Fluvial Remote Sensing for Science and Management

Fluvial Remote Sensing for Science and Management

Patrice E. Carbonneau
Durham University, Science site, Durham, UK

Hervé Piégay
University of Lyon, CNRS, France

ADVANCING RIVER RESTORATION AND MANAGEMENT

WILEY-BLACKWELL

A John Wiley & Sons, Ltd., Publication

This edition first published 2012
© 2012 by John Wiley & Sons, Ltd

Wiley-Blackwell is an imprint of John Wiley & Sons, formed by the merger of Wiley's global Scientific, Technical and Medical business with Blackwell Publishing.

Registered office: John Wiley & Sons, Ltd, The Atrium, Southern Gate, Chichester, West Sussex, PO19 8SQ, UK

Editorial offices: 9600 Garsington Road, Oxford, OX4 2DQ, UK
The Atrium, Southern Gate, Chichester, West Sussex, PO19 8SQ, UK
111 River Street, Hoboken, NJ 07030-5774, USA

For details of our global editorial offices, for customer services and for information about how to apply for permission to reuse the copyright material in this book please see our website at www.wiley.com/wiley-blackwell

Library of Congress Cataloging-in-Publication Data

CIP data has been applied for.

A catalogue record for this book is available from the British Library.

Wiley also publishes its books in a variety of electronic formats. Some content that appears in print may not be available in electronic books.

Cover image: Leonardo da Vinci
A map of the Arno west of Florence
1504
Pen and ink, green and blue wash, over black chalk and stylus
42.2 × 24.2 cm

Reproduced with permission from the Royal Collection Enterprises

Typeset in 9.25/11.5pt Minion by Laserwords Private Limited, Chennai, India.
Printed and bound in Singapore by Markono Print Media Pte Ltd.

First Impression 2012

Contents

Series Foreword

Advancing River Restoration and Management

The field of river restoration and management has evolved enormously in recent decades, driven largely by increased recognition of the ecological values, river functions, and ecosystem services. Many conventional river management techniques, emphasizing hard structural controls, have proven difficult to maintain over time, resulting in sometimes spectacular failures, and often degraded river environment. More sustainable results are likely from a holistic framework, which requires viewing the 'problem' at a larger catchment scale and involves the application of tools from diverse fields. Success often hinges on understanding the sometimes complex interactions among physical, ecological and social processes.

Thus, effective river restoration and management requires nurturing the interdisciplinary conversation, testing and refining our scientific theories, reducing uncertainties, designing future scenarios for evaluating the best options, and better understanding the divide between nature and culture that conditions human actions. It also implies that scientists better communicate with managers and practitioners, so that new insights from research can guide management, and so that results from implemented projects can in turn, inform research directions.

The series provides a forum for 'integrative sciences' to improve rivers. It highlights innovative approaches, from the underlying science, concepts, methodologies, new technologies, and new practices, to help managers and scientists alike improve our understanding of river processes, and to inform our efforts to better steward and restore our fluvial resources for more harmonious coexistence of humans with their fluvial environment.

G. Mathias Kondolf,
University of California, Berkeley

Hervé Piégay
University of Lyon, CNRS

Foreword

Images and maps have provided visual support to fluvial sciences and river management for centuries. Perhaps the earliest case in point, renaissance master Leonardo Da Vinci (1452–1519) frequently used his artistic talents in order to illustrate his scientific observations of fluid motion. His notes, reflections and observations on water, mostly contained in the *Codex Leicester* (Da Vinci, c1510), contain remarkably accurate drawings of wave erosion and water flow in both meanders and confluences. Da Vinci was also greatly concerned by the control and management of rivers and as a result the *Codex Leicester* contains detailed observations and drawings of water flowing around man-made obstacles. Most striking to the modern eye is the accuracy of Da Vinci's technical drawings and illustrations. Flow lines around bridge piers and other obstacles are painstakingly drawn to scale and into patterns giving an impressive level of process representation. This detailed sketching recorded the observations which often provided the foundations for scientific reasoning. In short, Da Vinci used a visual media (i.e. an image) as a tool to encode information on fluvial processes.

Furthermore, Da Vinci's scientific inquiry was not confined to the small scale observation of local flow phenomena. His illustrations of the Arno River which flows through Florence, demonstrate once again his acute sense of observation. For example, the cover art for this volume shows a sketch of the Arno and Mugnone rivers. To fluvial scientists and practitioners, this early visual representation of a river from an aerial perspective has a few fascinating features. The accurately rendered fluvial forms almost give the viewer the impression of an image rather than a map. Da Vinci clearly has a good understanding of channel features at a range of scales and these were rendered faithfully in his sketches. We can see small scale meanders, confluences and bifurcations. At larger scales, we can see clear distinctions between the main channel, secondary channels and inactive, possibly dry, channels. Furthermore, Da Vinci has clearly defined the boundaries of the braided band thus giving the viewer the impression of three distinct land-cover types in this landscape: the wetted channels, the active braided band and the vegetated area. Whilst it would obviously be exaggerated to consider Da Vinci as a pioneer of remote sensing, his work nevertheless provides one of the earliest and best known examples of visual data (*i.e.* sketches) used to convey technical information about fluvial landscapes and flow processes with a view towards both fundamental science and management.

Da Vinci relied on his well trained powers of observation in order to faithfully reproduce, among other, fluvial forms and features. However, the invention of photography in the late nineteenth century triggered a fundamental reversal of this process. With the photograph as an acceptable, or at least approximate, representation of geometric reality, the viewer of the image becomes the observer and information about the photographed scene can be acquired without direct physical presence or contact. In short, the photograph becomes a data acquisition method. Furthermore, with the contemporary invention of air travel in the late nineteenth century, it is not surprising that many early aeronauts and photographers collaborated and gave birth to aerial photography. In 1855, Gaspard-Félix Tournachon, most often referred to by his pseudonym 'Nadar', patented the concept of using aerial photographs for surveying and mapping. After three years of experimentation, in 1858, Nadar took the very first aerial photograph, a moment which was caricatured by Honoré Baumier in 1863 (Figure 1). Nadar had clearly understood the future potential of imagery as a source of information. Indeed, imagery and remote sensing have now become standardised data acquisition approaches with far reaching applications in all the physical and environmental sciences.

Inspired by these early thinkers who pioneered the use of visual representations in fluvial sciences and mapping, this edited volume will examine the most recent applications and uses of imagery and image-derived information in river sciences and management. Our goal is to present some key highlights of nearly two decades of research during which the use of image data has emerged along with important advances in sciences and technologies which

H. DAUMIER

NADAR

élevant la photographie à la hauteur de l'Art.

Figure 1 Caricature by Honoré Daumier of the first aerial photograph taken by Gaspard-Félix Tournachon (Nadar) in 1858.

are fundamentally enhancing our ability to characterise river geometry, sediment calibre, water characteristics, vegetation type, vegetation dynamics, ice dynamics, flooding, organisms, social value, etc and, ultimately, allows managers to provide practical recommendations. Our intended audience is a non-specialist one, this volume seeks to serve as an accessible entry point to both river managers and students who are looking for a condensed reference text capable of answering basic questions and explaining some of the more fundamental concepts. It is our hope that this volume will allow readers to determine if image-based approaches are suitable to their needs and thus encourage them to pursue the wider literature on the topic.

The Editors,
Patrice E. Carbonneau and Hervé Piégay

List of Contributors

Adrien Alber, University of Lyon, CNRS, France

Imtiaz Ali, Universitéy de Lyon, Bron cedex, France; Optics Labs, Islamabad, Pakistan

Tristan Allouis, Irstea, UMR TETIS, Montpellier, France

Jean-Stéphane Bailly, AgroParisTech, UMR TETIS and UMR LISAH, Montpellier, France

Giuliano Di Baldassarre, Department of Hydroinformatics and Knowledge Management, UNESCO-IHE, Delft, The Netherlands

Paul D. Bates, School of Geographical Sciences, University of Bristol, Bristol, UK

Normand Bergeron, Institut National de Recherche Scientifique, Centre Eau Terre et Environnement, Québec, Canada

Walter Bertoldi, Dipartimento di Ingegneria Civile e Ambientale, Universita degli studi di Trento, Italy

Mélanie Bertrand, Irstea, Unité de Recherche ETNA, Saint-Martin-d'Hères, and University of Lyon, CNRS, France

Thomas Buffin-Bélanger, Département de biologie, chimie et géographie, Université du Québec à Rimouski, Rimouski, Québec, Canada

Patrice E. Carbonneau, Department of Geography, Durham University, Science site, Durham, UK

Keith A. Cherkauer, Department of Agricultural and Biological Engineering, Purdue University, West Lafayette, IN, USA

Yann Le Coarer, Irstea, Unité de Recherche, HYAX, Aix-en-Provence, France

S. Corgne, LETG - Rennes COSTEL, CNRS, Université Rennes 2, Place Recteur Henri le Moal, France

Marylise Cottet, University of Lyon, CNRS, France

Simon Dufour, LETG - Rennes COSTEL, CNRS, Université Rennes 2, Place Recteur Henri le Moal, France

Robert Dunford, Environmental Change Institute, Oxford University Centre for the Environment, Oxford, UK

Russel N. Faux, Watershed Sciences, Inc., Corvallis, OR, USA

Denis Feurer, IRD, UMR LISAH, Montpellier, France

Mark A. Fonstad, Department of Geography, University of Oregon, Eugene, OR, USA

Philippe Frey, Irstea, Unité de Recherche ETNA, St-Martin-d'Hères, France

Alan R. Gillespie, Department of Earth and Space Sciences, University of Washington, Seattle, WA, USA

David Graham, Department of Geography, Loughborough University, Leicestershire, UK

Stan Gregory, Department of Fisheries and Wildlife, Oregon State University, Corvallis, OR, USA

Rebecca N. Handcock, Commonwealth Scientific and Industrial Research Organization, Floreat, WA, Australia

Alexandre Hauet, EDF - DTG - CHPMC, Toulouse, France

George Heritage, JBA Consulting, The Brew House, Wilderspool Park, Grenalls Avenue, Warrington, UK

Dave Hulse, Department of Landscape Architecture, University of Oregon, Eugene, OR, USA

Paul J. Kinzel, USGS, Geomorphology and Sediment Transport Laboratory, Golden, CO, USA

Eric Lajeunesse, Institut de Physique du Globe de Paris, Sorbonne Paris Cité, Paris, France

Andy Large, School of Geography, Politics and Sociology, Newcastle University, UK

J. Wesley Lauer, Department of Civil and Environmental Engineering, Seattle University, Seattle, WA, USA

Yves-Francois Le Lay, University of Lyon, CNRS, France

Carl J. Legleiter, Department of Geography, University of Wyoming, Laramie, WY, USA

Jérôme Lejot, University of Lyon, CNRS, France

Angela Limare, Institut de Physique du Globe de Paris, Sorbonne Paris Cité, Paris, France

Bruce J. MacVicar, Department of Civil and Environmental Engineering, University of Waterloo, Ontario, Canada

W. Andrew Marcus, Department of Geography, University of Oregon, Eugene, OR, USA

David C. Mason, Environmental Systems Science Centre, University of Reading, UK

François, Métivier, Aix-Marseille Université, France

Kristell Michel, University of Lyon, CNRS, France

Etienne Muller, Université de Toulouse; CNRS, INP, UPS; EcoLab (Laboratoire Ecologie Fonctionnelle et Environnement), France

Doug Oetter, Department of History, Geography, and Philosophy, Georgia College and State University, Milledgeville, GA, USA

Hervé Piégay, University of Lyon, CNRS, France

Stephen Rice, Department of Geography, Loughborough University, Leicestershire, UK

Anne Rivière-Honegger, University of Lyon, CNRS, France

Anne-Julia Rollet, University of Caen Basse-Normandie, Géophen, LETG UMR 6554 CNRS, France

Pierre Sagnes, Université Claude Bernard Lyon 1, Villeurbanne Cedex, France

Guy J-P. Schumann, School of Geographical Sciences, Bristol, UK

Menno Straatsma, University of Twente, Enschede, The Netherlands

Michal Tal, Aix-Marseille Université, France

Jing Tan, Department of Agricultural and Biological Engineering, Purdue University, West Lafayette, IN, USA

Klement Tockner, Leibniz-Institute of Freshwater Ecology and Inland Fisheries and Institute of Biology, Freie Universität Berlin, Germany

Christian E. Torgersen, U.S. Geological Survey, Forest and Rangeland Ecosystem Science Center, School of Environmental and Forest Sciences, University of Washington, Seattle, WA, USA

Laure Tougne, Universitéy de Lyon, Bron cedex, France

Elise Wiederkehr, Université of Lyon, CNRS, France

Wonsuck Kim, University of Texas, Austin, TX, USA

1

Introduction: The Growing Use of Imagery in Fundamental and Applied River Sciences

Patrice E. Carbonneau[1] and Hervé Piégay[2]

[1]Department of Geography, Durham University, Science site, Durham, UK
[2]University of Lyon, CNRS, France

1.1 Introduction

Earth observation now plays a pivotal role in many aspects of our lives. Indeed, hardly a day goes by without some part of our lives relying on some form of remote sensing. Weather predictions, mapping and high level scientific applications all make intensive use of imagery acquired from satellites, aircraft or ground-based remote sensing platforms. This form of data acquisition which relies on the reflection or emission of radiation on a target surface is now well accepted as a standard approach to data acquisition. However, the fields of river sciences and remote sensing have operated independently during much of their respective histories. Indeed remote sensing practitioners generally consider streams as linear, or perhaps network, entities in the landscape. In contrast, river scientists such as fluvial geomorphologists, lotic and riparian ecologists, with their focus on the internal structure of rivers and the processes which create these structures, often have a much more localised but three dimensional view of river systems. Nevertheless, both modern fluvial geomorphology and ecology are increasingly recognising that we need to reconcile these viewpoints. In a seminal paper, Fausch et al. (2002) discuss the scientific basis for this reconciliation. These authors argue that natural processes, both biotic and abiotic, frequently operate on larger spatial scales and longer time scales than traditional river sciences and management. Consequently, the authors argue that localised, non-continuous, sampling of small scale river processes, forms and biota leads to a fundamental scale mismatch between the processes under scrutiny and our data collection. Fausch et al. (2002) therefore argue that river sciences and management must begin to consider and sample river catchments (i.e. watersheds) at larger scales and that these units must be considered more explicitly as holistic system.

The need to study and sample river catchments as holistic systems naturally leads to the use of remote sensing as a basic methodology. Remotely sensed data and imagery is indeed the only approach which could conceivably give continuous data over entire catchments (Mertes, 2002; Fonstad and Marcus, 2010). However, in the 1990s and early 2000s, existing remote sensing acquisition hardware and analysis methods were neither tailored nor very suitable to the needs and interests of river scientists and managers. Mertes (2002) presented a review of remote sensing in riverine environments at the turn of the century. At that time, any data with sub-metric spatial resolution was considered of 'microhabitat' scale. Consequently, riverine features identified by remote sensing in the late twentieth century were generally of hectametric or kilometric scales. However, developments in the early twentieth century proceeded at a rapid pace and our ability to resolve fine details in the landscape has dramatically

Fluvial Remote Sensing for Science and Management, First Edition. Edited by Patrice E. Carbonneau and Hervé Piégay.
© 2012 John Wiley & Sons, Ltd. Published 2012 by John Wiley & Sons, Ltd.

improved in the last decade (see Chapter 8 and Marcus and Fonstad (2008) for a comprehensive review). Therefore, publications on the remote sensing of rivers have dramatically increased and 'Fluvial Remote Sensing' (FRS) is emerging as a self-contained sub-discipline of remote sensing and river sciences (Marcus and Fonstad, 2010). Moreover, the technical progress accomplished in the past two decades of research in FRS means that this sub-discipline of remote sensing has now begun to make real contributions to river sciences and management and the appearance of a volume on the topic is therefore timely. Our aim with this edited volume is to give readers with a minimal background in remote sensing a concise text that will cover the broadest possible range of potential applications of Fluvial Remote Sensing and provide contrasted examples to illustrate the capabilities and the variety of techniques and issues. Readers will notice when consulting the table of contents that we take a very broad view of 'remote sensing'. In addition to more conventional remote sensing approaches such as satellite imagery, air photography and laser scanning, the volume includes a wider range of applications where image and/or video data is applied to support river science and management. This chapter will set the context of this volume by first giving a very brief introduction to remote sensing and by discussing the evolution of journal publications in fluvial remote sensing approaches and river management. Finally, we will give a brief outline of the volume.

1.2 Remote sensing, river sciences and management

1.2.1 Key concepts in remote sensing

Here we will introduce some key remote sensing concepts which will help us illustrate and contextualise fluvial remote sensing as a sub-discipline. However, this introduction is not meant as a foundation text in remote sensing and we refer the reader in need of some fundamental material to classic remote sensing textbooks such as Lillesand et al. (2008) or Chuvieco and Alfredo (2010).

Remote sensing has a multitude of definitions. In broad terms, 'remote sensing may be formally defined as the acquisition of information about the state and condition of an object through sensors that are not in physical contact with it' (Chuvieco and Alfredo, 2010). This type of broad definition does not place any restriction on the type of interactions that occur between the target and the sensor. According to this definition, echo-sounding devices

such as sonar which use acoustic energy in order to detect objects in a fluid media such as air or water should be considered as remote sensing. However it should be noted that references to remote sensing usually apply to the collection of information via electromagnetic energy such as visible light, infrared light, active laser pulses, etc. Remote sensing is then generally divided in two broad categories: active or passive remote sensing. This description refers to the source of radiation. Passive remote sensing relies on externally emitted sources of radiation whilst active remote sensing relies on internally generated and emitted radiation. The best-known example of active remote sensing is RADAR (Radio Detection And Ranging) which uses radio waves to establish the position of objects in the vicinity of the sensor. More recently, lasers have been used in active remote sensing to give birth to LiDAR (Light Detection And Ranging) technology. LiDAR technology is rapidly becoming the method of choice for the generation of topography from ground based and airborne platforms and is the focus of Chapters 7 and 14 of this volume.

The key parameter exploited by active remote sensing has always been the time elapsed between the emission of a radiation pulse and it's detected return. As a result, active remote sensing uses a narrow and finite portion of the electromagnetic spectrum. For example, typical LiDAR technology uses infrared lasers with a wavelength of 1024 nm and radar relies on radio waves with wavelengths of 1–10 cm. Passive sensors, which rely on an external source of radiation (usually the sun), make a much more comprehensive usage of the electromagnetic spectrum. This is the type of remote sensing which is familiar to all of us because our visual system uses solar radiation to detect features in our surroundings. Table 1.1 presents a simplified form of the electromagnetic spectrum. This table gives the common names and categories of radiation as we move, from left to right, from the very short wavelengths of high energy cosmic radiation to the very long wavelengths of lower energy micro-waves and radio waves. Generally speaking, the majority of passive remote sensing sensor devices applied to earth observation uses radiation in the visible and infrared portions of Table 1.1. Given that the electromagnetic spectrum has a continuous range of frequencies (i.e. radiation wavelength is not intrinsically discreet), their detection and quantification relies on sensors that can detect incident radiation within a specified, finite, range of wavelengths. The most basic example of this would be greyscale (black and white) imagery where the brightness of a point on the photograph is proportional to the total amount of visible

Table 1.1 Simplified Electromagnetic Spectrum table (Modified from Ward et al., 2002).

Wavelength (λ)	<0.01 nm	0.01 to 1 nm	1 nm to 0.4 µm	0.4 to 0.7 µm			0.7 to 3 µm	3 to 8 µm	8 to 15 µm	15 µm to 1 mm	1 mm to 1 m				>1 m
Name	Cosmic Rays	X-Rays	Ultraviolet	Visible (Optical)			Infrared				Micro-waves				
				blue	green	red	near	middle	thermal	far	K-band: 1.1-1.4 cm	X: 2.4-3.75 cm	C: 3.75-7.5 cm	L: 15-30 cm	Radio

radiation, with frequencies ranging from approximately 0.4 to 0.7 microns, received by the sensor (e.g. the camera film). A further example would be standard colour photography. In this case, it would clearly be impossible to have a near infinite number of detectors each sensitive to a specific wavelength in the continuous visible spectrum. The solution which was therefore adopted in the early days of colour photography was to emulate human vision and to re-create colour by first sampling radiation in three distinct areas of the spectrum: red, green and blue (Lillesand et al., 2008). Within each of these primary colour bands, the total amount of radiation incident upon the sensor is recorded. Therefore for the red band, the sensor detects all the radiation with frequencies between approximately 0.6 and 0.7 microns. For the green band the sensor detects all the radiation from approximately 0.5 and 0.6 microns and for the blue band, detectable wavelengths range from 0.4 to 0.5 microns. It should be noted that the term 'band' mentioned earlier is one of the most fundamental in the remote sensing vocabulary. Formally, a 'spectral band' is a finite section of the electromagnetic spectrum, recorded and stored in a raster data layer. In the examples above, a greyscale image is a one band image and a colour image is a three band image. The term 'multispectral' therefore refers to a remote sensing approach or dataset which has several bands. Strictly speaking, colour photography, with its three bands in red, green and blue, can be considered as multispectral imagery. However, many authors and practitioners reserve the term 'multispectral' for datasets which have at least four spectral bands with one of the bands usually covering the infrared portion of the spectrum. It should be noted that the number of available bands is not the only important characteristic of a remotely sensed image. Potential applications

of remotely sensed data are often limited and one might even say, defined, by four additional parameters: spectral resolution, spatial resolution, temporal resolution and, to a lesser extent, radiometric resolution.

The concept of spectral resolution is closely related to the concept of a spectral band. It relates to the width, expressed in linear units of radiation wavelength (nm or µm), of the spectral bands of the imaging device. A clear distinction must therefore be made between the number of bands measured by a sensor which determines the range of radiation wavelengths that is sampled and the width (or narrowness) of an individual band which determines the sensors sensitivity to specific spectral features. Arguably the most classic example of the use of spectral features in remote sensing is the detection of vegetation. In healthy green vegetation, chlorophyll absorbs over 90% of incident radiation within the visible spectrum, albeit with a slightly lesser absorption and higher reflection in green wavelengths, which explains the colour of vegetation. However, in the infrared wavelengths, vegetation is a strong reflector. Sensors designed to detect vegetation, such as the classic Thematic Mapper sensor mounted on Landsat satellites, therefore try to exploit these differences by sampling red light (0.63–$0.69\,\mu m$) which is strongly absorbed by vegetation and near infrared light (0.76–$0.90\,\mu m$) which is strongly reflected. Note the relatively narrow width, in spectral terms of these bands. Our ability to accurately detect vegetation from remote sensing therefore depends not only on increasing the number of bands beyond the visible spectrum, but also on an improvement of the spectral resolution. If we follow this line of thought to its logical conclusion, we realise that it would be desirable to produce a sensor with a very high number of bands each with a very narrow bandwidth.

Such sensors are called 'Hyperspectral' and can have hundreds or even thousands of bands with resolutions as small as 0.002 μm. Whilst such hyperspectral sensors have huge potential, their usage in river sciences has been relatively limited and most of the progress in fluvial remote sensing rests on standard colour imagery with the conventional three bands of <u>R</u>ed, <u>G</u>reen and <u>B</u>lue (hence the term RGB imagery) which equates to a relatively coarse spectral resolution of approximately 0.2 μm.

One key advantage of widely available colour imagery is its very high spatial resolution. One of the most fundamental descriptors of remote sensing data, spatial resolution refers to the ground footprint of a single image pixel on real ground. This distance is generally quoted as a linear unit with the underlying assumption that the pixels are square. The spatial resolution of a dataset will define the smallest object that can be identified. Whilst there is no absolute rule for the number of pixels required to define a simple object (e.g. a boulder), our experience has shown that a minimum of 5X5 pixels are required in order to get an approximation of the object shape whilst 3X3, or even 2X2, pixels are required to establish to presence of an object of undefined shape in the image.

In parallel with spatial resolution, temporal resolution refers to the elapsed time between repeated imagery. Repeated image sampling has been somewhat less exploited in fluvial remote sensing. While studies of large rivers based on satellite imagery have been able to exploit the regular revisit frequency of orbital sensors (Sun et al., 2009; Frankl et al., 2011), airborne data is not acquired with the same regularity and studies reporting change based on airborne data are much less frequent. As a result, substantial progress remains to be made in terms of monitoring rivers and examining changes occurring at the smaller spatial resolutions that can be detected with airborne remote sensing. However, repeated imagery, including video imagery, has been successfully used at smaller scales for laboratory studies (see Chapter 13) and reach based studies (see Chapters 15 and 16). Furthermore, a largely un-exploited archive or terrestrial and airborne archival imagery exists for many parts of the world which does indeed include riverine areas. If issues such as image georeferencing (spatial positioning of the imagery), and image quality can be addressed (see Chapter 8), then these images could provide a very important source of data sometimes dating as far back as the nineteenth century.

The final parameter, radiometric resolution is easily confused with spectral resolution. Here the term 'radiometric' refers to the recording of data in the sensors memory. When radiation reaches a device, the intensity of radiation must be converted to some proportional brightness scale which can then be represented on an image. In the case of digital devices, this proportional brightness is termed the Digital Number (DN). The digital number is the dimensionless actual value of the pixel that can be seen if the image is accessed with image processing software. Typically, these pixel values are scaled to increasing powers of 2. For example, standard RGB imagery contains three bands, each of which has pixel values ranging from 0 to 255. These 256 possible values arise from data storage in an '8 bit' binary format meaning that each DN value is coded with 8 binary digits with possible values of 0 or 1 thus leading to $2^8 (256)$ possible values for the image pixels. However, more advanced sensors and satellites will frequently use higher 'bit-depths' of 11 or 12 bits thus leading to a wider range of 2048 (2^{11}) or even 4096 (2^{12}) DN values. This higher number of DN values can help in resolving finer differences in image brightness. In river sciences, radiometric resolution can be an important parameter when trying to measure river properties through the water interface (Legleiter et al., 2009).

In summary, from the point of view of an end-user, the fundamental properties of a remote sensing data acquisition system can be described by four key parameters: Spatial resolution, spectral resolution, temporal resolution and radiometric resolution. Spatial resolution is often considered as the primary parameter as it defines the size of the smallest object which can be resolved on the ground. Spectral resolution can be crucial in identifying certain materials, such as chlorophyll, based on their reflection of light as a function of the wavelength of the incident light. Temporal resolution is obviously crucial in change detection studies. Finally, radiometric resolution, often called 'bit-depth', defines the amount of information devoted to the storage of each image pixel. Higher radiometric resolutions allow for the recording of smaller differences in image brightness.

1.2.2 A short introduction to 'river friendly' sensors and platforms

A remote sensing 'platform' is simply the physical support which carries the 'sensor' that does the actual data collection. We have illustrated four classic and new platforms in Figure 1.1. This distinction between platform and sensor is not always clear, especially in the field of satellite remote sensing. For example, the TERRA satellite platform carries both the MODIS and ASTER sensor. However, the commercial term 'QuickBird' is used to describe both

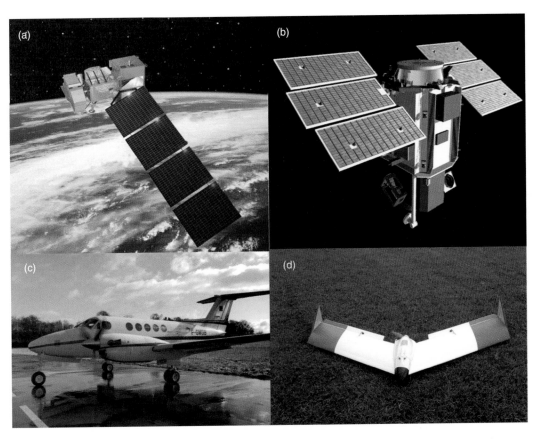

Figure 1.1 Typical Remote Sensing Platforms. a) Landsat-7 satellite (15m spatial resolution), b) QuickBird-2 satellite (61 cm spatial resolution), c) Full sized fixed wing aircraft operated by the French *Institut Géographique National* (commonly 0.5 m spatial resolution). Copyright IGN – France, d) Ultralight UAS system (1m total wingspan) operated by Durham University, UK.

the satellite and sensor. In the field of airborne remote sensing the distinction is usually clearer since a given sensor can usually be mounted on a range of fixed wing aircraft or helicopters.

Unsurprisingly, there is currently an abundance of remote sensing images and products. Finding a starting point and locating an appropriate data source and/or acquisition method can therefore be quite a daunting process. Here we give a short description of remote sensing data sources most likely to be of use in the context of fluvial sciences and river management. Many river managers are still under the impression that fluvial remote sensing is not an appropriate tool for river environments. This is a reasonable viewpoint if we consider the most classic and widely known remote sensing data: Landsat imagery. With spatial resolutions of typically 15 m or 30 m, Landsat images only sample river outlines accurately for very large rivers. Clearly, such imagery has little to offer a manager

or scientist needing to characterise a small stream with widths below 50 m. However, there has been remarkable technological progress in imaging which has now made images with resolutions of less than 1 m available globally. Several satellites now offer image resolutions below 1 m and low altitude airborne colour photography is now capable of resolutions as low as 2–3 cm. The availability of such data, offering a 100-fold improvement in spatial resolution when compared to classic Landsat, has been an important driver of methodological progress in fluvial remote sensing (see Marcus and Fonstad, 2008).

For readers who are unfamiliar with the topic, Table 1.2 gives a very brief summary of a few key satellites and platforms which are likely to be of interest to river scientists and managers. We have also included some older platforms that may be of lesser interest in a modern context but which nevertheless often appear in publications. This list is far from complete or exhaustive. Our aim is merely

Table 1.2 Common Satellite/Platforms with key characteristics.

Sensor/Platform	Launch Date	Spatial Resolution (at Nadir)	Temporal Resolution	Spectral Bands
MODIS/Terra	Dec. 1999	250 m (bands 1–2) 500 m (bands 3–7) 1000 m (bands 8–36)	16 days	36 bands from the visual to infrared and thermal
ASTER/Terra	Dec. 1999	15 m (bands 1–3) 40 m (bands 4–9) 90 m (bands 10–14)	16 days	14 bands from the visual to infrared and thermal
ETM+/ Landsat-7	Apr. 1999	15 m Panchromatic 30 m (bands 1–5 and 7) 60 m (band 6)	18 days	8 bands: Panchromatic, 3 visual, 2 infrared, 2 thermal
SPOT-5	May 2002	2.5 m Panchromatic 10 m (bands 1–3) 20 m (band 4)	2-3 days	5 bands: Panchromatic, 2 visual (no blue), infrared, thermal
Ikonos	Sept. 1999	82 cm Panchromatic 3.2 m Multispectral	3 days	5 bands: Panchromatic, 3 visual, infrared
QuickBird	Oct. 2001	65 cm Panchromatic 2.62 m Multispectral	2.5 days	5 bands: Panchromatic, 3 visual, infrared
WorldView-1	Sept. 2007	50 cm Panchromatic	1.7 days	1 band: Panchromatic
WorldView-2	Oct. 2009	50 cm Panchromatic 1.85 m Multispectral	1.1 days	9 bands: Panchromatic, 6 visual, 2 infrared,
GeoEye	Sept. 2008	50 cm Panchromatic 1.65 m Multispectral	2.1 days	5 bands: Panchromatic, 3 visual, infrared
Air Photography	N.A.	Variable. Typically 2 to 50 cm.	≈1 day	Variable. Typically standard colour. Most types of instruments available.
Unmanned Aerial systems (UAS)	N.A.	Variable. Typically 2 to 50 cm.	<1 day	Variable. Typically small format RGB digital cameras. Other instruments available on large UAS.

to suggest a few data acquisition options and justify these suggestions with the appropriate data characteristics. The first point to note is the variable spatial resolution, for each sensor, when images are acquired in panchromatic mode (i.e. greyscale) and multispectral mode. It should always be remembered that when satellite image vendors quote a sub-metric spatial resolution, they are referring to panchromatic imagery. At the time of writing, no satellite platform in earth orbit can acquire multispectral imagery with sub-metric resolutions. A possible substitute for high resolution imagery is called 'pan-sharpened' imagery. In a pan-sharpened image, the sub-metric resolution image is fused with the multispectral images. This transformation uses the brightness values in the panchromatic band to weigh the interpolation of the lower resolution multispectral bands. The result is a multispectral or colour image with the same resolution as that of the panchromatic image. Another interesting point to note about

spatial resolutions is the apparent 50 cm limitation which seems to have been reached in the more recent satellites. In fact, the GeoEye in Table 1.2 satellite is capable of producing 41 cm greyscale imagery and the Worldview-2 satellite can acquire at 46 cm. However, US regulations prohibit these companies from delivering data in the public domain with spatial resolutions below 0.5 m and therefore the images are resampled before delivery to the customer. Unfortunately, it seems that for the foreseeable future, satellite image spatial resolutions will be blocked at 50 cm. In terms of temporal resolutions, these satellites can all revisit a site within a few days. From the perspective of fluvial sciences, this makes them well suited to seasonal monitoring. In terms of spectral resolutions, the basic array of bands for a so-called 'multispectral' satellite image has long been four bands in Red, Green, Blue and Near Infrared. Many satellites in Table 1.2 conform to this standard and have three spectral bands in the visible range with an additional band in the infrared which is generally intended for vegetation. However, the recently launched WorldView-2 satellite proposes a marked improvement in spectral terms with eight bands with widths of 40 to 70 nm in the visible range with two bands in the near-infrared. This recently available imagery has not yet been applied to small rivers and holds much potential.

For users interested in studying or managing very small rivers with metric scale widths, even the best currently available satellite image may still be insufficient. In such cases, airborne remote sensing should be considered. The final two entries in Table 1.2 are meant to give a broad, preliminary, indication of the potential of airborne remote sensing (see Chapters 2, 5, 7, 8, 9 and 11 for further discussions). Airborne remote sensing is obviously a very wide topical area. Here we present only two broad types of acquisition platforms: air photography from conventional aircraft and Unmanned Aerial systems. Traditional air photography is now widely available from both the private sector and government agencies. In addition to colour imagery, traditional aircraft can be used to mount a range of instruments which have been shown to be useful in river sciences. For example, Fausch et al. (2002) present high resolution temperature acquired from a fixed wing aircraft and Marcus et al. (2003) show how hyperspectral data can provide a rich database of information which significantly surpasses the limits of standard RGB imagery. In terms of spatial resolution, aerial photography generally fills the niche below satellite imagery. The temporal resolution of air photos is obviously not as rigid as that of a satellite which is bound in an elliptical orbit around the earth. In theory, an aircraft can be mobilised very frequently and visit a site at least once a day. However, potential users should be aware that in practice, this is very rarely possible. Government agencies only very rarely commission repeat flights of an area at intervals smaller than one year. Similarly, private sector companies can sometimes have the availability for repeat flights within a year although our experience has been that this is very difficult for a specific rivers owing to cost and logistic constraints. Unmanned Aerial Systems (UAS) can free users from these logistic constraints by giving the opportunity for managers and scientists to operate their own aircraft. UAS exist in a very wide range of sizes and purposes. In fact some UAS, for example the Global Hawk and Ikhana systems operated by NASA, are in essence full sized, pilotless, aircraft. However, of particular interest here is the ever growing range of small, toy-sized, UAS available on the civilian commercial market. These systems are easy to pilot and come equipped with small format digital cameras and onboard navigation hardware which often allows for fully automated flight and data acquisition. These small aircraft can fly at very low altitudes and therefore can deliver very high resolution imagery. Their small size makes them very easy to deploy at high temporal resolutions. At the time of writing, publications using UAS data are relatively rare in river sciences (but see Dunford et al., 2011). However, this new technology is prompting much excitement in the river sciences community and the publication record can be expected to grow in the coming years.

1.2.3 Cost considerations

Most users considering remotely sensed data will probably turn to free data sources in the first instance. Classic Landsat data is freely downloadable from the United States Geological Service (USGS) via their EarthExplorer website (earthexplorer.usgs.gov). Whilst the resolution is low, this data can still provide some initial insights for medium to large rivers. For smaller rivers, most users will likely turn to free online mapping services like Google Earth which displays very good quality imagery, often with sub-metric resolutions. Google corporation purchases this imagery from a range of airborne and satellite sources (some in Table 1.2) and makes them freely viewable online. However, users cannot download full, raw, image products from Google Earth. Therefore, in the majority of cases, the purchase of data will still be required. The costs of such purchases are obviously a crucial consideration. Whilst these are quite variable across the full range of data types, sensors and platforms, we give here a basic summary which is not specific to any single company or service provider and which will hopefully provide the reader with some initial estimates.

In the case of satellite imagery, there are two important, broad, distinctions. First, is a new image required? Satellite image providers maintain full archives of all previously acquired images. These archived images are sold at discounted costs which range from 10–20 US\$ per km^2. However, if a new image is required, the purchase of a new acquisition will increase the cost to at least 20–80 US\$/km^2. The second factor in satellite image cost is the level of pre-processing. The cost estimates above are for basic standard imagery. However, image providers offer pre-processing services which range from improved image quality in terms of position, geometry and radiometry to the full production of Digital Terrain Models (DTMs). These levels of processing will obviously increase the cost, sometimes in excess of 100 US\$/km^2. Readers should also note that a minimum area must always be purchased. This is typically in excess of $20 \, km^2$ which therefore places the minimum cost of a single, high resolution, satellite image in the vicinity of 2000 US\$.

In the case of airborne imagery, costs are also quite variable. Dugdale et al. (2010) cite a cost of £150/km (approximately 250 US\$/km) for the acquisition of 3 cm airborne imagery. This would however be in addition to an initial mobilisation cost required to get the aircraft to the mission locality. Typically, in the case of small rivers with lengths below 100 km and widths below 100 m, surveys of full river lengths in order to acquire sub-decimetric resolution colour imagery will probably cost 10 000 to 25 000 US\$. However, many national agencies maintain image archives for their territories. These are generally of a much lower resolution, typically 25–50 cm. However, their cost is much lower. Government agencies, particularly in the US, will often provide these free of charge. Even when not freely available, the cost is roughly 10% of the cost of a new survey. Small UAS are generally affordable for most organisations. Depending on the size, level of automation and imaging equipment of the craft in question, costs can range from roughly 5000 US\$ to 30 000 US\$. These make them affordable options for 'do-it-yourself' remote sensors. However, prospective UAS pilots should take careful notice of national airspace regulations. Airspace regulations in most western nations now have specific regulations pertaining to UAS. The spirit of most UAS airspace usage regulations is that small, light weight, UAS operated in non-urban areas, at low altitudes (below 400 ft or 120 m) and within line of sight of the pilot are allowed. This situation is generally suitable to most river applications thus making UAS a good option for river study and management in the US and Europe. However, we strongly encourage readers to consult specific regulatory agencies before purchasing a UAS since regulations will vary across the globe and may change rather rapidly. Furthermore, many regions of the world do not allow any type of UAS operations. For example, in India, airborne photography, both from UAS and full aircraft, is strictly reserved to military uses. Readers considering airborne photography of any kind should therefore always check the regulatory framework for their intended field site.

1.3 Evolution of published work in Fluvial Remote Sensing

The past decade has clearly seen remarkable contributions to methodological aspects of fluvial remote sensing. As discussed in later chapters of this volume, river scientists now have a wide range of remote sensing and image based methods capable of quantifying the biotic and abiotic aspects of river environments. This progress has been reflected in academic publications and here we focus on a bibliometric survey in order to analyse the evolution of Fluvial Remote Sensing (FRS). The ISI Web of Science (WOS) database was used to provide a summary in international peer-reviewed scientific journals and conferences. Different searches were carried out based on a set of technical key-words, such as 'Remote sensing', 'imagery/image', 'photogrammetry/photography', 'video' combined with specific thematic key-words describing our geographical objects such as 'river', ' stream', 'fluvial channel', 'fluvial geomorphology', 'floodplain' and 'riparian'. We decided to reject the term 'river basin', which we found was used for catchment or regional scale hydrology, an observation in itself. We also rejected the terms 'video stream' and 'image stream' which are used purely for video technologies. The term 'channel' must also be used with caution since it can be used in the purely technical sense of a radiometric channel or video channel. From this request,[1] 224 references are specifically related to our topic. Of the 224 references, 200 have an abstract. In a second search phase, we introduced the terms 'management', 'restoration', 'maintenance', but also 'planning'. We did the second request[2] on the title for these additional keywords, the others being searched in the topics to reassemble more papers 12 only were then identified.

As a first order analysis, if we consider the pace of publications, we find that 1 to 3 papers were published every year between 1976 and 1996, 7 to 9 papers per year were published between 1997 and 2001, increasing to 11 to 14 per year from 2001 to 2006 and finally surpassing 30 per year since then with a maximum of 37 in 2010. This increase in the number and pace of publications is in itself a good indicator of the accelerating pace of progress in this sub-discipline of remote sensing. In order to pursue

---

[1]Exact request done in May 2011 : Title = (Remote sensing OR image OR imagery OR photogr* OR video) AND Title = (river* OR stream OR streams OR fluvial channel* OR fluvial geomorphology OR floodplain OR riparian) NOT Title = (basin* OR catchment OR watershed OR "video stream*" OR "image stream") = 333 References listed but only 224 were really in the scope of the discipline.

[2]Exact request done in August 2011 : Title = (Remote sensing OR imagery OR image OR photogr* OR video) and Topics = (river OR stream OR fluvial) and Topics = (management OR restoration OR maintenance OR planning OR conservation).

the bibliometric analysis in more detail, we considered three elements: authorships and journals, platforms and sensors and topical areas of study.

1.3.1 Authorships and Journals

First authorship is dominated by the USA (36%) and the UK (12%). However, a set of countries are quite well invested in this domain such as Australia (9%), China (9%), France (6%), Canada (6%), Holland (3.6%) and India (4.5%) (Figure 1.2). Many of these countries have active satellite remote sensing programs. If we compare these results to a broad WOS search with the single term 'rivers' (>100 000 papers) or 'river management' (>15 000 papers), the UK (5.7% and 7.7% of papers respectively), or India (2.5% and 2.0% of papers respectively) are significantly stronger in Fluvial Remote Sensing whereas USA is slightly stronger (31% and 34%) as well as France (5.0% and 4.3%), Australia (4.7 and 8%) and China (11% and 7%), Holland (2.3% and 3.8) and Canada

(6.8% and 6.4) are similar. Germany has weak research in this domain compared to its scientific weight in river and river management research (5.2% and 4.8%), similar to Japan (4.4% and 2.3%).

The papers dealing with Fluvial Remote Sensing were published in 91 journals. 81% of these journals only published one or two manuscripts (Figure 1.3). Among the remaining 19%, specialised journals in geomatics and remote sensing such as *International Journal of Remote Sensing* and *Remote Sensing of Environment* are the most popular (respectively 7.5 and 10% of the manuscripts). The thematic journals *Earth Surface Processes and Landforms* and *Geomorphology* are almost as attractive as these specialised journals. They are followed by the *Journal of Hydrology* and the *Journal of the American Water Resources Association*. In the field of ecology, the remote sensing papers are published in a large set of ecological journals none of which is devoted exclusively to remote sensing. Overall, 33% of papers found in our search are published in Geomatics/Remote Sensing Journals, 17% in Ecology/Biology, 16% in Earth Sciences, 13% in Hydrology, 9% in Water Environment, 6% in Ocean Environment, 5% in Environment and 1% in Agriculture.

1.3.2 Platforms and Sensors

Within our search results, papers based on satellite data are slightly more frequent than aerial/airborne data with 34% of papers referring to 'satellite' against 27% to 'aerial/airborne' (Figure 1.4a). Landsat is the most frequently used satellite platform (21%) following by Terra (16%) and Spot (7.5%). In terms of satellites capable of delivering imagery with spatial resolutions at or below a meter, Quickbird is more popular than Ikonos, but both are still quite infrequently used in the literature (respectively 5.5% and 2% of manuscripts). Envisat and Formosat are cited only in a very few papers. The Shuttle Radar Topography Mission was mentioned in the abstracts of two contributions. The terms 'UAV' or 'drone' do not appear in any of the abstracts. The terms 'blimp', 'balloon' and 'Unmanned' in one manuscript each and 'helicopter' in four of the 200 abstracts.

When considering sensors, we observe a range of equipment used, from spacecraft imagers such as ASTER or MODIS to ground or airborne equipment covering a large part of the electromagnetic spectrum in both passive and active modes (Figure 1.4b). If we combine satellite imagery (both panchromatic and three-band colour), film based archival photography, ground based photography

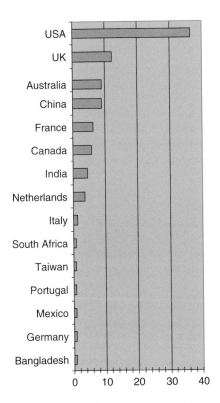

Figure 1.2 Distribution of the manuscripts according to the laboratory citizenship of the first author (in % of the studied papers).

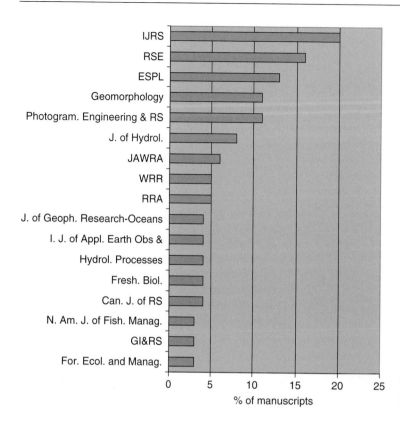

Figure 1.3 Distribution of the manuscripts within the journals having published more than two manuscripts in the WOS (in % of the studied papers).

and contemporary digital airborne photography, we find that the traditional camera (either film or digital) is still the most commonly used sensor (13,5%). LiDAR (Light Detection And Ranging), RADAR and TIR (Thermal InfraRed) sensors are also well cited with respectively 13.5% 11% and 9% of manuscripts. Spaceborne sensors such as ASTER (Advanced Spaceborne Thermal Emission and Reflection Radiometer), MODIS (Moderate Resolution Imaging Spectroradiometer) and MERIS (Medium-spectral Resolution Imaging Spectrometer) are also cited. Airborne hyperspectral imagers such as CASI (Compact Airborne Spectrographic Imager) are less frequently cited (3%). 'Terrestrial remote sensing' is also cited with devices such as TLS (Terrestrial Laser Scanning, 0.5%), LSPIV (Large Scale Particle Image Velocimetry, 3,5%) and ground-based video (6%).

We also explored the temporal trend of the platforms/sensors used for the most frequent (Figure 1.5). Two relative references were used, the 15 000 manuscripts focused on river management and stored in the WOS, and the 200 papers studied without distinguishing any method. These two cumulated curves show the WOS database prior to 1990 is not very rich and the steep

trend we observed in recent years is also partly due to the database structure itself. When looking at the relative cumulated curves per year for the different platforms/sensors, two groups can be observed: Pioneer platforms/sensors such as photograph and Infra-Red for which the median year is 2000 and their use seems to decrease a bit after, and new sensors such as TIR (median year 2005), SAR/Radar (median year 2006) but also LiDAR (median year 2008). Airborne/Aerial data, Landsat and video seem to follow the general trend in term of publications. However, Landsat seems more popular in the 1998–2004 period and its relative use is decreasing.

1.3.3 Topical Areas

An examination of the abstracts revealed that FRS is contributing to a large set of topics that we can group into three broad areas (Figure 1.6). First, the drive for a better science base in management decisions has seen remote sensing applied to ecological and habitat studies aiming to identify land-use types, specific habitat types and biotopes (37% of papers). Second, investigations in water sciences which are related to the fields of water

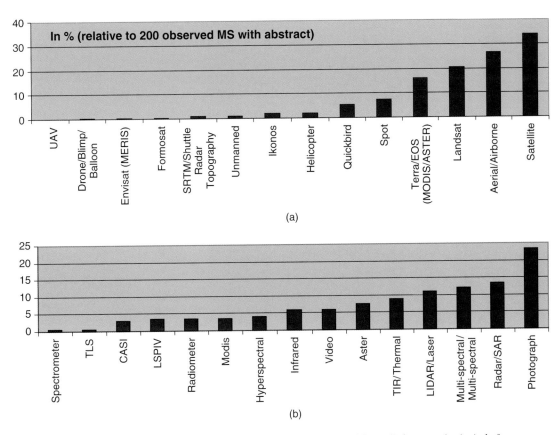

Figure 1.4 Frequency of terms within the abstracts of the 200 manuscripts (in % of the studied manuscripts): a) platforms, b) sensors.

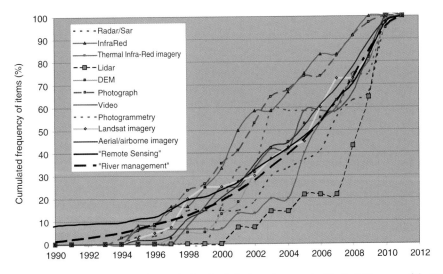

Figure 1.5 Cumulative frequency curve (in % of papers) of each of platforms/sensors cited in the 200 papers of the Web of Science dealing with rivers and remote sensing. We compared the temporal evolution of frequency for the terms "Remote sensing" and "River management" in the 200 studied manuscript within the whole Web of Science dataset.

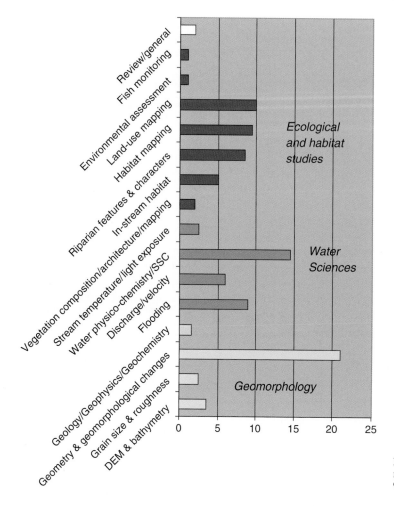

Figure 1.6 Review of abstracts of 203 manuscripts to identify the main topics (in % of the studied manuscripts).

chemistry, hydraulics and hydrology and which study specific topics such as plumes, pollution, suspended sediment concentrations, stream temperature, flooding, discharge and velocity are seeing an increasing dependence on FRS (33.5% of papers). Third, with the aim of improving and often of up-scaling the data acquisition process in traditional fluvial geomorphology, inquiries relating to regional and tectonic settings, bank erosion monitoring and decadal channel shifting, geomorphic changes, channel geometry, bathymetry, grain size, have all seen an increasing use of remote sensing data (27% of papers).

Figure 1.7 presents box and whisker plots showing the publication periods for the five most frequent topics shown on the vertical axis of Figure 1.6. This shows that some of the topics have emerged fairly recently such as SSC & Water Chemistry for which most of the

papers were published between 2008 and 2009 on habitat mapping and riparian features, whereas others are more popular all along the studied period such as flooding and geomorphic changes.

In addition to the timing of publications, we briefly examined the abstract content for these five topics in order to identify the main research thrusts within each area. In the case of habitat mapping, vegetation mapping is by far the most dominant application of remote sensing. Studies range from native vegetation assessment to the identification of invasive species. Satellite platforms are the major source of data but airborne platforms seem increasingly utilised. We also find a few published works using underwater video in order to characterise fish and animal behaviour. However, these studies of fish behaviour and/or habitat are actually rare which indicates

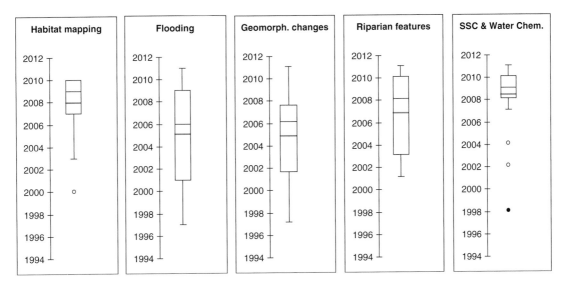

Figure 1.7 Distribution of the occurrence of the five most frequent topics (shown in Figure 1.6) within the set of 200 studied papers. Box plots provide the two deciles and quartiles with a black line indicating the median and the red line showing the average.

that progress remains to be made in the applications of FRS to the mapping and characterisation of stream biota.

Most of the papers dealing with flooding focused on the use of the synthetic aperture radar (which can sense through cloud cover) in order to map flooding extent in near real time at both coarse and fine spatial resolutions. This application uses both spaceborne platforms (ERS-1, RADARSAT-1) and airborne platforms. Additionally, Landsat TM is used to determine inundation from a range of flows because of its temporal capacity to cover areas repeatedly. This topic area also makes heavy use of topographical data derived from remotely sensed sources in order to identify peaks, troughs and slopes in flood affected areas. At large scales, the Shuttle Radar Topography Mission (SRTM) DEM is commonly used. At smaller scales, LiDAR is increasingly used to provide high resolution, high accuracy topographic height and even bathymetry (i.e. water depths). LiDAR also has the advantage of measuring vegetation height, which can be converted to friction coefficients. Generally speaking, these flooding studies employ this range of FRS tools in order to provide baseline data which is then fed into hydraulic and/or hydrologic models.

In the geomorphic change topic, most of the contributions focused on channel changes at a decadal scale based on repeated aerial photos or satellite imagery (e.g. SPOT or Landsat) in order to understand bank or delta erosion, meander migration rates and sediment production.

There is also a good volume of published work on the spatial organisation of fluvial landforms or reaches and the factors controlling them, notably geology, tectonics and riparian vegetation which have often been conducted over very long reaches (catchment and sometimes sub-continental scales). Other papers also explored smaller scale, in-channel morphological changes such as bars, channel branches, considering their sizes, their forms and the associated land cover attributes. At these smaller scales, human pressures such as gravel mining and urbanisation have also been discussed in the literature. In the case of these smaller scale studies, air photo or satellite imagery remains the norm. However, one contribution used Synthetic aperture radar (SAR) imagery for monitoring the changing forms of braided rivers over a short time scale. This is likely a reflection of the technical progress in SAR technology. Finally, fluvial geomorphology seems to be the field where most methodological developments are occurring. Here we find a significant body of published works demonstrating the use of both passive and active remote sensing in order to characterise channel width, channel depth, riparian vegetation and sediment characteristics. In terms of data sources, this area is dominated by standard photography and LiDAR (both terrestrial and airborne).

Abstracts found with the keywords 'Riparian Features' were quite varied in content. However, in common with the habitat mapping topic, vegetation identification remains as a dominant application of remote sensing. Here we find applications of LiDAR, colour and multi-spectral data aimed at identifying the composition and land-uses of the riparian zones along with their temporal dynamics. Traditional image classification of these datasets remains the principal method. However, a few papers did mention newly developed object-based classification methods. We also find that the scale of the studies in this category varies quite widely from studies focusing on bankside vegetation a few meters or tens of meters away from the channel to studies examining the entire catchment of large rivers.

Studies of river water chemistry and suspended sediment concentrations (SSC) are well established in oceanic sciences. They are also well established in large river science with some early work taking advantage of the Landsat program (Aranuvachapun and Walling, 1988). However, in the context of the smaller, so-called 'normal', rivers which are the focus of this book, remote sensing of water quality publications is scarce. Within our search results, water-quality papers were dominated with estuarine and large river studies at the interface between oceanic and fluvial sciences. The rationale behind most of these studies is to replace expensive and labour intensive ground-based field monitoring with multispectral or hyperspectral remote sensing data in order to perform what is in essence 'remote spectroscopy'. The key focus is the study of river plumes in terms of sediment load

and pollution load. The parameters which are directly measured in these studies are turbidity (i.e. water clarity) and organic matter concentrations (i.e. presence of chlorophyll). These metrics can then be used as proxies for other parameters such as pollution load and salinity. The most commonly used sensors are the ETM+ (Landsat), MODIS and the Advanced Land Imager (ALI) which is a multispectral sensor mounted on NASA's Earth Observation-1 (EO-1) satellite.

1.3.4 Spatial and Temporal Resolutions

Finally, the abstracts were used to examine the range of temporal and spatial resolutions in use within our abstract database (Figure 1.8a). Most of the contributions are based on spatial resolutions of 10–50 m confirming the use of satellite imagery. Coarser resolutions are less frequent. Metric and sub-metric resolution mainly based on airborne imagery are also very common, reaching 25% of papers. Ground-based remote sensing (here we assume decimetric or centimetric resolutions even if not specified in the abstract) is also a field which is well explored within 10% of the contributions. When combining topical areas and the spatial resolutions, a χ^2 test shows they are dependant ($p < 0.0001$) (Table 1.3). Discharge and fish monitoring are based on ground remote-sensing, whereas DEM and bathymetry use very high resolution (>1 m) data. Riparian features and land-use mapping also used very high resolution (1 m) data mainly based

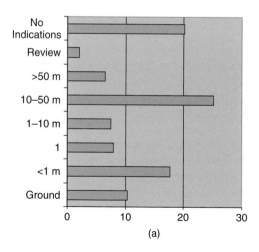

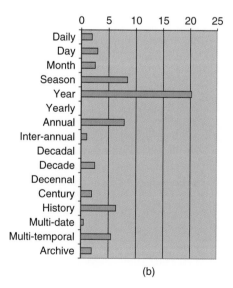

(a)   (b)

Figure 1.8 Frequency of terms within the abstracts of the 200 manuscripts (in % of the studied manuscripts) dealing with (a) spatial and (b) temporal framework.

Table 1.3 A posterior contribution of each of the cells to the χ^2 test testing the independence between the classes of broad topics and spatial resolution.

Topic	Spatial Resolution					
	Ground	1	<1 m	1–10 m	10–50 m	>50 m
DEM & bathymetry	(+) NS	(−) NS	**(+)*****	(−) NS	(−) NS	(−) NS
Discharge/velocity	**(+)*****	(−) NS	(−) NS	(−) NS	(−) NS	(−) NS
Environmental assessment	(−) NS	(−) NS	(−) NS	(−) NS	(+) NS	(−) NS
Fish monitoring	**(+)*****	(−) NS	(−) NS	(−) NS	(−) NS	(−) NS
Flooding	(−) NS	(−) NS	(+) NS	(−) NS	**(+)*****	(−) NS
Geometry & geomorphological changes	(−) NS	(−) NS	(+) NS	(−) NS	(+) NS	(+) NS
Grain size & roughness	(+) NS	(+) NS	(+) NS	(−) NS	(−) NS	(−) NS
Habitat mapping	(−) NS	(+) NS	(−) NS	**(+)****	(−) NS	(−) NS
Land-use mapping	(−) NS	**(+)****	(+) NS	(−) NS	(+) NS	(−) NS
Riparian features & characters	(−) NS	**(+)****	**(+)****	(+) NS	(−) NS	(−) NS
Stream temperature/light exposure	(−) NS	(−) NS	(+) NS	(+) NS	(−) NS	(−) NS
Vegetation composition/architecture/ mapping	(−) NS	(+) NS	(−) NS	(−) NS	(−) NS	(−) NS
In-stream habitat	(+) NS	(−) NS	(+) NS	(−) NS	(−) NS	(+) NS
Water physico-chemistry/SSC	(−) NS	(−) NS	(−)*	(−) NS	(+) NS	**(+)*****

(+): *Positive association*
(−): *Negative association*
NS: *not significant at* $\alpha = 0.1$
*: *significant at* $\alpha = 0.1$
**: *significant at* $\alpha = 0.05$
***: *significant at* $\alpha = 0.01$

Figure 1.9 Location of the detailed examples shown in the different chapters of the book: public site on Google Earth.
Source: http://maps.google.fr/maps/ms?msid=215028322631048652408.0004a7dd26c4dfd045d2c&msa=0 © 2012 Google.

on high resolution satellite images or airborne photos. Current habitat mapping publications used a lower resolution platform such as Spot or Landsat. Flooding and water physico-chemistry are often based on coarser resolution images.

Temporal resolutions were often difficult to find and/or not explicitly defined in the abstracts. We therefore separated the abstracts into several categories which imply a certain resolution timescale rather than exact quantitative values (Figure 1.8b). The terms 'year' and 'annual' are the most frequent (respectively 20% and 8%) but 'early' or 'inter-annual' are less common. The term 'season' is quite often cited (8% of MS) as well. 'Decade' and 'century' but also 'day' or 'daily' also occurred occasionally. The terms 'multi-temporal' or 'historic' concerns 5.5 to 6% of MS. Interestingly, we see that 47.5% of papers mention temporal resolution terminology. This obviously shows the importance of monitoring work in remote sensing. However, it also illustrates the importance and persistence of satellite data as a source of data acquisition in river sciences. Despite the lower resolution, the reliable availability of satellite imagery at predictable time intervals is a major advantage which could very well explain the past, current and future importance of satellite data in fluvial remote sensing.

1.3.5 Summary

This survey of published literature in FRS illustrates some key points about this sub-discipline of remote sensing. Our database search revealed that over the last 35 years traditional satellite data was the major data source employed by fluvial remote sensing studies. We found that a surprisingly high proportion of published work used traditional remote sensing data such as Landsat, ASTER and even MODIS (Brodie et al., 2010; Liu et al., 2010; An et al., 2011). The legacy of traditional satellite remote sensing can also be seen in the very high number of publications which focus on vegetation characterisation/quantification (Laba et al., 2010; Bertoldi et al., 2011). This trend has continued well into the twenty-first century with airborne data remaining in second place and, despite being capable of higher spatial resolutions, not yet overtaking spaceborne data in the published literature. The causes for this are difficult to establish with certainty. However, the availability of reliable repeat (multi-temporal) imagery from satellite sources is a likely factor. Furthermore, we believe that the dominance of satellite-based publications also shows that classic river sciences and

management studies have not made heavy use of remote sensing since spaceborne data is rarely suited to the spatial and temporal scales which characterise river processes. In the papers we surveyed, only a small minority examined classic river science topics such as fluvial bedforms and channel topography, sediment calibre and dynamics (especially in the gravel to boulder size range) and river fauna. However, within our search results, we can clearly see the impact of recent published works aimed at developing remote sensing technology and methods which are tailored to river sciences and capable of providing data acquisition strategies that are well suited to river science investigations. Advances in imaging technology which now allow for centimetric imagery from the air (Carbonneau et al., 2004; Forzieri et al., 2010) and decimetric imagery from space (Zhang et al., 2004; Johansen et al., 2010), new LiDAR technology which is customised to river environments (Kinzel et al., 2007) and processing methods designed to extract a range of features of interest to river sciences (Carbonneau, 2005; Jordan and Fonstad, 2005; Buscombe and Masselink, 2009), have all radically improved our capability to characterise the fluvial forms and processes mentioned above. Given time we expect this progress to change the overall profile of publications in fluvial remote sensing. We would therefore hope that an identical bibliometric survey conducted in 2020 would yield a significantly enhanced list of publications where the line between traditional river sciences and traditional remote sensing has become blurred or even invisible.

1.4 Brief outline of the volume

The volume is divided into three main sections. First, we present a series of six chapters with a slightly more theoretical perspective on the 'Spectrum of Remote Sensing Techniques and their Applications'. Chapter 2 explores the basic rationale for using remote sensing in river environments. Starting from the question of 'What can we see?' this chapter explores the possibilities and limitations of Fluvial Remote Sensing. Chapter 3 follows this topic with a discussion on the basic physics which underpins the application of remote sensing to river environments. The following chapters then begin addressing specific elements of technical progress. Chapter 4 discusses hyperspectral (very high spectral resolution) remote sensing, while Chapter 5 deals with thermal imagery, which is clearly of importance in the context of changing climates and potentially warming rivers. Chapter 6 deals with FRS methods applied to another emerging impact of changing hydrologic cycles: flooding. Chapter 7 deals with LiDAR technology and it's specific application to river environments. This first section is followed by a section of five chapters with a more applied perspective and which focuses on 'Hyperspatial to catchment-scale imagery'. Chapter 8 defines and discusses the concept of 'Hyperspatial' imagery. Chapter 9 presents an extensive habitat mapping based on hyperspatial imagery. Chapter 10 examines how high resolution imagery can be used to go beyond classic characterisation and predict the evolution of riparian vegetation. Similarly, Chapter 11 presents image-based characterisation approaches which extend beyond local study areas and can be applied to long reaches or even entire networks. Finally, Chapter 12 examines the uses of remote sensing in predicting the land-use changes of entire catchments (i.e. watersheds). In the third and final section of the book, we examine the increasing use of ground-based (terrestrial) remote sensing methods in river sciences. Chapter 13 considers the uses of image-based data acquisition in indoor flume experiments. Chapter 14 examines the application of ground-based LiDAR, usually called 'Terrestrial Laser Scanners' (TLS) to river sciences. Chapter 15 focuses on oblique and vertical ground-photos which can provide millimetric spatial resolution for grain size or grain morphometry at a very high temporal resolution. These approaches provide powerful tools for small-scale process monitoring. The final three chapters still use imagery as their primary data source but they represent a definite departure from areas which are normally considered as within the remit of remote sensing. Chapter 16 discusses the uses of videography in river monitoring works. Chapter 17 discusses the uses of imagery in the study of small individual lotic organisms. Finally, Chapter 18 examines the use of photo-questionnaires in the assessment of the societal value of rivers and associated restoration works. Practical conclusions close the volume in Chapter 19. This volume therefore introduces the scope of research already achieved and shows that the techniques now available can be the basis for further exciting developments in the next few years ensuring the field of Fluvial Remote Sensing is poised to achieve more significant contributions. Locations of case-studies for the different chapters are also available on line so as to provide opportunities for readers to see in more detail size, geometry and characters of the rivers and field sites discussed in the volume (Figure 1.9).

References

An, S.Q., Zhao, D.H., Cai, Y., Jiang, H., Xu, D.L., and Zhang, W.G. 2011. Estimation of water clarity in Taihu Lake and surrounding rivers using Landsat imagery. *Advances in Water Resources*, **34**(2), 165–173.

Aranuvachapun, S., and Walling, D.E. 1988. Landsat-mss radiance as a measure of suspended sediment in the lower Yellow River (Hwang Ho). *Remote Sensing of Environment*, **25**(2), 145–165. 10.1016/0034-4257(88)90098-3.

Bertoldi, W., Gurnell, A.M., and Drake, N.A. 2011. The topographic signature of vegetation development along a braided river: Results of a combined analysis of airborne lidar, color air photographs, and ground measurements. *Water Resources Research*, **47**. 10.1029/2010WR010319.

Brodie, J., Schroeder, T., Rohde, K., Faithful, J., Masters, B., Dekker, A., Brando, V., and Maughan, M. 2010. Dispersal of suspended sediments and nutrients in the Great Barrier Reef lagoon during river-discharge events: conclusions from satellite remote sensing and concurrent flood-plume sampling. *Marine and Freshwater Research*, **61**(6), 651–664.

Buscombe, D., and Masselink, G. 2009. Grain-size information from the statistical properties of digital images of sediment. *Sedimentology*, **56**(2), 421–438.

Carbonneau, P.E. 2005. The threshold effect of image resolution on image-based automated grain size mapping in fluvial environments. *Earth Surface Processes and Landforms*, **30**(13), 1687–1693.

Carbonneau, P.E., Lane, S.N., and Bergeron, N.E. 2004. Catchment-scale mapping of surface grain size in gravel bed rivers using airborne digital imagery. *Water Resources Research*, **40**(7). 10.1029/2003WR002759.

Chuvieco, E., and Alfredo, H. 2010. *Fundamentals of Satellite Remote Sensing*, CRC Press, Taylor and Francis, Boca Raton. 448 pp.

Dunford, R., Hervoue, A., Piegay, H., Belletti, B., and Tremelo, M.L. 2011. Analysis of post-flood recruitment patterns in braided-channel rivers at multiple scales based on an image series collected by unmanned uerial vehicles, ultra-light aerial vehicles, and satellites. *Giscience & Remote Sensing*, **48**(1), 50–73.

Fausch, K.D., Torgersen, C.E., Baxter, C.V., and Li, H.W. 2002. Landscapes to riverscapes: Bridging the gap between research and conservation of stream fishes. *Bioscience*, **52**(6), 483–498.

Fonstad, M.A., and Marcus, W.A. 2010. High resolution, basin extent observations and implications for understanding river form and process. *Earth Surface Processes and Landforms*, **35**(6), 680–698.

Forzieri, G., Moser, G., Vivoni, E.R., Castelli, F., and Canovaro, F. 2010. Riparian vegetation mapping for hydraulic roughness estimation using very high resolution remote sensing data fusion. *Journal of Hydraulic Engineering-Asce*, **136**(11), 855–867.

Frankl, A., Nyssen, J., De Dapper, M., Haile, M., Billi, P., Munro, R.N., Deckers, J., and Poesen, J. 2011. Linking long-term gully and river channel dynamics to environmental change using repeat photography (Northern Ethiopia). *Geomorphology*, **129**(3-4), 238–251.

Johansen, K., Phinn, S., and Witte, C. 2010. Mapping of riparian zone attributes using discrete return LiDAR, QuickBird and SPOT-5 imagery: Assessing accuracy and costs. *Remote Sensing of Environment*, **114**(11), 2679–2691.

Jordan, D.C., and Fonstad, M.A., 2005, Two-Dimensional mapping of river bathymetry and power using aerial photography and GIS on the Brazos river, Texas, *Geocarto International*, **20**(3), 1–8.

Kinzel, P.J., Wright, C.W., Nelson, J.M., and Burman, A.R. 2007. Evaluation of an experimental LiDAR for surveying a shallow, braided, sand-bedded river. *Journal of Hydraulic Engineering-Asce*, **133**(7), 838–842.

Laba, M., Blair, B., Downs, R., Monger, B., Philpot, W., Smith, S., Sullivan, P., and Baveye, P.C. 2010. Use of textural measurements to map invasive wetland plants in the Hudson River National Estuarine Research Reserve with IKONOS satellite imagery. *Remote Sensing of Environment*, **114**(4), 876–886.

Legleiter, C.J., Roberts, D.A., and Lawrence, R.L. 2009. Spectrally based remote sensing of river bathymetry. *Earth Surface Processes and Landforms*, **34**(8), 1039–1059.

Lillesand, T.M., Keifer, R.W., and Chipman, J. 2008. *Remote Sensing and Image Interpretation* 6th Ed., John Wiley and Sons. 804 pp.

Liu, D.Z., Chen, C.Q., Gong, J.Q., and Fu, D.Y. 2010. Remote sensing of chlorophyll-a concentrations of the Pearl River Estuary from MODIS land bands. *International Journal of Remote Sensing*, **31**(17-18), 4625–4633.

Marcus, W.A., and Fonstad, M.A. 2008. Optical remote mapping of rivers at sub-meter resolutions and watershed extents. *Earth Surface Processes and Landforms*, **33**(1), 4–24.

Marcus, W.A., and Fonstad, M.A. 2010. Remote sensing of rivers: the emergence of a subdiscipline in the river sciences. *Earth Surface Processes and Landforms*, **35**(15), 1867–1872.

Marcus, W.A., Legleiter, C.J., Aspinall, R.J., Boardman, J.W., and Crabtree, R.L. 2003. High spatial resolution hyperspectral mapping of in-stream habitats, depths, and woody debris in mountain streams. *Geomorphology*, **55**(1–4), 363–380.

Mertes, L.A.K. 2002. Remote sensing of riverine landscapes. *Freshwater Biology*, **47**(4), 799–816.

Sun, Z.Y., Ma, R., and Wang, Y.X. 2009. Using Landsat data to determine land use changes in Datong basin, China. *Environmental Geology*, **57**(8), 1825–1837.

Ward, J.V., Robinson, C.T., and Tockner, K. 2002. 'Applicability of ecological theory to riverine ecosystems.' In: *International Association of Theoretical and Applied Limnology, Vol 28, Pt 1, Proceedings*, R.G. Wetzel, ed., E Schweizerbart'sche Verlagsbuchhandlung, Stuttgart, 443–450.

Zhang, J.Q., Xu, K.Q., Watanabe, M., Yang, Y.H., and Chen, X.W. 2004. Estimation of river discharge from non-trapezoidal open channel using QuickBird-2 satellite imagery. *Hydrological Sciences Journal-Journal Des Sciences Hydrologiques*, **49**(2), 247–260.

2 Management Applications of Optical Remote Sensing in the Active River Channel

W. Andrew Marcus[1], Mark A. Fonstad[1] and Carl J. Legleiter[2]

[1]Department of Geography, University of Oregon, Eugene, OR, USA
[2]Department of Geography, University of Wyoming, Laramie, WY, USA

2.1 Introduction

As a potential user of remote sensing for river monitoring and analysis, you may find yourself wondering 'what can be measured, mapped, and/or modeled with remote sensing?'; 'will clients accept results based on remote sensing?'; and 'why use remote sensing rather than classical field methods?' These are all critical questions, and answering them correctly determines whether a remote sensing approach will substantially benefit a project or detract from the project's success.

This chapter addresses these questions in the context of *passive optical imagery* of the *active river channel*. 'Passive' refers to the measurement of light occurring naturally in the environment – reflected solar energy. This is in contrast to 'active sensors' such as radar and LiDAR, which emit a pulse of energy and record the return of that energy. Active sensors are explored in later chapters. 'Optical' refers to the dominant wavelengths of light originating from the sun: blue, green, red and near-infrared wavelengths. The blue, green and red wavelengths are 'visible light' that we can see with our eyes. The near-infrared is invisible to the human eye, but is still a major component of sunlight. Passive optical imagery is something we all have seen since childhood – most photos are passive optical imagery, as are panchromatic, colour, and near-infrared air photos. This chapter discusses imagery where the data are broken into narrow, discrete wavelengths of light, as with hyperspectral sensors. It also presents information on broad-spectrum imagery where many wavelengths are bundled together, as with panchromatic photography that encompasses all visible wavelengths to form a single image representing overall brightness.

The chapter also focuses on the 'active channel', which is the low flow channel plus adjacent areas that are free of vegetation and subject to scour or deposition under typical hydrologic conditions. The active channel includes the submerged channel, unvegetated mid-channel islands, chutes, and exposed bars. The active channel has been the focus of a great deal of remote sensing research since the mid-1990s, a period that coincides with the increased availability of centimeter- to meter-resolution digital imagery. Prior to that time, research on remote sensing of rivers was generally limited to film-based aerial photos or digital multispectral satellite imagery. Studies of the active channel with digital multispectral data were limited because the pixel size of satellite imagery was too large. A 30 or 80 m Landsat pixel from the 1970s or 1980s, for example, would cover the entire active channel plus large portions of the bank and floodplain in all but the largest lowland rivers.

The vegetated floodplain is not included in our definition of the active channel and is not discussed in this chapter. The optical techniques for mapping riparian vegetation on floodplains are, for the most part, the same techniques used to map vegetation, regardless of setting. These methods are extensively discussed in many text books, articles, and on-line sources and are already widely used for resource management (for overviews see Dieck and Robinson, 2004; Jensen, 2007). In contrast, techniques specifically focused on the active channel are sufficiently new that they are not yet widely known to the management community.

Evaluating whether to use remote sensing requires knowing which river features can be measured, mapped, and/or modeled in this manner. The first part of this chapter therefore reviews active channel features that can be mapped with passive optical imagery, the general nature of the remote mapping techniques, and some of the limitations specific to each application. A subsequent section summarises issues that are common to many remote sensing applications (e.g., accuracy assessment). The chapter concludes with a discussion of factors to consider when determining whether or not to use remote sensing rather than, or in addition to, other available techniques.

Finally, the focus throughout the chapter is on parameters that can be monitored and mapped using remote sensing. These mapping applications may be an endpoint in their own right, but are often just the starting point for management applications related to modeling, planning, and active intervention in the stream. We briefly mention some management applications such as habitat assessment and flood planning, but for the most part we encourage the reader to think about, and imagine, the many uses to which the remote sensing-based measurement and maps can be put.

2.2 What can be mapped with optical imagery?

What can you measure and map with optical images? At one level, the answer to this question is simple: anything you can see with the naked eye is potentially 'mappable' with optical imagery. But the answer does not stop there; hyperspectral and multispectral imagery can detect features at resolutions and wavelengths not visible to the human eye. When looking at a clear-water stream, one therefore can intuitively determine which features might be mappable with remote sensing (muddy streams where the bottom cannot be seen are generally poor candidates for remote sensing of features within the water column

Figure 2.1 Variations in depth (and other river features) can be easily detected with the naked eye. In this example along the Trinity River, California, variations in depth are immediately apparent as variations in water darkness. Likewise, glides, riffles and pools (i.e. biotypes) can be distinguished by variations in depth and surface turbulence, and variations in sediment size are apparent in the shallow water portions of the stream. River features that are visible to the naked eye are features that, in theory, can be measured and mapped with optical remote sensing.

or on the bed). In Figure 2.1, for example, one can gauge which areas are deeper by the darkness of the water. Likewise, the human eye can easily pick out variations in sediment size, pieces of wood, algae on rock surfaces, surface turbulence, and so forth. All of these are therefore good candidates for mapping via remote sensing. In addition, multi- or hyperspectral data might be able to pick out more subtle variations in turbidity, depth, and turbulence.

Alas, life is rarely so simple. Our eyes and brains process a remarkable amount of information on the fly: brightness, colour, texture, shape, shadow, size, spatial context, rate of movement, and so on. In contrast, remote sensing algorithms typically use just one, or maybe two of these factors at a given time. Often, therefore, a remote sensing-based, image processing approach misses the subtle identifying characteristics that are readily detected by the human brain. What can be 'seen' with remote sensing therefore often differs in subtle ways from what the eye-brain combination can detect. Yet sometimes the different perspectives provided by remote sensing (especially with multispectral and hyperspectral imagery) coupled with trained users can detect more than the brain-eye combination can see (e.g., Legleiter and Goodchild, 2005).

Much of the previous research on remote sensing of rivers has focused on using imagery to map the same features that humans observe and map in the field. The bulk of the following section summarises this kind of work. However, some of the more interesting advances in remote sensing of rivers will probably result from abandoning these anthropic constraints and letting the images 'speak for themselves;' we briefly discuss some of these prospects in the conclusion of this chapter. Likewise, many of the most exciting advances in remote sensing of rivers are coming from devices that do not use the visible and near-infrared wavelengths with which people are most familiar. Radar, LiDAR, and thermal imagery and combinations of these with optical imagery are opening up a wide range of new remote sensing applications for river management; the reader should be sure to review those chapters to understand the full spectrum of potential applications.

The following summary of optical image applications highlights recent studies that have achieved the highest accuracies. Reviews by Mertes (2002) and Gilvear and Bryant (2003) provide more information on the history of different remote sensing sensors and applications in rivers, while Marcus and Fonstad (2008) and Feurer et al. (2008) focus on reviews of optical imagery only. Marcus (2012) provides more detail on recent advances in using passive and active instruments to measure the hydraulic environment of rivers. In the following discussion of potential applications we provide an overview of how various techniques work, identify major limitations specific to those applications, and briefly summarise theoretical and practical limitations generic to all applications (e.g. image quality, resolution, shadow, logistics).

2.3 Flood extent and discharge

Documenting areas of flood inundation is critical for emergency response and floodplain planning. Mapping flood extents also is useful for identifying locations where efforts to restore or maintain stream habitat and biodiversity could be targeted. Finally, even when rivers are not flooding, repeat mapping of inundated area provides insight into discharge variability over time as well as an early warning of flood or drought hazards in remote regions. The importance of remote sensing for tracking flow variability and floods is particularly important in areas without hydrologic monitoring systems or where access to such data are limited (Brakenridge et al., 2005).

Aerial photos have long been used to document flood extent, especially during the peak of major floods when access to the river is limited and potentially dangerous. Some of the earliest applications of satellite imagery were for measuring flood extents (Smith, 1997), but the 17-day repeat cycle of the Landsat satellite did not allow monitoring of change during one flood event. Subsequently, the advent of satellites that covered the same location on a daily basis enabled monitoring of floods with very coarse (1 km) spatial resolution data in large rivers (Barton and Bathos, 1989). More recently, 1 to 5-m resolution satellite imagery provides a mapping tool in smaller streams. In contrast to satellites, however, aircraft can fly beneath cloud cover and capture flood events that might be missed by satellite imagery.

Because water is generally visible to the human eye on imagery, it is possible to trace or manually digitise the wetted channel extent. Automated approaches for mapping flow extent and discharge using satellite imagery are reviewed by Smith (1997). Mapping the extent of water is typically done with the shortwave infrared bands. Water is a strong absorber in these wavelengths and therefore very dark on images compared to other features. It is our experience that automated approaches with meter or cm resolution imagery can confuse water with dark shadow and require manual editing to ensure reasonable results. At coarser resolutions, however, the shadows are mixed with other features in the pixel (e.g., trees) and the spectral signal is distinct from water. The task of separating water from other features is also simplified if one has change-over-time imagery. In this case, the difference between dry and inundated conditions for a given pixel is sufficiently large to enable change detection algorithms to accurately map surface inundation using a variety of sensors, including the Landsat Thematic Mapper (Kishi et al., 2001), AVHRR (Sheng et al., 2001), and the EO-1 hyperspectral sensor (Ip et al., 2006). Smith (1997, p. 1429) notes that there has been little change since the 1970s in how inundated areas are mapped and that the approach is 'now considered operational.'

Discharge can be estimated directly from an inundated area if there is a nearby ground-based gauging station and a correlation can be established between discharge and flood extent. The highest accuracy discharge measurements using this approach are achieved if 'gauging reaches' are identified where the wetted width varies dramatically with changes in discharge (Brakenridge et al., 2005), a response that characterises many braided rivers but is more difficult to find in single thread channels. To avoid confusion, it is worth noting that large changes in width in response to discharge variations is the opposite of what is considered ideal for ground-based gauging stations, where it is preferable to have minimal variation in width in response to fluctuations in discharge. Space-based estimates of this sort are particularly valuable in areas where gauging stations are being discontinued. Correlations established in this manner can be extended to other rivers of similar morphology. Alternatively, flood volume and depth can be derived by coupling flood inundation maps with digital elevation models, which can be coupled with hydrologic models to estimate discharge (Smith, 1997). In a related application, remote sensing-based measurements of flood extent can be used to calibrate and validate numeric models of flooding (Bates, 2004).

Aerial flights and satellite imagery will doubtlessly remain useful remote sensing tools for flow monitoring. Cloud and tree cover are major obstacles to these approaches, however, because the sensor cannot see the water surface. Researchers and monitoring programs therefore are increasingly turning to active radar imagers, which penetrate forest and cloud cover and allow water to be distinguished from other features as well as providing surface elevation data for broad areas (Alsdorf et al., 2007; Schumann et al., 2009).

2.4 Water depth

The importance of water depth to monitoring, mapping and modeling river habitats has generated considerable research interest in measuring depth from optical imagery. As long as the water is clear enough to see to the bottom, there are three general approaches that provide relatively accurate depth estimates (Table 2.1).

The easiest depth mapping technique is the correlation approach, where the brightness of the image (i.e., the pixel value) at a number of locations is correlated to field measurements of depth at the same locations. The regression equation derived from this correlation is then applied to the remainder of the image to estimate water depths. There are a number of variations to the correlation approach. The highest accuracies (Table 2.1) are achieved by using multiple regressions where more than one image band (e.g. red, green and blue bands) are correlated to depth (e.g., Gilvear et al., 2007; Lejot et al., 2007). Marcus et al. (2003) achieved high accuracies using 128-band hyperspectral imagery that covered the visible and shortwave infrared wavelengths. They first ran a principal components analysis on the water portion of the image to remove noise and reduce the dimensionality of the spectral signal, then ran a step-wise regression

for each biotype (riffle, pool, glide, etc.) to derive depths that were specific to different surface turbulence regimes within the stream. The major limitation specific to the correlation approaches is that water depths must be measured in the field at the same time as imagery is collected to avoid variations in discharge and channel shape that could modify the relationship between pixel values and depth.

Physically-based models provide an alternative to simple correlation approaches. These models are based on the physics of how light moves through water, as is summarised in Chapter 3 and reviewed by Legleiter et al. (2004; 2009). From a management perspective, the most readily applicable of these models are those that avoid the need for field teams to collect field data at the time-of-flight, which removes a major logistical constraint. Fonstad and Marcus (2005) developed a hydraulically assisted bathymetry (HAB) model that couples equations describing light attenuation by the water column and the hydraulics of open channel flow with data on discharge, slope and channel width to map depths throughout a stream (Figure 2.2). Their physical modeling technique does not require field crews or data collection specific to the project. The discharge data can come from nearby gauging stations, slopes can come from maps or other sensors, and width can be measured from the imagery. Moreover, the model is sufficiently simple that the mathematical formulae used to compute depth estimates can be implemented in a spreadsheet. Because the HAB technique does not require field measurements, it can be used with historical imagery so long as discharge data are available for a nearby site.

Some researchers have expressed concern, however, that simple models like HAB do not consider the effects of variables like turbidity, substrate size and colour, surface turbulence, algae on rocks, and other factors that could potentially complicate depth mapping. In an extensive theoretical and empirical experiment, however, Legleiter et al. (2009) demonstrated that these other factors can be accounted for and accurate depth estimates achieved by using the natural log of an appropriate band ratio (Table 2.1). A band ratio is simply one band divided by another, and typically the green band value is divided by the red band value for the same pixel. Differences in pixel values due to sediment colour, sediment size, vegetation, and other substrate characteristics are, in effect, normalized by band ratios, leaving the depth signal as the primary factor driving variations in pixel values. Legleiter et al. (2009) also demonstrated that depth maps derived from such ratios show remarkable resiliency across a range

Table 2.1 Accuracies for depth measurements. Accuracy measures represent the highest accuracies achieved using optical imagery to map bathymetry. Detailed explanations of the methods are in the cited references. Updated and modified from Marcus and Fonstad (2008).

Analytical Approach	Location	Imagery				Accuracy		Author(s)
		Platform & Monitor	Spatial Resolution	Spectral Range (nm)	# of Bands Used	Metric Reported	Optimal Values Achieved	
Multiple regression of spectra	Rhône River, France	Paraglider drone – digital camera	5 cm, 14 cm	Visible	3 (r,g,b)	r^2, measured vs. observed	0.53 all substrates 0.81 alluvial substrate 0.90 vegetated substrate	Lejot et al., 2007
Multiple regression of pc bands	Lamar River, WY, USA	Helicopter – PROBE-1	1 m	400–2400	128	r^2, measured vs. observed	0.67 in high gradient rifles, 0.99 in glides	Marcus et al., 2003
Physical model – field tested	Brazos River, TX, USA	Aircraft – digital camera	20 cm	Visible	3 (r,g,b)	r^2, measured vs. observed	0.77	Fonstad and Marcus, 2005
Physical model – field tested	Soda Butte Creek & Lamar River, WY, USA	Ground-based hyperspectral spectro-radiometer	0.21 to 1.00 m	400-900	various sensor configurations evaluated	r^2, measured vs. observed	0.79 to 0.98 evaluated across wide range of conditions	Legleiter et al., 2009
Photogrammetry plus physical model	South Saskatchewan River, Canada	Aircraft – film-based or digital camera	4 cm	Visible	1	Mean error: Std deviation of error:	−0.025 to 0.18 m ±0.18 to ±0.31 m	Lane et al., 2010

Figure 2.2 Depth map for the McKenzie River above Springfield, Oregon, developed using the true color air photo to the left and the HAB-2 technique (Fonstad and Marcus, 2005). Darker tones indicate deeper water, except where shadows obscure the water. Depths in this reach vary from zero to approximately 1.2 m. Pixel resolution is ~10 cm.

of spatial and spectral resolutions and environmental conditions. One finding of particular note is that the best depth estimates usually occurred when one of the ratio bands was centered around 710 nm, due to the strong absorption of near-infrared wavelengths by water, which makes the remotely sensed signal highly responsive to subtle variations in depth.

Finally, in addition to correlation approaches and physical models, classical photogrametry with stereo pairs can provide high resolution, accurate results. Application of these techniques to determine water depth is complicated, however, by the need to know the height of the water surface above the bed surface that is being mapped. Westaway et al. (2000, 2001) solved this by identifying water surface elevations at the channel edge and extending these water surfaces to the remainder of the channel. More recently, Lane et al. (2010) coupled photogrammetric depth estimation techniques with optical physics to achieve accurate depth measurements in areas covered by only one photo (Table 2.1). To accomplish this, they developed depth-reflectance relations in areas of stereo coverage and then applied these relationships to reflectance values in areas covered by only one photo.

Water clarity is the key limitation to all techniques for measuring water depth with optical imagery. The maximum depth that can be remotely measured is the maximum water depth to which the light can penetrate and return to the surface and be detected by the sensor, which varies with water column optical properties, wavelength, instrument sensitivity, and substrate composition (Legleiter et al., 2004).

2.5 Channel change

Managers and researchers have long used aerial photographs to map channel boundaries, bars, floodplain cover, erosion and other features (Gilvear and Bryant, 2003). Channel change maps are important for documenting flood hazard, erosion hazard, and changes in habitat diversity, as well as for understanding the causes of those changes. The concept behind using remote imagery to create two dimensional planimetric channel change maps is simple: features visible on air photos can be traced or digitised and transferred to map coordinates, and if images are available from different dates, the maps can be overlain to document change over time.

A major component of channel change often relates to the introduction of human features such as groynes and weirs. These features often stand out clearly on vertical images because of their shape and texture, which make them jump out to the naked eye, and because of their different composition relative to surrounding materials, which enables their detection through automated, spectrally-based techniques. However, when covered by vegetation or buried in sediments, remote sensing may well miss features of this nature.

Since the 1980s the use of remotely sensed optical imagery for channel change mapping has exploded, largely because of the widespread availability of Geographic Information Systems (GIS). GIS made it much easier to digitise river features from aerial or satellite imagery, georectify and reproject the features to a common coordinate system, conduct a wide variety of spatial and change-over-time analyses, and produce maps from the results. The techniques for mapping river features with remote imagery are standard to any GIS or remote sensing software, and many college students now have the skills necessary to carry out the work.

The ease of mapping with GIS, however, has led to a certain degree of complacency regarding the results. Too often change detection mapped from optical imagery is considered to be 'real', when in fact the change may be the result of cumulative errors. Scale distortions in the original imagery, poor selection of ground control points used to georeference the image, and the algorithms used to transform the image to a specific coordinate system all generate errors. These errors can be substantial and are compounded when attempting to detect change over time (Hughes et al., 2006). Simply put, just because it looks good does not mean that it is good. Analyses of channel change should include error assessment and use of stable control points.

Error of this nature can be significantly reduced using orthorectification techniques, which correct for planform distortion caused by topography and sensor geometry. The orthorectification techniques, however, are more costly and complex. They require additional data layers on elevation, use more expensive software, need more time to set up, and call for more specialised personnel.

In addition to GIS, an exciting advance in optical remote sensing of rivers is the three dimensional mapping of channels and channel change (this can also be done with active sensors, as is discussed in elsewhere in this volume). Previously, mapping elevation changes in channels has been a challenge because it requires high accuracy to capture channel bed elevation changes that are often subtle (Brasington et al., 2003). It is also requires measuring elevations of both the submerged and emergent parts of the channel throughout the entire active channel.

Several approaches have been used to solve these problems with optical imagery. Westaway et al. (2000, 2001) modified classical photogrammetric techniques to account for the effects of refraction. In addition, Westaway et al. (2003) developed an alternative approach that uses classical photogrammetry to derive topographic information for exposed surfaces and a correlation-based method of estimating water depth within the wetted channel from the image data.

More recently, Lane et al. (2010) extended these techniques so that they can be used with historical imagery. In their procedure, overlapping images and photogrammetric principles are used to derive local elevations inside and outside the channel. A filter then identifies which photogrammetrically derived water elevations are likely to be valid based on the certainty with which corresponding points can be identified on the two photos. The valid water depths are used to develop a relation between image brightness and depth that is applied to the rest of the channel to map depths. These depths in turn can be subtracted from the elevation at the dry surface of the channel margin to determine bed elevation. Lane et al. found that the method provided accurate depth estimates and could be used to detect vertical changes of 0.40 m or more with a 67% confidence of change (Table 2.1). Although more complex than this brief description would suggest, the Lane et al. methodology should ultimately be applicable anywhere: (1) tie points can be identified on historical imagery that can be surveyed in the present day; (2) the image scale is sufficiently fine to detect depth changes; and (3) the imagery of the submerged areas displays a range of brightness that can be correlated with depth. Marcus (2012) provides more in-depth summaries of the techniques developed by Westaway et al. and Lane et al.

As with channel change in the horizontal dimension, it is critical to consider error when estimating changes in the vertical dimension. For example, the 67% confidence of correctly detecting a vertical change of 0.40 m in the system examined by Lane et al. (2010) is reassuring at one level, but also a sobering reminder that vertical changes typical of many small streams will be missed or inaccurately portrayed using the photogrammetric approach. Brasington et al. (2003) and Lane et al. (2003) provide guidelines on how to evaluate vertical error and its implications for detection of vertical change. Likewise, one can determine the potential range of depths that can be mapped and the precision of bathymetric mapping techniques by modeling the optical physics under varying stream conditions (e.g. Legleiter et al., 2009).

2.6 Turbidity and suspended sediment

Turbidity, in its formal optical definition, refers to the amount of attenuation and back scattering of light due to suspended solids and dissolved load (Davies-Colley and

Smith, 2001). Turbidity in and of itself is an important component of water quality that controls light availability for photosynthetic organisms. Turbidity is also an important control on other remote sensing applications in rivers; for example, depth estimates from optical imagery are far less likely to be reliable where the bottom is obscured (Figure 2.3). Finally, because turbidity is related to sediment concentrations, its measurement can provide information on timing of sediment movement, variability in sediment sources, and (when coupled with discharge data) sediment load, although some approaches move directly from pixel values to suspended sediment concentrations without reference to turbidity (e.g. Pavelsky and Smith, 2009).

There is an extensive literature describing the use of optical imagery to monitor turbidity and, to a lesser degree, suspended sediment concentrations. When suspended sediments are remotely mapped, it is usually accomplished via remote mapping of the turbidity. A correlation between turbidity and ground-based

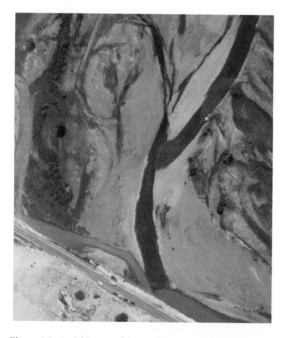

Figure 2.3 Aerial image of the confluence of Soda Butte Creek (left) and the Lamar River (right), Wyoming, USA. The turbidity in Soda Butte Creek prevents light penetration of the water column, which in turn limits measurement of depths, substrates size, or other features below the water surface. If ground measurements are available to calibrate image reflectance to turbidity, however, the imagery can be used to map turbidity and suspended sediment concentrations in the two streams.

measurements of suspended sediment concentrations then generates an estimate of suspended sediment concentrations. Because optical remote sensing images the upper portion of the water column where suspended sediment concentrations are usually lower, remote sensing approaches may underestimate depth-integrated suspended sediment concentrations.

The vast majority of literature on turbidity and sediment concentrations focuses on oceans, lakes and estuaries. The research that does exist for rivers focuses almost entirely on large systems like the Amazon (Mertes et al., 1993), Yellow (Aranuvachapun and Walling, 1988) or Yangtze Rivers (Liu et al., 2008), where turbidity, suspended sediment concentrations, and surface characteristics of the streams are relatively constant over long reaches. In contrast, a large host of factors in smaller systems alter the relation between turbidity, suspended sediment concentrations, and reflected light over short distances. Surface turbulence, water depth, substrate colour and size, aquatic vegetation, sun angle relative to the water, and local variations in sediment colour all can alter the signal received by the sensor independent of any actual changes in turbidity. Different inputs and variable timing of flow from tributaries also can alter the correlations between turbidity and sediment concentrations. In short, it remains difficult to accurately map turbidity and suspended sediment concentrations in small and medium streams with remote sensing.

The majority of approaches for estimating turbidity or sediment concentrations use field measurements to establish empirical relations between image values, turbidity, and suspended sediment. Mertes et al. (2002) provide an overview of image processing techniques for mapping suspended sediment concentrations. Presently, there are no widely accepted standard approaches for setting up these empirical relationships. Most researchers use some combination of red and/or near infrared bands to map turbidity, although green bands can be of use, especially when chlorophyll concentrations strongly influence turbidity. The relationships set up between sediment concentrations or turbidity and spectral signatures include spectral mixture analysis (Mertes, 1993), simple logarithmic relations (Aranuvachapun and Walling, 1988), normalised difference relations (Han et al., 2006), simple regression (Wang et al., 2009), multiple regression (Shibayama et al., 2007), General Additive Models (Bustamante et al., 2009) and neural networks (Teodoro et al., 2008). As of now, researchers seem inclined to choose the method that generates the best fit to the field data without having strong theoretical arguments for choosing one method over another. Moreover, simple methods often seem to generate results that are nearly as accurate as more complex approaches. Bustamante et al. (2009), for example, found that the simple Water Turbidty Index of Yagamata et al. (1988) yielded an r^2 value of 0.75 for measured versus modeled turbidity values, while a far more complex General Additive Model yielded an r^2 of 0.79. Similarly, Wang et al. (2009) found that one variable regression using band 4 (near infrared) of Landsat provided excellent estimates of suspended sediment concentrations in the Yangtze River (Table 2.2).

The wide range of techniques for estimating turbidity with optical imagery indicates the need for a more general theoretical basis for turbidity measurement. Dekker et al. (1997) reviewed the optical theory that underlies turbidity mapping by remote sensing and identified several issues of particular importance. Light interacts with turbid waters in a non-linear manner, making it hard to develop empirical relations that can be transferred from one system to another. Sun angle relative to the surface and sensor can also strongly alter turbidity estimates. The many factors controlling turbidity and its reflected signal have hindered development of physical models that can be applied without local calibration data. Mertes et al. (1993) developed an approach based on spectral mixture analysis that holds promise, but even their technique required calibration to laboratory data. For the foreseeable future, managers therefore will have to rely on establishing empirical relations between local turbidity and the sensor signal. Empirical approaches of this nature can produce accurate results, but are difficult to implement in remote rivers or during flood conditions, which are often the periods of greatest interest to managers.

2.7 Bed sediment

Mapping bed sediment size is important for documenting in-stream habitat for fish, macroinvertebrates and other organisms, for characterising flow resistance for hydraulic and flood inundation models, and for modeling sediment transport and channel stability. There is a substantial body of literature on *ground-based* optical measurement of sediment size where the pixel resolution is far smaller than the sediment size. In this case the individual particles can be seen with the naked eye on the imagery, so the focus shifts to automating the procedure in order to delineate particle boundaries and measure particle axes (e.g. Raschke and Hryciw, 1997; Graham et al., 2005). More recently, terrestrial laser scanning (an active sensing

system) has been used to map exposed sediment sizes at reach scales (Hodge et al., 2009). These approaches to ground-based mapping of sediment size are useful at the scale of an individual plot, bar, or reach but are not feasible over longer lengths of stream where thousands to millions of photos or ground-based surveys might be required to cover the entire area.

It is only recently that *airborne* optical imagery has been available at sufficiently fine spatial resolutions to measure sediment size over long lengths of stream. The coarser resolution of these photos compared to ground images, however, limits what they can detect.

Rather than trying to measure individual grains as is done with ground photos, approaches for measuring sediment size with aerial imagery use the image semivariance, a statistical technique that characterises the 'graininess' or texture of an image. The concept is simple. Areas with larger sediment sizes have more shadows cast by the large clasts and therefore have a more heterogeneous texture. In contrast, surfaces with much finer sediments have less of a size difference between clasts, have less shadowing effect, and are more homogenous in appearance. These variations in image texture within a given window (e.g. 35×35 pixels) can be measured by a two-dimensional variogram (among other techniques), with higher values indicating more brightness variation from one pixel to the next. Carbonneau et al. (2004, 2005) discovered that the values from a two dimensional variogram are linearly related to grain size for both dry and submerged sediments. A linear regression between field measurements of the median particle size, D_{50}, and the two dimensional semi-variance of the image for the same locations was used to develop equations that could then be applied to the remainder of the image to accurately estimate D_{50} (Table 2.2). This technique can also be used to map other percentiles of the sediment grain size distribution (D_{16}, D_{84}, etc.) so long as: (a) the window size is large enough to get a stable semivariance signal; (b) the sediment patches are relatively uniform at the scale of the window; and (c) the grain size fraction (e.g. D_{50}) is larger than the image resolution (Carbonneau et al., 2005). Using this technique with 3 cm resolution imagery, Carbonneau et al. were able to continuously map sediment size along 80 km of the St. Marguerite River in Quebec.

Assuming that the sediment can be seen through the water, image resolution is the biggest limitation on measuring sediment size. The general rule of thumb is that the smallest size of sediment that can be mapped with *black and white* or *three band colour* imagery is equal to the pixel resolution of the image (Carbonneau, 2005).

Table 2.2 Accuracies of sediment and turbidity measurements. Accuracy measures represent the highest accuracies achieved using optical imagery to map bathymetry. Detailed explanations of the methods are in the cited references. Updated and modified from Marcus and Fonstad (2008).

Feature	Analytical Approach	Location	Imagery				Accuracy		Author(s)
			Platform & Monitor	Spatial Resolution	Spectral Range (nm)	# of Bands Used	Metric Reported	Optimal Values Achieved	
Turbidity	Simple regression of spectra	Tampa Bay estuary, FL, USA	Satellite (MODIS)	250 m	620–670	1	r^2, measured vs. observed	0.73	Chen et al. 2007
	Generalized Additive Model	Guadalquivir River, Spain	Landsat-5 TM Landsat-7 ETM+	30 m	450 – 2350 evaluated	varies with setting	r^2, measured vs. observed	0.79	Bustamante et al., 2009
Suspended sediment	Regression	Yangtze River, China	Landsat-7 ETM+	30 m	770–860	1 (band 4)	r^2, measured vs. observed	0.86 to 0.87	Wang et al., 2009
Bed material grain size	Spatial variogram	Sainte Marguerite River, Quebec, Canada	Helicopter – digital camera	3 cm	Visible	3 (r,g,b)	r^2, measured vs. observed	0.80 for D_{50}	Carbonneau et al., 2004
	Spatial variogram	Sainte Marguerite River, Quebec, Canada	Gantry – digital camera	0.3 mm	Visible	1 (gray scale)	r^2, measured vs. observed	0.93 for distinguishing sand from larger clasts	Carbonneau et al., 2006
	Linear unmixing	Ribble Estuary, Lancashire, UK	Daedalus 1268 Airborne Thematic Mapper	1.75 m	Visible to Short-wave IR	10	r^2, measured vs. observed	0.79 for % clay 0.60 to 0.84 for % sand Best in drier conditions	Rainey et al., 2003

For example, an image with 10 cm resolution can be used to map sediments of 10 cm or larger. In turn, this means the image cannot be used to accurately characterise the full distribution of sediment sizes (unless the finest sediments are 10 cm in size), because the smaller sediments are not being 'seen' by the sensor. Managers applying the technique with panchromatic or colour imagery will almost certainly need to charter special flights to collect imagery at the resolution of the sediment size of interest. For example, Carbonneau et al. (2004, 2005) chartered a helicopter to fly at 155 m above the river to acquire 3 cm resolution imagery.

If multispectral or hyperspectral imagery are available it may be possible to map features smaller than the nominal image pixel size. Rainey et al. (2003), for example, used spectral mixture analysis to map fine sediment sizes in an estuary. This type of 'pixel unmixing' is accomplished based on knowledge of the spectral characteristics of the different pure end members present within a pixel. For example, it is easy to estimate how much black and white paint are mixed together to create a grey colour in a single pixel; the darker the grey, the larger the proportion of black paint. Rainey et al. (2003) used the optical theory that small spaces between fine grains act as blackbody cavities. The black body behaviour fills in the spectra for each pixel to a degree proportional to the grain size, enabling estimates of the proportion of sand and mud, even in pixels that were 1.75 m in size (Table 2.2). Unfortunately, the optical theory only works well in relatively dry sediments. As of yet, there are no techniques for measuring sediment smaller than the pixel size in submerged sediments. This means that issues such as sand embeddedness cannot at present be characterized with airborne imagery.

2.8 Biotypes (in-stream habitat units)

Biotypes, also called 'in-stream habitats', 'micro-habitats' or 'morphologic units', refer to features such as riffles, pools, glides, and exposed bars. From an optical perspective, these features vary in a number of important ways that can be captured by remote sensing. Optical variations associated with depth, surface turbulence, substrate size, and vegetation associated with the units enable differentiation of features that are essentially composed of the same material – water. Likewise, the differences in composition between exposed bars, water and vegetation make it easy to manually map bars if the image resolution is appropriate. Automated mapping of bars with classification

algorithms is also relatively straightforward, provided the features are not in deep shadow.

Biotypes are receiving increasing attention from stream managers. In Europe, biotype mapping is a mechanism for defining biodiversity in streams under the Water Framework Directive (Dodkins et al., 2005). In the United States, many agencies use in-stream habitats to characterise stream health. A number of schemes exist for defining the different kinds of biotypes, leading to confusion around definitions; one person's glide can be another person's run. A good summary of the different biotype classification schemes and their overlap is provided in Table 1 of Milan et al. (2010).

Despite their growing importance, there is relatively little research on mapping biotypes with remotely sensed optical imagery. Researchers have achieved good results mapping in-stream habitats using supervised classification, a classical technique that is included on all remote sensing software. To apply this technique, the user first identifies a number of pixels, called training sites, on the image that are characteristic of each class of interest (e.g., riffles, pools, glides). The algorithm then maps other pixels on the image that have spectral signals similar to the training sites. Remotely sensed biotype maps can be more precise than ground-based maps. Ground surveyors will often lump many small features into a larger adjacent features (e.g., lumping low velocity stream margins into the riffle that dominates that stream reach). If the pixel size is small enough, the imagery will differentiate these many smaller extent variations.

Legleiter et al. (2002) and Marcus (2002) found that mapping accuracy is sensitive to the number of spectral bands used for classification. Simply put, more bands are better. Hyperspectral imagery (discussed in another chapter in this volume) thus provides better results than multispectral imagery (Marcus, 2002), and imagery that includes short wave infrared is better than imagery that only spans the visible wavelengths (Legleiter, 2003). The spatial resolution of imagery relative to the size of the channel being mapped is also a crucial consideration. Marcus et al. (2003) documented higher mapping accuracies with 1-m imagery in a 5th order stream than in a 4th order stream, which in turn had higher accuracies than in a 3rd order stream (Table 2.3). The drop in accuracy results from the 'mixed pixel' problem, which occurs when one pixel encompasses multiple features. In the stream context, a 1-m pixel in a third order stream is more likely to include portions of two units (e.g., a glide and pool) than in the 5th order stream, where the biotypes are much larger.

Table 2.3 Accuracies for biotype, algae and wood mapping. Accuracy measures represent the highest accuracies achieved using optical imagery to map bathymetry. Detailed explanations of the methods are in the cited references. Updated and modified from Marcus and Fonstad (2008).

Feature	Analytical Approach	Location	Imagery				Accuracy		Author(s)
			Platform & Monitor	Spatial Resolution (nm)	Spectral Range (nm)	# of Bands Used	Metric Reported	Optimal Values Achieved	
Biotypes	Principal component reduction followed by supervised maximum likelihood classification	Lamar River, WY, USA	Helicopter – PROBE-1	1	400–2400	128	Producer's accuracy	86%	Marcus et al., 2003
	Principal component reduction followed by co-kriging	Lamar River, WY, USA	Helicopter – PROBE-1	1	400 to 2400	128	Producer's accuracy	94%	Goovaerts, 2002
Algae	Matched filter	Cache Creek, WY, USA	helicopter	1	400 to 2400	128	User's accuracy	75%—intentionally overestimated # algae sites to make sure none were missed	Marcus et al., 2001
	Regression	Swan River, Australia	CASI (spectral mode with 288 bands) on aircraft	Not reported	423–946	1 at 750 nm	r^2, measured vs. observed	total cells 0.94 dinoflagellates, 0.44 chlorophyll-a, 0.77 chlorophyta 0.70 diatoms 0.59 chrytophta 0.95	Hick et al., 1998
Wood	Threshold filter on multiple pc bands	Lamar River, WY, USA	Helicopter-PROBE-1	1	400 to 2400	128	Producer's accuracy	85%	Marcus et al., 2003
	Threshold filter on 1 pc band	Unuk River, AK, USA	Aircraft – digital camera	0.45	Visible	3 (r,g,b)	Overall user's accuracy	89%	Smikrud and Prakash, 2006

Several researchers have developed alternative approaches to supervised classification. Maruca and Jacquez (2002) achieved good results using a spatially agglomerative cluster technique, and Goovaerts (2002) achieved high accuracies using a geostatistical co-kriging approach (Table 2.3). Legleiter and Goodchild (2005) examined the potential of a 'fuzzy' approach to stream classification that provides a more realistic representation of the gradual transitions between similar habitat units. Their work revealed far more complexity in the stream patterns than is captured in the simple either/or dichotomy of biotype classification. None of these approaches, however, are presently available in existing software packages. Special programming skills are therefore needed to implement them, making them less accessible to the management community.

Regardless of the approach, no physical models yet exist that allow remote mapping of biotypes in the absence of field data. Because the spatial pattern of biotypes can change with variations in discharge, personnel must be in the field at or near the time of image acquisition to map biotypes. The field maps are then used to select pixels to train the classification algorithms. Given that field mapping of biotypes is necessary regardless of mapping approach, using remote sensing only makes sense if the spatial extent of mapping far exceeds what a survey team can readily achieve while in the field.

2.9 Wood

Wood in rivers plays a major role in forming and altering stream habitats in many streams. Wood can force sediment deposition and transport, alter in-stream habitats and stream morphology, provide organic debris for macroinvertebrates, and create shelter for fish. Emplacing wood is a major tool used in stream restoration. Promoting wood accumulation is often a central goal of riparian management strategies (e.g. Fox and Bolton, 2007; Jochem et al., 2007).

In theory, remote detection of wood in river channels or on exposed bars should be relatively straightforward. At its simplest level, if sufficiently fine resolution imagery is available, one can see the wood on the images and map it manually. Even from an automated perspective, detection of wood should be relatively simple. Mapping depth, turbidity, channel change, and biotypes all require detecting variations, often subtle, in one feature type (water).

In contrast, mapping wood, at least against a background of water or sediment, requires detecting the difference between features with distinctly different compositions.

Automated mapping of wood in and along rivers can, in fact, be relatively simple if one has hyperspectral imagery or fine resolution, high quality colour imagery Marcus et al. (2003) and Smikrud and Prakash (2006) both used similar approaches to detect wood with relatively high accuracies (Table 2.3). Both sets of researchers first calculated principal components from their hyperspectral imagery to isolate the spectral signatures of different features. They then applied a matched filtering technique to the principal component images to detect wood. The matched filter operates by using some wood pixels within the image to train an algorithm that then finds similar pixels elsewhere in the image. A major advantage of the matched filtering technique is that there is no need for field teams if some wood can be seen clearly on the imagery. Principal component transformations are available in all remote sensing software and matched filtering is included in an increasing number of these packages.

If hyperspectral imagery is available, the matched filter will almost certainly detect wood in areas where it cannot be seen with the naked eye on the same image. In this case, the algorithm is 'unmixing' the pixel to detect locations where wood makes up only a portion of the pixel. This can lead to confusion on the part of users, who may believe the classification is showing wood where it does not exist. In addition, finding pieces of wood smaller than a pixel may cause problems if the user is seeking to map wood of a given size or larger. For example, many fluvial wood studies only seek to document wood that is 1 or 2 m in length and at least 10 cm in diameter – large enough to affect the flow. Hyperspectral imagery may detect much smaller wood, making it difficult to determine which pieces are of sufficient size to alter flow dynamics. Pixel unmixing approaches or shape detection algorithms might provide a solution to this issue, but no research has yet been attempted along these lines.

2.10 Submerged aquatic vegetation (SAV) and algae

SAV and algae are important to ecosystem health, providing food and cover for a wide range of species, removing toxins from water and sediments, and stabilising stream beds. On the other hand, an over abundance of SAV or the presence of certain species can be damaging to ecosystem health, water quality, and human structures. Problems associated with SAV and algal blooms range from eutrophication to acute toxicity associated with blue green algae to fouling of engineering works. Remote mapping of SAV and algae for management applications should thus seek to provide more than a simple percent cover map; ideally, it would also provide information on species composition, biomass, and physiology.

Almost all SAV research has focused on lakes, deltas, wetlands, estuaries and coastal waters. As with turbidity, where research also has centered on large water bodies, focusing on these environments has not significantly advanced remote sensing of SAV in small stream systems, where local-scale variations can confound the spectral signal. Work from large water bodies thus provides a guide to mapping SAV and algae, but should not be transferred directly to smaller streams without additional study and validation.

Silva et al. (2008) review remote sensing and plant physiology issues to consider in mapping SAV. A central issue in mapping SAV is the relatively strong absorption of optical wavelengths by water. An identical plant therefore looks different to the sensor when it is emergent, just beneath the surface, or more deeply submerged. Similarly, changes in plant structure, age, and reflectance confuse the identification of plants and complicate estimation of biomass.

Researchers thus use field measurements that incorporate plant- and location-specific variations to develop regressions that use individual bands, band ratios, or principal components to predict biomass (Silva et al., 2008). Models of this sort have yielded R^2 values of 0.79 (Armstrong et al., 1993) to 0.85 (Zhang, 1998) for comparisons of measured and estimated SAV biomass in large relatively stationary water bodies. Regression-based estimates reach a plateau, however, beyond which biomass continues to increase without a corresponding change in the spectral reflectance – the signal becomes saturated (Figure 2.4). In addition to biomass, remote sensing has been used to map SAV community type, chlorophyll concentration (Peñuelas et al. 1993), photosynthetic efficiency (Peñuelas et al. 1997, 1993), and foliar chemical composition (LaCapra et al. 1996).

Hyperspectral data are useful for separating SAV and algal chlorophyll signals (Williams et al. 2003) and identifying invasive species (Underwood et al., 2006). Even the additional spectral information, however, does not entirely overcome the complex signals generated by variable turbidity, water depths, and plant physiology (Hestic et al., 2008). Regardless of sensor type or platform, the variability in results among research projects and the potential complexities in mapping SAV indicate that – from a management perspective – this application is still in a developmental rather than an operational phase.

Figure 2.4 Submerged aquatic vegetation (SAV), Browney Brook, County Durham England, and algae along a side channel of the Tummel River, Scotland. The distinct spectra of chlorophyll relative to water and substrate enables mapping of general locations of SAV and algae, although separation of species and mapping of parameters such as biomass is more problematic.

As with SAV, the literature on algae focuses on large water bodies. Mapping of marine algae is so well established that satellites have been launched largely to monitor chlorophyll concentrations and map phytoplankton (e.g., the Coastal Zone Color Scanner in 1978). High spatial resolution imagery can provide good measurements of algal blooms over time in lake settings (Hunter et al., 2008). Hoogenboom et al. (1997) and Quibell (1991) describe some of the basic optics to consider in mapping algae with remote imagery.

Research on algae mapping in rivers suggests significant potential for this application. Hick et al. (1998) found high correlations between the 750 nm band and a number of algal parameters (Table 2.3) and concluded that only three to four bands were needed to map algal blooms over large areal extents, provided that in-water calibration data are available. Marcus et al. (2001) used hyperspectral imagery and pixels from an algae filled pool to train a matched filter to find similar sites throughout a backcountry region. In a subsequent survey of the stream, 75% of the sites they classified as algae had algae (Table 2.3) and led to the discovery of four previously unknown amphibian sites. This application demonstrated the value of remote sensing for exploratory purposes in inaccessible regions.

sensing to show that classical downstream hydraulic geometry relations do not provide accurate predictions of depth. This finding has significant implications for stream restoration, which frequently uses hydraulic geometry to estimate target depths (and other parameters) for naturalising streams. Ongoing work by the authors and others is examining the potential to use parameters derived from remote sensing to document habitat impacts of low-head dams, model water quality, and target river reaches for restoration. The list of potential applications is expanding rapidly, especially if one includes work that is examining fusion of optical data with LiDAR or radar, which can enable extraction of other hydraulic variables such as water surface slopes, velocity, and Froude numbers. Carbonneau et al. (2011), for example, used topographic information derived from radar imagery to obtain water surface slopes and used optical imagery to measure wetted widths and water depths, then combined these data to derive stream power and velocities every meter along 16 km length of a stream. Because they mapped water depth and substrate size for every square meter along the stream, they were also able to derive continuous metrics of spawning habitat suitability for the entire stream.

2.11 Evolving applications

The applications discussed above are ones for which there has already been a substantial body of work. Several evolving lines of research are also worth tracking depending on management needs. Forecasting and detection of ice breakup on large rivers is receiving increased attention (Pavelsky et al., 2004; Morse and Hicks, 2005, Kääb and Prowse, 2011). This application has the potential to become a major tool in inaccessible sub-polar and polar rivers, just as detection of floods has become one of the established applications of remote sensing. There is also an emerging effort to use remote sensing measurements from streams to map derived variables such as stream power (Jordan and Fonstad, 2005; Carbonneau et al., 2011), although most of these require merging active and passive sensors.

Moving beyond technique development, researchers are now beginning to apply the remote sensing methods discussed above to better understand rivers, although this work is just now beginning to appear in the published literature. Marcus and Fonstad (2008, 2010), for example, used continuous data on depths derived from remote

2.12 Management considerations common to river applications

Our discussion so far has focused on reasons for using remote sensing, potential applications in river settings, and some of the potential complications specific to individual applications (e.g., problems associated with depth measurement). But there are also issues common to almost any effort to use remotely sensed data to obtain river information on turbidity, SAV, or whatever. These issues are discussed in detail in Marcus and Fonstad (2008). Rather than repeat their work, we summarise their discussion in Table 2.4 and focus here on issues of particular relevance to stream managers. In doing so, we assume that the manager is working with a remote sensing professional who has addressed issues of data acquisition, quality and analysis. In this circumstance, the manager will be given remote sensing results, probably in the form of maps and, preferably, accuracy assessments. The manager is then faced with interpreting those results, presenting them to other users, and determining whether and to what extent this information can be incorporated into management plans.

Table 2.4 Common issues in river remote sensing. The issues highlighted below commonly arise in the context of using optical remote sensing to map river variables. The table does *not* summarize issues that are common to remote sensing across all applications, such as atmospheric and geometric corrections, standard image processing issues, or large data storage requirements.

Category	Common Issues in River Applications	Potential Solutions
Logistics: see Aspinall et al., 2002; Marcus and Fonstad, 2008	*Timing of flights:* Aircraft-based imagery cannot be acquired when variable needs to be measured	– acquire your own aircraft – use satellite imagery – use existing imagery for historical analysis
	Timing of field data: Field data collected under conditions different than those under which the image data were acquired	– schedule field data collection during periods of relatively stable stream conditions (e.g. low flow period)
	Location: Field sites inaccessible or too dangerous to collect ground validation data	– substitute data from similar settings – collect data at less dangerous time (e.g. use post-flood indicators of depths) – model data to simulate plausible values
	Expertise: Requires personnel with significant expertise in remote sensing	– contract with existing experts – use existing web-based data sets (e.g. for flooding) – seek out automated approaches
Optical environment: see Legleiter et al., 2009; Marcus and Fonstad, 2008	*Obstructed view of river:* from above	– use low elevation platforms (e.g. hand held balloon, drones) – use active device (LiDAR, radar)
	Turbidity: blocks light penetration of water column	– use ground-based measurements (e.g. total station surveys, sonar, ground-based radar)
	Shadows: create different radiance values over identical features	– mask out shadowed areas – develop algorithms for shadow removal
	Sun-target-sensor geometry: obscures features (e.g. reflections, sun glint)	– plan image acquisition to avoid unfavorable conditions – use algorithms to normalize lighting conditions
	Local variations: turbulence, substrate color, SAV, etc. generate different reflectances for a given measure (e.g. depth)	– use band ratios to normalize for variations – stratify image to measure variable separately in each category
Imagery & ground data: see Aspinall et al., 2002; Legleiter et al., 2002, 2009; Marcus et al., 2003	*Location precision:* must be high to match small features on imagery and in streams	– map directly to the imagery – set up benchmark/targets to tightly co-registered imagery to ground surveys
	Spatial resolution: may not be sufficient to detect small stream features	– acquire imagery with high spatial resolution for smaller streams – use pixel unmixing for features with distinct spectra (e.g. wood)
	Spectral coverage: Narrow spectral resolution over a broad spectral range sometimes needed to separate features (e.g. biotypes, SAV) that have similar compositions	– use hyperspectral imagery – use mapping algorithms that do not rely to such a large extent on between-spectra variations (e.g. semi-variograms, kriging)

2.13 Accuracy

Remote sensing results are almost always less than 100% accurate when compared to ground data. In fact, the large majority of the *highest* accuracies achieved with remote sensing of rivers range between about 75 and 90% (Tables 2.1, 2.2 and 2.3). Accuracies are limited to less than 100% due to the previously discussed obstacles that are specific to individual applications, as well as a number of issues common to most optical remote sensing projects. These 'generic' issues are summarised in Table 2.4 and discussed in more detail by Aspinall et al. (2002), Legleiter et al. (2002, 2009) and Marcus and Fonstad (2008). Figure 2.5 shows how the optical environment can vary over short distances with viewing angle and location, highlighting some of the issues identified in Table 2.4.

In addition, certain river settings do not work well for optical remote sensing. In particular, high energy and small headwater streams tend to have the view of the water column obscured by turbulent flow, in-channel features like boulders and wood, and overhanging vegetation. In addition, the small size of these streams means that very fine resolution imagery is necessary to capture the fine scale of variations in the stream (e.g., the rapid transition from a step to a pool in a step-pool system).

Do these levels of uncertainty mean that remote sensing results are too fraught with error to be useful for management applications? The simple answer is 'no' – for several reasons.

First, the metrics classically used to assess 'accuracy' of remote sensing data might not be entirely appropriate (Marcus et al., 2003, Legleiter et al., 2011). Standard methods of characterising accuracy in remote sensing assume the ground data are correct. Differences between ground-based and remote sensing results thus are assumed to represent error in the remote sensing. But what if the remote sensing data is actually more reliable and informative than the ground data? Marcus (2002) and Legleiter et al. (2002), for example, argued that their remote sensing maps of biotypes were more accurate than their field data. This was because surveyors on the ground combined large sections of river into one unit (e.g., a riffle), even if 'mini-glides' were present within the riffle. In contrast, the high spatial resolution imagery would also map most of that same unit as a riffle, but also map some pixels as glides depending on local variations in surface turbulence and depth. In this case the remote sensing imagery is probably more *precise* in its mapping of fine resolution features. The determination as to whether the remote sensing map is more or less *accurate* depends on whether you are a detail-oriented 'splitter' or a 'lumper' focused on the big picture; others may be more comfortable representing this kind of natural variability using fuzzy approaches, as described by Legleiter and Goodchild (2005). Similar arguments can be made regarding remote sensing maps of wood and bed sediment size.

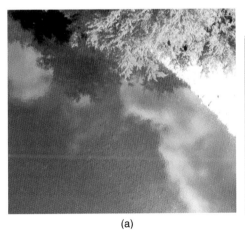 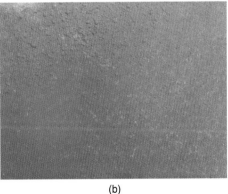

(a)   (b)

Figure 2.5 Photos from a bridge over the Garry River below Killecrankie, Scotland, demonstrating how the optical environment can change dramatically over short distances (Table 2.4). (a) Looking upstream, portions of the river are obscured by trees, shadows alter the lighting in some areas, and reflections obscure features. (b) Looking downstream from the same bridge at approximately the same photo scale, the different viewing angle enables the camera to readily captures variations in depth and substrate color and size.

Second, sometimes ground based measurements are more precise, but the broad areal coverage provided by the imagery makes remotely sensed maps more globally accurate. Marcus and Fonstad (2008) make this case with the example of field surveys of cross sections, which are precise and very accurate at that one location. Yet those cross sections are poor predictors of depth a short distance away from the survey location. The ± 15 cm precision of the remote sensing depths at all locations in the stream thus provides better *global* bathymetric accuracy than the high *local* accuracy of the cross section survey. Similar arguments can be made for SAV, turbidity, and algae, all of which require relatively time consuming local surveys that are representative of values at or near the sample site.

Third and finally, remote sensing maps are clearly the most accurate alternative when they are the only alternative. This occurs when historic data are needed or where sites are inaccessible or too dangerous for field surveys. Reconstruction of historic channel change, mapping of algae in roadless areas, and mapping of floods and ice breakup all represent cases discussed above where remote sensing might provide the only viable mapping alternative.

None of the points above are intended to imply that remote sensing is always the most accurate or the preferred alternative. Rather, our point is that accuracy measures must be considered in the context of the management needs for local precision, global accuracy, and the range of alternatives available for gathering data.

2.14 Ethical considerations

Remote sensing of rivers can raise ethical issues not typically encountered with classical ground-based data. Some of these issues are addressed in guidelines regarding professional conduct (ASPRS, 2007), data sharing as it relates to national security (Federal Geographic Data Committee, 2005), and the use of remote sensing to benefit people of all backgrounds and economic levels (United Nations General Assembly, 1986). Many countries also have a history of legal precedent regarding privacy and the use of remote imagery. We do not attempt to provide an in-depth discussion of these issues, but rather, raise some concerns that are particularly relevant to river managers.

The use of remote sensing raises social equity issues, potentially providing users who have access to remote sensing technology an advantage in accessing the river resource. This situation is nothing new; timber and mineral companies, for example, have used remote sensing

for decades to acquire information they do not share with other parties. However, because rivers are often a publicly owned and managed resource and are connected to up- and down-stream users, the issue of privileging a set of technologically savvy users over other residents of the basin is particularly problematic with rivers. If a small number of users have information on the best fish habitat, gravel supplies, reaches with high quality water, or other factors, they can potentially exploit the river resource to their own benefit, but not necessarily to the benefit of others living within the same watershed. Whether this is viewed as good or bad depends on the perspective of the individuals and society.

One solution to the concern regarding private exploitation is to make all remote sensing results publicly available, an approach advocated by the United Nations (United Nations General Assembly, 1986). Yet allowing this degree of access raises concerns regarding intellectual property and environmental protection. Particularly with high spatial resolution imagery, the maps may reveal the locations of sensitive resources (e.g., habitat for an endangered species), potentially jeopardising the very resource one is trying to protect (Marcus and Fonstad, 2008). Managers may therefore want to consider aggregating the data to a coarser spatial resolution, as is done with the U.S. Census, where results are reported at the block level but not the household level. Alternatively, one could make the locations blurry or indistinct, obscuring the map coordinates to prevent resource exploitation. But these approaches invite the same criticism leveled before, where a privileged elite have access to information that others cannot share. Competing social goals thus leave managers having to grapple with such issues on a case-by-case basis.

As with any technology, the potential for use and misuse of river remote sensing is broad. The importance of water to human society and the connectivity of rivers make the potential benefits and costs all the greater in rivers. It will largely be up to river managers, who have the broadest access to information about rivers and their user populations, to contemplate and develop uses for remote sensing that most benefit rivers and the communities that depend upon them.

2.15 Why use optical remote sensing?

The reasons for using remote sensing for river management are similar to those reasons for considering remote sensing in any setting. Among the most important considerations are which features can be remotely mapped (discussed in the first part of the chapter), image availability and cost; the spatial extent, coverage, scale and resolution of imagery; the need to collect repeat measurements; logistical constraints; the availability of digital data; the availability of multispectral data; and accuracy. In rivers, the potential to protect personnel from dangerous field situations can be a major consideration as well. The tradeoffs among these various factors need to be carefully attended to before a project commences: managers who assume remote sensing will meet their mapping and monitoring needs may be disappointed and upset.

From a management perspective, logistics and cost considerations often drive the decision as to whether to use remote sensing. In turn, the spatial extent of the study and the number of measurements needed over time are typically primary factors driving logistics and cost. Assuming that the variable in question can be mapped with remote sensing, the general rule is that the larger the spatial coverage and the greater the number of measurements needed over time, the more attractive the remote sensing option becomes. For example, consider the common scenario where information on channel depths is needed for flood inundation modeling. If the modeling involves only one relatively short river reach at one point in time, surveyors can often measure the necessary cross sections with relative ease in just one day. In contrast, use of remote sensing would require acquisition of remote imagery, specialised software, and expert personnel, making this alternative cost prohibitive. However, if the intent is to collect cross sections throughout the entire river, then the cost of remote sensing becomes increasingly reasonable relative to mobilising field crews for prolonged periods of time. Even in the case of one time, local data collection, the remote sensing approach could be cost effective if existing imagery were available for the appropriate date and trained personnel were at hand. The one time application of remote sensing makes even more sense in large or fast-flowing rivers where field data collection could place personnel at risk.

Moreover, the preceding example undersells the potential utility of remote sensing. Field measurements typically provide point or transect data. In contrast, remote sensing imagery provides continuous measurements of the entire channel over long distances, assuming the channel can be seen by the sensor. In the context of flood inundation analysis, remote sensing does more than just survey isolated cross-sections as a field crew would; it also provides continuous depth measurements for the entire channel at the resolution of the imagery. This enables more detailed monitoring of depth changes and

more sophisticated modeling, which in turns enhances the potential to develop more accurate models of flood inundation.

Furthermore, unlike the field-based cross sections, the imagery is more versatile in its applications. Once the imagery has been acquired, depending on its spectral coverage and spatial resolution, it can potentially be used to map features such as riparian cover, macrophytes, sediment size, wood, or biotypes (glides, riffles, etc.). The digital image format also provides a permanent record that is useful for validation by independent parties and allows for the possibility of developing historic change maps for variables that cannot be detected using present approaches. If this sounds farfetched, consider that prior to the year 2000, the only variables that remote sensing imagery was used for were flood extent, depth, turbidity, and riffle/pool delineation; this list that is now much longer (Marcus and Fonstad, 2008).

Yet many of the positive attributes listed above are hypothetical. Most river managers do not know if they will need continuous coverage, monitoring of other variables, or more sophisticated modeling at a later date. What they do know is that they need a specific measurement *now*. Furthermore, techniques for remote sensing of rivers are still being developed; there are no standard approaches that have been endorsed by regulatory agencies or incorporated into readily available software packages. Finally, and perhaps most importantly, our personal experience suggests that projects requiring new image data must incorporate significant flexibility. In general, one wishes to acquire imagery for the right place at the right time with the right specifications (resolution, signal to noise ratio, etc.). We have considered ourselves lucky on occasions where we have achieved two of these three objectives. Because the costs are realised in the present, application of remote sensing for river management, especially if new data must be acquired, can be risky business.

Remote sensing becomes a reasonable alternative to ground-based field surveys when it can monitor or map the variables of interest and when it can do so on a cost effective or safer basis than ground-based techniques. The cost-benefit ratio varies with the factors discussed above and with the risk averseness of the management agency. What is certain is that remote sensing will become an ever more viable option in the future. Increasingly, river management goals extend beyond the active channel to include the floodplain and nearby landscapes, so that multiple users contribute to the purchase of image data. Moreover, the cost of imagery has generally decreased over the past decade, the availability of software and personnel who can implement remote sensing algorithms is growing, and methods are being more widely tested and accepted. Present trends suggest that the utility and cost effectiveness of remote sensing will only improve in the coming years.

References

Alsdorf, D., Bates, P.D., Melack, J., Wilson, M., and Dunne, T. 2007. Spatial and temporal complexity of the Amazon flood measured from space. *Geophysical Research Letters* **34**(8): Article number L08402.

American Society for Photogrammetry and Remote Sensing (ASPRS). 2007. *Code of Ethics*. http://www.asprs.org/membership/certification/appendix_a.html. Accessed 30 November 2007.

Aranuvachapun, S. and Walling, D.E 1988. Landsat-MSS radiance as a measure of suspended sediment in the Lower Yellow River (Hwang Ho). *Remote Sensing of Environment* **25**(2): 145–165.

Armstrong, R.A 1993. Remote sensing of submerged vegetation canopies for biomass estimation. *International Journal of Remote Sensing* **14**(3): 621–627.

Aspinall, R.J., Marcus, W.A., and Boardman, J.W. 2002. Considerations in collecting, processing, and analyzing high spatial resolution, hyperspectral data for environmental investigations. *Journal of Geographical Systems* **4**(1): 15–29.

Barton, I.J. and Bathols, J.M. 1989. Monitoring floods with AVHRR. *Remote Sensing of Environment* **30**(1): 89–94.

Bates, P. 2004. Remote sensing and flood inundation modeling. *Hydrologic Processes* **18**: 2593–2597.

Brakenridge, G.R., Nghiem, S.V., Anderson, E., and Chien, S. 2005. Space-based measurement of river runoff. *EOS: Transactions of the American Geophysical Union* **86**(19): 185–192.

Brasington, J., Langham, J., and Rumsby, B. 2003. Methodological sensitivity of morphometric estimates of coarse fluvial sediment transport. *Geomorphology* **53**(3/4): 299–316.

Bustamante, J., Pacios, F., Diaz-Delgado, R., and Aragones, D. 2009. Predictive models of turbidity and water depth in the Doñana marshes using Landsat TM and ETM+ images. *Journal of Environmental Management* **90**(7): 2219–2225.

Carbonneau, P.E. 2005. The threshold effect of image resolution on image-based automated grain size mapping in fluvial environments. *Earth Surface Processes and Landforms* **30**(13): 1687–1693.

Carbonneau, P.E., Lane, S.N., and Bergeron, N.E. 2004. Catchment-scale mapping of surface grain size in gravel bed rivers using airborne digital imagery. *Water Resources Research* **40**(7): Article No. W07202.

Carbonneau, P.E., Bergeron, N.E., and Lane, S.N. 2005. Automated grain size measurements from airborne remote sensing for long profile measurements of fluvial grain sizes. *Water Resources Research* **41**(11): Article No. W11426.

Carbonneau, P.E., Fonstad, M.A., Marcus, W.A., and Dugdale, S. 2011. Making riverscapes real. *Geomorphology*. **137**(1): 74–86.

Davies-Colley, R.J. and Smith, D.G. 2001. Turbidity, suspended sediment, and water clarity: A review. *Journal of the American Water Resources Association* **37**(5): 1085–1101.

Dekker, A.G., Hoogenboom, H.J., Goddijn, L.M., and Malthus, T.J.M. 1997. The relationship between inherent optical properties and reflectance spectra in turbid inland waters. *Remote Sensing Reviews* **15**: 59–74.

Dieck, J.J and Robinson, L.R 2004. Techniques and Methods Book 2, Collection of Environmental Data, Section A, Biological Science, Chapter 1, General classification handbook for floodplain vegetation in large river systems: U.S. Geological Survey, Techniques and Methods 2 A–1, 52 p.

Federal Geographic Data Committee. 2005. *Final Guidelines for Providing Appropriate Access to Geospatial Data in Response to Security Concerns*. Federal Geographic Data Committee: Reston, VA. http://www.fas.org/sgp/othergov/fgdc0605.pdf. Accessed 30 November 2007.

Fonstad, M.A. and Marcus, W.A. 2005. Remote sensing of stream depths with hydraulically assisted bathymetry (HAB) models. *Geomorphology* **72**(1–4): 107–120.

Feurer, D., Bailly, J.S., Puech, C., LeCoarer, Y., and Viau, A. 2008. Very high resolution mapping of river immersed topography by remote sensing. *Progress In Physical Geography* **32**(4): 1–17.

Fox, M. and Bolton, S. 2007. A regional and geomorphic reference for quantities and volumes of instream wood in unmanaged forested basins of Washington State. *North American Journal of Fisheries Management* **27**(1): 342–359.

Gilvear, D.J. and Bryant, R. 2003. Analysis of aerial photography and other remotely sensed data. In *Tools in Fluvial Geomorphology*, Kondolf, G.M. and Piegay, H. (eds). Wiley: London, 133–168.

Gilvear, D.J., Hunter, P., and Higgins, T. 2007. An experimental approach to the measurement of the effects of water depth and substrate on optical and near infra-red reflectance: a field-based assessment of the feasibility of mapping submerged instream habitat. *International Journal of Remote Sensing* **28**(10): 2241–2256.

Goovaerts, P. 2002. Geostatistical incorporation of spatial coordinates into supervised classification of hyperspectral data. *Journal of Geographical Systems* **4**(1): 99–111.

Graham, D.J., Reid, I., and Rice, S.P. 2005. Automated sizing of coarse-grained sediments: image-processing procedures. *Mathematical Geology* **37**(1): 1–28.

Graham, D.J., Rice, S.P., and Reid, I. 2005. A transferable method for the automated grain sizing of river gravels. *Water Resources Research* 41(W07020): doi:10.1029/2004WR003868.

Han, Z., Jin, Y-Q., and Yun, C-X. 2006. Suspended sediment concentrations in the Yangtze River estuary retrieved from the CMODIS data. *International Journal of Remote Sensing* 27(19): 4329–4336.

Hestir, E.L., Khanna, S., Andrew, M.E., Santos, M.J., Viers, J.H., Greenberg, J.A., Rajapakse, S.S., and Ustin, S.L. 2008. Identification of invasive vegetation using hyperspectral remote sensing in the California Delta ecosystem. *Remote Sensing of Environment* 112: 4034–4047.

Hick, P., Jernakoff, P., and Hosja, W. 1998. Algal bloom research using airborne remotely sensed data: Comparison of high spectral resolution and broad bandwidth CASI data with field measurements in the Swan River in Western Australia. *Geocarto International* 13(3): 19–28.

Hodge, R.A., Brasington, J., and Richards, K.S. 2009. Characterisation of grain-scale fluvial morphology using TLS. *Earth Surface Processes and Landforms* 34: 954–968.

Hoogenboom, H.J., Volten, H., Schreurs, R., and de Haan, J.F. 1997. Angular scattering functions of algae and silt: an analysis of backscattering to scattering fraction. Ackleson, S.G. and Frouin, R.J. (eds). Ocean Optics XIII, SPIE vol. **2963**: 392–400.

Hughes, M.L., McDowell, P.F., and Marcus, W.A. 2006. Accuracy assessment of georectified aerial photographs: Implications for measuring lateral channel movement in a GIS. *Geomorphology* 74: 1–16.

Hunter, P.D., Tyler, A.N., Gilvear, D.J., and Willby, N.J. 2008. The spatial dynamics of vertical migration by Microcystis aeruginosa in a eutrophic shallow lake: A case study using high spatial resolution time-series airborne remote sensing. *Limnology and Oceanography* 53: 2391–2406.

Ip, F., Dohm, J.M., Baker, V.R., Doggett, T., Davies, A.G., Castano, B., Chien, S., Cichy, B., Greeley, R., and Sherwood, R. 2006. Flood detection and monitoring with the Autonomous Sciencecraft Experiment onboard EO-1. *Remote Sensing of Environment* 101(4): 463–481.

Jensen, J.R. 2007. *Remote Sensing of the Environment: An Earth Resource Perspective*, 2nd ed., Upper Saddle River, NJ: Prentice Hall, 592 pages.

Jochem, K., Daniel, H., Muhar, S., Gerhard, M., and Preis, S. 2007. The use of large wood in stream restoration: experiences from 50 projects in Germany and Austria. *Journal of Applied Ecology* 44(6): 1145–1155.

Jordan, D.C. and Fonstad, M.A. 2005. Two-dimensional mapping of river bathymetry and power using aerial photography and GIS on the Brazos River, Texas. *Geocarto* 20(3): 1–8.

Kääb, A. and Prowse, T. 2011. Cold-regions river flow observed from space. *Geophysical Research Letters* 38(L08403): doi: 10.1029/2011gl047022.

Kishi, S., Song, X., and Li, J. 2001. Flood detection in Changjiang 1998 from Landsat-TM data. *Space Technology* 20(3): 99–105.

LaCapra, V.C., Melack, J.M., Gastil, M., and Valeriano, D. 1996. Remote sensing of foliar chemistry of inundated rice with imaging spectrometry. *Remote Sensing of Environment* 55(1): 50–58.

Lane, S.N., Westaway, R.M., and Hicks, D.M. 2003. Estimation of erosion and deposition volumes in a large, gravel bed, braided river using synoptic remote sensing. *Earth Surface Processes and Landforms* 28(3): 249–271.

Lane, S.N., Widdison, P.E., Thomas, R.E., Ashworth, P.J., Best, J.L., Lunt, I.A., Sambrook Smith, G.H., and Simpson, C.J. 2010. Quantification of braided river channel change using archival digital image analysis. *Earth Surface Processes and Landforms* 35: 971–985

Legleiter, C.J. 2003. Spectrally driven classification of high spatial resolution, hyperspectral imagery: A tool for mapping in-stream habitat. *Environmental Management* 32(3): 399–411.

Legleiter, C.J. and Goodchild, M.F. 2005. Alternative representations of in-stream habitat: classification using remotely sensed data, hydraulic modeling, and fuzzy logic. *International Journal of Geographical Information Science* 19(1): 29–50.

Legleiter, C.J., Marcus, W.A., and Lawrence, R. 2002. Effects of sensor resolution on mapping in-stream habitats. *Photogrammetric Engineering and Remote Sensing* 68(8): 801–807.

Legleiter, C.J., Kinzel, P.J., and Overstreet, B.T. 2011. Evaluating the potential for remote bathymetric mapping of a turbid, sand-bed river: 2. Application to hyperspectral image data from the Platte River. *Water Resources Research* 47(W09532): doi: 10.1029/2011wr010592.

Legleiter, C.J., Roberts, D.A., and Lawrence, R.L. 2009. Spectrally based remote sensing of river bathymetry. *Earth Surface Processes and Landforms* 34: 1039–1059.

Legleiter, C.J. and Roberts, D.A. 2005. Effects of channel morphology and sensor spatial resolution on image-derived depth estimates. *Remote Sensing of Environment* 95: 231–247.

Legleiter, C.J., Roberts, D.A., Marcus, W.A., and Fonstad, M.A. 2004. Passive remote sensing of river channel morphology and in-stream habitat: physical basis and feasibility. *Remote Sensing of Environment* 93(4): 493–510.

Lejot, J., Delacourt, C., Piégay, H., Fournier, T., Trémélo, M.L., and Alleman, P. 2007. Very high spatial resolution imagery for channel bathymetry and topography from an unmanned mapping controlled platform. *Earth Surface Processes and Landforms* 32(11): 1705–1725.

Marcus, W.A. 2002. Mapping of stream microhabitats with high spatial resolution hyperspectral imagery. *Journal of Geographical Systems* 4(1): 113–126.

Marcus, W.A. 2012. Remote sensing of the hydraulic environment in gravel-bed rivers. In M. Church, P. Biron, A. Roy (eds.), *Gravel-bed rivers: Processes, tools, environments.* Chichester, John Wiley and Sons. p. 261–285.

Marcus, W.A., Aspinall, R., Boardman, J., Crabtree, R., Despain, D., Halligan, K., and Minshall, W. 2001. Validation of

High-Resolution Hyperspectral Data for Stream and Riparian Habitat Analysis. Annual Report (Phase 2) to NASA EOCAP Program, Stennis Space Flight Center, Mississippi, 127 p. plus figures.

Marcus, W.A and Fonstad, M.A. 2008. Optical remote mapping of rivers at sub-meter resolutions and watershed extents. *Earth Surface Processes and Landforms* 33: 4–24.

Marcus, W.A., Legleiter, C.J., Aspinall, R.J., Boardman, J.W., and Crabtree, R.L. 2003. High spatial resolution, hyperspectral (HSRH) mapping of in-stream habitats, depths, and woody debris in mountain streams. *Geomorphology* 55(1–4): 363–380.

Maruca, S. and Jacquez, G.M. 2002. Area-based tests for association between spatial patterns. *Journal of Geographical Systems* 4(1): 69–83.

Mertes, L.A.K. 2002. Remote sensing of riverine landscapes. *Freshwater Biology* 47: 799–816.

Mertes, L.A.K., Dekker, A., Brakenridge, G.R., Birkett, C., and Le'tourneau, G. 2002. Rivers and lakes. In S.L. Ustin (ed.), Natural Resources and Environment, Manual of Remote Sensing, John Wiley and Sons, New York.

Mertes, L.A.K., Smith, M.O., and Adams, J.B. 1993. Estimating suspended sediment concentrations in surface waters of the Amazon River wetlands from Landsat Images. *Remote Sensing of Environment* 43: 281–301.

Milan, D., Heritage, G., Large, A., and Entwistle, N. 2010. Mapping hydraulic biotopes using terrestrial laser scan data of water surface properties *Earth Surface Processes and Landforms* 35(8): 918–931.

Morse, B. and Hicks, F. 2005. Advances in river ice hydrology 1999–2003. *Hydrological Processes* 19(1): 247–263.

Pavelsky, T.M. and Smith, L.C. 2004. Spatial and temporal patterns in Arctic river ice breakup observed with MODIS and AVHRR time series. *Remote Sensing of Environment* 93(3): 328–338.

Pavelsky, T.M. and Smith, L.C. 2009. Remote sensing of suspended sediment concentration, flow velocity, and lake recharge in the Peace-Athabasca Delta, Canada. *Water Resources Research* 45(W11417): doi:10.1029/2008WR007424.

Peñuelas, J., Gamon, J.A., Griffin, K.L., and Field, C.B. 1993. Assessing community type, plant biomass, pigment composition and photosynthetic efficiency of aquatic vegetation from spectral reflectance. *Remote Sensing of Environment* 46: 110–118.

Peñuelas, J., Filella, I., and Gamon, J.A. 1997. Assessing photosynthetic radiation-use efficiency of emergent aquatic vegetation from spectral reflectance. *Aquatic Botany* 58: 307–315.

Rainey, M.P., Tyler, A.N., Gilvear, D.J., Bryant, R.G., and McDonald, P. 2003. Mapping intertidal estuarine sediment grain size distributions through airborne remote sensing. *Remote Sensing of Environment* 86(4): 480–490.

Quibell, G. 1991. The effect of suspended sediment on reflectance from freshwater algae. *International Journal of Remote Sensing* 12: 177–182.

San Miguel-Ayanz, J., Vogt, J., De Roo, A., and Schmuck, G. 2000. Natural hazards monitoring: Forest fires, droughts, and floods-The example of European pilot projects. Survey Geophysics 21: 291–305.

Schumann, G., Di Baldassarre, G., and Bates, P.D. 2009. The utility of spaceborne radar to render flood inundation maps based on multialgorithm ensembles (Part 2). *IEEE Transactions on Geoscience and Remote Sensing* 47(8): 2801–2807.

Sheng, Y., Gong, P., and Xiao, Q. 2001. Quantitative dynamic flood monitoring with NOAA AVRR. *International Journal of Remote Sensing* 22: 1709–1724.

Shibayama, M., Kanda, K., and Sugahara, K. 2007. Water turbidity estimation using a hand-held spectropolarimeter to determine surface reflection polarization in visible, near and short-wave infrared bands. *International Journal of Remote Sensing* 28(16): 3747–3755.

Silva, T.S.F., Costa, M.P.F., Melack, J.M., and Novo, E.M.L.M. 2008. Remote sensing of aquatic vegetation: theory and applications. *Environmental Monitoring and Assessment* 140: 131–45.

Smith, L.C. 1997. Satellite remote sensing of river inundation area, stage, and discharge: A review. *Hydrologic Processes* 11: 1427–1439.

Teodoro, A.C., Goncalves, H., and Veloso-Gomes, F. 2008. Statistical techniques for correlating total suspended matter concentration with seawater reflectance using multispectral satellite data. *Journal of Coastal Research* 24(4) SUPPL. (2008 07 01): 40–49.

Underwood, E., Ustin, S., and DiPietro, D. 2003. Mapping nonnative plants using hyperspectral imagery. *Remote Sensing of Environment* 86: 150–161.

United Nations General Assembly. 1986. *Principles Relating to Remote Sensing of the Earth from Space*, A/RES/41/65. http://www.un.org/documents/ga/res/41/a41r065.htm. Accessed 30 November 2007.

Wang, J.J., Lu, X.X., Liew, S.C., and Zhou, Y. 2009. Retrieval of suspended sediment concentrations in large turbid rivers using Landsat ETM+: an example from the Yangtze River, China. *Earth Surface Processes and Landforms* 34(8): 1082–1092.

Westaway, R., Lane, S.N., and Hicks, D.M. 2003. Remote survey of large-scale braided rivers using digital photogrammetry and image analysis. *International Journal of Remote Sensing* 24: 795–816.

Westaway, R., Lane, S.N., and Hicks, D.M. 2001. Airborne remote sensing of clear water, shallow, gravel-bed rivers using digital photogrammetry and image analysis. *Photogrammetric Engineering and Remote Sensing* 67: 1271–81.

Westaway, R., Lane, S.N., and Hicks, D.M. 2000. Development of an automated correction procedure for digital photogrammetry for the study of wide, shallow gravel-bed rivers. *Earth Surface Processes and Landforms* 25: 200–26.

Williams, D.J., Rybicki, N.B., Lombana, A.V., O'Brien, T.M., and Gomez, R.B. 2003. Preliminary investigation of submerged aquatic vegetation mapping using hyperspectral remote sensing. *Environmental Monitoring and Assessment* 81: 383–392.

Yamagata, Y., Wiegand, C., Akiyama, T., and Shibayama, M. 1988. Water turbidity and perpendicular vegetation indices for paddy rice flood damage analyses. *Remote Sensing of Environment* **26**(3): 241–251.

Yuan, L. and Zhang, L-Q. 2007. The spectral responses of a submerged plant Vallisneria spiralis with varying biomass using spectroradiometer. *Hydrobiologia* **579**: 291–299.

Zhang, X. 1998. On the estimation of biomass of submerged vegetation using Landsat thematic mapper (TM) imagery: A case study of the Honghu Lake, PR China. *International Journal of Remote Sensing* **19**(1): 11–20.

3

An Introduction to the Physical Basis for Deriving River Information by Optical Remote Sensing

Carl J. Legleiter[1] and Mark A. Fonstad[2]

[1]Department of Geography, University of Wyoming, Laramie, WY, USA
[2]Department of Geography, University of Oregon, Eugene, OR, USA

3.1 Introduction

Remote sensing affords considerable potential to not only foster significant advances in our scientific understanding of river systems but also facilitate monitoring and management of these crucial resources. The chapters in this book illustrate the range of ways in which this technology has been applied to diverse fluvial settings. This spectrum of applications can be expected to expand as new sensors, new methods, and new uses of remotely sensed data are developed in the coming years (Marcus and Fonstad, 2008). While the successes documented in this volume (Marcus et al., Chapter 2) and in the primary literature (reviewed by Feurer et al., 2008) justify a degree of optimism regarding these possibilities, a certain amount of caution is required as well. The key to efficient, effective use of remote sensing techniques in riverine environments is an appreciation of the physical processes that govern the interaction of light and water in stream channels. These processes, collectively referred to as radiative transfer, enable the attributes of interest to managers to be inferred from various kinds of image data. At the same time, the mechanics of radiative transfer, along with certain sensor characteristics, impose fundamental limitations as to what kind of information

can be derived from river images, as well as the reliability of that information. The goal of this chapter is to provide an overview of the physical processes that underlie the application of remote sensing to rivers. Our objective is to equip prospective users of image data with knowledge of these processes and thus prepare them to make informed, rational assessments of what can and cannot be achieved via remote sensing. Such knowledge, we believe, is the key to realising this technology's widely recognised but largely untapped potential to contribute to river management.

Remote sensing of rivers is a novel field of inquiry that remains in an early stage of development relative to remote mapping of coastal environments, where sophisticated, physics-based approaches are now favored (Lee et al., 2001; Mobley et al., 2005; Lesser and Mobley, 2007). Stream studies, in contrast, tend to rely upon more empirical techniques. Typically, image pixel values are related to ground-based measurements of channel attributes – depth, suspended sediment concentration, bed material grain size, etc. – to establish statistical correlations. The resulting regression equations are then applied to pixels throughout the image to map the quantities of interest. This approach has been implemented successfully (e.g., Winterbottom and Gilvear, 1997; Lejot et al., 2007, compilation of results in Chapter 2) but is

Fluvial Remote Sensing for Science and Management, First Edition. Edited by Patrice E. Carbonneau and Hervé Piégay.
© 2012 John Wiley & Sons, Ltd. Published 2012 by John Wiley & Sons, Ltd.

limited in several important respects. To develop such correlations, field data must be collected at the same time the image is acquired, a requirement that negates one of the primary advantages of *remote* sensing. Ground-based measurements also must be associated with specific image pixels, placing a premium on accurate geo-referencing of image data, which can be difficult for images with relatively coarse spatial resolution. Moreover, the resulting calibrations are scene and sensor-specific and cannot be readily extended to other locations, nor to images of the same river acquired at other times, due to spatial and temporal variations in solar radiation and differences in sensor characteristics. Notably, however, more recently developed techniques could help to overcome these limitations by incorporating hydraulic (Fonstad and Marcus, 2005) or photogrammetric (Lane et al., 2010) information to calibrate image-derived depth estimates when field data are not available.

Similarly, a more physics-based approach built upon radiative transfer theory could also help to alleviate some of these issues by providing greater generality, but such techniques are not yet available to the fluvial community. At present, remote sensing of rivers typically involves a certain degree of empiricism. Nevertheless, even regression-based methods must have some underlying physical basis, or meaningful correlations with field data could not be established. The strength of these correlations, and the extent to which accurate maps of river attributes can be derived from them, thus depends on the physics governing the interaction of light and water, even if these processes are not explicitly considered during image analysis. For this reason, we maintain that regardless of position along the continuum from purely empirical to fully physics-based, an understanding of radiative transfer processes is essential.

This chapter provides an introduction to these processes, and focuses specifically on passive optical remote sensing of river bathymetry. Estimating water depth from image data was one of the earliest applications of remote sensing technology to aquatic environments (e.g., Lyzenga, 1978), and is perhaps the most mature use of such methods in a fluvial context. Moreover, in the shallow, relatively clear-flowing streams of interest to many scientists and managers, efforts to characterize other channel attributes must account for the influence of a water column of variable thickness, or depth. A discussion of the physical basis for bathymetric mapping thus encompasses many of the principles underlying other applications as well. The material presented in this chapter is a river-centric distillation of more general treatments of

remote sensing concepts (Schott, 1997), radiative transfer theory and numerical modeling (Mobley, 1994), optical properties of various types of water bodies (Bukata et al., 1995), and recent advances in our understanding of the unique radiative transfer processes operating in shallow water (e.g., Mobley and Sundman, 2003; Dierssen et al., 2003); the interested reader is referred to these publications for additional detail. Similarly, more thorough discussion of the topics covered herein can be found in our earlier work (Legleiter et al., 2004; Legleiter and Roberts, 2005; Legleiter et al., 2009; Legleiter and Roberts, 2009).

By focusing our attention on passive optical remote sensing of the active river channel, we have chosen to neglect other, complementary technologies and approaches. For example, an active form of remote sensing, Light Detection And Ranging (LiDAR), has become the preferred method of topographic measurement for research on surface processes (Slatton et al., 2007), and newly developed, water-penetrating green LiDAR systems could enable precise, simultaneous mapping of both subaerial topography and wetted channel bathymetry (Bailly et al., Chapter 7; Kinzel et al., 2007; McKean et al., 2009). Nor do we address thermal remote sensing, mapping of riparian vegetation, or efforts to quantify water quantity and quality via remote sensing; coverage of these topics is available elsewhere.

This chapter is organised to first provide some background on general remote sensing concepts and radiative transfer theory and then focus more specifically on the optical properties of the fluvial environment and the ways in which channel characteristics can be inferred from measurements of reflected solar energy. The following sections:

1) introduce quantities used to characterise light fields and provide an overview of the radiative transfer processes operating in shallow streams;

2) describe the spectral characteristics of river channels, including the water surface, optically significant constituents of the water column, and bed and bank materials;

3) illustrate how information on river morphology can be derived from image data by separating the remotely sensed signal into component parts and evaluating their relative magnitudes; and

4) summarise the effects of sensor characteristics such as spectral, spatial, and radiometric resolution on the accuracy and precision of image-derived depth estimates. The chapter concludes by evaluating the prospects for more physics-based remote sensing of rivers, discussing

some additional applications, and identifying future challenges and research needs.

3.2 An overview of radiative transfer in shallow stream channels

In essence, passive optical remote sensing of rivers involves measuring energy – *electromagnetic radiation* that originated from the sun, propagated through the Earth's atmosphere, and interacted with the fluvial environment in various ways. Some proportion of the solar energy incident upon a river channel is reflected upward and travels through the atmosphere to an air- or space-borne sensor, where a light-sensitive detector records the amount of energy received. Multispectral or hyper-spectral instruments partition this energy into several, distinct ranges of wavelengths, called bands, and thus provide information on the distribution of energy across the spectrum. For remote sensing of rivers, optical data spanning the visible and near-infrared wavelengths from 400 nm to approximately 1100 nm ($1 \, \text{nm} = 1 \times 10^{-9} \, \text{m}$) are most useful; longer wavelengths are strongly absorbed by pure water. In any case, measurements of reflected solar energy are the basic data from which river information is derived. It is important to note, however, that although remote sensors record the total amount of energy incident upon them, they cannot distinguish whence that energy came – the river channel of interest, adjacent terrestrial features, or the atmosphere. Consequently, only a portion of the at-sensor signal is directly related to the channel characteristics of interest, and some understanding of the various ways in which electromagnetic energy interacts with riverine environments is needed to decompose and interpret this composite signal.

A useful concept in this context is that of the *image chain*, articulated by Schott (1997). The image chain approach involves viewing the remote sensing process as a series of interdependent, related events or steps that lead to a final output – an image of a river or some data product derived therefrom. Importantly, because each link along the image chain results from previous links and influences subsequent links, this approach provides a means of identifying potential limiting factors. As stated by Schott (1997, p. 14), 'if we study and understand the chain of events associated with a particular output image or product, we will be better able to understand what we have (i.e., what the product tells us or means), what we don't have (i.e., the limitations or confidence levels associated with the output product), and where those

limitations were introduced (i.e., where are the weak links in the image chain)'. This approach thus provides a convenient, insightful framework for examining the various radiative transfer mechanisms operating in fluvial environments, the measurement of reflected solar energy by remote detectors, and the conversion of image data into information on channel attributes. We thus begin our discussion by introducing some physical quantities used to characterise electromagnetic energy. We then use these quantities to describe the chain of processes by which sunlight interacts with stream channels to produce images of rivers.

3.2.1 Quantifying the light field

Electromagnetic radiation is one of the dominant forms by which energy is transferred through the environment; measurement of this radiation is the basis for remote sensing. Elementary particles of light called *photons* travel as waves, each moving at the speed of light and having a specific *wavelength* (distance between two consecutive peaks or troughs) and corresponding *frequency* (number of waves traveling past a fixed point in one second). Wavelength (λ, in m), frequency (v, in s^{-1} or Hertz), and the speed of light c are related as

$$c = \lambda v \qquad (3.1)$$

where c has a value of 2.9979×10^8 m/s in a vacuum. The amount of *radiant energy* associated with a single photon, denoted by q and expressed in units of joules (J), is directly proportional to the frequency at which the photon oscillates and inversely proportional to wavelength:

$$q = hv = h\frac{c}{\lambda} \qquad (3.2)$$

where $h = 6.6256 \times 10^{-34}$ J·s is Planck's constant. The important implication of these expressions is that shorter-wavelength photons, such as those comprising blue light, have a higher frequency and thus a greater amount of energy than longer wavelength, lower-frequency photons in the near-infrared. The total energy Q encompassed by a beam of light is a function of the number of photons present and their wavelengths, or, equivalently, frequencies:

$$Q = \sum q_i = \sum_{i=1} n_i h\frac{c}{\lambda_i} = \sum_{i=1} n_i h v_i \qquad (3.3)$$

where the summation is over all the wavelengths of light present in the beam and n_i is the number of photons of each wavelength.

Table 3.1 Radiometric and reflectance quantities used to quantify light fields. Note that each of these quantities can be expressed per unit wavelength to provide spectral information, in which case the units would include a per nanometer (nm^{-1}) factor as well.

Quantity	Symbol	Units	Unit Symbol
Radiant energy	Q	Joule	J
Radiant flux	Φ	Watt = Joule/second	W
Radiant flux density: irradiance	E	Watt/square meter	Wm^{-2}
Radiance	L	Watt/square meter/steradian	$Wm^{-2}\,sr^{-1}$
Irradiance (volume) reflectance	R	dimensionless	N/A
Bidirectional reflectance distribution function (BRDF)	r_{BRDF}	per steradian	sr^{-1}

Building upon this concept of radiant energy, we can define several other radiometric quantities relevant to remote sensing, summarised in Table 3.1. First, consider the rate at which the energy present in our light beam is propagating. The amount of radiant energy transferred per unit time is called the ***radiant flux*** or ***radiant power***, assigned the symbol Φ, and expressed in units of J/s, or Watts (W). A related quantity of fundamental interest is the rate at which this radiant flux is delivered to a surface.

This radiant flux density is referred to as ***irradiance*** and is denoted by E, with units of Wm^{-2}. The irradiance thus represents the amount of radiant energy received by a surface per unit time and per unit area. As illustrated in Figure 3.1, the actual amount of energy received also depends on the angle at which the light beam strikes the surface. This ***angle of incidence*** is given the symbol θ and is measured relative to a line perpendicular to the surface; $\theta = 0°$ indicates a beam striking the surface from directly

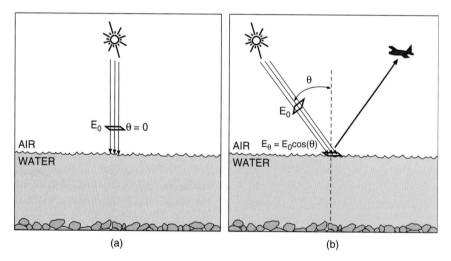

(a)   (b)

Figure 3.1 Projected area effects on the irradiance E incident upon a surface. (a) Surface is oriented perpendicular to the incident radiation, and the irradiance is simply the radiant flux per unit area E_0. (b) Surface is oriented at an angle relative to the incident radiation. The angle of incidence θ is measured relative to a line drawn perpendicular to the surface. In this case, the irradiance is reduced because the flux is projected, or spread, over a larger area and is given by $E_\theta = E_0 \cos(\theta)$. Figure adapted from Schott, J.R. *Remote Sensing: The Image Chain Approach*. New York, Oxford University Press, 1997.

overhead (Figure 3.1a), whereas larger values of θ like that shown in Figure 3.1b imply more glancing, oblique angles of approach. The irradiance incident upon a surface is given by **Lambert's cosine law**:

$$E_\theta = E_0 \cos(\theta) \qquad (3.4)$$

where E_0 is the irradiance onto a surface oriented perpendicular to the direction in which the light beam is propagating. For larger incidence angles, the energy encompassed by the beam is projected, or spread, over a lager surface area, resulting in smaller amount of energy per unit area and thus a smaller irradiance (Figure 3.1b). The important implication of Equation (3.4) is that the amount of radiant energy incident upon a river, some fraction of which will be reflected and ultimately recorded by a remote sensor, depends on the incidence angle, which in turn depends on the time of day, time of year, geographic location, degree of cloud cover, and surface slope and aspect. For example, images acquired late in the day, when θ is large, will have a weaker energy signal than data acquired when the sun is higher in the sky. Similarly, local variations in streambed orientation relative to the sun dictate that some areas of the bottom will receive less energy than others. If the instrumental noise associated with an imaging system is assumed constant, those times and places for which irradiance is reduced will have lower signal-to-noise ratios, and information derived therefrom will be less reliable.

Irradiance provides spatial information about the flux of radiant energy but not angular or directional information. This radiometric quantity is an integrated measure in that the irradiance includes energy incident upon a surface from all directions encompassed by a hemisphere above that surface: incidence angles from 0 to 90° and all **azimuths** (i.e., compass directions from 0 to 360°). For remote sensing, however, we are interested in the amount of energy arriving at a detector from a specific direction. To obtain this type of directional information, the irradiance is normalised to a unit **solid angle**, which is a three-dimensional angular measure. To understand the concept of a solid angle, first consider an angle θ on a two-dimensional plane like that shown on the left side of Figure 3.2. We can define such an angle as the ratio of the length l of an arc bracketed by two lines originating from the centre of a circle to the radius r of the circle: $\theta = l/r$. For example, an angle encompassing the

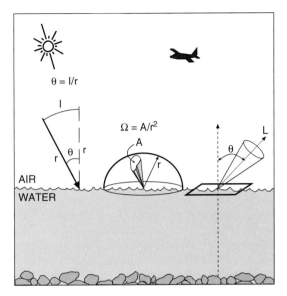

Figure 3.2 Geometric principles involved in the concept of radiance. On the left is an angle θ on a two-dimensional plane, defined as the ratio of the length l of an arc bracketed by two radii of a circle of radius r: $\theta = l/r$. By analogy with this two-dimensional case, the three-dimensional solid angle shown in the center of the figure is defined as the ratio of the surface area A encompassed by a range of incidence angles and azimuths to the square of the radius r of the sphere: $\Omega = A/r^2$. On the right side of the figure is an illustration of radiance, L, defined as the amount of radiant energy per unit time per unit area perpendicular to the direction of photon travel per unit solid angle. Figure adapted from Schott, J.R. *Remote Sensing: The Image Chain Approach*. New York, Oxford University Press, 1997.

entire circle has an arc length equivalent to the circumference of the circle, $l = 2\pi r$; dividing this arc length by the circle's radius yields the angle in radians – there are $2\pi r/r = 2\pi$ radians in a circle. For the three-dimensional case depicted in the centre of Figure 3.2, we can define a solid angle analogously by dividing the area of a spherical surface encompassed by a specific set of incidence angles and azimuths by the square of the radius of the sphere

$$\Omega = \frac{A}{r^2} \qquad (3.5)$$

where A is the area of the surface and Ω denotes the solid angle. The units of a solid angle are called steradians (sr), and there are $4\pi r^2/r^2 = 4\pi$ steradians in a sphere.

Now, using the concept of a solid angle to account for the directional structure of a three-dimensional light field, we can define another radiometric quantity central to remote sensing of rivers: *radiance* is the amount of radiant energy per unit time per unit area perpendicular to the direction of photon travel per unit solid angle (Figure 3.2). Radiance thus has units of $Wm^{-2}sr^{-1}$ and is given the symbol L. Unlike irradiance, which varies as the inverse square of the distance between the source of radiant energy and the surface receiving the energy, it can be shown that the radiance along a beam of light is constant over distance (Schott, 1997). Radiance thus has the important advantage of being independent of both distance and geometric effects; illumination and viewing angles are accounted for by normalising the radiant energy to a unit solid angle. Because of these unique properties, radiance measurements made under a range of illumination conditions, at various heights above the Earth's surface, and from different view angles can all be directly compared to one another. For these reasons, radiance is the quantity actually measured by remote sensors. Also, note that each of the radiometric terms defined above can be normalised to a unit wavelength to provide additional information on the distribution of radiant energy across the spectrum. Passive optical remote sensing is thus based upon measurements of a fundamental physical quantity: the spectral radiance $L(\lambda)$, reported in units of $Wm^{-2}sr^{-1}\,nm^{-1}$.

The spectral radiance measured by a remote detector depends on the amount of radiant energy incident upon a surface and the proportion of that energy that is reflected into the sensor's field of view. Where and when the surface of interest, a river channel in our case, receives a smaller amount of irradiance, the amount of reflected radiance will necessarily be smaller as well, even if the properties of the surface remain the same. A more general description of the innate characteristics of a surface or water body can be obtained by normalising the measured, upwelling spectral radiance by the incident, downwelling radiation. To provide such information, we can relate the amount of energy reflected away from a surface (in various directions) to the amount of energy received by the surface (from various directions) using quantities analogous to the radiometric terms in Table 3.1. These quantities are generally referred to as reflectance, and two particular forms of reflectance are especially relevant to remote sensing of rivers.

First, the ***irradiance reflectance*** is defined as

$$R(\lambda) = \frac{E_u(\lambda)}{E_d(\lambda)} \tag{3.6}$$

Here, $E_u(\lambda)$ refers to the upwelling spectral irradiance, which represents the radiant flux density integrated over the entire upper hemisphere, with light arriving from various directions weighted in proportion to the cosine of the incidence angle (Equation 3.4) to account for the projected area effect illustrated in Figure 3.1. The downwelling spectral irradiance, denoted by $E_d(\lambda)$, is defined similarly for the lower hemisphere. Simply put, $R(\lambda)$ is the ratio of the amount of energy traveling upward away from a surface to the amount of energy impinging upon the surface, irrespective of the directional structure of the light field. In other words, $R(\lambda)$ is the proportion of the total energy incident upon a surface that is reflected back away from that surface. A perfectly absorbing black object would have $R(\lambda) = 0$, and brighter objects would have $R(\lambda)$ values approaching 1. In an aquatic context, we can define an irradiance reflectance for any location above or within the water column. Within the water, $R(\lambda)$ is referred to as the volume reflectance and will depend on the optical properties of pure water and other suspended or dissolved materials. The volume reflectance can also vary with depth, a dependence made explicit by the notation $R(z, \lambda) = E_u(z, \lambda)/E_d(z, \lambda)$, where z denotes depth below the water surface.

A second, somewhat more complex reflectance quantity is the ***bidirectional reflectance***, which, as the name implies, does account for the directional structure of the light field. Intuition tells us that such directional effects are important, particularly in rivers, where the water surface can act as a mirror-like reflector for certain illumination and viewing angles, producing strong sun glint. Nearby sandbars, in contrast, appear to have a similar brightness regardless of sun angle and viewing direction. The smooth, glassy water surface shown in Figure 3.3a is an example of a *specular* reflector – all of the radiation reflected from the surface is directed in the exact opposite direction from the incident light ray. Sandbars behave more like **Lambertian** or diffuse reflectors, with a similar amount of reflectance in all directions (Figure 3.3a). Other objects might appear brighter when viewed from the specular or backscatter directions and darker from grazing angles. These variations in visual brightness can be quantified by considering the magnitude of the radiance emanating from a surface

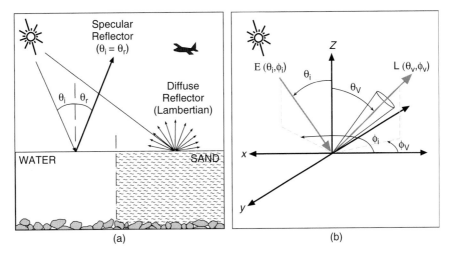

Figure 3.3 Bidirectional reflectance concepts. (a) A level water surface can act as a specular reflector, with all of the incident radiation reflected in the exact opposite direction. (b) A sandbar acts as a diffuse or Lambertian reflector, with a similar amount of radiance in all directions. (c) The bidirectional reflectance distribution function (BRDF) describes reflectance values for all combinations of illumination and viewing geometry. These geometries are specified by an incidence angle θ_i and azimuth ϕ_i for the downwelling irradiance and the view angle θ_v and azimuth ϕ_v for the direction in which the upwelling radiance is measured. The BRDF is defined by Equation (3.7) in the text. Figure adapted from Schott, J.R. *Remote Sensing: The Image Chain Approach*. New York, Oxford University Press, 1997.

in each direction as a probability distribution – what is the likelihood that a photon incident upon the surface will be scattered in a particular direction? Guided by this concept, we can define a ***bidirectional reflectance distribution function (BRDF)***, illustrated in Figure 3.3b. This distribution describes reflectance values for all combinations of illumination and viewing geometry. The BRDF is defined as the ratio of the radiance scattered into the direction described by the view angle θ_v and view azimuth ϕ_v to the downwelling irradiance arriving from an incidence angle of θ_i and an azimuth of ϕ_i:

$$
\begin{aligned}
&r_{BRDF}(\theta_i, \phi_i, \theta_v, \phi_v, \lambda) \\
&= \frac{L(\theta_v, \phi_v, \lambda)}{E_d(\theta_i, \phi_i, \lambda)} = \frac{dL(\theta_v, \phi_v, \lambda)}{L(\theta_v, \phi_v, \lambda) \cos(\theta_i)\, d\Omega(\theta_i, \phi_i)}
\end{aligned}
$$

$$(3.7)$$

The last expression on the right explicitly indicates that the BRDF uses the projected horizontal area of the incident radiance, as seen in the $\cos(\theta_i)$ factor in the denominator. The BRDF thus has units of sr^{-1}. Although these directional effects can be important (Mobley et al., 2003), in practice, surfaces are typically assumed to behave as Lambertian reflectors. For a perfectly

Lambertian, 100% reflectant surface, $L(\lambda) = E_d(\lambda)/\pi$ and the BRDF for such a surface thus reduces to $1/\pi$. A bidirectional reflectance can thus be converted to the more common and intuitive irradiance reflectance by assuming Lambertian behaviour and multiplying by π.

3.2.2 Radiative transfer processes along the image chain

For passive optical remote sensing, the first link in the image chain is the sun, the source of radiant energy. The top of Earth's atmosphere receives an irradiance of $1367\,Wm^{-2}$, an average, spectrally-integrated value known as the solar constant. Upon entering the atmosphere, this radiation is modified as photons interact with air molecules and aerosols through various ***absorption*** and ***scattering*** mechanisms. Importantly, these mechanisms depend on wavelength, the type and concentration of certain atmospheric constituents, and the path length radiation must traverse to reach the surface, which is primarily a function of the solar incidence angle. A full discussion of atmospheric effects is beyond the scope of this chapter, and more complete treatments are available elsewhere (Bukata et al., 1995; Schott, 1997). For our purposes, the key concept to bear in mind is that remote sensing of rivers involves viewing one optical medium (water in a stream channel) through a second optical medium (Earth's atmosphere) that influences the amount and spectral distribution of radiant energy incident upon the river. Moreover, absorption and scattering within the atmosphere impart a directional structure to the downwelling light stream, summarised in terms of a *sky radiance distribution*. This distribution describes how the radiant flux is partitioned within the hemisphere above the surface. Typically, more energy arrives from that portion of the sky centered around the sun and smaller, but still significant, amounts of energy arrive from other directions. We can therefore divide the total downwelling spectral irradiance $E_d(\lambda)$ into two parts: 1) a direct component, associated with the solar beam itself; and 2) a diffuse component that consists of skylight scattered within the atmosphere and possibly energy reflected from nearby objects.

For those photons that propagate through the atmosphere to the river, the next link in the image chain consists of interactions that occur at the *air-water interface*. Radiance is more useful for examining these processes than is irradiance because the former quantity includes directional information whereas the latter does not. Consider a beam of downwelling spectral radiance in air $L_{da}(\lambda)$, impinging upon the water surface at an incidence angle of θ_a as shown in Figure 3.4a. A portion of this beam will be *reflected* from the water surface, with the reflected beam $L_{ra}(\lambda)$ traveling away from the surface at an angle of θ_{ra} measured relative to a line drawn perpendicular to the water surface. The remainder of the downwelling beam is *transmitted* through the air-water interface and into the water column. This transmitted beam $L_{tw}(\lambda)$ is *refracted*, or bent, as it passes from air to water due to the different speeds at which light travels through the two media. The in-water incidence angle θ_w of the transmitted beam can be determined from *Snell's law*:

$$n_a \sin \theta_a = n_w \sin \theta_w, \qquad (3.8)$$

where $n_a = 1.000$ and $n_w = 1.333$ are the *indices of refraction* of air and water, respectively. Accounting for refraction via Equation (3.8) is critical because θ_w determines both the in-water path length transmitted photons must travel to reach the streambed and their angle of incidence upon the bed, which in turn influences the bottom reflectance according to the substrate's BRDF.

We now turn our attention to that portion of the incident beam that is reflected from the water surface. For the simple case of a perfectly *level water surface*, the angle at which the reflected beam travels away from the air-water interface is the same as the angle at which the incident beam approached the surface. The angle of incidence equals the angle of reflection, and the water surface behaves as a specular reflector. The reflectance ρ

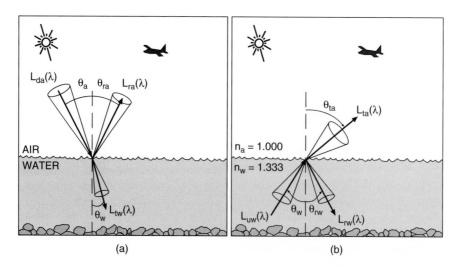

(a) (b)

Figure 3.4 Interaction of radiant energy with the air-water interface. (a) A portion of the incident light beam is reflected from the water surface and the remainder is transmitted through the air-water interface and refracted as the light beam enters the water column. (b) A portion of the light beam approaching the air-water interface from below experiences internal reflection and the remainder is refracted and dispersed over a larger solid angle as it enters the atmosphere. See text for details. Figure adapted from Bukata, R.P., J.H. Jerome, et al. *Optical Properties and Remote Sensing of Inland and Coastal Waters*. Boca Raton, FL, CRC Press, 1995.

from a level water surface is essentially independent of wavelength and is given by **Fresnel's formula**:

$$\rho(\theta_i, \theta_r) = 0.5 \left[\frac{\sin(\theta_i - \theta_r)}{\sin(\theta_i + \theta_r)} \right]^2$$
$$+ 0.5 \left[\frac{\tan(\theta_i - \theta_r)}{\tan(\theta_i + \theta_r)} \right]^2 \qquad (3.9)$$

where θ_i and θ_r are the angles of incidence and refraction, respectively. The magnitude of the surface-reflected radiance is thus $L_{ra}(\lambda) = \rho L_{da}(\lambda)$. For incidence angles up to $30°$, the Fresnel reflectance from a level water surface is essentially a constant value of $\rho = 0.02-0.03$. For larger incidence angles (i.e., as the sun sinks in the sky), ρ increases gradually, does not exceed 0.1 until $\theta_i > 65°$, and then rises rapidly to 1 at a grazing angle of incidence.

Although a level water surface provides a convenient starting point and a useful first approximation for the surface-reflected radiance, this simple geometry is clearly not realistic. Water surfaces in rivers are often highly irregular, with a temporally and spatially variable topography dictated by flow hydraulics and surface turbulence. For such non-uniform water surfaces, alternative approaches must be used to characterise radiative transfer processes at the air-water interface. This is not to say that the physics pertaining to a level surface are no longer pertinent; instead, these principles are applied to a probability distribution of water surface orientations. A stochastic approach of this kind has been developed in marine settings (Mobley, 1994), where probability distributions of wave slopes can be related to wind speed (Cox and Munk, 1954). These models have also been applied to rivers as a surrogate for flow-related surface turbulence. The primary effect of water surface irregularity is to increase surface-reflected radiance (Legleiter et al., 2004, 2009). In essence, rougher water surfaces appear brighter because more of the incident radiation is reflected, which reduces the amount of energy transmitted across the surface and into the water.

Continuing with our image chain analogy, the next link is the water column itself. Here, the downwelling light stream is attenuated as photons are absorbed and scattered by pure water and other optically significant materials. Both absorption and scattering interactions act to reduce the intensity of the transmitted radiance, denoted by $L_{tw}(\lambda)$ (Figure 3.4a); scattering processes also modify the directional structure of $L_{tw}(\lambda)$. The farther photons propagate through the water column, the more likely they are to be absorbed or scattered, implying that radiance varies as a function of path length (i.e., depth).

This concept is expressed via **Beer's law**, which relates the radiant flux $\Phi(0, \lambda)$ entering an absorbing medium to the radiant flux $\Phi(r, \lambda)$ at a distance r into the medium:

$$\Phi(r, \lambda) = \Phi(0, \lambda)e^{-a(\lambda)r} \qquad (3.10)$$

where $a(\lambda)$ is an **absorption coefficient** with units of m^{-1}. The radiant flux thus decreases exponentially with distance traveled through the water column as photons are absorbed.

The radiant flux is also reduced due to scattering, at a rate quantified by the **scattering coefficient** $b(\lambda)$. A closely related quantity, the **scattering phase function**, is illustrated schematically in Figure 3.5 and describes how scattering interactions redirect photons onto new paths and thus provides directional information. For example, by integrating the scattering phase function over the

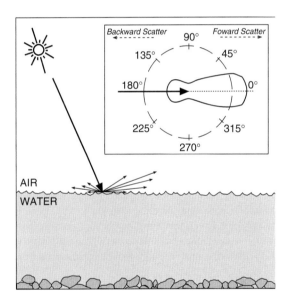

Figure 3.5 Schematic, two-dimensional illustration of the scattering phase function, which describes how scattering interactions redirect photons onto new paths. For the scattering center shown in the inset, the phase function is the probability that a scattering event will cause the scattered photon to travel in a specific direction, with an angle of $0°$ indicating no change in direction and an angle of $180°$ implying that the photon was scattered back into the direction from which it came. A forward scattering probability can be defined by integrating the phase function over angles from $0°$ to $90°$ and $270°$ to $0°$ (the right half of the unit circle). The back-scattering probability is the integral of the phase function over angles from $90°$ to $270°$ (the left half of the unit circle). The scattering phase function illustrated here is strongly biased in the forward direction. Figure adapted from Schott, J.R. *Remote Sensing: The Image Chain Approach*. New York, Oxford University Press, 1997.

hemisphere ahead of the incident flux (the right side of Figure 3.5 inset), we can determine a forward scattering probability for the photons in the incident beam. Similarly, integrating over the hemisphere trailing the incident beam (left side of Figure 3.5 inset) yields a back-scattering probability. Scattering in natural waters tends to be strongly biased toward the forward direction – that is, most scattered photons continue to propagate in a similar direction – but back-scattering is largely responsible for imparting a volume reflectance, defined by Equation (3.6), to the water column.

Attenuation of the downwelling light stream thus results from both absorption and scattering. The total attenuation coefficient $c(\lambda)$ is defined as the fraction of radiant energy removed from an incident beam per unit distance due to the combined effects of both types of processes:

$$c(\lambda) = a(\lambda) + b(\lambda) \qquad (3.11)$$

These three coefficients are ***inherent optical properties*** of the water column in the sense that they do not depend on how they are measured nor how the water is illuminated. Because inherent optical properties are independent of the spatial distribution and intensity of the incident radiation, these quantities could be directly related to certain attributes of interest, such as the suspended sediment concentration. Remote sensing, however, involves observing rivers under a range of illumination conditions, so we must instead rely on ***apparent optical properties***, which are used to normalise for the effects of external environmental conditions, such as changes in the incident illumination. Inferring water column characteristics from remotely sensed data thus relies heavily upon relationships between apparent and inherent optical properties (Bukata et al., 1995).

Attenuation, in many ways, forms the basis for interpreting measurements of radiance that has interacted with the water in a stream channel. As a result, one apparent optical property of particular interest is the ***irradiance attenuation coefficient***. This quantity, denoted by $K(\lambda, z)$, can be derived from Beer's law and is defined as the change in the spectral irradiance, denoted by $\Delta E(\lambda, z)$ corresponding to a unit change in depth z at an initial depth z:

$$K(\lambda, z) = -\frac{1}{E(\lambda, z)} \left[\frac{\Delta E(\lambda, z)}{\Delta z} \right] \qquad (3.12)$$

This is an apparent optical property because, unlike $c(\lambda)$, $K(\lambda, z)$ depends on the directional structure of

the radiance distribution comprising $E(\lambda, z)$ and is thus commonly referred to as the diffuse attenuation coefficient. We must also consider the angle at which light is propagating through the aquatic medium in order to calculate irradiance at a particular depth using a depth-averaged attenuation coefficient $K(\lambda)$. In this case, Beer's law can be re-written as

$$E(\lambda, z) = E(\lambda, 0^-)e^{-\overline{K(\lambda)}z} \qquad (3.13)$$

where $E(\lambda, 0^-)$ is the spectral irradiance just beneath the water surface.

In shallow stream channels, however, the attenuation of light is more complex than Equation (3.13) would appear to imply. Strictly speaking, Beer's law is not appropriate for these environments because the attenuation of the downwelling flux differs from that of the upwelling flux due to the presence of a reflective bottom that acts as a source of radiant energy (Dierssen et al., 2003). Because irradiance can be readily partitioned into downwelling and upwelling components, separate attenuation coefficients for the downwelling and upwelling irradiance – $K_d(\lambda, z)$ and $K_u(\lambda, z)$, respectively – can be defined by analogy with Equation (3.12). By making such a distinction, Legleiter et al. (2004) demonstrated that whereas $K_d(\lambda, z)$ was essentially independent of depth, $K_u(\lambda, z)$ varied considerably within the water column. Although the downwelling flux decreases monotonically with depth, the upwelling flux can actually increase with depth as more and more photons reflected from the bottom contribute to $E_u(\lambda, z)$. These effects clearly complicate the definition of a single, representative attenuation coefficient. In practice, however, the two-way trip photons must make to return from the streambed to the water surface is accounted for by using $2K_d(\lambda)$ as an ***effective attenuation coefficient***. This approach has been shown to provide reasonable results for a range of plausible stream conditions (Legleiter et al., 2009; Legleiter and Roberts, 2009).

One of the most critical links along the image chain is the manner in which those photons that do propagate through the full depth of the water column interact with the streambed. Some proportion of the downwelling irradiance impinging upon the bed, at depth z_b, will be reflected upward and begin a second traverse of the water column. Both of the reflectance concepts described in Section 3.2.1 apply here, and we can define an irradiance reflectance for the bottom, $R_b(\lambda)$, using Equation (3.6). A more complete description of the reflectance properties of the streambed would account for directional structure by considering the BRDF of the substrate. To fully characterize the BRDF, however, radiance measurements must be made for a range of source and sensor orientations. The difficulty of collecting such data, even in terrestrial settings, dictates that BRDF's are only available for a small number of materials (e.g., Zhang et al., 2003). Typically, the substrate is assumed to behave as a Lambertian surface (i.e., reflectance is the same for all view directions); radiative transfer simulations performed by Mobley et al. 2003 indicated that replacing a more complex BRDF with a uniform, Lambertian BRDF with the same irradiance reflectance entailed errors of less than 10%. In practice, then, $R_b(\lambda)$ is the primary factor that determines what fraction of the radiant flux reaching the bed will begin propagating upward toward a remote detector. This **bottom reflectance**, or **albedo**, varies with wavelength and depends on the characteristics of the sand, gravel, vegetation, or other materials comprising the streambed (Section 3.3.3).

Upon reflection from the streambed, photons embark on a second traverse of the water column, along which they are subject to the same absorption and scattering processes that affect the downwelling light stream. Because the substrate behaves, at least approximately, as a Lambertian surface, the angular structure of the upwelling flux is more uniform than that of the downwelling radiation incident upon the bed. As a result, the upwelling flux will be attenuated more rapidly as it propagates back toward the air-water interface (Dierssen et al., 2003). Moreover, these directional effects dictate that the attenuation of upwelling photons reflected from the bed differs from the attenuation of photons scattered upward within the water column (Maritorena et al., 1994). These two factors give rise to the distinction between $K_d(\lambda, z)$ and $K_u(\lambda, z)$. Fully accounting for these effects involves complex radiative transfer modeling, and most applications have relied upon an effective attenuation coefficient to summarise the influence of the water column (e.g., Philpot, 1989). This assumption over-simplifies the radiative transfer process, however, and more rigorous approaches that distinguish between the downwelling and upwelling light streams have been developed by the coastal remote sensing community and could be applied to rivers as well.

Whether reflected from the bottom or scattered within the water column, upwardly mobile photons approaching the air-water interface from below are subject to the same processes as the downwelling solar radiation approaching the interface from above. The interaction of the upwelling light stream with the water surface differs in some important ways, however, as illustrated in Figure 3.4b. Whereas the Fresnel reflectance of air-incident rays from the water

surface remained small for incidence angles up to 65°, rays approaching the air-water interface from below experience greater Fresnel reflectance at smaller incidence angles. In fact, because the index of refraction of air is less than that of water ($n_a < n_w$), a critical angle of incidence exists for which the angle of refraction from water into air is 90°, implying that the incident beam will be refracted back into the water. For a typical n_w of 1.333, this critical angle, determined from Snell's law (Equation 3.8), is 48.36°. Any upwelling photon with an in-water incidence angle greater than 48.36° will not be transmitted through the air-water interface; instead, **total internal reflection** occurs. As a result, all of the radiance recorded by an airborne detector is derived from an underwater cone with a half-angle of 48.36°. Moreover, because the radiant flux contained within a given solid angle below the air-water interface is spread into a larger solid angle above the water surface (Figure 3.4b), the radiance from within the stream channel is reduced by a factor of $1/n_w^2 \approx 0.563$ upon entering the air (Bukata et al., 1995). These effects can be summarised by saying that light can much more easily get 'into the water' than 'out of the water' (Mobley, 1994). This is one of the main reasons why remote sensing of rivers involves the measurement of relatively small amounts of reflected solar energy.

In any case, for those photons that do manage to escape the water column, the next link along the image chain is a second trip through the atmosphere. These photons experience additional absorption and scattering interactions that modify the **water-leaving radiance** $L_w(\lambda)$ *en route* to the sensor. These effects are summarised in terms of a **path transmittance** $T(\lambda)$ and **path radiance** $L_p(\lambda)$. The former quantity, $T(\lambda) < 1$, is a multiplicative coefficient that accounts for the losses that occur as radiance is transmitted from the river channel to the sensor. The latter quantity, $L_p(\lambda)$, makes an additive contribution to the radiance received by the sensor as light is scattered into the detector's field of view by the atmosphere or objects near the channel. Again, a full treatment of atmospheric effects and their correction is beyond our scope, and the interested reader is referred to remote sensing texts (e.g., Schott, 1997). For our purposes here, a few key points are worth noting. First, because water bodies have such low reflectance, the energy signal associated with the fluvial features of interest can be quite small relative to extraneous sources of radiance, such as the atmosphere or adjacent, typically much brighter terrestrial surfaces. For these reasons, remotely sensed data acquired under hazy conditions can be of limited value, and in many cases deriving useful river information from images will require some degree of atmospheric correction. Even for airborne remote sensing, which involves a shorter stream-sensor path length than satellite data, the atmosphere exerts a significant influence on the water-leaving radiance that must be accounted for. Also note that the atmosphere affects the downwelling irradiance incident upon the river.

The critical implication of the preceding discussion of the image chain is that the radiance measured above a river channel comes from a number of different sources and has been influenced by a number of different radiative transfer processes. These diverse energy pathways can be summarised by expressing the total, at-sensor upwelling spectral radiance $L_T(\lambda)$ as the sum of four components:

$$L_T(\lambda) = L_B(\lambda) + L_C(\lambda) + L_S(\lambda) + L_P(\lambda) \qquad (3.14)$$

These components are illustrated in Figure 3.6. The bottom-reflected radiance $L_B(\lambda)$ consists of photons that have interacted with the streambed. $L_B(\lambda)$ thus depends not only on depth, due to the exponential attenuation of light with distance traveled through the water column, but also on bottom type, which determines the **bottom contrast** between the reflectance of the streambed

and the volume reflectance of the water column. If a study's objectives are to map bathymetry or substrate composition, $L_B(\lambda)$ is the radiance component of interest and the other terms in Equation (3.14) must somehow be accounted for. Next, $L_C(\lambda)$ represents radiance from the water column itself and encompasses those photons that entered the water but were scattered into the upward hemisphere before reaching the bed. $L_C(\lambda)$ is the radiance component most useful for measuring concentrations of suspended sediment or other suspended or dissolved materials. The combination of the first two terms in Equation (3.14) constitutes the water-leaving radiance $L_W(\lambda) = L_B(\lambda) + L_C(\lambda)$, which includes all of the photons that have interacted with the river channel in a potentially informative manner. The surface-reflected radiance $L_S(\lambda)$ is comprised of photons that never entered the water and are thus unaffected by depth, bottom type, or water column optical properties. Similarly, path radiance $L_P(\lambda)$ scattered into the sensor's field of view by the atmosphere contributes to $L_T(\lambda)$ but is essentially independent of the river attributes of interest. This partitioning of the radiance signal into component parts, some of which provide useful river information and some of which do not, is an important concept to keep in mind in any application of remote sensing to rivers.

3.3 Optical characteristics of river channels

With this overview of the image chain comprising remote sensing of rivers, we now shift our attention to the spectral characteristics of some important features of the resulting images. Each of the radiative transfer processes and surface interactions described above depends, to some extent, on wavelength. Radiance measurements in a number of spectral bands thus provide information that can be used to identify and remove extraneous contributions to the at-sensor radiance signal and gain additional leverage for inferring channel attributes from remotely sensed data. Panchromatic images are readily available (e.g., historical aerial photography) and often have high spatial resolution (e.g., commercial satellites featuring a high-resolution panchromatic band and several multispectral bands with somewhat larger pixels). These gray-scale data can be used to map channel planform, lateral migration, and, to a lesser extent, variations in water depth (e.g. Winterbottom and Gilvear, 1997). Incorporating spectral information, however, allows channel morphology to be measured with greater confidence and opens up

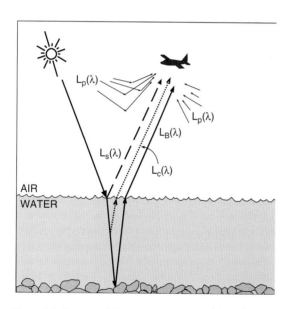

Figure 3.6 Conceptual diagram of the components of the total at-sensor radiance signal $L_T(\lambda)$. Radiance reaching the sensor originates from (1) reflectance from the streambed, $L_B(\lambda)$; (2) volume reflectance within the water column, $L_C(\lambda)$; (3) surface reflectance from the air-water interface $L_S(\lambda)$; and (4) scattering within the atmosphere, $L_P(\lambda)$. Adapted from Bukata, R.P. et al. (1995) *Optical Properties and Remote Sensing of Inland and Coastal Waters*. Boca Raton, FL: CRC Press, p. 54.

the possibility of examining other river attributes, such as substrate composition or suspended sediment concentration. Many airborne sensors and commercial satellites provide multispectral data, most often consisting of blue, green, red, and near-infrared bands, and more advanced hyperspectral instruments measure radiance in a larger number of narrower bands. We briefly examine the effects of sensor spectral resolution in Section 3.4.3, and hyperspectral remote sensing is considered in greater detail in Chapter 4 (Fonstad). Here, we focus on the spectral characteristics of the river channels themselves.

3.3.1 Reflectance from the water surface

For most applications, reflectance from the water surface is an extraneous source of radiance that confounds mapping of channel attributes such as depth, substrate type, and water column optical properties. As discussed above, the magnitude of the surface-reflected radiance can be determined from Fresnel's Equation (3.9) for a level water surface and remains small (0.02–0.03) as long as the solar incidence angle is not too large. Radiative transfer simulations indicate that more realistic, irregular water surfaces have greater reflectance (Legleiter et al., 2004, 2009). The spectral shape of the surface-reflected radiance $L_s(\lambda)$ also depends on surface roughness. For a level water surface, $L_s(\lambda)$ consists of reflected diffuse skylight and thus increases slightly at shorter wavelengths that experience greater scattering within the atmosphere. For the special case of a view direction opposite the direct solar beam, specular reflection results in an $L_s(\lambda)$ spectrum that literally mirrors the solar spectrum – pure sun glint. For rougher water surfaces, the orientation of individual surface facets becomes more variable, a greater number of facets reflect light from the brighter, near-sun portion of the sky, and $L_s(\lambda)$ begins to resemble the solar spectrum rather than the background sky.

An important property of reflection from the air-water interface is that all wavelengths are affected to a similar degree – that is, surface reflectance can be approximated as spectrally flat. When $L_S(\lambda)$ is normalised by the incident flux the resulting surface reflectance $R_S(\lambda)$ is nearly equal for all wavelengths, $R_S(\lambda) \approx R_s$. The magnitude of R_S is greater for rougher water surfaces, however. Although surface reflectance can increase the radiance measured above a river channel, this increase is purely additive and does not modify the spectral shape of the radiance signal because all wavelengths are affected approximately equally. Importantly, for those portions of the spectrum for which the water-leaving radiance $L_W(\lambda)$ is small

(i.e., the near-infrared), $L_S(\lambda)$ can comprise a large proportion of $L_T(\lambda)$, overwhelming the other components in Equation (3.14). Because $L_S(\lambda)$ is unrelated to the channel attributes of interest, this situation can complicate, if not preclude, mapping of these attributes. In other words, the noise associated with water surface reflectance can drown out the signal associated with radiance that has interacted with the water column and possibly the streambed (Legleiter et al., 2009).

Images of rivers are often adversely affected by sun glint due to strong reflections from the water surface. This problem can be mitigated by carefully planning data acquisition so as to avoid unfavorable combinations of illumination and viewing geometry. Due to the highly variable nature of water surface topography in rivers, however, sun-dominated surface reflectance will always be present to some extent. Because such reflectance is spectrally flat, radiance measured in wavelengths for which $L_W(\lambda)$ is assumed negligible can be attributed entirely to reflection from the water surface, along with the ubiquitous contribution from the atmosphere. Subtracting this radiance component across the spectrum could thus serve to remove the contribution of $L_S(\lambda)$ to $L_T(\lambda)$ and help to isolate the water-leaving signal of interest. In fact, this procedure is used to correct for sun glint in oceanographic remote sensing (Hooker et al., 2002; Hochberg et al., 2003). Two issues that arise in a fluvial context imply that such an approach is not directly applicable to rivers, however. First, the texture of the water surface, and hence the magnitude of $L_s(\lambda)$, is dictated by local flow hydraulics and is thus highly spatially variable, implying that a single surface-reflectance correction factor cannot be subtracted uniformly across an image. Second, $L_W(\lambda)$ cannot be assumed negligible, even in the near-infrared, due to the shallow depths and highly reflective substrates typical of rivers. Radiance observations from longer, shortwave-infrared wavelengths, where $L_W(\lambda)$ would be negligible even for very shallow waters, could be used for this purpose, but the multispectral sensors used in most stream studies lack such bands. For these reasons, detection and removal of water-surface reflectance remains difficult.

3.3.2 Optically significant constituents of the water column

The optical properties of the water column are a crucial consideration in any effort to derive information on the wetted portion of a river channel from remotely sensed data. Even for attributes such as channel morphology or bottom type that do not appear to be directly related to these properties, the water column cannot be ignored. The optical characteristics of the water column determine how radiant energy is modified as it propagates to and from the streambed. For applications focused on the water conveyed within the channel rather than the shape and composition of the channel boundary, optical properties provide the critical link between radiance measurements and concentrations of suspended sediment, chlorophyll, or dissolved organic matter. In either case, a useful approach to characterising the water column, detailed by Bukata et al. (1995), is to consider the composite, or bulk, inherent optical properties as the additive consequence of the *specific inherent optical properties* of the suspended and/or dissolved organic and/or inorganic materials in the water column, as well as pure water itself. For example, we can express the absorption coefficient $a(\lambda)$, a *bulk inherent optical property* of the water column as a whole, as the sum of the absorption coefficients, denoted by $a_i(\lambda)$, associated with each of n components, weighted by their concentrations x_i:

$$a(\lambda) = \sum_{i=1}^{n} x_i a_i(\lambda) \qquad (3.15)$$

Analogous expressions can be used to describe the scattering and back-scattering coefficients as well. The $a_i(\lambda)$ in Equation (3.15) represent the amount of absorption attributable to specific components of the water column; similarly, $b_i(\lambda)$ describe scattering by specific components. Specific inherent optical properties of this kind are referred to as *optical cross-sections* and have units of area per unit mass (i.e., m^2g^{-1}). Multiplying an optical cross-section by a concentration in gm^{-3} thus yields an absorption or scattering coefficient with units of m^{-1}. Legleiter et al. (2004, 2009) used optical cross-sections to model the optical properties of a simple two-component water column consisting of pure water and suspended sediment and quantify the effects of suspended sediment concentration on the water-leaving radiance. Conversely, if optical cross-sections for the components of interest are known and the bulk inherent optical properties of the water column can be derived from remotely sensed data, this information could be used to determine the concentration of each component. This approach is described in greater detail by Bukata et al. (1995), who focus on coastal waters and lakes. In principle, similar methods could be applied to fluvial environments, but in shallow streams the influence of the bottom could complicate such efforts.

Whether the attribute of primary interest pertains to the streambed or to the water column, one optically

significant component must always be considered: pure water. The absorption spectrum of pure water, denoted by $a_w(\lambda)$ is relevant to many fields of study, and numerous attempts have been made to measure this fundamental quantity with a high degree of accuracy. The most widely used data on $a_w(\lambda)$ for visible and near-infrared wavelengths were compiled by Pope and Fry (1997) and Smith and Baker (1981) and are shown in Figure 3.7a. Absorption of light by water is weakest in the blue and green, and photons of these wavelengths can thus penetrate considerable distances into a (pure) water body. As wavelength increases into the red and especially the near-infrared, $a_w(\lambda)$ increases by an order of magnitude, implying that photons in this portion of the spectrum will not propagate through the water column with nearly as much ease as their shorter-wavelength counterparts. Importantly, the spectral shape of $a_w(\lambda)$ dictates that red and near-infrared bands are most useful for estimating water depth in shallow stream channels with depths of 1 m or less. The strong absorption by pure water over this spectral range implies that small changes in depth will correspond to relatively large changes in radiance. This sensitivity allows for precise depth estimates in shallow water, but strong absorption also leads to saturation of the radiance signal at greater depths. Beyond a certain point, an additional increase in depth produces a very small, potentially undetectable, decrease in radiance because the vast majority of photons have already been absorbed. To map bathymetry across a broad range of depths, information from across the visible spectrum is thus necessary. Similarly, because the absorption spectrum of pure water dictates which wavelengths are most likely to propagate to the bed, $a_w(\lambda)$ influences which bands are useful for mapping bottom type. Scattering by pure water is strongest in the blue and decreases with wavelength but is insignificant relative to suspended sediment (Figure 3.7b).

The optical properties of suspended sediment, chlorophyll, and dissolved organic matter are best summarised in terms of their optical cross-sections, shown in Figure 3.7 (Bukata et al., 1995). For suspended sediment in particular, data on optical cross-sections are sparse and reflect natural variability in particle size, shape, and lithology. In practice, one of a handful of existing optical cross-sections is assumed to be representative of the area of interest. Nevertheless, some general observations can be made. Absorption by chlorophyll and dissolved organic matter tends to be strongest at shorter wavelengths, decrease with increasing wavelength, and be of a similar magnitude as absorption by suspended sediment. Absorption by suspended sediment is greatest in the blue and green and generally decreases with wavelength, although the data of Bukata et al. (1995) suggest an increase in absorption in the red. In all cases, however, absorption by suspended sediment, chlorophyll, or dissolved organic matter remains an order of magnitude less than absorption by pure water. Scattering by dissolved organic matter is insignificant, and the scattering cross-section for chlorophyll is spectrally flat and considerably smaller in magnitude than that for suspended sediment, illustrated for four different materials in Figure 3.7b. The radiative transfer modeling of (Legleiter et al., 2004, 2009) demonstrated that higher concentrations of suspended

sediment produce greater scattering and increase the volume reflectance of the water column, and thus $L_C(\lambda)$, throughout the blue and green. As concentrations increase, the range of wavelengths affected by these scattering interactions extends farther into the red.

An important result of these studies is that the visible spectrum can be partitioned, at least conceptually and for a given set of conditions, into: 1) a scattering-dominated regime at shorter wavelengths, where suspended sediment concentration is a primary control on the water-leaving radiance; and 2) an absorption-dominated regime, where $L_W(\lambda)$ is mainly sensitive to variations in water depth due

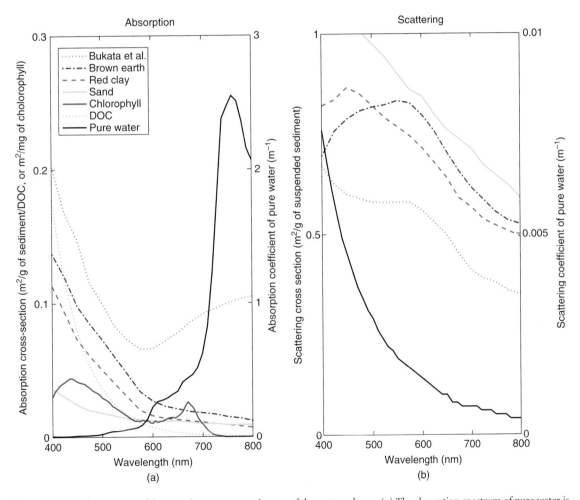

Figure 3.7 Optical properties of the most important constituents of the water column. (a) The absorption spectrum of pure water is plotted on the right vertical axis, based on the data of Pope and Fry (1997) from 400–720 nm and Smith and Baker (1981) from 730–800 nm. The absorption cross-sections for four different types of suspended sediment, chlorophyll, and dissolved organic carbon (DOC) are plotted on the left vertical axis and are based on data compiled in the Hydrolight radiative transfer model (Mobley and Sundman, 2001) and Equation (5.7) of Bukata et al. (1995) for DOC. (b) Scattering cross-sections for suspended sediment. Based on data from Hydrolight (Mobley and Sundman, 2001).

to large values of $a_w(\lambda)$ over this range. In the scattering-dominated regime, radiance increases with depth due to a greater number of scattering interactions in a thicker water column. Conversely, radiance decreases with depth in the absorption-dominated regime because more photons are absorbed within a thicker water column. At the transition between the two regimes is a wavelength for which the radiance is equal for all depths; this transition shifts toward longer wavelengths as suspended sediment concentration increases. Remote sensing of shallow water bathymetry and bottom type thus relies primarily upon red and near-infrared wavelengths, whereas efforts to infer concentrations of suspended sediment, chlorophyll, or dissolved organic matter will primarily make use of the blue and green portions of the spectrum. Note that these results were based on radiative transfer simulations for which pure water and suspended sediment were the only optically significant constituents of the water column. The presence of dissolved organic matter or other substances can modify the optical properties of the water column, for example. In general, all wavelengths are always affected by both absorption and scattering to some degree.

3.3.3 Reflectance properties of the streambed and banks

If the effects of the water column are somehow accounted for (i.e., via an effective attenuation coefficient), remote sensing techniques can be used to map channel bathymetry and/or substrate composition. For either of these applications, the reflectance properties of the streambed are an important consideration because the bottom-reflected radiance $L_B(\lambda)$ depends on both depth and bottom type. Even if water column optical properties are uniform at the reach scale, the relationship between depth d and $L_B(\lambda)$ can vary due to local variations in the irradiance reflectance of the streambed, $R_B(\lambda)$. Conversely, bathymetric variation can complicate efforts to estimate $R_B(\lambda)$ and map different substrate types. In other words, estimates of water depth ideally would be informed by knowledge of substrate composition, whereas mapping of bottom type would proceed based on knowledge of depth. In practice, of course, neither depth nor substrate type is known *a priori*. Inferring either attribute from remotely sensed data thus represents an under-determined problem, but spectral information can provide additional leverage.

An important concept in this context is the bottom contrast between the streambed, with an irradiance reflectance $R_B(\lambda)$, and the overlying water column, with a volume reflectance $R_C(\lambda)$. If the bed is brighter than the water – that is, if $R_B(\lambda) > R_C(\lambda)$ – the bottom is detectable as an increase in radiance relative to deep water; shallower depths correspond to higher radiances. This scenario is typical of clear water and highly reflective substrates at red and near-infrared wavelengths, where radiative transfer is dominated by pure water absorption. The opposite case, where the bed is actually darker than the overlying water column and $R_B(\lambda) < R_C(\lambda)$, implies that depth and radiance are inversely related; shallower water appears darker than deeper water. In this case, the presence of a relatively dark substrate at a small depth truncates the water column and reduces the amount of radiance scattered by suspended sediment or other materials. This scenario occurs at shorter blue and green wavelengths dominated by scattering, particularly when suspended sediment concentrations are high. A third possibility is $R_B(\lambda) \approx R_C(\lambda)$, implying that the bed is indistinct from the water itself, making the bottom effectively invisible and precluding estimation of depth. For the conditions examined by Legleiter et al. (2004, 2009), this scenario occurred from 550–600 nm at the transition from scattering- to absorption-dominated radiative transfer; at the wavelength of this crossover, radiance was independent of depth.

Because bottom reflectance, along with water column optical properties, determines the nature of the relationship between depth and radiance, some knowledge of different bottom types is useful for mapping bathymetry. If the objective is to characterise the spatial distribution of various substrates, knowledge of their spectral characteristics is critical – these characteristics determine the utility of remotely sensed data for distinguishing among streambed features. Reflectance spectra for several different bottom types measured along a gravel-bed river via ground-based spectroscopy are shown in Figure 3.8a (Legleiter et al., 2009). Certain substrates, such as limestone bedrock, are much brighter than other, darker materials, such as gravel. Moreover, some bottom types have very distinctive spectral shapes, such as periphyton, which is a type of algae that adheres to sediments comprising the streambed. This spectral diversity is conducive to substrate classification via remote sensing, and the pronounced differences in overall brightness and spectral shape among these three features imply that limestone, gravel, and periphyton could be distinguished from one another.

Substrate spectral variability could undermine efforts to map bathymetry, however. Standard depth retrieval

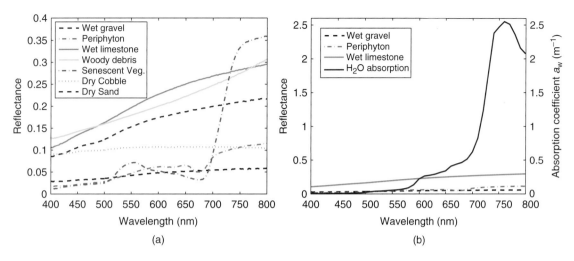

Figure 3.8 (a) Reflectance spectra for various in-stream and terrestrial features observed along a gravel-bed river. (b) Spectral differences in bottom reflectance are minor relative to spectral differences in attenuation by the water column, which is dominated by absorption by pure water. Data on the absorption coefficient for pure water $a_w(\lambda)$ are from Pope and Fry (1997) and Smith and Baker (1981). Reproduced from Legleiter, C.J. et al. (2009) Spectrally based remote sensing of river bathymetry. In: *Earth Surface Processes and Landforms*, pp. 1039–1059, with permission from John Wiley & Sons, Inc.

algorithms tend to yield under-estimates where the river flows over bright limestone and over-estimates in areas of darker gravel; more sophisticated spectrally-based methods are more robust to these effects (e.g., Lee et al., 2001; Mobley et al., 2005). Although few field observations of bottom reflectance in natural river channels have been made, the most extensive data set of which we are aware indicated that bottom reflectance was actually quite homogeneous in a fairly typical gravel-bed river (Legleiter et al., 2009). One hundred and thirty-nine ground-based reflectance measurements are summarised in Figure 3.9, which strongly resembles the periphyton spectrum in Figure 3.8a, implying that algal coating of the streamed was pervasive during late summer. This result is fortuitous for estimating depth because variation in bottom reflectance apparently need not be a concern, at least in this environment. Streambed photographs acquired along with the field spectra featured diverse lithologies and a range of particle sizes, but these grain-to-grain differences were not significant at the scale of the reflectance measurements (0.21 m field of view). The most salient feature of these spectra was a strong absorption band (i.e., decrease in reflectance) centered at 675 nm due to absorption by algal chlorophyll. Strong spectral signals of this kind will be most evident at shallow depths, where water column exerts less of an influence, and more subdued in deeper water.

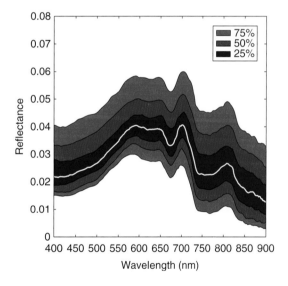

Figure 3.9 Field spectra measured along a gravel-bed river by Legleiter et al. (2009). The thick white line represents the median of 139 spectra and the solid areas encompass the indicated percentage of the distribution of reflectance measurements. These spectra represent a range of depths but have a similar spectral shape, indicating that bottom reflectance is fairly uniform and dominated by periphyton. Reproduced from Legleiter, C.J. et al. (2009) Spectrally based remote sensing of river bathymetry. In: *Earth Surface Processes and Landforms*, pp. 1039–1059, with permission from John Wiley & Sons, Inc.

Also shown in Figure 3.8a are reflectance spectra for various bank materials encountered along the same river (Legleiter et al., 2009). Even for studies focused specifically on in-channel attributes such as depth and substrate type, the banks cannot be ignored. Along the margins of the channel, the water-leaving radiance signal of interest will inevitably be mixed with radiance from adjacent terrestrial features. The severity of this problem depends on the dimensions of the channel relative to the image pixel size and also on the spectral contrast between submerged areas and various bank cover types. This issue was examined in detail by Legleiter and Roberts (2005), who attempted to 'unmix' near-bank pixels using reflectance spectra of pure aquatic and terrestrial end-members. The field spectra in Figure 3.8a indicate that dry sediments are brighter than their submerged counterparts but still relatively dark, particularly for coarser cobble, which was essentially spectrally flat. Finer sand-sized sediment was brighter and had a more pronounced increase in reflectance with increasing wavelength; woody debris had a similar spectral shape but somewhat higher reflectance than sand. The most spectrally distinct bank cover type was riparian vegetation, with low reflectance in the blue and red due to chlorophyll absorption and much higher reflectance in the near-infrared. The much greater reflectance of vegetation than water at these wavelengths makes this bank cover type spectrally distinct from the channel. Legleiter and Roberts (2005) thus found that pixels along vegetated banks could be unmixed successfully. Conversely, gravel bar surfaces were much more difficult to distinguish due to their spectral similarity to the wetted channel. These results imply that separating the aquatic signal of interest from the composite radiance from a mixed pixel will be more challenging for certain portions of the riverine landscape than for others. Under some circumstances, such as shallow, braided channels with numerous, unvegetated gravel bars, even identifying the water's edge could be difficult. For bathymetric mapping applications, an important consequence of terrestrial contamination is the common occurrence of negative depth estimates for shallow water near the banks. Typical, inverse relationships between depth and radiance result in negative depth estimates when applied to mixed pixels that include gravel, sand, or especially riparian vegetation. The development of improved methods of characterising channel margins remains an important challenge in the remote sensing of rivers.

3.4 Inferring river channel attributes from remotely sensed data

With this overview of the pertinent radiative transfer processes and optical properties, we are now prepared to examine how river information can be derived from remotely sensed data. We have chosen to focus our discussion on one of the more mature applications of remote sensing to fluvial systems, the estimation of water depth from passive optical image data. This topic has received considerable research attention due to the fundamental role of spatially distributed data on flow depths in geomorphic investigations, habitat assessments, and monitoring activities. Moreover, depth is perhaps the most tractable river attribute to infer from image data, and several other applications, such as substrate mapping, are facilitated by, if not dependent upon, bathymetric information.

3.4.1 Spectrally-based bathymetric mapping via band ratios

Although numerous depth retrieval methods have been applied to rivers with varying degrees of success, a semi-empirical approach based on a simple spectral band ratio has been shown to have a number of important advantages. The physical basis of this technique has been examined in detail elsewhere (Legleiter et al., 2004; Legleiter and Roberts, 2005; Legleiter et al., 2009), and only a summary of these findings is presented here. Recall our discussion of Equation (3.14), which separated the total, at-sensor radiance signal $L_T(\lambda)$ into its component parts. The component of primary interest for depth retrieval is the bottom-reflected radiance $L_B(\lambda)$, which depends not only on depth d but also on the irradiance reflectance of the streambed $R_B(\lambda)$. Some means of distinguishing variations in depth from variations in the brightness and/or composition of the substrate is thus necessary. The basic principle behind band ratio-based bathymetric mapping is that spectral differences in $R_B(\lambda)$ are much smaller than the order-of-magnitude spectral differences in the rate at which light is attenuated by the water column (Figure 3.8b). As a result, the bottom reflectance in two bands, λ_1 and λ_2, is similar for different bottom types, and the ratio $R_B(\lambda_1)/R_B(\lambda_2)$ is relatively unaffected by substrate heterogeneity (e.g., Dierssen et al., 2003). In contrast, for clear-flowing streams the rate of attenuation primarily depends on the absorption coefficient of pure water $a_w(\lambda)$ and thus exhibits a strong increase with wavelength through the red and near-infrared. Therefore, as depth increases, the radiance $L_T(\lambda_2)$ measured in a longer-wavelength band experiencing greater attenuation decreases more rapidly than the radiance $L_T(\lambda_1)$ in the band with weaker attenuation. Consequently, the band ratio $L_T(\lambda_1)/L_T(\lambda_2)$ increases as depth increases and is not strongly influenced by changes in bottom reflectance. The ratio calculation also accounts for variations in the irradiance incident upon sloping streambeds because these topographic effects influence the numerator and denominator bands in the same way. In addition, both bands are affected to a similar degree by reflectance from the water surface, which is spectrally flat and thus cancels in the ratio. Taking the natural logarithm of the band ratio to account for the exponential attenuation of light with distance traveled through the water column thus provides an image-derived quantity useful for bathymetric mapping (Legleiter et al., 2009). Importantly, however, this simple algorithm requires ground-based measurements to calibrate the relationship between the band ratio and water depth. More recent, radiative transfer-based techniques do not require such tuning and have largely replaced band ratios for remote sensing of shallow marine environments (e.g., Lee et al., 2001; Mobley et al., 2005). Similarly, in the context of rivers, new methods based on hydraulic (Fonstad and Marcus, 2005) and photogrammetric (Lane et al., 2010) principles can now be used to calibrate image-derived depth estimates without requiring simultaneous field data. Development of robust, flexible bathymetric calibration procedures remains an area of active research.

3.4.2 Relative magnitudes of the components of the at-sensor radiance signal

Equation (3.14) indicates that the radiance measured by a remote detector is the sum of four components: 1) $L_B(\lambda)$, the bottom-reflected radiance of primary interest for depth retrieval or substrate mapping; 2) $L_C(\lambda)$, radiance scattered within the water column before reaching the bed; 3) $L_S(\lambda)$, radiance reflected from the water surface before interacting with the water column or substrate; and 4) $L_P(\lambda)$, radiance scattered within the Earth's atmosphere. Considering the relative magnitudes of these four components can provide some insight as to the conditions under which remotely sensed data can be used to

quantify river attributes such as bathymetry and bottom type, which are directly related only to $L_B(\lambda)$. A scaling analysis of this kind was conducted by Legleiter et al. 2009, who supported their arguments by performing radiative transfer simulations, collecting ground-based reflectance measurements, and producing bathymetric maps from hyperspectral image data. This study indicated that the bottom-reflected radiance is the dominant component of $L_T(\lambda)$ in shallow, clear-flowing rivers with depths on the order of tens of cm, water columns dominated by pure water absorption rather than scattering by suspended sediment, and highly reflective substrates, provided that the illumination and viewing geometry are favourable (small $L_S(\lambda)$) and that atmospheric effects do not overwhelm the aquatic signal of interest (small $L_P(\lambda)$). Under these circumstances, the other radiance components can be considered negligible – that is,

$$L_B(\lambda) \gg L_C(\lambda) + L_S(\lambda) + L_P(\lambda).$$

More specifically, the radiance contribution from the bottom will exceed that from the water column by an increasing amount as depth decreases, as bottom reflectance increases, and as absorption within the water column predominates over scattering. Where, when, and in which wavelengths these conditions are met, water depth d is linearly related to the image-derived quantity X given by

$$X = \ln\left[\frac{L_T(\lambda_1)}{L_T(\lambda_2)}\right] \approx \ln\left[\frac{L_B(\lambda_1)}{L_B(\lambda_2)}\right]$$

$$= [K(\lambda_2) - K(\lambda_1)]\, d \qquad (3.16)$$

$$+ \ln\left[\frac{R_B(\lambda_1) - R_C(\lambda_1)}{R_B(\lambda_2) - R_C(\lambda_2)}\right] + A$$

In this expression subscripts denote the two spectral bands used to compute the ratio, $K(\lambda)$ is the effective attenuation coefficient, $R_B(\lambda)$ is the irradiance reflectance of the streambed, $R_C(\lambda)$ is the volume reflectance of the water column, and A is a constant that accounts for the downwelling irradiance incident upon the river, transmission and reflection at the air-water interface, and transmission losses within the atmosphere. This equation describes a straight line with a slope given by the difference in the effective attenuation coefficient between the two bands, and X increases with d if $K(\lambda_2) > K(\lambda_1)$. The intercept of the line accounts for the bottom contrast between the streambed and water column, as well as the other factors included within the constant A.

The only quantity in Equation (3.16) that varies on a pixel-by-pixel basis within an image is the one of primary interest, the flow depth d. For a reach without tributary inputs or other sources of suspended sediment, the inherent optical properties of the water column can be assumed homogeneous, and these properties determine the values of $K(\lambda)$ and $R_C(\lambda)$. Substrate composition might vary within a reach, if not on a pixel scale, but the ratio $R_B(\lambda_1)/R_B(\lambda_2)$ remains approximately constant across bottom types. These other factors therefore have relatively little influence on X, implying that this image-derived quantity primarily depends on depth and is thus useful for bathymetric mapping. In practice, depth maps are produced by regressing field measurements of depth against X values for the corresponding image pixels and applying the resulting calibration equation throughout the image. The radiative transfer simulations, field spectra, and image processing reported by Legleiter et al. 2009 provided both theoretical and empirical validation of the critical assumptions underlying this ratio-based approach.

3.4.3 The role of sensor characteristics

In addition to the physical processes that govern the interaction of light and water and thus dictate the optical properties of river channels, the characteristics of the sensors themselves are also an important consideration. The technical specifications of an imaging system play a key role in determining which river attributes can be remotely mapped, and with what degree of accuracy and precision. For the most part, the physics that control the upwelling spectral radiance must simply be accepted, although the timing of data acquisition can be specified so as to avoid unfavourable circumstances such as high turbidity or strong sun glint. Faced with these largely immutable physical constraints, prospective users of remotely sensed data can improve their chances of deriving useful river information by selecting instrumentation appropriate for a particular project's objectives. Certain types of sensors will be better able to provide image data from which the attributes of interest can be reliably inferred than other kinds of instruments. Financial and logistical considerations always play a role, of course, but in planning a remote sensing mission one would like to specify three principal sensor characteristics: 1) *spatial resolution*, which is typically equated with the edge dimension of an image pixel on the ground and depends on sensor optics and flying height; 2) *spectral resolution*, which refers to the number of wavelength bands in which radiance measurements are made, as well as the location and width of these bands; and 3) *radiometric resolution*, which describes the sensor's ability to detect small changes in radiance and depends on the manner in which the continuous upwelling radiance signal is converted to discrete, digital image data. In the following paragraphs, we examine each of these sensor characteristics and discuss their influence on the utility of remotely sensed data for specific applications.

Perhaps the most obvious decision made in planning the acquisition of remotely sensed data, or identifying appropriate existing data sets, is that of spatial resolution, or pixel size. In essence, the image data must be sufficiently detailed to allow the user to 'see' the river features of interest. The ratio of mean channel width to image pixel size thus represents a fundamental constraint. If this ratio is on the order of one, even detecting the channel could be problematic; this is often the case for small- to medium-sized rivers less than 50 m wide m and moderate-resolution satellite systems (e.g., Landsat's 30 m pixels). A number of commercial satellites now provide much greater spatial resolution, often with a sub-meter panchromatic band and multispectral bands with pixel sizes of 2 m or less; current examples include Ikonos, QuickBird, GeoEye, and WorldView-2.

Airborne systems provide greater flexibility because the flying height can be adjusted so as to achieve a desired pixel size, and some low-altitude deployments achieve resolutions on the order of a few cm (e.g., Carbonneau et al., 2004; Lejot et al., 2007). Acquiring such high resolution data can mitigate the problem of mixed pixels along the channel banks, although some degree of mixing is inevitable. With smaller pixels, the area over which such mixtures occur will be reduced, as will the area subject to negative depth estimates along shallow channel margins. Within the channel proper, ambiguity due to sub-pixel variations in depth and bottom reflectance can largely be avoided with high spatial resolution data. These detailed images also enable more traditional photo interpretation, and features such as woody debris will often be clearly visible. Such high resolution data are not without drawbacks, however. Other factors to consider include the practicality of flying at low altitudes, the small area covered by individual images along a flight line, the massive volume of data involved, and the effort required to geo-reference and analyse large numbers of images.

For many applications, then, a somewhat coarser resolution, with pixels on the order of 1–4 m, might be more appropriate for large-scale mapping and monitoring.

In this, more typical situation, reliable depth estimates can be obtained using the ratio-based method described above. Legleiter and Roberts (2005) showed that this approach is robust to the effects of sub-pixel variations in depth and bottom reflectance. This study also described how spectral mixture analysis could be used to disaggregate aquatic and terrestrial radiance contributions to mixed pixels along the banks, which will be more prevalent in coarser-resolution data. A key point made by Legleiter and Roberts (2005) is that the ability to characterise channel morphology depends in a complex and spatially variable manner on the spatial resolution of the sensor and the dimensions of the channel features of interest. As a result, a pixel size that is adequate for one reach might not be sufficient for other, more morphologically complex portions of the river.

Although the radiance upwelling from a river channel comprises a continuous spectrum, remote sensing instruments typically integrate the spectral radiance over a few discrete bands. The number, width, and location of these bands determine how faithfully spectral variations in the original radiance signal are reproduced by image data and thus dictate the extent to which spectral information can be used to infer channel attributes. High spectral resolution is potentially quite valuable because several important optical characteristics of rivers vary appreciably with wavelength. Most notably, the reflectance of the streambed and the absorption and scattering processes that determine the rate of attenuation within the water column have well-defined spectral shapes. For bathymetric mapping applications, the availability of a large number of narrow spectral bands enables selection of wavelengths that are highly sensitive to depth, typically due to strong absorption by pure water, but unaffected by variations in bottom reflectance and scattering within the water column.

Legleiter et al. (2009) introduced a simple method, called *optimal band ratio analysis* or OBRA, for identifying wavelength pairs that satisfy these requirements. This technique exploits high spectral resolution data by computing the image-derived quantity X defined by Equation (3.16) for all possible combinations of bands. The resulting X values are then regressed against field measurements of depth and the band combination that yields the highest R^2 is deemed optimal for depth retrieval. OBRA of both field spectra and simulated data from a radiative transfer model indicated that integrating the radiance signal over broader spectral bands did not significantly degrade the predictive power of X vs. d regression equations, implying that high spectral resolution data are not necessarily

essential for accurate bathymetric mapping. When field spectra were re-sampled to match the bands of specific multispectral sensors, R^2 values were somewhat lower, however, suggesting that the wavelength position of the bands might be more important than their widths.

For other applications focused on substrate mapping, high spectral resolution could be more important, if not critical. Distinguishing among bottom types requires sufficient spectral information to resolve subtle differences between streambeds. For example, periphyton exhibits a diagnostic chlorophyll absorption feature at 675 nm that can be used to detect algal presence, but only if the imaging system provides sufficient spectral detail to resolve this feature. Moreover, higher spectral resolution could allow the effects of depth and bottom reflectance to be be disentangled with greater confidence where heterogeneous substrates might act to complicate depth retrieval. Hyperspectral data also create opportunities to apply more sophisticated approaches such as spectral mixture analysis, which could be especially helpful along the banks. For management applications where bathymetric mapping is the primary objective, the robust performance of ratio-based depth retrieval implies that more readily available, multispectral data will often be adequate, however.

A third sensor characteristic, radiometric resolution, is not visually apparent in the same way as spatial resolution and does not affect the structure of an image data set in an obvious manner as does spectral resolution. Radiometric resolution can thus be easily overlooked, but this fundamental property of an imaging system should be a primary consideration in any remote sensing investigation. For example, the precision with which depths can be estimated and the maximum depth that can be detected are both largely determined by sensor radiometric resolution, which describes an instrument's ability to distinguish subtle variations in the amount of upwelling radiance. The critical concept to bear in mind is that although the original radiance signal is continuous, image data are recorded as digital numbers. Converting the continuous signal to a discrete form necessarily entails some loss of information. In the context of rivers, this principle implies that a change in depth, bottom reflectance, or volume reflectance can only be detected if the corresponding change in radiance exceeds the fixed amount of radiance corresponding to one digital number.

Continuing with our focus on depth retrieval, this reality of remote sensing dictates that truly continuous bathymetric maps cannot be derived from digital image data. Instead, each depth estimate is associated with a bathymetric contour interval. The width of these contour intervals, which can be defined as the sum of the smallest detectable increase in depth and the smallest detectable decrease in depth, depends on the radiometric resolution of the imaging system and the brightness contrast between the bottom and the water column. If the reflectance of the streambed and the optical properties of the water column are assumed constant, bathymetric precision (i.e., contour interval width) can be treated as a function of sensor characteristics. A useful metric in this regard is the **noise-equivalent delta radiance** $\Delta L_N(\lambda)$, which is most simply defined as the change in radiance equivalent to one digital number, or the inverse of the number of possible discrete values. For example, a 12-bit imaging system would have a smaller $\Delta L_N(\lambda)$ proportional to $1/2^{12} = 0.00024 \, \mathrm{Wm}^{-2}\mathrm{sr}^{-1}$ (integrated over the sensor band pass) than a less sensitive instrument that records 8-bit data and thus enables only $2^8 = 256$ possible values. In practice, $\Delta L_N(\lambda)$ also depends on instrumental and environmental signal-to-noise characteristics and a more inclusive definition would account for these other factors as well (e.g., Giardino et al., 2007). In any case, this approach can be used to characterise the inherent uncertainty associated with image-derived depth estimates. Building upon earlier work by Philpot (1989), Legleiter et al. (2004) showed that the magnitude of this uncertainty increases as depth increases, as bottom reflectance decreases, and as scattering predominates over absorption. This study demonstrated that relatively precise depth estimates, with uncertainties of <5 cm, could be achieved in shallow water with highly sensitive instruments but that uncertainties of >30 cm would be associated with deeper water and less sensitive detectors.

Closely related to the precision of depth estimates is the dynamic range of depth retrieval. Philpot (1989) reasoned that the maximum detectable depth occurs where the difference between the measured at-sensor radiance and the radiance from a hypothetical infinitely deep water body is equal to the radiance corresponding to one digital number. The bottom contrast between the water column and substrate also exerts an important control in this context. For the conditions examined by Legleiter et al. (2004), the maximum detectable depth ranged from approximately 0.5 m in strongly absorbing near-infrared wavelengths for a sensor with a relatively large $\Delta L_N(\lambda)$ to over 5 m in less strongly absorbing green wavelengths for a more sensitive detector with a smaller $\Delta L_N(\lambda)$. These results highlighted a tradeoff between precision and dynamic range. Depth estimates are most precise in the near-infrared, where strong absorption by

pure water implies that even small changes in depth correspond to large changes in radiance. The strength of absorption also dictates that at greater depths the radiance signal will saturate, with additional increases in depth producing very small changes in radiance that will not be detectable by many imaging systems.

Legleiter and Roberts (2009) incorporated these concepts into a **forward image model** that allows depth retrieval accuracy and precision to be examined for a particular river of interest and a given sensor configuration. An application of the forward image model to a gravel-bed river is shown in Figure 3.10, which indicates considerable variation in bathymetric accuracy and precision within the channel. In essence, this approach involves simulating an image from the streambed up by combining information on bed topography, bottom reflectance, and water column optical properties with numerical models describing radiative transfer in the water column and atmosphere. The forward image model can thus be used to assess, a priori, the utility of image data for specific applications in river research and management. This study demonstrated that the reliability of image-derived depth estimates is strongly dependent on channel morphology and thus varies spatially. Sensor spatial resolution was the primary control on bathymetric accuracy, with depths tending to be underestimated in pools (i.e., yellow tones along the left side of Figure 3.10a), while mixed pixels along shallow channel margins also made near-bank depth estimates unreliable. For example, the bright red tones along the lower (south) end of the mid-channel bar indicate that large under-predictions of depth occurred where pixels along the water's edge included relatively bright, sandy sediment. Similarly, bathymetric precision was determined mainly by sensor radiometric resolution, with broader bathymetric contour intervals also occurring in deeper water (i.e., green tones along the left side of Figure 3.10b). This type of analysis can temper one's enthusiasm regarding the potential for remote sensing of rivers, but carefully considering the limitations as well as the capabilities associated with this technology will ultimately lead to more effective use of remote sensing in fluvial environments.

Another important point with regard to sensor characteristics is that the three types of resolution discussed above are not independent of one another, but rather intimately connected. Essentially, remote sensing involves collecting photons and sorting them into different 'bins' based on where they came from (spatial resolution) and the wavelength at which they travel (spectral resolution). In order for the photons collected in one of these bins to

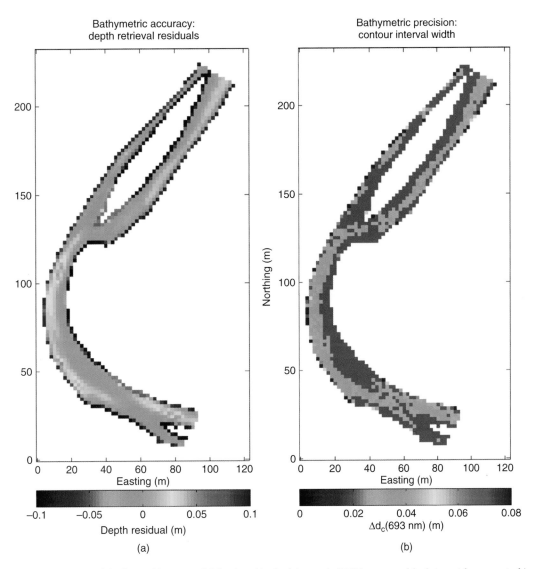

Figure 3.10 Application of the forward image model developed by Legleiter et al. (2009) to a gravel-bed river. A hyperspectral image was simulated using topographic survey data for the channel reach of interest, technical specifications of the Airborne Imaging Spectrometer for Applications (AISA), and radiative transfer models of the water column and atmosphere. The simulated image has 2 m pixels and consists of 12-bit data. (a) Bathymetric accuracy is expressed in terms of depth retrieval residuals, or errors relative to the known input bathymetry. (b) Bathymetric precision is expressed in terms of contour interval widths Δd_c, defined for the band centered at 693 nm. Reprinted from *Remote Sensing of Environment*, 113, Legleiter, C.J. & Roberts, D.A., A forward image model for passive optical remote sensing of river bathymetry, pp. 1025–1045, Copyright 2009, with permission from Elsevier.

'count', in the sense that they are represented in the final, digital image data, they must be sufficient in number to 'fill up' the bin and induce a transition from one digital number to the next; the depth of these bins is related to the sensor's radiometric resolution. The signal of interest is the population of photons that have interacted with the river and reached the sensor, but the fidelity with which

this signal is recorded is determined by the manner in which the imaging system collects and bins these photons. This binning analogy highlights the tradeoffs that must be made in selecting appropriate instrumentation for a particular remote sensing project, with some compromise reached between spatial detail, spectral discrimination, and radiometric sensitivity (Legleiter et al., 2004).

For example, high spatial resolution corresponds to small spatial bins, implying that the bins must encompass a broader range of wavelengths in order for them to be filled to the depth set by the sensor's radiometric resolution. Conversely, subdividing photons into narrow spectral bands typically implies an increase in pixel size. As a result, satellite or airborne sensors with very high spatial resolution tend to be multispectral, with only three or four bands, whereas hyperspectral instruments tend to have larger pixel sizes. Radiometric resolution determines bin depth and thus represents a key constraint, often dictated by the signal to noise properties of the system electronics. In any case, these sensor characteristics are a crucial consideration in any application of remote sensing to rivers. Prospective users of this technology must understand what levels of spatial, spectral, and radiometric resolution are required for different types of studies. For example, an investigation of channel change might benefit from a highly sensitive detector with 12-bit radiometric resolution that enables very precise depth estimates, whereas an effort to map different algal species might prefer greater spectral resolution that allows distinctive chlorophyll absorption features to be distinguished. To make effective use of remote sensing techniques in river management, project objectives must be clearly defined and prioritised so that instrumentation and/or existing data sources can be selected accordingly.

3.5 Conclusion

This chapter has provided an overview of the physical basis for remote sensing of rivers. Various radiometric quantities were defined and used to describe the radiative transfer processes by which light and water interact in shallow stream channels. We employed an image chain analogy to link these processes to one another, discuss the influence of the Earth's atmosphere, and partition the radiance signal recorded by a remote detector into a set of components. A particular application, estimation of water depth, was emphasised throughout the chapter and used to illustrate how river information can be derived from remotely sensed data. Finally, the effects of sensor characteristics – spatial, spectral, and radiometric resolution – on the nature and reliability of this information was discussed. In keeping with the chapter's principal objective, we have stressed the importance of understanding the physical processes underlying remotely sensed data, even when these data are primarily analysed using empirical approaches.

Whereas traditional techniques that rely upon statistical correlations between ground-based measurements and image pixel values produce scene-specific results, a more physics-based approach could provide greater generality and reduce the need for *in situ* observations. Significant progress toward this end has already been made in coastal and shallow marine environments. For example, Mobley et al. (2005) used a numerical radiative transfer model to create lookup tables containing spectra for a range of depths, bottom types, and water column optical properties. By matching these spectra to hyperspectral image data, Lesser and Mobley (2007) were able to map these attributes over a broad area and to a high degree of accuracy, without extensive field data. Remote sensing of rivers has not yet reached such an advanced stage, but additional effort toward this goal is justified by the advantages of a physics-based approach: (1) the amount of field data required for calibration could be greatly diminished, with a smaller number of ground-based measurements used primarily for validation; (2) more flexible, generic mapping algorithms could be developed and applied to larger spatial extents and/or archival image data to examine river dynamics over an expanded range of scales; and (3) the accuracy and precision of image-derived river information can be assessed *a priori* within a forward image modeling framework that represents physical processes along the image chain (Legleiter and Roberts, 2009). Ultimately, we hope to see continued progress toward more physics-based remote mapping of fluvial systems.

In the interim, there is potential to leverage existing monitoring programs to obtain the ground reference data needed to implement traditional, empirical approaches to remote sensing of rivers. For example, repeat cross-section surveys conducted regularly for geomorphic monitoring purposes might be coordinated with the acquisition of remotely sensed data and thus used to calibrate image-derived depth estimates. Exploiting the synergy between current, ground-based data collection activities and remote sensing campaigns could serve as a bridge for extending spectrally-based bathymetric mapping over longer stream segments and on a more routine basis. Such an effort could facilitate various management objectives and help to build confidence in the use of remote sensing techniques in future monitoring programs. In any case, we believe that even for practitioners using empirical approaches, some appreciation of the underlying radiative transfer processes is critical.

An improved understanding of these processes could enable a number of innovative applications, beyond the bathymetric mapping emphasised in this chapter. One obvious extension is quantitative mapping of different bottom types on the basis of their reflectance properties. Image-derived depth estimates, along with information on optical properties, could be used to account for the influence of the water column, estimate bottom reflectance, and thus delineate various substrate types (e.g., Dierssen et al., 2003). For example, one could potentially distinguish among various grain size patches (e.g., gravel vs. sand vs. fine sediments) or different species of submerged aquatic vegetation or algae; quantitative estimates of biomass might be possible as well. A similar approach could be used to identify particular features of interest, such as bank failure, which could be expressed as a distinct spectral signal of water flowing over blocks of grassy sod. More sophisticated techniques, such as spectral mixture analysis or lookup-table based methods similar to those employed by Lesser and Mobley (2007) would facilitate the development of such novel applications. Additional prospects worthy of attention include improved characterisation of various aspects of water quality, such as concentrations of suspended sediment, chlorophyll, and dissolved organic matter, and inference of flow hydraulics from remote observations of surface-reflected radiance.

We anticipate a number of future challenges and research needs that must be addressed in order to advance remote sensing of rivers. For example, we know little about the spectral variability of different substrate and streambank materials, and this dearth of information justifies a concerted effort to build representative spectral libraries from a range of fluvial environments. Similarly, additional data on optical cross-sections are needed to improve parametrisation of water column optical properties in radiative transfer models. Sun glint is pervasive in many river images and is present to some extent in almost all cases, but effective techniques for detecting and removing surface-reflected radiance have yet to be developed. This problem is important because sun glint can preclude inference of channel attributes of interest. Given the relatively small amounts of radiance that leave river channels and the increasing availability of high spatial resolution satellite image data, improved, practical methods of atmospheric correction are also necessary. Finally, before remote sensing of rivers can become a viable, operational tool for monitoring and management, significant effort must be invested in the development of efficient, readily available software tools that facilitate practical implementation of various image processing techniques.

The very existence of this volume is testament to the current utility and future potential for remote sensing of rivers. In order to fully realise this considerable potential, we as a community of river-oriented managers and scientists must first develop confidence in image-derived river information. This confidence must be justified. Along with the capabilities afforded by remote sensing technology, we must bear in mind the inherent limitations associated with these data. To reiterate the premise of this chapter, an understanding of the physical processes that both enable and limit the application of remote sensing to rivers is critical to the effective use of this powerful tool. Improving our knowledge of these processes will allow us to identify appropriate uses of remote sensing techniques and to define realistic expectations of what can and cannot be achieved using these methods. Such a critical perspective, informed by an appreciation of the underlying physics, will enable us to more efficiently derive river information from remotely sensed data and more effectively use this information in a range of fluvial applications.

3.6 Notation

λ = Wavelength

ν = Frequency

c = Speed of light

q = Radiant energy of a single photon

h = Planck's constant

Q = Total radiant energy in a beam of light

n_i = Number of photons at a particular wavelength

Φ = Radiant flux

E = Irradiance

θ = Angle of incidence

l = Arc length

r = Radius of circle of sphere; distance

Ω = Solid angle

A = Surface area on sphere used to compute a solid angle; constant in Equation (3.16)

L = Radiance

$R(\lambda)$ = Irradiance reflectance

$E_u(\lambda)$ = Upwelling spectral irradiance

$E_d(\lambda)$ = Downwelling spectral irradiance

z = Vertical position (depth) within the water column

z_b = Vertical position (depth) at bottom of water column

θ_v = Viewing angle

$\phi_v =$ View azimuth

$\theta_i =$ Illumination angle (incidence angle)

$\phi_i =$ Illumination azimuth

$r_{BRDF}(\lambda) =$ Bidirectional reflectance distribution function (BRDF)

$\theta_w =$ In-water incidence angle after refraction

$n_a, n_w =$ Refractive index of air, water

$\theta_r =$ Angle of refraction

$\rho =$ Fresnel reflectance from the air-water interface

$a(\lambda) =$ absorption coefficient

$b(\lambda) =$ scattering coefficient

$c(\lambda) =$ total attenuation coefficient

$K(\lambda, z) =$ diffuse irradiance attenuation coefficient

$\overline{K(\lambda)} =$ depth-averaged diffuse irradiance attenuation coefficient

$K_d(\lambda, z) =$ diffuse irradiance attenuation coefficient for downwelling irradiance

$K_u(\lambda, z) =$ diffuse irradiance attenuation coefficient for upwelling irradiance

$L_W(\lambda) =$ water-leaving radiance

$T(\lambda) =$ transmittance of the atmosphere

$L_P(\lambda) =$ path radiance from the atmosphere

$L_T(\lambda) =$ total at-sensor radiance

$L_B(\lambda) =$ bottom-reflected radiance

$L_C(\lambda) =$ radiance from the water column

$L_S(\lambda) =$ surface-reflected radiance

$x_i =$ concentration of ith optically significant constituent of the water column

$a_i(\lambda) =$ absorption coefficient of ith optically significant constituent of the water column

$b_i(\lambda) =$ scattering coefficient of ith optically significant constituent of the water column

$a_w(\lambda) =$ absorption coefficient of pure water

$R_b(\lambda) =$ bottom reflectance of the streambed

$R_c(\lambda) =$ volume reflectance of the water column

$d =$ water depth

$X =$ log-transformed band ratio (Equation 3.16)

$R^2 =$ coefficient of determination for regression

$\Delta L_N(\lambda) =$ noise-equivalent delta radiance

References

Bukata, R. P., Jerome, J.H., Kondratyev, K.Y., and Pozdnyakov, D.V.: 1995 *Optical Properties and Remote Sensing of Inland and Coastal Waters*, CRC Press, Boca Raton, FL.

Carbonneau, P.E., Lane, S.N., and Bergeron, N.E. 2004 Catchment-scale mapping of surface grain size in gravel bed rivers using airborne digital imagery, *Water Resources Research* **40**(W07202), doi:10.1029/2003WR002759.

Cox, C. and Munk, W. 1954 Measurement of the roughness of the of the sea surface from photographs of the sun's glitter, *Journal of the Optical Society of America* **44**(11), 838–850.

Dierssen, H.M., Zimmerman, R.C., Leathers, R.A., Downes, T.V., and Davis, C.O. 2003 Ocean color remote sensing of seagrass and bathymetry in the Bahamas Banks by high-resolution airborne imagery, *Limnology and Oceanography* **48**(1), 444–455.

Feurer, D., Bailly, J.-S., Puech, C., Le Coarer, Y., and Viau, A.A. 2008 Very-high-resolution mapping of river-immersed topography by remote sensing, *Progress in Physical Geography* **32**(4), 403–419.

Fonstad, M.A. and Marcus, W.A. 2005 Remote sensing of stream depths with hydraulically assisted bathymetry (HAB) models, *Geomorphology* **72**(1–4), 320–339.

Giardino, C., Brando, V.E., Dekker, A.G., Strombeck, N., and Candiani, G. 2007 Assessment of water quality in Lake Garda (Italy) using Hyperion, *Remote Sensing of Environment* **109**(2), 183–195.

Hochberg, E.J., Atkinson, M.J., and andrefouet, S. 2003 Spectral reflectance of coral reef bottom-types worldwide and implications for coral reef remote sensing, *Remote Sensing of Environment* **85**(2), 159–173.

Hooker, S.B., Lazin, G., Zibordi, G., and McLean, S. 2002 An evaluation of above- and in-water methods for determining water-leaving radiances, *Journal of Atmospheric and Oceanic Technology* **19**(4), 486–515.

Kinzel, P.J., Wright, C.W., Nelson, J.M., and Burman, A.R. 2007 Evaluation of an experimental LiDAR for surveying a shallow, braided, sand-bedded river, *Journal of Hydraulic Engineering* **133**(7), 838–842.

Lane, S.N., Widdison, P.E., Thomas, R.E., Ashworth, P.J., Best, J.L., Lunt, I.A., Smith, G.H.S., and Simpson, C.J. 2010 Quantification of braided river channel change using archival digital image analysis, *Earth Surface Processes and Landforms* **In press**.

Lee, Z., Carder, K.L., Chen, R.F., and Peacock, T.G. 2001 Properties of the water column and bottom derived from airborne visible infrared imaging spectrometer (AVIRIS) data, *Journal of Geophysical Research-Oceans* **106**(C6), 11639–11651.

Legleiter, C.J. and Roberts, D.A. 2005 Effects of channel morphology and sensor spatial resolution on image-derived depth estimates, *Remote Sensing of Environment* **95**(2), 231–247.

Legleiter, C.J. and Roberts, D.A. 2009 A forward image model for passive optical remote sensing of river bathymetry, *Remote Sensing of Environment* **113**(5), 1025–1045.

Legleiter, C.J., Roberts, D.A., and Lawrence, R.L. 2009 Spectrally based remote sensing of river bathymetry, *Earth Surface Processes and Landforms* **34**(8), 1039–1059.

Legleiter, C.J., Roberts, D.A., Marcus, W.A., and Fonstad, M.A. 2004 Passive optical remote sensing of river channel morphology and in-stream habitat: Physical basis and feasibility, *Remote Sensing of Environment* **93**(4), 493–510.

Lejot, J., Delacourt, C., Piégay, H., Fournier, T., Trémélo, M.-L. and Allemand, P. 2007 Very high spatial resolution imagery for channel bathymetry and topography from an unmanned mapping controlled platform, *Earth Surface Processes and Landforms* **32**(11), 1705–1725.

Lesser, M.P. and Mobley, C.D. 2007 Bathymetry, water optical properties, and benthic classification of coral reefs using hyperspectral remote sensing imagery, *Coral Reefs* **26**(4), 819–829.

Lyzenga, D.R. 1978 Passive remote-sensing techniques for mapping water depth and bottom features, *Applied Optics* **17**(3), 379–383.

Marcus, W.A. and Fonstad, M.A. 2008 Optical remote mapping of rivers at sub-meter resolutions and watershed extents, *Earth Surface Processes and Landforms* **33**(1), 4–24.

Maritorena, S., Morel, A., and Gentili, B. 1994 Diffuse-reflectance of oceanic shallow waters-influence of water depth and bottom albedo, *Limnology and Oceanography* **39**(7), 1689–1703.

McKean, J., Nagel, D., Tonina, D., Bailey, P., Wright, C.W., Bohn, C., and Nayegandhi, A. 2009 Remote sensing of channels and riparian zones with a narrow-beam aquatic-terrestrial LiDAR, *Remote Sensing* **1**(4), 1065–1096.

Mobley, C.D. 1994 *Light and Water: Radiative Transfer in Natural Waters*, Academic Press, San Diego.

Mobley, C.D. and Sundman, L.K. 2001 *Hydrolight 4.2 User's Guide*, 2 edn, Sequoia Scientific, Redmond, WA.

Mobley, C.D. and Sundman, L.K. 2003 Effects of optically shallow bottoms on upwelling radiances: Inhomogeneous and sloping bottoms, *Limnology and Oceanography* **48**(1), 329–336.

Mobley, C.D., Sundman, L.K., Davis, C.O., Bowles, J.H., Downes, T.V., Leathers, R.A., Montes, M.J., Bissett, W.P., Kohler, D.D.R., Reid, R.P., Louchard, E.M., and Gleason, A. 2005

Interpretation of hyperspectral remote-sensing imagery by spectrum matching and look-up tables, *Applied Optics* **44**(17), 3576–3592.

Mobley, C.D., Zhang, H., and Voss, K.J. 2003 Effects of optically shallow bottoms on upwelling radiances: Bidirectional reflectance distribution function effects, *Limnology and Oceanography* **48**(1), 337–345.

Philpot, W.D. 1989 Bathymetric mapping with passive multispectral imagery, *Applied Optics* **28**(8), 1569–1578.

Pope, R.M. and Fry, E.S. 1997 Absorption spectrum (380–700 nm) of pure water. 2. Integrating cavity measurements, *Applied Optics* **36**(33), 8710–8723.

Schott, J.R. 1997 *Remote Sensing: The Image Chain Approach*, Oxford University Press, New York.

Slatton, K.C., Carter, W.E., Shrestha, R.L., and Dietrich, W. 2007 Airborne laser swath mapping: Achieving the resolution and accuracy required for geosurficial research, *Geophysical Research Letters* **34**(L23S10), doi:10.1029/2007GL031939.

Smith, R.C. and Baker, K.S. 1981 Optical properties of the clearest natural waters (200–800 nm), *Applied Optics* **20**(2), 177–184.

Winterbottom, S.J. and Gilvear, D.J. 1997 Quantification of channel bed morphology in gravelbed rivers using airborne multispectral imagery and aerial photography, *Regulated Rivers: Research & Management* **13**(6), 489–499.

Zhang, H., Voss, K.J., Reid, R.P., and Louchard, E.M. 2003 Bidirectional reflectance measurements of sediments in the vicinity of Lee Stocking Island, Bahamas, *Limnology and Oceanography* **48**(1), 380–389.

4

Hyperspectral Imagery in Fluvial Environments

Mark J. Fonstad

Department of Geography, University of Oregon, Eugene, OR, USA

4.1 Introduction

One of the most conspicuous river tools on the rise during the past decade has been the use of hyperspectral imagery to characterise riverine environment components such as aquatic habitats, bathymetry, and water quality. Hyperspectral remote sensing, also known as imaging spectroscopy, involves the use of instruments that record electromagnetic radiation as narrow contiguous spectra, often dispersing light into several dozen or even hundreds of wavebands (often called either 'bands' or 'channels' by the hyperspectral remote sensing community), as opposed to traditional cameras or multispectral imagers that capture radiation as broad and often discrete spectral bands (3–20 bands at most). Scientific articles extolling the virtues of hyperspectral imaging have shown the improved capability of this type of remote sensing for mapping instream habitats, classifying riparian vegetation, and extracting geometric information such as water depth. As river managers planning applied river projects, an important question should be the costs versus the benefits of including hyperspectral imaging in a river survey. This chapter focuses on that question.

The three colours sensed by the human eye, blue, green, and red (of which all brain-processed colours are derived), are actually each a mixture of several ranges of wavelengths that centre on each of the three colours. In optics, we refer to this phenomenon as being 'wide bands' of light. The bands actually overlap considerably; in some cases the eye can register something as more than one colour even though it is a single wavelength. In contrast, hyperspectral instruments are designed so that each 'colour' or 'channel' collected are recording very narrow band of wavelengths. This narrow, precisely-tailored range of wavelengths for each band is as important as the number of bands in distinguishing hyperspectral imagery from broad-band passive optical imagery. This band narrowness means that very specific information about materials, such as biophysical and biochemical properties, can be correlated and connected with very specific optical wavelengths. This specificity is what makes hyperspectral remote sensing so powerful.

As different hyperspectral instruments have different spectral resolutions, wavelength numbers and ranges, platforms, and advantages/disadvantages, this chapter strives to be deliberately vague about the exact boundary between what is multispectral imagery (described in detail elsewhere in this volume) and hyperspectral imagery. The typical rule-of-thumb is that hyperspectral imagery has at least dozens of bands of light being collected. Another way of defining hyperspectral imagery is the idea that the word 'hyper' can mean 'too many'; hence, hyperspectral imagery is imagery that contains 'too many' colours. This may sound like a bad thing (and having a large amount of data does have its problems), but what having 'too many' colours really means is that you have too many colours to worry about the technical quantity limitations usually constraining colour remote sensing of environments. A more practical way to think of this concept is thinking about classifying a river scene. If you wish to classify instream river habitats into six different categories, then the number of colour bands to do the job

Fluvial Remote Sensing for Science and Management, First Edition. Edited by Patrice E. Carbonneau and Hervé Piégay.
© 2012 John Wiley & Sons, Ltd. Published 2012 by John Wiley & Sons, Ltd.

reasonably well may be in the neighbourhood of 6 – 10 bands (perhaps). The actual number is based on how spectral unique the various feature classes are and how these unique modes line up with specific bands. With only three bands, it becomes more and more difficult to separate these six categories using three colour bands alone. With 128 bands from one particular hyperspectral system, however, a researcher is freed from having to limit the number of 'categories' or 'units' because of the band quantity limitation. Instead, decisions about categorisation potentially can be made based on a physical, functional role, rather than a technical limitation. Maybe this river truly has 18 functional habitats, and wouldn't it be nice if we had so much data that we could pull these 18 classes out almost automatically? On the flip side, many hyperspectral bands covary with one another to a large degree; there is a large amount of data redundancy in many hyperspectral images of river environments. It is therefore not possible to really treat these example 128 bands as truly separate 128 pieces of information; their true 'data dimensionality' is less than 128 bands would suggest.

This chapter is organised into four sections. The first describes the nature of hyperspectral data, how it is collected, and how hyperspectral data are related to river environments. The second section discusses some of the advantages of hyperspectral imagery compared with other types of remote sensing in the assessment of rivers. The third section discusses many of the logistical and optical limitations of hyperspectral imagery that may hinder its use in applied river management. The fourth main section looks at some of the image processing techniques used to extract river information from hyperspectral data.

4.2 The nature of hyperspectral data

Spectrometers have been used as part of remote sensing for more than a century. In astronomy, for example, light collected from distant planets, stars, and galaxies with a telescope was dispersed into constituent wavelengths with a prism or diffraction grating. Dark lines that appeared at specific wavelengths were signatures of atoms or molecules. These early users of optical spectrometry were not seeing true hyperspectral imagery, however; the spectrometer was yielding a one-dimensional spectrum from only one portion of the telescope's field of vision. Imaging spectroscopy, the collection of simultaneous images, one for each of many different wavelengths,

came much later. Spectroscopy remains a vital tool in ground-based and spaceborne astronomy.

Imaging spectrometry as an earth observation tool began in earnest in the mid-1980s (Goetz et al., 1985). The AIS (Airborne Imaging System) developed by NASA in 1983 was the first to be used from high-altitude airborne platforms. The first research quality 'workhorse' hyperspectral sensor is known as AVIRIS (Figure 4.1), Airborne Visible Infrared Imaging Spectrometer (Green, 1994), which is still in current use. Many similar such airborne instruments such as CASI (Compact Airborne Spectrographic Imager) and HyMAP (Hyperspectral Mapper) now are operational in both private and public realms. The lower spatial resolution that normally is required in order to increase the spectral resolution of instruments has hampered spaceborne hyperspectral development. Today, the Hyperion spaceborne instrument on NASA's EO-1 satellite has 220 spectral bands and 30 m resolution. It is currently the only true spaceborne hyperspectral instrument, and it is a prototype mission with a limited operational lifetime.

Normal multispectral imaging instruments, such as a digital camera or the Quickbird spaceborne instrument, often are designed in a fashion known as a 'framing

Figure 4.1 The upper portion shows the AVIRIS sensor aboard a twin-engine Otter airplane on its way to image portions of Yellowstone National Park; the lower image is a closeup of the sensor system.

camera'. In this design, the light that is passed through a camera's optics is converted to electrical signals at the back of the camera in a two-dimensional array of sensors. Often these sensors are designed so that some elements only respond to certain wavelengths of light, hence the red, green, and blue light that is measured in a normal digital camera. More advanced multispectral instruments use a variety of methods to measure additional wavelengths of light, including firing several cameras, filtered for different wavelengths, simultaneously over a target.

Hyperspectral imaging requires a redesign of the way an instrument collects and measures light. The huge number of wavelengths desired means that splitting light into dozens of filtered cameras is very unwieldy or impossible. Instead, two instrument designs are primarily used for hyperspectral imaging, both quite different from a typical framing camera. In the *whiskbroom* linear (or cross-track) array, such as that used in NASA's AVIRIS hyperspectral instrument, a moving mirror allows light from points on the ground into the sensor, where a diffraction grating is used to spread out the spectrum of colours onto a linear array of detectors that record the intensity of light in different wavelengths. The number of array elements becomes the number of wavelengths measured by the detector, and therefore the number of channels generated through the imaging. Basically, each instantaneous measurement yields a spectrum for one place on the ground. The spatial pattern of the data collection is a zig-zag pattern perpendicular to the plane's flight path, with the forward movement of the plane yielding another spatial dimension of data. The zig-zags are then converted to two-dimensional images through a spatial discretisation process, and the various spectra collected at each instant are converted to brightness values in different layers. The resulting product is known as a hyperspectral datacube (Figure 4.2a). Whiskbroom sensors have the advantage of having one detector measuring the intensity of the dispersed radiation; therefore quantitative comparisons between pixels are relatively simple. The disadvantage is that the complex movement of the mirror coupled with the complex movements of the airplane can induce non-uniform mapping that can make spatial measurements problematic without tight geometric control.

The other hyperspectral design commonly used is known as a *pushbroom* (or along-track) array. Instead of a moving mirror, a narrow slit oriented perpendicular to the flight path allows light into the instrument. The light is scattered into various wavelengths through a diffraction grating, and the resulting two-dimensional dataset (one dimension of space, one-dimension of spectra) falls

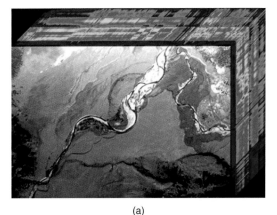

(a)

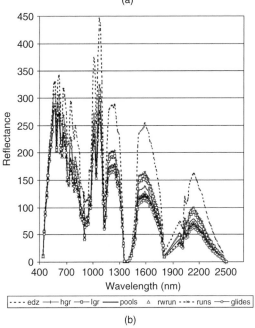

(b)

Figure 4.2 (a) A hyperspectral datacube (from the Probe-1 instrument) showing a true colour image on the top of the stack, and the various wavelength images along the sides, (b) example spectra from various fluvial biotype features including eddy drop zones (edz), high gradient riffles (hgr), low gradient riffles (lgr), rough water runs (rwrun), as well as pools, runs, and glides. (b) Reproduced with permission from Marcus, W.A. (2002) Mapping of stream microhabitats with high spatial resolution hyperspectral imagery. Journal of Geographical Systems, 4, 113–126.

upon a two-dimensional CCD array, which records the data. Sequential frames build up a dataset of cross-track, along-track and spectral information. These data are then reprocessed into hyperspectral images with many wavelength bands by aligning and merging successive image frames. The advantage of pushbroom arrays is that they are simpler and cheaper, and the rigid slot means that there are no complicated movements during each frame taken (as there is with the whiskbroom scanner). It also increases the exposure time and, therefore, helps counteract the issues caused by the low signal to noise ratio of hyperspectral data. The disadvantage is that each column of pixels in the instrument array is now a separate sensor, and through time these columns might yield slightly different brightness values even when flown over uniform materials. This can produce 'bad lines' (lines that are consistently brighter or darker than neighbouring lines) that can complicate image processing.

The coastal zone and benthic mapping community have had a long period of development of optical analysis of water and bottom properties, and this has yielded several approaches to mapping such environments with specifically designed and processed sensors (Ackleson, 2003). Both empirical and theoretical research in these environments showed that broad spectral bands provide limited information for mapping several distinct types of materials in the water column simultaneously, and hyperspectral-type imagers are thus preferred in many situations (Kutser et al., 2003). The success of hyperspectral approaches in ocean environments suggested that river applications might also be successful, if airborne imaging spectrometers could be flown at low altitude, yielding high spatial resolution imagery needed for the spatial heterogeneity of most rivers.

Mertes (2002) and Marcus and Fonstad (2008) provide an overview of the history, physical nature, and uses of the optical remote sensing of rivers. Most spectral analyses used in river remote sensing have been applied to multispectral imagery with a low number of broad bands. While these have been successful in projects such as mapping suspended sediment concentrations (Mertes, 1993), the low number of bands limits the types of algorithm than can be applied and, more importantly, limits the number of properties of the river that can be extracted from imagery. This issue is known as underspecification, a limitation hyperspectral data excel at overcoming.

Earlier multispectral scanner imagers had been used to characterise river water depths (Lyzenga, 1978; Lyon and Hutchinson, 1995) and a combination of water depths and bottom sediment type (Lyon, Lunetta, and Williams, 1992). In an attempt to broaden the types of river environments that could be characterised through remote sensing, Marcus (2002) was one of the first to directly apply hyperspectral imaging to stream environments, focusing on the mapping of in-stream aquatic habitats in

Yellowstone National Park. Other research using different instruments and techniques soon followed, and these will be discussed in later sections.

One set of interesting results arose early. Legleiter et al. (2002) compared the ability of multispectral and hyperspectral imagery and various spatial resolutions to correctly classify in situ stream habitats, and found that high-resolution, hyperspectral imagery significantly out-performed other imagery. They also found that many of the errors that were observed in the comparison of the classified hyperspectral imagery were in fact not errors, but situations where the field mapping (the so called 'ground truth') was more in error than the classified imagery.

4.3 Advantages of hyperspectral imagery

Compared with traditional multispectral imagery comprising a small number of bands, hyperspectral imagery has many profound advantages. The most general advantage is that there are typically far more spectral bands available (Figure 4.2b, 4.2c) than different classes of features being mapped (for example, water habitats, or types of vegetation species). This situation stands in stark contrast to most remote sensing situations with multispectral imagery, which are underspecified. While this issue can sometimes require advanced algorithms (such as spectral unmixing, fuzzy classification, physically-based atmospheric normalisation, regression trees, and artificial neural networks) to produce information-rich maps, it allows users to precisely tune techniques to identify very precise types of objects, such as individual species-level identification of riparian vegetation.

As one example, having hyperspectral data allows a user to choose exactly which pair of channels will best map the relative water depths in clearwater streams (Legleiter et al., 2009); multispectral band ratios lack this precision and generally have less accurate water depth map results. Additionally, the large number of channels means that, once the imagery has been radiometrically calibrated, it is possible to use optical-physical relationships to extract continuously-varying (rather than classified) water parameters without the need for in situ ground data. Such a process has been used for several years in the shallow ocean water community, but it is just starting to be used by river researchers.

Several studies have shown that hyperspectral imagery provides superior river habitat mapping accuracy than habitat maps derived from multispectral imagery (for example, Legleiter et al., 2002). This classification advantage is partly based on the number of channels for exact types, but also because of the ability to better categorise mixtures; a more difficult problem for multispectral (underspecified) imagery. Most published work on this aspect of hyperspectral data in fluvial systems has used standard supervised classification techniques, the same or similar to those used in undergraduate remote sensing classes. The high number of channels, however, has raised the possibility that classes derived from the imagery itself (unsupervised classification) may be better at mapping rivers than standard supervised methods for classifying river habitats on the ground. This unsupervised approach to mapping rivers is a form of data mining, and many techniques developed for standard remote sensing topics (such as self-organising maps, a type of artificial neural network algorithm) could be imported to river remote sensing using hyperspectral data that has been done previously.

One of the chief advantages of hyperspectral data, as noted already, is its ability to quantify mixtures of types. This type of process, often called 'soft classification', can yield a variety of new ways of measuring rivers. One typical example of soft classification, linear unmixing, uses pixels covered by a single pure feature (also known as an endmember for example a 'pure' water habitat such as a pool) as a baseline, and then analyses all other imaged pixels in order to extract what proportions of each pixel is made of the various habitats – in other words, it is a subpixel mapping approach. A separate soft classification approach treats mixtures not as combinations of pure habitats, but rather shows how 'impure' habitats are; this approach is known as fuzzy classification (Legleiter and Goodchild, 2005). Soft classification approaches tend to yield more realistic data products compared with standard 'hard' classification approaches. They are also more comparable to continuous-data ground validation methods in the field.

4.4 Logistical and optical limitations of hyperspectral imagery

The single biggest downside to utilising hyperspectral imaging in fluvial environments is its cost. Hyperspectral work on rivers has been almost entirely based on airborne imaging, and hired hyperspectral flights can be expensive. An economy of scale exists if larger areas are to be mapped; only part of the total costs are associated with flying time, and larger river areas might be cost-effective

when compared to traditional river mapping. For very large rivers (>30 m width minimum), it may be possible to use spaceborne hyperspectral instruments such as NASA's Hyperion (on the experimental EO-1 satellite) at relatively low cost. Future spaceborne hyperspectral instruments, such as EnMap and HyspIRI, will have similar 30–50 meter ground resolutions, though their lowered repeat times and increased global coverage will add to the number of rivers that can be examined from space using hyperspectral imaging. The problem of pixel mixing usually means that a river must be at least a few pixels wide in order for high quality information to be extracted; therefore, spaceborne hyperspectral platforms do not, as yet, allow most of the world's rivers to be imaged. For most rivers, existing hyperspectral datasets with the necessary resolution to image river environments are few, and the cost can be high. Most projects desiring hyperspectral imagery generally hire outside consulting firms to produce new hyperspectral imagery. This sometimes can cost tens of thousands of dollars (US). Some governmental programs can provide limited hyperspectral imagery at lowered cost. NASA AVIRIS (Airborne Visible/Infrared Imaging Spectrometer), for example, flies hyperspectral imagery in support of US government programs at reduced prices. Additionally, the AVIRIS program has a competitive program to provide one free flight line to a graduate student per year, subject to the flight schedule of that year. But these opportunities are often competitive in nature and projects can therefore not control the timing and logistics of their flight. Buying a well-calibrated, good quality hyperspectral scanner that is useful for river remote sensing is in the low hundreds of thousands of dollars (US), similar to the cost of buying a LiDAR airborne system. The paucity of agencies and companies that fly contract imagery also makes the imaging difficult to control in terms of exactly when things are flown; this in turn can affect the quality of the imaging. A few days can turn a clear water stream to a muddy morass. Time will tell if the number of hyperspectral instruments flying will increase and the costs will decrease.

Most students of river remote sensing do not have extensive hyperspectral imaging skills. While many techniques of hyperspectral digital image processing are similar to those in the multispectral arena, some of the best uses of hyperspectral data require advanced skills. These include physically-based approaches for separating various water and channel characteristics and advanced statistical algorithms for mapping mixtures of classes such as Spectral Angle Mapper (SAM) and linear unmixing. Specially-trained remote sensing students often have these skills, but most are not familiar with the three-dimensional river environment. This knowledge gulf has been a problem for the development of the science of the remote sensing of rivers. Finding technically competent hyperspectral operators is thus an important consideration when choosing to use hyperspectral imagery in a river project.

Because imaging spectrometers break up incoming light into a large number of channels, the number of photons coming into each channel is much smaller than would be entering into a multispectral instrument in the same situation. While hyperspectral instruments often have a high signal to noise ratio (to try to compensate for this low number of photons), the main effect has been to require the instrument designers to provide larger pixel sizes. The spatial (ground) resolution of airborne hyperspectral imagers is therefore often significantly larger than typical high-quality photographs. Centimeter-to-decimeter resolution digital imaging is now becoming quite straightforward for normal aerial photography. Hyperspectral imagery is typically meter-resolution or larger. The lower number of photons per channel can also reduce the maximum detectable depth in shallow clearwater environments (Legleiter et al., 2009), but this effect can be offset by using algorithms that use several channels.

The primary uses of hyperspectral imagery in the river environment to date have been (a) classification of river habitats (Marcus et al., 2003), (b) production of water depth maps (Legleiter and Roberts, 2005), (c) water quality/suspended sediment concentration mapping (Karaska et al., 2004), and (d) riparian cover and submerged vegetation identification (Williams et al., 2003). In the past decade, the increase in the spatial resolution of aerial photography (to decimeter or centimeter scales) has allowed this branch of river remote sensing to compete with hyperspectral imaging in habitat mapping, by substituting high-resolution textural information for the increased number of channels in hyperspectral imagery. As one example, larger grain size mapping is now typically done using textural approaches on high-resolution photography rather than spectral analysis of hyperspectral data. Also, LiDAR developments in the past decade have included the development of water penetrating LiDAR for depth mapping. Unfortunately, it is not totally clear which instruments are currently best for mapping various fluvial environments in a given situation; logistics may provide a better reason for choosing one technology over another at the present time. In principle, the fusion of hyperspectral and LiDAR approaches might be very advantageous to fluvial research and management. At present, however, the cost of such fusions has negated their joint use in all but a tiny number of studies.

Many hyperspectral instruments operate in the whiskbroom (cross-track) configuration; therefore an image pixel's data is collected a tiny moment later than the pixel next to it. As an airplane is moving during that momentary instant, the individual pixels are usually not in a standard uniform rectangular array (with respect to the ground) as are most air photos. In the case of most extensive aircraft roll, pitch, yaw, and vibration, pixels may not line up well (Figure 4.3). Such imagery also needs to be orthorectified (just like aerial photography) to correct for distortions such as topography and cross-track distance distortion. The issue also exists with pushbroom scanning systems, though to a slightly lesser degree. Rectification, therefore, is a significant issue the project planners may need to examine. More recent innovations have automated some of these procedures, though these automations are not yet standardised throughout the industry.

Because hyperspectral imagery is, by design, divided up into a large number of narrow spectral channels, some of these channels can be very sensitive to non-river effects, such as atmospheric water vapour. Other channels may be sensitive to imaging geometry, creating considerable illumination differences from the middle to edge of a flight line. Standard digital image processing techniques (such as band ratioing) may alleviate some of these effects, but more severe issues may require advanced techniques such as physically-based optical approaches. While there are somewhat standardised methods for atmospherically normalising ('atmospheric correction') many multispectral sensors (such as Landsat), these approaches were often not designed for the spatial and spectral sensitivity of the hyperspectral scanner. Atmospheric optical models such as MODTRAN and FLAASH allow automated atmospheric correction, though they rely on (often reasonable) assumptions about the atmospheric structure and composition that may not be precisely known for the time and location of study. The best approach, though expensive, to such correction is to make field spectroscopy measurements at various sites on the ground at the time of the flight. These can then be used to directly calibrate the various channels of the hyperspectral imagery. Field spectrometers are now fairly common pieces of equipment in remote sensing.

Another counterintuitive issue in using hyperspectral imaging is the issue of ground validation. Typically, remote sensing analyses, such as classification, is validated by going to various sites on the ground and recording

(a)

(b)

Figure 4.3 Scanning errors due to complex platform motions: (a) the helicopter platform, and (b) the resulting imagery before geometric correction.

the 'truth' and comparing this with the digital image estimates; this is often called 'ground truthing'. The issue is that in some cases, the hyperspectral image analyses can produce habitat maps and estimates of physical, biological, and chemical parameters that are so good, that they are actually better than the field measurements. This is particularly the case in areas of high spatial complexity and mixing. Therefore, it is often not correct to use the

classified field measurements as 'ground truth' (Legleiter et al., 2002; Marcus, 2002). Perhaps a better approach in some cases is to remove ground classification from the analysis and replace it with ground measurement of actual physical quantities (water depth, substrate size, vegetation cover percent, etc.). A field spectrometer is one way of directly comparing hyperspectral imagery with in situ optical properties of the water, while at the same time aiding in atmospheric correction (Figure 4.4). Another

Figure 4.4 Ground validation through in situ measurement of river optics using a field spectrometer. Such an approach can be used to compare the spectra of 'pure' stream classes to one another and to hyperspectral imagery, and it can also be used to aid in atmospheric correction.

option is to only do comparisons in areas of relative 'pristine' or 'pure' classes where the faith in the ground measurements is very high.

A more pragmatic issue with using hyperspectral imagery is the computer processing power and memory needed for digital image processing. Long hyperspectral flight lines are both large spatially and have hundreds of individual layers (the spectral channels). Any additional processing will create for such layers. This was a larger problem in the 'early days' of hyperspectral remote sensing of rivers (the 1990s). Though the problem is fairly simply dealt with today with careful computer planning, the issue can become serious in cases where many orthorectified datasets are joined together.

The final problem with hyperspectral use in fluvial environments is not really a problem, but rather a matter of project specification. There are very few situations today where we have long-term historical hyperspectral data that we can compare with modern hyperspectral data. If change detection over long time spans is what is required for a particular project, hyperspectral imaging is perhaps better supplanted by more traditional imagery.

4.5 Image processing techniques

The conversion of hyperspectral imagery into maps showing important fluvial environments quantitatively has been the primary area of research in the field during the past fifteen years. While there are many different processing techniques, they can be divided into two broad groups. The first are statistical-empirical techniques; these include algorithms that compare image data to reference data (ground or lab-based), and these comparisons allow generalisations that can be applied to entire images. The second group is the physically-based approaches. Physically-based radiative transfer models such as Hydrolight use knowledge of the absorption, scattering and backscattering to model light propagation through the medium of interest (e.g. a water column). This second approach has the advantage of not necessarily needing ground reference data, but comes at the price of being more technical and complex, requiring specialised software (such as Hydrolight), and may require more experienced personnel. It also has been used more rarely in the fluvial environment, though it is much more standard in the ocean water science community. Some techniques attempt to combine both statistical and physically-based methods together. Unfortunately for the river manager community, there is no broad consensus

on exactly which techniques ought to be used in a given fluvial situation.

Statistical-empirical techniques include such methods as supervised classification (automated classification based on user-input training targets), unsupervised classification (automated production of mapped clusters or categories based on statistical differentiation), and regression (statistically-constructed continuous functions between an image variable and a ground variable). More advanced statistical-empirical approaches may use a number of such techniques in sequence or combination.

The most commonly used statistical-empirical methods in river remote sensing are supervised classification methods. While there are several different numerical approaches to supervised classification, all basically try to find separation rules that group different pixels with similar spectral signatures with spectrally similar targets input as 'true' targets on the ground by a researcher. One such approach, maximum likelihood classification, is very commonly used with multispectral data, and is also commonly used by practitioners with hyperspectral data as well. Maximum likelihood supervised classification (Marcus, 2002) yielded producer's accuracies of 85%–91% for different in-stream habitats using hyperspectral image data. Mapping of these habitats in Yellowstone National Park (Figure 4.5) included features such as pools, glides, and riffles and used 128 channel hyperspectral imagery at 1m spatial resolution. Further work with this same imagery (Marcus et al., 2003) used a combination of maximum likelihood classification and principle components analysis (PCA) to map in-stream habitats over a wider range of scales, as well as map woody debris occurrence and water depths. Calibration depths were measured in the field, and these were regressed (using multiple linear regression) against several PCA component images to produce water depth maps with R^2 values of 0.28 to 0.99 depending on the size of the stream and on the habitat type. These researchers were able to map woody debris by using a Matched Filter technique trained on field maps of fluvial wood. While overall accuracies using this technique were quite high (overall accuracies of 83%), it was apparent that the technique was working better than the validation statistics suggested, because the imagery was picking up wood that was missed by field crews because of the small size of some of the pieces. Techniques such as PCA are very useful in showing visually the high data dimensionality associated with hyperspectral imagery, and in visually separating various riverscape elements (Figure 4.6). Extending hyperspectral imagery and supervised classification to estuary and tidal areas,

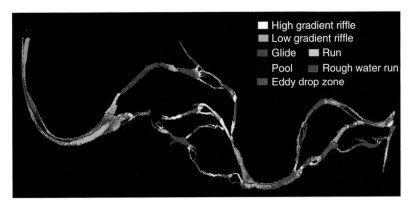

Figure 4.5 Supervised classification used to construct an instream habitats map of a portion of the Lamar River in Yellowstone National Park with the Probe-1 sensor.

Gilvear et al. (2004) used CASI imagery to identify river morphologies as well as many artificial features of interest.

More advanced supervised classification methods such as Spectral Angle Mapper (SAM) can sometimes take fuller advantage of the wealth of spectral information in hyperspectral imagery better than traditional supervised classification techniques. This is not always the case, however; Jollineau and Howarth (2008) compared SAM and maximum likelihood classification using CASI imagery over wetlands and found SAM to not be as good as the more traditional approach. Geostatistical techniques or techniques based on spatial feature recognition can often be used for supervised classification, and sometimes do some better than many spectral-only techniques. Harken and Sugumaran (2005), for example, compared SAM to an object-oriented (i.e. spatial feature recognition) supervised classification algorithm and found the object-oriented approach to be significantly more accurate using high-spatial resolution data.

Although less commonly used in the hyperspectral remote sensing of rivers, unsupervised classification is another powerful technique to extract fluvial environments from imagery. Legleiter (2003) used canonical discriminate analysis to show that unsupervised classification explained much more of imaged stream habitat variability compared with supervised approaches based on in situ field measurements. More advanced unsupervised techniques, such as artificial neural networks and self-organising maps, have yet to be employed by river practitioners.

Empirical regression between continuous water variables (depth, water clarity, etc.) and imagery is a very common use for multispectral and hyperspectral image processing. The work of Marcus et al. (2003) is one straightforward use of multiple linear regression to predict water depth given hyperspectral imagery and some ground measurements of depth for training. Moving from fairly small rivers to large rivers, estuaries, and bays, Chen et al. (2007a) used regression-based approaches to estimate Tampa Bay turbidity from MODIS reflectance. They also estimated both water clarity and Secchi disk depth through regression of MODIS imagery against ground measurements (Chen et al., 2007b). While MODIS does not have the number of bands associated with most hyperspectral sensors (Table 4.1), it does have enough to effectively use many of the same techniques commonly used in hyperspectral remote sensing.

In principle, all river remote sensing imagery should first be processed to remove the effects of the atmosphere; these can obscure and complicate the optical signals leaving the river. Some statistical-empirical methods are fairly robust and do not require extensive atmospheric effect removal. However, such atmospheric correction becomes absolutely crucial if reference spectra are to be used in comparison with imagery in order to extract water variables, rather than using in-situ calibration data (Aspinall et al., 2002). There are many approaches to removing the effects of the atmosphere; some require the collection of in situ data, some rely on physically-based models of the atmosphere, and some use specific in-scene information (Reinersman et al.,1998; Mustard et al., 2001). Atmospheric processing of hyperspectral imagery is particularly crucial in the case of rivers, because water vapour above the river environment can significantly alter the optical signals in a number of wavelengths, as can other effects including dust, haze, smoke, and a variety of other issues.

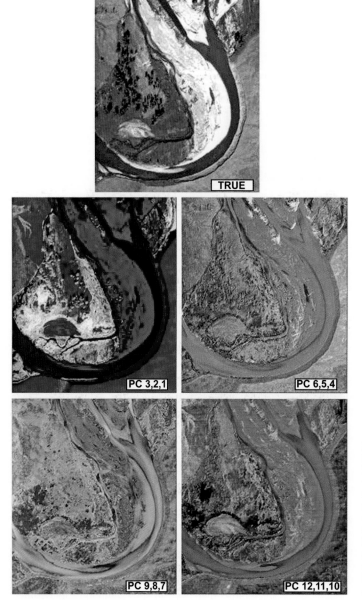

Figure 4.6 An example of Principle Components Analysis (PCA) processed hyperspectral image components. Starting at the top is the 'True-Color' image. The second line shows the PC components 3-2-1, and also the PC components 6-5-4. The last line shows the PC components 9-8-7 and 12-11-10. This image is intended to convey the broad dimensionality latent in the hyperspectral imagery. As such, the statistical properties of each new image (for example variable loadings on each component) have been excluded. Original imagery courtesy of Andrew Marcus.

Removal of atmospheric effects and conversion to reflectance (rather than instrument brightness values) is required for some advanced statistical-empirical techniques, such as those that match reference spectra against measured image spectra in order to classify types or to measure proportional fractions of types within each image pixel. As one example, comparison of airborne hyperspectral imagery of the Potomac River to reference spectra yielded submerged vegetation maps (Williams et al., 2003).

Modified spectral unmixing, a technique based on multiple linear regression, can allow mapping of both depth and subpixel composition of aquatic benthos (Hedley and Mumby, 2003). A somewhat similar approach by Huguenin et al. (2004) characterised depth, suspended chlorophyll, suspended sediments, and coloured

Table 4.1 Comparison of a small set of example hyperspectral sensors.

Acronymn	Platform	# of Channels	Spectral Range (nm)	Typical Spatial Res
ASIA Eagle	Airborne	244	400–970	Variable by height
AVIRIS	Airborne	224	400–2500	Variable by height
CASI	Airborne	288	400–1000	Variable by height
HYDICE	Airborne	210	400–2500	Variable by height
HYMAP	Airborne	126	450–2500	Variable by height
HYPERION	Spaceborne	220	400–2500	30 m
IMSS	Airborne	400+	400–800	Variable by height
PROBE-1	Airborne	128	400–2500	Variable by height

dissolved organic carbon in the nearshore water environment. The same algorithm as Hugeunin's was used by Karaska et al. (2004) to characterise the Neuse River, North Carolina, using the AVIRIS hyperspectral scanner. The authors used the technique to map variations in suspended chlorophyll, suspended minerals, coloured dissolved organic carbon, and turbidity. Reference spectra can come from laboratory measurements of water environmental variables, or through careful field spectroscopy studies, such as those done by Gilvear et al. (2007) to identify river water depth and substrate type. Spectral unmixing does not have to be a purely statistical-empirical approach, it can be combined with physically-based approaches to take optical physics into account.

Physically-based approaches recognise that a large quantity of the light reaching an instrument has been modified in several ways and by several environmental components other than the target of interest. By understanding the nature of these components and their interactions, it becomes possible to separate out the various effects, quantify them, and perhaps map them. More importantly, taking a physically-based approach allows researchers to use physical law to predict why light does various things in different water environments, and therefore we can predict ahead of time what the results of certain river remote sensing projects might be. The key advantage of using a physically-based approach as that because the derived models have a physical basis they are more robust and less site-specific than those derived using empirical methods. This does not, however, guarantee that they will work better for a specific mapping purpose.

While many of the optical laws and reference spectra for various river parameters are known, the use of these to back-calculate conditions for a given image is somewhat complex; this problem can be solved by so-called inversion methods. For example, Lee et al. (1999) provide an early shallow coastal reflectance model/inversion model to solve for water depth and various water properties using hyperspectral data, with no field data other than measured reflectance required. Another inversion algorithm for in-water absorption and scattering coefficients is suggested by Gould et al. (2001).

Mobley's Hydrolight software (Mobley, 1999; Mobley and Sundman, 2001), a radiative transfer model based on earlier theoretical and applied research on optics (Mobley, 1994) as well as later research (Mobley, 1999; Mobley and Sundman, 2003; Mobley et al., 2003) provided a strong optical physical basis for understanding the nature of light in water, and has been used extensively in the ocean community to understand water parameters under various lighting conditions, including depth in the shallow water environment. Legleiter et al. (2004) and Legleiter and Roberts (2005) used this general approach to predict the effects of channel morphology and sensor spatial resolution on imager-derived depth estimates and other habitat variables. Among other findings, they found that a careful chosen ratio of red and green wavelengths is a stable correlate to bottom depth. Further work by Legleiter et al. (2009) found that depth estimates could be predictably improved by a technique called optimal band ratio analysis (OBRA). If hyperspectral image data are available, OBRA produces the best set of band ratios necessary to extract water depth from imagery. Optimal band ratios only provide relative depth maps, and other methods must be added to convert these relative maps to absolute depth maps (Fonstad and Marcus, 2005). Legleiter and Roberts (2009) continued this line of research, producing a 'forward image model' (FIM) that predicts how well a given sensor will work for mapping a given set of stream habitats. Such forward models have a long history in the remote sensing of ocean habitats (for example, Maritorena et al., 1994), and there is definitely room for further advances of this type in the remote sensing of rivers; it would be extremely useful for river managers to know the likely quality levels of their data before they are contracted.

4.6 Conclusions

Hyperspectral data are not a panacea for remote river studies. At their best, they are still quite expensive and logistically complex. At their worst, the effects of the atmosphere and other factors can make interpretation very difficult and misleading. For these reasons alone, caution needs to be exercised when considering hyperspectral imaging for river monitoring. The apparent realism and flexibility of hyperspectral tools is considerable, and experience is needed to know how to analyse these data to provide real river information rather than apparently real yet false information.

There are, unfortunately, no hyperspectral instruments in orbit that have the ground resolution needed to map most of the world's small to medium-sized river environments. While some airborne hyperspectral imaging can yield very high resolution imagery, they cannot approach the cm-scale resolution that some digital cameras can provide; such resolutions are necessary for some river analyses such as extracting particle size information. Low-flying remote control aircraft with the ability to carry hyperspectral sensors and be used by civilians over most river areas do not seem to become likely in the near future. Most of all, hyperspectral imaging is only available for the last few years at best, and it usually is only available once an investigator pays for a tasked mission. As such, long-term decadal change detection via hyperspectral methods is not yet a reality. There are not yet agreed-upon standards for using hyperspectral data in river analysis, for example the many methods for radiometric calibration of these images for rivers have not been tested against each other systematically.

The advantages that come with well-used hyperspectral imagery are, however, immense. The large number of narrowly-defined image bands provides the perfect data source for modern image processing algorithms. Specific river variables, such as water depth, are strongly related to particular wavelengths, and these can be singled out from hyperspectral data and used to provide optimum information extraction. Modern optical methods, coupled with hyperspectral data and spectral libraries, can in principle yield quantitative river environment information without the need for in situ data measurement. The multitude of layers allows for realistic classification of river habitats, whether these classifications are supervised or unsupervised. For those larger projects that have the money and need for a great deal of remotely-collected river data, hyperspectral data are second to none, and are likely quite cost-effective when compared with ground measurements.

The future of hyperspectral imaging of river environments is very bright. Most importantly, the number of hyperspectral sensors and organisations flying these sensors is growing very rapidly. This has the long-term effect of reducing costs. Computational power and data storage, once major problems with using large hyperspectral imagery, are quickly becoming non-issues. Hyperspectral data can be used to test the usefulness and veracity of more modest instruments (such as three-band aerial imagery) to extract river instruments. The work by the ocean optics community and a handful of river researchers are quickly determining the uses and limitations of hyperspectral imagery in detecting various water environments, and doing so in a quantitative, replicable way that provides a level of quality control necessary in the profession uses of such imagery. Hyperspectral remote sensing of rivers is beginning to be combined with other data types, such as with LiDAR data (Hall et al., 2009). The combination of hyperspectral, LiDAR, and high-resolution aerial imagery to provide synoptic views riverscape constituents should become feasible in the near future. Whether riparian or fluvial, hyperspectral remote sensing of riverscapes is one of the crowning jewels of river mapping.

Acknowledgments

I would like to acknowledge the efforts of Andrew Marcus and Carl Legleiter, who contributed images, valuable discussions, and unyielding field camaraderie.

References

Ackleson, S.G. 2003. Light in shallow waters: A brief research review. *Limnology and Oceanography*, 48(1, part 2), 323–328.

Aspinall, R.J., Marcus, W.A., and Boardman, J.W. 2002. Considerations in collecting, processing, and analyzing high spatial resolution hyperspectral data for environmental investigations. *Journal of Geographical Systems*, 4, 15–29.

Chen, Z., Hu, C., and Muller-Karger, F. 2007. Monitoring turbidity in Tampa Bay using MODIS/Aqua 250-m imagery. *Remote Sensing of Environment*, 109, 207–220.

Chen, Z., Muller-Karger, F., and Hu, C. 2007. Remote sensing of water clarity in Tampa Bay. *Remote Sensing of Environment*, 109, 249–259.

Fonstad, M.A. and Marcus, W.A. 2005. Remote sensing of stream depths with hydraulically assisted bathymetry (HAB) models. *Geomorphology*, 72(1–4), 107–120.

Gilvear, D.J., Hunter, P., and Higgins, T. 2007. An experimental approach to the measurement of the effects of water depth and substrate on optical and near infra-red reflectance: a field-based assessment of the feasibility of mapping submerged instream habitat. *International Journal of Remote Sensing*, 28(10), 2241–2256.

Gilvear, D., Tyler, A., and Davids, C. 2004. Detection of estuarine and tidal river hydromorphology using hyper-spectral and LIDAR data: Forth estuary, Scotland. *Estuarine Coastal and Shelf Science*, 61(3), 379–392.

Goetz, A., Vane, G., Solomon, J.E., and Rock, B.N. 1985. Imaging spectrometry for Earth remote sensing. *Science*, 228(4704), 1147–1153.

Gould, R.W., Arnone, R.A., and Sydor, M. 2001. Absorption, scattering; and remote-sensing reflectance relationships in coastal waters: Testing a new inversion algorithm. *Journal of Coastal Research*, 17, 328–341.

Green, R.O. 1994. AVIRIS Operational Characteristics. Pasadena, CA: Jet Propulsion Lab, 10 pp.

Hall, R.K., Watkins, R.L., Heggem, D.T., Jones, K.B., Kaufmann, P.R., Moore, S.B., and Gregory, S.J. 2009. Quantifying structural physical habitat attributes using LIDAR and hyperspectral imagery. *Environmental Monitoring and Assessment*, 159, 63–83.

Harken, J. and Sugumaran, R. 2005. Classification of Iowa wetlands using an airborne hyperspectral image: a comparison of the spectral angle mapper classifier and an object-oriented approach. *Canadian Journal of Remote Sensing*, 31(2), 167–174.

Hedley, J.D. and Mumby, P.J. 2003. A remote sensing method for resolving depth and subpixel composition of aquatic benthos. *Limnology and Oceanography*, 48, 480–488.

Huguenin, R.L., Wang, M.H., Biehl, R., Stoodley, S., and Rogers, J.N. 2004. Automated subpixel photobathmetry and water quality mapping. *Photogrammetric Engineering and Remote Sensing*, 70(1), 111–125.

Jollineau, M.Y. and Howarth, P.J. 2008. Mapping and inland wetland complex using hyperspectral imagery. *International Journal of Remote Sensing*, 29(12), 3609–3631.

Karaska, M.A., Huguenin, R.L., Beacham, J.L., Wang, M., Jensen, J.R., and Kaufman, R.S. 2004. AVIRIS measurements of chlorophyll, suspended minerals, dissolved organic carbon, and turbidity in the Neuse River, North Carolina. *Photogrammetric Engineering and Remote Sensing*, 70(1), 125–133.

Kutser, T., Dekker, A.G., and Skirving, W. 2003. Modeling spectral discrimination of Great Barrier Reef benthic communities by remote sensing instruments. *Limnology and Oceanography*, 48, 497–510.

Lee, Z., Carder, K.L., Mobley, C.D., Steward, R.G., and Patch, J.S. 1999. Hyperspectral remote sensing for shallow waters: Deriving bottom depths and water properties by optimization. *Applied Optics*, 38, 3831–3843.

Legleiter, C.J. (2003. Spectrally driven classification of high spatial resolution, hyperspectral imagery: A tool for mapping in-stream habitat. *Environmental Management*, 32, 399–411.

Legleiter, C.J. and Goodchild, M.F. 2005. Alternative representations of in-stream habitat: classification using remote sensing, hydraulic modeling, and fuzzy logic. *International Journal of Geographic Information Science*, 19(1), 29–50.

Legleiter, C.J., Marcus, W.A., and Lawrence, R. 2002. Effects of sensor resolution on mapping in-stream habitats. *Photogrammetric Engineering and Remote Sensing*, 68, 801–807.

Legleiter, C.J. and Roberts, D.A. 2005. Effects of channel morphology and sensor spatial resolution on image-derived depth estimates. *Remote Sensing of Environment*, 95, 231–247.

Legleiter, C.J. and Roberts, D.A. 2009. A forward image model for passive optical remote sensing of river bathymetry. *Remote Sensing of Environment*, 113, 1025–1045.

Legleiter, C.J., Roberts, D.A., and Lawrence, R.L. 2009. Spectrally based remote sensing of river bathymetry. *Earth Surface Processes and Landforms*, 34, 1039–1059.

Legleiter, C.J., Roberts, D.A., Marcus, W.A., and Fonstad, M.A. 2004. Passive optical remote sensing of river channel morphology and in-stream habitat: Physical basis and feasibility. *Remote Sensing of Environment*, 93, 493–510.

Lyon, J.G. and Hutchinson, W.S. 1995. Application of a radiometric model for evaluation of water depths and verification of results with airborne scanner data. *Photogrammetric Engineering and Remote Sensing*, 61, 161–166.

Lyon, J.G., Lunetta, R.S., and Williams, D.C. 1992. Airborne multispectral scanner data for evaluating bottom sediment types and water depths of the St. Marys River, Michigan. *Photogrammetric Engineering and Remote Sensing*, 58, 951–956.

Lyzenga, D.R. 1978. Passive remote-sensing techniques for mapping water depth and bottom features. *Applied Optics*, 17, 379–383.

Marcus, W.A. 2002. Mapping of stream microhabitats with high spatial resolution hyperspectral imagery. *Journal of Geographical Systems*, 4, 113–126.

Marcus, W.A. and Fonstad, M.A. 2008. Optical remote mapping of rivers at sub-meter resolutions and watershed extents. *Earth Surface Processes and Landforms*, 33, 4–24.

Marcus, W.A., Legleiter, C.J., Aspinall, R.J., Boardman, J.W., and Crabtree, R.L. 2003. High spatial resolution hyperspectral mapping of in-stream habitats, depths, and woody debris in mountain streams. *Geomorphology*, 55, 363–380.

Maritorena, S., Morel, A., and Gentili, B. 1994. Diffuse-reflectance of oceanic shallow waters – Influence of water depth and bottom albedo. *Limnology and Oceanography*, 39, 1689–1703.

Mertes, L.A.K. 2002. Remote sensing of riverine landscapes. *Freshwater Biology*, 47, 799–816.

Mertes, L.A.K., Smith, M.O., and Adams, J.B. 1993. Estimating suspended sediment concentrations in surface waters of the Amazon River Wetlands from Landsat Images. *Remote Sensing of Environment*, 43, 281–301.

Mobley, C.D. 1994. Light and Water: Radiative Transfer in Natural Waters. *San Diego Academic Press*. 592 pp.

Mobley, C.D. 1999. Estimation of the remote-sensing reflectance from above-surface measurements. *Applied Optics*, 38, 7442–7455.

Mobley, C.D. and Sundman, L.K. 2001. Hydrolight 4.2 User's Guide. Redmond, WA. *Sequoia Scientific*. 88 pp.

Mustard, J.F., Staid, M.I., and Fripp, W.J. 2001. A semianalytical approach to the calibration of AVIRIS data to reflectance over water: application in a temperate estuary. *Remote Sensing of Environment*, 75, 335–349.

Rainey, M.P., Tyler, A.N., Gilvear, D.J., Bryant, R.G., and McDonald, P. 2003. Mapping intertidal estuarine sediment grain size distributions through airborne remote sensing. *Remote Sensing of Environment*, 86, 480–490.

Reinersman, P.N., Carder, K.L., and Chen, F.-I.R. 1998. Satellite-sensor calibration verification using the cloud-shadow method. *Applied Optics*, 37, 5541–5549.

Williams, D.J., Rybicki, N.B., Lombana, A.V., O'Brien, T.M., and Gomez, R.B. 2003. Preliminary investigation of submerged aquatic vegetation mapping using hyperspectral remote sensing. *Environmental Monitoring and Assessment*, 81, 383–392.

5

Thermal Infrared Remote Sensing of Water Temperature in Riverine Landscapes

Rebecca N. Handcock[1], Christian E. Torgersen[2],
Keith A. Cherkauer[3], Alan R. Gillespie[4], Klement Tockner[5],
Russel N. Faux[6] and Jing Tan[3]

[1]Commonwealth Scientific and Industrial Research Organization, Floreat, WA, Australia
[2]U.S. Geological Survey, Forest and Rangeland Ecosystem Science Center, School of Environmental and Forest Sciences, University of Washington, Seattle, WA, USA
[3]Department of Agricultural and Biological Engineering, Purdue University, West Lafayette, IN, USA
[4]Department of Earth and Space Sciences, University of Washington, Seattle, WA, USA
[5]Leibniz-Institute of Freshwater Ecology and Inland Fisheries and Institute of Biology, Freie Universität Berlin, Germany
[6]Watershed Sciences, Inc., Corvallis, OR, USA

5.1 Introduction

Water temperature in riverine landscapes is an important regional indicator of water quality that is influenced by both ground- and surface-water inputs, and indirectly by land use in the surrounding watershed (Brown and Krygier, 1970; Beschta et al., 1987; Chen et al., 1998; Poole and Berman, 2001). Coldwater fishes such as salmon and trout are sensitive to elevated water temperature; therefore, water temperature must meet management guidelines and quality standards, which aim to create a healthy environment for endangered populations (McCullough et al., 2009). For example, in the USA, the Environmental Protection Agency (EPA) has established water quality standards to identify specific temperature criteria to protect coldwater fishes (Environmental Protection Agency, 2003). Trout and salmon can survive in cool-water refugia even when temperatures at other measurement locations are at or above the recommended maximums (Ebersole et al., 2001; Baird and Krueger, 2003; High et al., 2006). Spatially extensive measurements of water temperature are necessary to locate these refugia, to identify the location of ground- and surface-water inputs to the river channel, and to identify thermal pollution sources.

Regional assessment of water temperature in streams and rivers has been limited by sparse sampling in both space and time. Water temperature has typically been measured using a network of widely distributed instream gages, which record the temporal change of the

Fluvial Remote Sensing for Science and Management, First Edition. Edited by Patrice E. Carbonneau and Hervé Piégay.
© 2012 John Wiley & Sons, Ltd. Published 2012 by John Wiley & Sons, Ltd.

Table 5.1 Comparison of conventional measurements and TIR remote sensing for regional assessment of water temperature in rivers and streams.

a)		Conventional Measurements	TIR Remote Sensing
Data acquisition	Advantages	• Measurements can be made at any point in the water column. • Limited technical expertise is needed to gather data. • Data can be obtained under most weather conditions including fog and cloud cover. • Continuous measurements are possible using data loggers. • Costs of collecting data can be low, depending on the number of instruments that must be deployed.	• An alternative to collecting validation data is to use existing networks of in-stream data loggers. **Satellite** • Capability for regional coverage, repeat monitoring with systematic image characteristics, and low cost. • Data can be gathered across multiple scales from local (e.g. upwelling ground-water) to regional (entire floodplains). **Airborne** • Can measure TIR images at fine pixel sizes suitable for narrower streams and rivers. **Ground** • Instruments are easy to deploy and validate *in situ*; requires physical access to the stream.
	Disadvantages	• Sparse sampling of T_k in space. • Gives limited information about the spatial distribution of water temperature. Data loggers can be destroyed or removed by vandalism or floods. • Data are collected only at point locations. Do not provide a view the entire thermal landscape of the river. • Temperature gauges are typically located in larger streams and rivers. • Calibration of thermometers is still necessary. • To collect spatially extensive measurements, it is necessary to deploy many personnel.	• Obtaining TIR images can be costly and complex, and temporally limited. • Care must be taken in interpretation of TIR data under off-nadir observation angles and with variable surface roughness (i.e. diffuse versus specular reflections). **Satellite** • TIR images may not be available due to cloud cover, limited duty cycle of platforms used to collect data (satellite orbits, or availability of aircraft). **Airborne** • Generally acquired over narrow swath widths covering small areas compared to satellite data. • Acquisition costs can be high, especially if multiple overlapping scan lines are needed to create a mosaic. **Ground** • Can only view the water from specific locations along the stream. • Observation angles need to be chosen carefully to reduce the effects of reflections from objects along the river bank.

b)		Conventional Measurements	TIR Remote Sensing
Data Processing	Advantages	• Standard data storage and processing techniques can be used (knowledge of the hydrological system is still necessary).	• For applications in which having a non-absolute temperature is useful, non-radiometrically corrected TIR images can be used to assess relative spatial patterns within a single image. • Validation is not required for applications that only need relative temperatures.
	Disadvantages		• Interpretation of TIR image data to determine water temperature can be complex and expensive, and requires trained technical expertise. • Care must be taken to interpret TIR images within their terrestrial and aquatic context. • Radiometric correction is necessary to accurately retrieve quantitative temperatures from TIR data accurately, but this can be time-consuming and expensive. • For data acquired from aircraft, changes in the stability of the aircraft as it flies can require complex and costly post-processing of images.

Table 5.1 (continued).

c)		Conventional Measurements	TIR Remote Sensing
Applications	Advantages	• T_k can be measured directly, which is both of interest biologically and applicable to management objectives.	• Repeatable, spatially extensive, and systematic measurements. • Can quantify spatial patterns of water temperature in streams, rivers, and floodplains at scales ranging from less than 1 m to over 100 km. • Can view the entire thermal landscape of the river, not just point locations. • Consistent data source for entire floodplains and can be used to calibrate stream temperature models. • TIR image data and concurrent visible and NIR images (where available) can be used to assess both the water surface and adjacent riparian areas. • Repeat flights can be used to assess habitat degradation.
	Disadvantages	• Difficult to collect spatially extensive data to use to calibrate stream temperature models for entire watersheds.	• T_r is measured at the surface layer of the water and may not be representative of T_k in the water column, which is of interest biologically. • Trade-off between pixel-size (i.e. to identify spatial patterns and reduce mixing with bank materials) and the cost of conducting broad-scale surveys.

bulk, or kinetic, temperature of the water (T_k) at specific locations. For example, the State of Washington (USA) recorded water quality conditions at 76 stations within the Puget Lowlands eco region, which contains 12,721 km of streams and rivers (Washington Department of Ecology, 1998). Such gages are sparsely distributed, are typically located only in larger streams and rivers, and give limited information about the spatial distribution of water temperature (Cherkauer et al., 2005).

Although hydrologists, ecologists, and resource managers are ultimately interested in T_k in the water column – because this is both biologically important and also the definition of temperature used for management purposes – measurements of radiant temperature (T_r) made at the water's surface using thermal infrared (TIR) remote sensing provide an attractive alternative to *in situ* measurement of T_k, if T_r measurements can be determined with suitable and known quality and detail. A key advantage of TIR remote sensing of T_r over conventional measurements of T_k is that it is possible to quantify spatial patterns of water temperature in rivers, streams, and floodplains, at multiple spatial scales throughout entire watersheds. However, remote sensing of water temperature can be time-consuming and costly due to the difficulties in obtaining images and the complexities of processing raw data to produce calibrated temperature maps. As will be explored in this chapter, understanding these benefits and limitations is necessary to determine whether thermal remote sensing of water temperature is suitable for water resource management applications (Table 5.1).

The goal of this chapter is to show how TIR measurements can be used for monitoring spatial patterns of water temperature in streams and rivers for practical applications in water resources management. We use the term 'water temperature' to refer specifically to water temperature of lotic systems ranging in size from streams to rivers. The chapter is divided into three parts. First, we examine the state of the science and application of TIR remote sensing of streams and rivers Section 5.2. Second, we explore the theoretical basis of TIR measurements of water temperature, data sources suitable for observing riverine landscapes, the required processing steps necessary to obtain accurate estimates of water temperature from TIR data, and the validation of such temperature estimates (Sections 5.3 to 5.6). Third, we show two examples of using TIR data to monitor water temperature in rivers of varying sizes (Sections 5.7 and 5.8). To illustrate the utility of TIR data for quantifying thermal heterogeneity over a range of spatial scales, we show very fine resolution (0.2–1 m) images of fine-scale hydrologic features such as groundwater springs and cold-water seeps. We also expand the scope to entire floodplains and river sections (1–150 km) to show characteristic patterns of lateral and longitudinal thermal variation in riverine landscapes. For TIR pixel sizes, we use the following terminology across a range of sensors and platforms: 'ultra-fine resolution' for pixel sizes of less than 1 m, 'very fine resolution' for pixel sizes of 1 to 5 m, 'fine resolution' for pixel sizes of 5 to 15 m, 'medium resolution' for pixel sizes of 15 to 100 m, and 'coarse resolution' for pixel sizes of greater than 100 m.

5.2 State of the art: TIR remote sensing of streams and rivers

The remote sensing of surface water temperature using measurements of emitted TIR radiation (3–14μm) can provide spatially distributed values of T_r in the 'skin' layer of the water (top 100μm). This is a well-established practice (Mertes et al., 2004), particularly in oceanography where daily observations of regional and global sea-surface temperature (SST) are made from satellites (Anding and Kauth, 1970; Emery and Yu, 1997; Kilpatrick et al., 2001; Parkinson, 2003). In the terrestrial environment, TIR remote sensing of surface water temperature initially focused on lakes (LeDrew and Franklin, 1985; Bolgrien and Brooks, 1992) and coastal applications such as thermal pollution from cooling water discharge from a nuclear power plant (Chen et al., 2003), but starting in the 1990s airborne TIR remote sensing has been conducted by government agencies over thousands of kilometers of rivers to monitor water quality, identify sources of cold-water inputs, and to develop spatially referenced river temperature models (Faux and McIntosh, 2000; Faux et al., 2001; Torgersen et al., 2001).

Applications of TIR technology to measure water temperature of rivers are diverse and have been employed in a wide variety of fluvial environments. Published examples of thermal maps can be found in the early 1970s (Atwell et al., 1971), and one of the earliest documented uses of TIR imaging to evaluate fish habitat in a river was by researchers in Australia, who identified cold groundwater inputs that were ostensibly important for the survival of rainbow trout (*Oncorhynchus mykiss*) in the Murray River (Hick and Carlton, 1991). The TIR images, collected from a fixed-wing aircraft mounted with a multispectral scanner, were particularly effective in the brackish sections of the Murray River where cool groundwater rose to the surface because it was less dense than saltwater. Subsequent work – like that in the Murray River – focused on thermal anomalies associated with wall-based channels, groundwater inputs, and thermal refugia important for salmon in the Pacific Northwest (USA) (Belknap and Naiman, 1998; Torgersen et al., 1999). The impetus for such work arose from the need to identify localised patches of cool water (e.g., Figure 5.1), but the utility of these data became even more apparent for assessing thermal diversity at broader spatial scales in the floodplain (e.g., Figures 5.2, 5.3, and 5.11) and longitudinally over tens of kilometers (Figure 5.4). Direct applications in fisheries continue to be conducted (Madej et al., 2006), but by far the most extensive use of TIR remote sensing has been by natural resource management agencies seeking to calibrate spatially explicit river temperature models for entire watersheds (Figure 5.4; Boyd and Kasper, 2003; Oregon Department of Environmental Quality, 2006). Prior to the availability of near-continuously sampled longitudinal water temperature data derived from airborne TIR remote sensing, discontinuities associated with groundwater inputs and hyporheic flow were very difficult to quantify empirically.

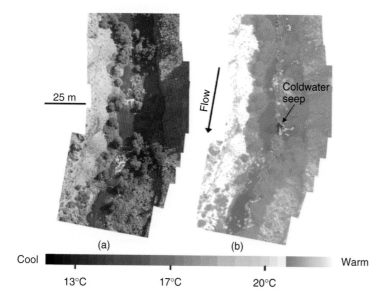

Figure 5.1 Natural-color (a) and airborne TIR (b) aerial images of cold-water seepage area in the Crooked River (Oregon, USA) in a high-desert basalt canyon (27 August 2002). The colored portion of the TIR temperature scale spans the approximate range in water surface temperature in the image; land and vegetation surface temperature are depicted in shades of gray. Lateral cold-water seeps, such as the one depicted above, are relatively small in area but provide important thermal refugia for coldwater fishes. (United States Bureau of Land Management, Dept. of Interior, USA; Watershed Sciences, Inc., Corvallis, Oregon, USA).

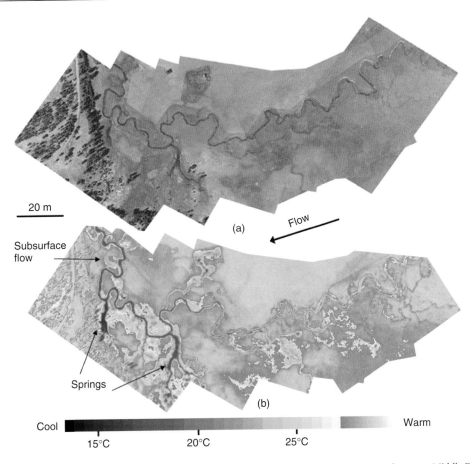

Figure 5.2 Natural-color (a) and airborne TIR (b) aerial images of groundwater springs flowing into the upper Middle Fork John Day River (Oregon, USA) in a montane meadow (16 August 2003). See Figure 5.1 for clarification of color and grayscale thermal classification. Complex subsurface hydrologic flow paths and areas of increased soil moisture adjacent to the wetted channel are revealed by lower TIR land and vegetation radiant temperature (United States Bureau of Reclamation, Dept. of Interior, USA; Watershed Sciences, Inc., Corvallis, Oregon, USA).

In the last decade, the increased awareness of TIR technology, combined with technological advances that have made TIR imaging systems more stable, portable, and affordable, has led to novel applications in riverine ecology. Both airborne- and ground-based approaches have proven highly effective for identifying and mapping the extent of very-fine resolution thermal heterogeneity associated with point sources, hyporheic flow, discharge patterns, and geothermal inputs within the river channel (Burkholder et al., 2008; Cardenas et al., 2008; Dunckel et al., 2009; Cardenas et al., in press). Other studies have utilised the entire swath width of TIR imaging systems to assess thermal variation beyond the river channel and across the floodplain and adjacent riparian areas (Rayne

and Henderson, 2004; Arrigoni et al., 2008; Smikrud et al., 2008; Cristea and Burges, 2009; Tonolla et al., 2010).

Recent developments in TIR remote sensing of rivers have expanded the area of interest beyond water surface temperature – but there is much to be learned from viewing the entire 'thermal landscape' of rivers, laterally, longitudinally, vertically, and temporally. The vertical and temporal dimensions of thermal diversity in riverine systems have just begun to be investigated with TIR remote sensing. The vertical dimension, or thermal stratification, is poorly understood in TIR remote sensing because measurements of radiant temperature are made only in the surface layer of the water (approximately top 10 cm), which may not be representative of T_k further down the water column (this will be expanded on in a

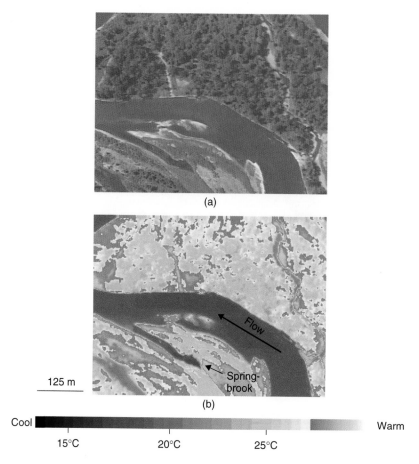

Figure 5.3 Natural-color (a) and airborne TIR (b) aerial images showing thermal heterogeneity in a complex floodplain of the Willamette River (Oregon, USA), which flows through a large, low-elevation agricultural valley (22 July 2002). See Figure 5.1 for clarification of color and grayscale thermal classification. Radiant water temperature varies laterally from the cooler and relatively homogeneous thalweg and main channel to warmer backwaters and disconnected channels. A springbrook is apparent where relatively cooler hyporheic flow emerges from the unconsolidated substratum of a large riverine island (Oregon Department of Environmental Quality 2006; Watershed Sciences, Inc., Corvallis, Oregon, USA).

later section). Where mixing in the water column occurs, cooler water can be detected at the surface, but few studies have determined *in situ* the necessary water velocities and fluvial morphology required to fully mix the vertical structure of the river. Further investigation of the vertical dimension may be conducted in winter when groundwater, which at this time is warmer than river water, is more likely to rise to the surface due to its lower density. Few studies have collected TIR images of rivers and streams in winter, but this area of inquiry holds much potential for quantifying surface water and groundwater interactions and identifying 'warm-water' refugia for fishes in cold regions (Tockner, 2006).

Comparisons of TIR images in rivers across seasons and years provides a means to assess changes in the thermal landscape associated with habitat degradation or to evaluate the effectiveness of floodplain restoration. The application of TIR remote sensing in restoration ecology of rivers is uncommon (for a notable exception see Loheide and Gorelick, 2006) but will likely gain momentum as rivers that were surveyed aerially in the 1990s are re-flown to monitor the effectiveness of management actions (e.g., channel modification and re-vegetation of riparian areas) at restoring thermal diversity in riverine landscapes. The following sections provide the technical context and practical applications of TIR remote

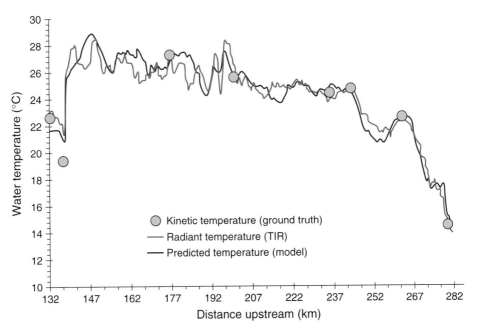

Figure 5.4 Longitudinal profile of water temperature in the upper Grande Ronde River (Oregon, USA) depicting radiant temperature acquired during an airborne FLIR overflight (20 August 1999), in-stream measurements of kinetic temperature, and calibrated model predictions. Distance upstream (x-axis) was determined from the river mouth (Oregon Department of Environmental Quality, 2010a, b).

sensing so that water resources managers and scientists can evaluate how this technology can be used both to address management needs in water quality assessment and biological conservation and also to further the understanding of hydrological processes and riverine ecosystems.

5.3 Technical background to the TIR remote sensing of water

This section focuses on the technical considerations necessary for informed planning and implementation of studies that use TIR remotely sensed images for monitoring streams and rivers. We therefore focus in this section, firstly, on the theoretical basis of the TIR remote sensing of water in general, and secondly, on the topics specific to the TIR remote sensing of riverine landscapes. We explicitly use the terminology of either the TIR remote sensing of *water* or of *rivers* to refer to whether the background applies to water in general, or to water in streams and rivers. A summary of the suggested processing required of TIR data to determine water temperature can be found

in Figure 5.5. The theory of thermal properties of natural materials is extensively covered in the literature, and for the thermal remote sensing of water, specifically, we recommend a good introductory text (e.g., Mather, 2004; Lillesand et al., 2008) or overview (e.g., Atwell et al., 1971; Prakash, 2000).

5.3.1 Remote sensing in the TIR spectrum

All materials with a temperature above 0 K emit radiation, and as described by Wien's Displacement Law, the hotter the object, the shorter the wavelength of its emitted radiation. For example, the sun's temperature is approximately 6000 K, and the sun emits its peak radiation in the visible part of the electromagnetic spectrum (0.4–0.8μm) to which the human eye is adapted. Remote sensing in the region of visible, near infrared (NIR) and mid-infrared radiation (<3μm) utilises reflected radiation. In contrast, the earth's ambient temperature is ~300 K and its peak radiation is emitted at the longer wavelength of 9.7μm. Thermal remote sensing captures radiation emitted in these longer wavelengths (3–1000μm). As TIR observations are strongly affected by radiation absorbed

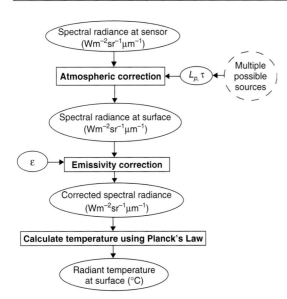

Figure 5.5 Flowchart summarizing the suggested processing of TIR data to determine stream temperature. See the Table of Abbreviations and text for definitions.

and emitted from water vapour, TIR applications focus on the 8–14µm region of the electromagnetic spectrum where atmospheric interference and contamination by solar radiation (in the 3–5µm region) is minimised.

5.3.2 The relationship between emissivity and kinetic and radiant temperature

A material's T_k is determined by the particular thermal characteristics of the material and its heat budget. In simplistic terms, this heat budget results from energy being absorbed, transmitted, and reflected, sometimes multiple times. For example, water has a high *thermal inertia*, which means that it changes temperature slowly as heat energy is added, a low *thermal conductivity*, which means that heat passes through it slowly, and a high *thermal capacity*, which means that it stores heat well.

Whereas the T_k of water can be measured directly using a thermometer that is in contact or immersed in the water, TIR remote sensing of water (using a radiant thermometer, or a ground-, air-, or space-borne imaging sensor) relies instead on indirect measurements of the radiation, emitted by the water body, to determine the water's radiant temperature, T_r. The amount and spectral distribution of this radiated energy is a combination of the material's T_k and its surface emissivity.

Emissivity (ε) defines the efficiency of a material at radiating energy compared to a blackbody. Emissivity values range from 0 to 1. A *blackbody* is a theoretical material with an emissivity of 1, which absorbs all incident radiation and re-emits it perfectly at all wavelengths. A *graybody* has an emissivity that is independent of wavelength λ and is <1. A *selective radiator* will have an emissivity that varies across wavelengths, and this characteristic can be used to identify the material based on its emissivity spectra. In the natural environment true blackbodies do not exist, which means that T_r will always be less than T_k.

Water can approximate a blackbody as its emissivity in the 8–14µm range is near 1 (Figure 5.6). There are many factors which influence water's actual emissivity. The emissivity of water can vary with the amount of suspended sediment (Salisbury and Aria, 1992), and dissolved minerals (e.g., as found in sea water). Features such as riffles and surface foam, which can roughen the water surface, will also alter the emissivity, as can physical characteristics such as wet rocks that have a different emissivity (closer to water) than dry rocks. As well as varying with wavelength, emissivity varies with the compound effects of both angular effects and the surface roughness. For large observation angles (>70° from nadir), rough water will have a higher emissivity than placid water at the same observation angle and will therefore appear warmer (Masuda et al., 1988). When the roughness of the water surface is constant, for small observation angles up to 30° there is a small decrease in spectrally variable emissivity and temperature (e.g., <0.1°C at 10µm for distilled water at 17.2 °C), but for observation angles >70° Fresnel reflection increases and the emissivity and T_r are significantly lower (Masuda et al., 1988; Ishiyama et al., 1995; Cuenca and Sobrino, 2004). In practical terms, providing the observation angles are within ~30° of nadir these effects are negligible.

The emissivity should therefore be chosen to match the specific water characteristics, and care taken in interpretation of TIR data under conditions which might change the emissivity. As the emissivity varies with wavelength, the emissivity should be matched to the band wavelength specifications of the sensor, and care should be taken when narrow-band temperatures are compared to temperatures obtained from sensors with different band characteristics.

The reflectivity of a material is the proportion of radiation incident upon it that is reflected back. There are a number of laws describing how energy is absorbed, transmitted and reflected, and for brevity we will only detail here with what is needed for applied TIR remote sensing

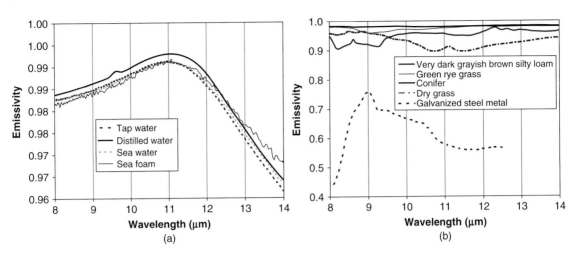

Figure 5.6 Emissivity in the 8–14 µm TIR region for (a) water and foam, and (b) other relevant materials. Note different scales for the x-axes ("Very dark grayish brown silty loam" [85P3707], green rye grass [grass], conifer [conifers], dry grass [drygrass], galvanized steel [0525UUUSTLb], tap water [TAPWATER], sea water [SEAWATER], and sea foam [FOAM] from the ASTER Spectral Library, Version 2.0 [Baldridge et al., 2009]; distilled water [WATER2_W] from the MODIS Emissivity Library, 2010).

of rivers. As described by Lillesand et al. (2008), for most natural materials that are opaque to thermal radiation, Kirchhoff's Law describes how an object's emissivity and absorptivity are equal. Due to the conservation of energy, this can be written as:

$$\varepsilon(\lambda) = 1 - \rho(\lambda) \tag{5.1}$$

where $\varepsilon(\lambda)$ is the emissivity of the material at a wavelength λ, and $\rho(\lambda)$ is the reflectivity of the material. This relationship describes why water, for example, has a very low reflectivity across all wavelengths, and a correspondingly high emissivity. Equation (5.1) can also be used to determine an object's emissivity if its reflectivity in the TIR is known.

Good sources of reflectivity and emissivity data are spectral libraries, such as the MODIS Emissivity Library (2010) and the ASTER Reflectivity Library (Baldridge et al., 2009). Spectral libraries have the advantage of being measured under controlled conditions. Although the spectral libraries cannot cover all variations in emissivity described above, standard spectra for distilled water are usually suitable for the accuracy requirements of many applications which require T_r, including the TIR remote sensing of rivers. Emissivity and reflectivity spectra from a library also generally cover a range in wavelengths, with measurements at fine spectral intervals (e.g., 1 nm), and, if needed, the spectral data can be mathematically convolved to match the specific spectral characteristics of the data acquired for a given application.

5.3.3 Using Planck's Law to determine temperature from TIR observations

The radiant energy incident on a surface (e.g., sunlight) is termed its irradiance, while the radiant energy exciting the surface is its radiant emittance. The radiation emitted by an object is often assumed to be *Lambertian*, or emitting equally at all angles, although in practice water can be roughened by wind and waves and behave in a non-Lambertian manner. The radiant emittance is usually not measured across the whole hemisphere, but only from a particular direction and solid angle (measured in steradians, sr). For an assumed Lambertian surface, the radiation leaving the surface is assumed to be uniformly distributed across the hemisphere above the surface (which subtends 2π sr). When measured at a particular wavelength, the units of spectral radiant emittance are $Wm^{-2}\mu m^{-1}$.

Planck's Law describes the non-linear relationship of the total radiant emittance from a blackbody, at a particular wavelength, to its temperature. When expressed per unit wavelength, Planck's Law in simplified form (Mather 2004) is as follows:

$$W(\lambda, T) = \frac{c_1}{\lambda^5 (e^{c_2/\lambda T} - 1)} \tag{5.2}$$

where $W(\lambda, T)$ is the total spectral radiant emittance (i.e. not per unit of solid angle) at a particular temperature per unit area of emitting surface at wavelength (λ) in meters (W m^{-2} m^{-1}), T is the object's temperature in Kelvin, c_1 is 3.7242×10^{-16} W · m^{-2}, and c_2 is 1.4388×10^{-2} m K. To obtain an object's brightness temperature (i.e. without correcting for emissivity), we can invert Equation (5.2) as follows:

$$T(\lambda, W) = \frac{c_2}{\lambda \ln \left(\frac{c_1}{\lambda^5 W} + 1 \right)} \quad (5.3)$$

To determine T_r rather than the brightness temperature from the measured TIR spectral radiance at a particular wavelength, we should first apply the emissivity correction (Equation 5.1) for W before calculating temperature using Equation 5.3. When implementing these equations, care needs to be taken with the units (e.g., wavelength is in m, not μm) and with the precision and number of significant digits for any computer software used in the calculation or else small errors in the calculation may be artificially magnified.

Note that while we describe here the general form of the corrections, they should be applied for individual bands and pixels, and adjusted according to the specific spectral characteristics of the sensor as described by its spectral response function. The band centre wavelength is generally used for calculations. An effective band centre for a sensor band can be determined by weighting all wavelengths within the defined sensor spectral band using for the band-specific spectral response function.

5.3.4 Processing of TIR image data

While some remote sensing data processing methods, such as geo-rectification, are common for all remote sensing data, the emitted nature of TIR remote sensing requires some different processing techniques compared to remote-sensing of reflected radiation. These special techniques will be described here, but for information on standard pre-processing, readers are referred to remote sensing textbooks (e.g., de Jong, et al., 2004; Mather 2004; Lillesand et al., 2008).

Required radiometric corrections of the data compensate for both the effect of the atmosphere on what is measured at the satellite, as well as between-image differences such as changes in the emissivity of the surface due to short-term factors such as wind blowing across the water surface. Radiometric corrections can be time-consuming and can be applied with different levels of processing depending on the application for which the data are to be used. To derive quantitative temperature values from raw TIR data requires either that a radiometric correction be applied to the data, or some form of other calibration data be used for an empirical correction (e.g., see Section 5.6.4 for an example with SSTs).

5.3.5 Atmospheric correction

While observations using TIR data focus on the atmospheric windows as previously described, water in the atmosphere between the water and the sensor is still one of the largest sources of error. For images acquired from a satellite in earth orbit, the sensors are recording at the top of the atmosphere (TOA) and therefore observe the radiation originally emitted from the earth's surface (i.e. L_g) after it has passed through the atmosphere. This emitted spectral radiance measured at the sensor (L_s) is influenced by many factors related to its path through the atmosphere and factors such as the viewing geometry of the sensor and sun. These factors can be summarised by two factors: an additive spectral radiance contribution (i.e. path radiance, L_p) resulting from upwelling spectral radiance contributed by the atmosphere, and a multiplicative factor (i.e. transmissivity, τ) which is due to the attenuation by atmospheric absorption and scattering of spectral radiance emitted by the surface and not reaching the sensor. The correction of L_s to determine L_g is as follows:

$$L_g(\lambda) = \frac{L_s(\lambda) - L_p(\lambda)}{\tau(\lambda)} \quad (5.4)$$

where:
L_g = land-leaving spectral radiance at a particular
        wavelength ($Wm^{-2}\mu m^{-1}sr^{-1}$)
L_s = sensor spectral radiance ($Wm^{-2}\mu m^{-1}sr^{-1}$)
L_p = path spectral radiance ($Wm^{-2}\mu m^{-1}sr^{-1}$)
τ = transmissivity (unitless)
λ = is the wavelength of the sensor (e.g., the band
        centre wavelength)

To determine L_g from L_s accurately it is essential to correct for atmospheric conditions as even on clear days there will be an effect from atmospheric gases and water vapour. Smoke, dust or haze can result in large effects. TIR radiation also cannot be sensed through clouds or fog, so standard remote sensing practices should be used to identify and mask these in the image. Once L_g has been determined from L_s using Equation 5.4, T_r can be determined using Equation 5.3.

L_p and τ can be determined for the specific image date using a radiative transfer model such as MODTRAN (Berk et al., 1989), 6S (Kotchenova, et al., 2006), or FLAASH (Adler-Golden et al., 1999) to calculate all aspects of scattering and transmission of radiance through the atmosphere. However, such models are time-consuming and require input data which may not be available real-time, such as vertical profiles of atmospheric water vapour. Some of these parameters can be estimated from standard atmospheres or models. For example in a study of TIR remote sensing of streams and rivers in the USA (Handcock et al., 2006), a radiosonde was launched on a weather balloon concurrent with sensor overpasses, and L_g and τ were determined for each image using the Penn State/NCAR mesoscale model (MM5) (Dudhia, 1993). The MM5 column atmospheric water vapour or radiosonde profile water vapour was used as an input to radiative transfer modeling (MODTRAN 4.0, Berk et al., 1989; Ontar Corporation, 1998) to determine the spectrally varying atmospheric τ and L_p to correct each image (Kay et al., 2005).

While critical to accurately retrieving temperature from TIR data, atmospheric correction can be time consuming and expensive. An alternative is to use non-atmospherically corrected water temperature data to assess relative spatial patterns within a single image. The

assumption is made that if a single uniform atmospheric correction had been applied across the whole image then its quality only affects the absolute accuracy and not the uncertainty of remotely sensed river-temperature measurements. The resulting relative temperatures can be used along with image interpretation of the river with its terrestrial and aquatic context for applications where knowing absolute temperatures is not critical. For example, to identify hydrological features such as seeps (e.g., Figure 5.1), confluences of streams or rivers with contrasting temperatures (e.g., Figure 5.7), as well as thermal refugia or possible pollution sources. In some situations having absolute temperature is still essential, such as mapping thermal characteristics of fish habitat.

5.3.6 Key points

- Interpretation of TIR image data to determine water temperature can be complex. However, understanding

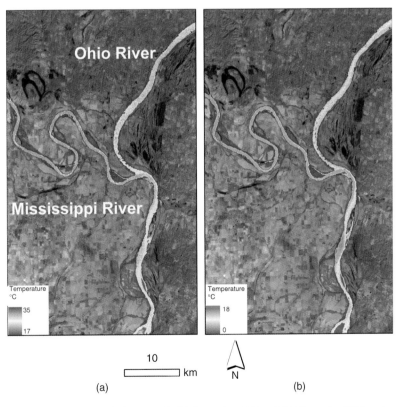

(a)   (b)

Figure 5.7 Confluence of the Mississippi and Ohio rivers in (a) September 2001 and (b) November 2001 as viewed in Landsat 7 ETM+ images. Monitoring bulk river temperature in this region is complicated by the mixing of cooler water from the Mississippi River and warmer water from the Ohio River, which continues to affect downstream temperature for more than 10 km.

this theoretical background is necessary for choosing whether the required tools and technical skills are available to process and interpret the images.

• Emissivity should be chosen to match the specific water and sensor characteristics. Spectral libraries provide a good source of emissivity and reflectivity data for most applications. Care must be taken in interpretation of TIR data under conditions that might affect the emissivity, such as large observation angles and rough water.

• To reduce errors that can occur at large observation angles, images should be chosen to be near nadir ($<30°$). When multiple images are obtained they should have similar observation angles.

• While critical to accurately retrieving quantitative temperature from TIR data, radiometric correction can be time consuming and expensive. An alternative is to use non-radiometrically corrected TIR images to assess relative spatial patterns within a single image, but this will limit the applications for which the data can be used to those not requiring absolute temperature information.

• TIR radiation cannot be sensed through clouds or fog, so standard remote sensing practices should be used to identify and mask these out of the image before quantifying water temperature.

• See also Table 5.1.

5.4 Extracting useful information from TIR images

Once the TIR image data have been processed to determine T_r, it is still necessary to extract information specific to the thermal application. In this section we first discuss how to calculate a representative water temperature. We then examine how both the size of the river relative to the pixel size of the TIR imaging sensor, and the near-band environment, influences the accuracy of extracted water temperature. We note that care is required for interpretation of TIR images within their terrestrial and aquatic context; a trained operator is required to reduce errors associated with image interpretation. A detailed examination of this topic is beyond the scope of this chapter, therefore we illustrate this complexity with examples.

5.4.1 Calculating a representative water temperature

Thermal applications usually require fine resolution TIR data to map thermal pollution sources or locate thermal refugia. Extracting a longitudinal water temperature profile can be complex because temperature from the stream or river centre may not be representative spatially across the stream. This necessitates approaches such as weighted averages (Cristea and Burges, 2009) or median filtering (Handcock et al., 2002).

Many TIR imaging sensors used for measuring water temperature are designed to have multiple spectral bands located at different wavelengths. These wavelengths are typically determined from the TIR emissivity spectra useful for geological applications (e.g., Gillespie et al., 1984; Bartholomew et al., 1989). When available, multiple bands have an advantage for checking the accuracy of image processing when ground-based temperatures are available, because of the physical constraint of there being only one true temperature. The band or bands with the least amount of instrument noise and atmospheric effects can then be selected to calculate the final image temperatures. Alternatively, multiple bands can be averaged to reduce noise due to atmospheric or sensor differences and provide a better estimate of the actual temperature (Handcock et al., 2006).

5.4.2 Accuracy, uncertainty, and scale

The issue of spatial scale is critical to the remote sensing of rivers using TIR data, as the combination of river width and pixel size will determine whether it is possible to distinguish the river from the bank at the desired levels of accuracy and uncertainty with the TIR imaging sensor. The accuracy (bias) of a TIR measurement can be compared to a known *in situ* reference value used for validation, while its uncertainty (precision) is the repeatability of measurements. The radiometric precision of a TIR sensor is described by its NEΔT, or 'noise-equivalent change in temperature,' which is the minimum difference in temperature that the sensor can resolve as a signal from the background noise.

Rivers often have a complex morphology of channels, boulders, shallow areas, gravel bars, islands and in-river rocks, and vary greatly in hydrological and hydraulic characteristics such as ground-water inputs, water depth, water velocity and turbulence fluctuations. Handcock et al. (2006) quantified the accuracy and uncertainly related to the TIR remote sensing of river temperature across multiple spatial scales, imaging sensors, and platforms, and showed that when the water was resolved by less than three pure water pixels of a well-mixed river, the measurements had low accuracies and high uncertainties.

In practice, it can be difficult to find three pure water pixels as the edge pixels are frequently contaminated by bank material. TIR remote sensing of water involves a trade-off between having a pixel size fine enough to identify spatial patterns and reduce mixing with bank materials, and coarse enough so that the cost of flying large areas is not prohibitive. TIR satellite-based images generally cover large areas (e.g. ASTER images cover a ground area of $3600\,km^2$ and Landsat ETM$^+$ $\sim 34200\,km^2$), but their pixel sizes are commonly too coarse to resolve the river channel, except for the widest rivers. However their low cost, capability for regional coverage, and the potential for repeat monitoring with systematic image characteristics, make satellite-based TIR imaging sensors attractive if the pixel size is suitable for the river size. Airborne- and ground-based platforms produce TIR images with finer pixel sizes suitable for narrower rivers. Ground based TIR imagery can also be limited by the difficulty in locating sites suitable for the operation of the TIR sensor at an elevation above the water to resolve significant amounts of the stream channel. Further discussion of the characteristics of different sensor platforms can be found in Section 5.5.

For pixels that are not pure, such as for edge pixels that are a mixture of the bank and water, either a single emissivity needs to be assumed for the pixel, resulting in additional error for the temperature estimation for the pixel, or some method needs to be used to unmix the pixel. For example, the pixel can be unmixed using a spectral mixture analysis (Gillespie, 1992; Gustafson et al., 2003; Sentlinger et al., 2008, Gu et al., 2008).

5.4.3 The near-bank environment

The near-bank environment includes objects with a wide variety of temperatures and emissivities. It includes bark, branches, grasses, leaves, soil, sand, and rocks. Depending on the season, the time of day, and whether the near-bank objects are shaded, they may be warmer or cooler than the water. For example, for summer day-time observations, sun-lit rocks and woody debris exposed during low water levels could be warmer than the water. In comparison, a gravel-bar shaded by the trees could be cooler than adjacent water and would be confused with cold water inputs, as could riparian vegetation cooled by evapo-transpiration. These examples illustrate the sort of complexities found in image interpretation of TIR images in a riverine landscape.

Emitted radiation from the near-bank environment of vegetation and rocks can reach the sensor directly or be scattered from the river surface as the result of emission, reflection, and multiple-scattering of emitted

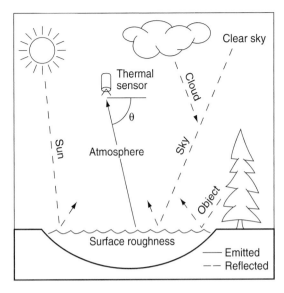

Figure 5.8 Sources of emitted and reflected TIR radiation in thermal remote sensing of rivers and streams (Torgersen et al., 2001, Figure 1). Reprinted from *Remote Sensing of Environment*, **76**(3), Torgersen, C.E. et al., Airborne thermal remote sensing for water temperature assessment in rivers and streams, pp. 386–398, Copyright 2001, with permission from Elsevier.

radiation (Figure 5.8). TIR radiation emitted from near-bank objects may pass directly into the path of the sensor, resulting in erroneous image interpretation, particularly when the bank material's temperature is very different from the water temperature. This is particularly problematic for ground-based TIR imaging sensors observing the water from the bank with a large observation angle (e.g. Handcock et al., Figure 7b, 2004; Cardenas, et al., Figure 2, 2008). TIR radiation emitted from near-bank objects may also be scattered or reflected from other surfaces into the sensor, both of which will increase the measured T_r. This multiple scattering effect is small compared to the temperature of the water. For example, for a tree with a temperature of 25 °C and water at 17.2 °C there is a calculated increase in the observed water temperature of 0.10 °C–0.65 °C, respectively, depending on the viewing geometry (Handcock et al., 2006). When the bank vegetation is tall compared to the width of the river it may not be possible to find regions in the river centre that are not contaminated by scattered radiation emitted by the bank vegetation. In situations where it is difficult to obtain pure-water pixels, higher resolution imagery may be needed, and more than 10 pixels may need to be sampled from the imagery to resolve T_r accurately (Torgersen et al., 2001).

5.4.4 Key points

- Depending on the TIR sensor and platform that is used, the available TIR data must be matched to the size of the river being monitored; as a general rule, when the channel width is resolved by fewer than three adjacent pure water pixels in a well-mixed river, the measurements will have low accuracies and high uncertainties. In most situations, more than 3 pixels will be needed. Similar accuracy problems would be expected when resolving other water bodies.
- TIR radiation emitted from near-bank objects may pass directly into the path of the sensor, resulting in erroneous image interpretation, particularly when the bank material's temperature is very different from the water temperature.
- Having multiple spectral bands is an advantage where there is a single known temperature, which can be used as a check of the accuracy of the data processing.
- Care must be taken to interpret images within their terrestrial and aquatic context.
- See also Table 5.1.

5.5 TIR imaging sensors and data sources

In this section, we discuss sources of TIR data, focusing on image-based sensors rather than point-based sensors such as radiant thermometers. Point-based sensors are discussed in the next section in relation to TIR validation. There are many sensors for measuring TIR data, and their availability and refinements is changing rapidly. This chapter does not attempt to provide a comprehensive overview. Instead, some examples of particular sensors are used to illustrate the characteristics typical of particular types of sensors. Common to all sensor platforms is the issue of calibration. Some TIR imaging sensor systems have on-board calibration sources, typically a hot and a cold calibration source. Other systems record only relative values of emitted radiation, or digital numbers (DN), and must be calibrated (see Section 5.6 for methods of measuring *in situ* water temperature).

5.5.1 Ground imaging

There are many examples of ground-based imaging systems that provide an array of TIR measurements. When choosing a TIR imaging sensor, specifications are usually given as to its pixel size and precision. Pixel sizes from such imaging systems will be determined by the distance between the sensor and the water, and are typically <1 m. Many ground-based TIR imaging systems are available, including forward looking infrared (FLIR) imagers (Rogalski and Chrzanowski, 2002) that can be used in both ground-based and airborne surveys. High quality systems have NEΔT values of 0.1 °C or better.

When using ground-based TIR imaging sensors, care must be taken with the viewing geometry of the sensor relative to the water and the surroundings. In particular, tall near-bank objects such as trees will scatter emitted radiation onto the water, and some of this radiation will be reflected back into the field of view (FOV) of the imaging sensor (see Section 5.4.3 for more details). Data from ground-based imaging sensors such as a FLIR can be calibrated (e.g., Handcock et al., 2006) using shielded and stirred water targets at different temperatures. Linear regression is used to relate raw image values (DN) to T_r measurements made using a hand-held broadband radiometer.

5.5.2 Airborne imaging

As airborne TIR imaging sensors are widely used for monitoring of water temperature in riverine environments, we will cover the topic in some detail. Airborne TIR imaging sensors can be mounted on either fixed-wing aircraft or on a helicopter which may be manned or unmanned. The resulting pixel size of the TIR data (instantaneous field of view, or IFOV) is a function of the distance from the water surface (determined by the height that the platform flies), the sensor characteristics, and the optical FOV of the sensing system. The height of the platform and optical FOV also determines the ground footprint of the resulting image.

The narrow swath widths of TIR images from airborne platforms – typically from a few kilometers for fine resolution images, to a few 10s or 100s of meters for very-fine resolution images – reduces their ability to capture long stream reaches as the channel winds in and out of the preferred straight-line flight pattern. To best manage the issues of edge pixels (discussed in Section 5.4.2) in airborne TIR image acquisition, a balance must be struck between the resolution of the sensor, the desired resolution of the image, the width and sinuosity of the stream channel, and the altitude of the aircraft. Finer pixel sizes may be obtained from the same sensor when the aircraft is flown at lower altitudes, however, as the number of times the aircraft is required to circle around and line-up on a new straight-line for acquisition of the stream channel, the longer and more expensive the image acquisition will become.

Another important consideration for airborne data acquisition is that the images do not provide a truly synoptic assessment, or 'snapshot', of water temperature at a specific time if the images are collected sequentially along the river course. Thus, diurnal changes in water temperature should be considered in planning airborne data collection. See further discussion of this topic in Section 5.5.2.1 on helicopter versus fixed-wing aircraft, and Section 5.6 on the validation of temperature measurements.

The selection of an airborne platform/TIR-imaging-sensor depends on project-specific details such as the river characteristics (size, sinuosity, etc.), temporal constraints, desired spatial and thermal resolutions and accuracies, and map accuracy specifications. TIR imaging systems have evolved continually with advances in technology, and there are a number of TIR imaging sensors available on the market suitable for use on airborne-platforms. These TIR imaging sensors have unique technical characteristics such as physical size, temperature resolution, integration times, detector types and sizes, and optics. However, there are also common features that make these TIR imaging sensors suitable for TIR remote sensing of water.

TIR imaging systems must be able to store raw DN that can be converted either internally or during post-processing to a measure of radiant energy. Manufacturers differ in how this is accomplished, but in most cases the detectors are calibrated in the laboratory environment against a black-body source and this information is stored (either internally or externally) as a conversion curve that is unique to the sensor. Because airborne remote sensing typically involves collecting sequential frames, another important sensor system characteristic is its ability to retain internal radiometric consistency throughout the data acquisition. Although conditions such as ambient temperature change, the TIR imaging system must be able to control or minimise internal drift such that frame-to-frame measurements are consistent. TIR imaging sensor manufacturers accomplish this in a number of ways such as using internal temperature references or cooling mechanisms which retains stability in the detector array. Finally, the TIR imaging system must account for radiometric distortion due to variability in individual detector response and lens optics (in the case of frame based TIR imaging sensors). This is referred to as uniformity correction and can be accomplished either internally or during the post processing.

A wide variety of TIR imaging sensors have been used for airborne applications. For example, in one study (Handcock et al., 2006) the MODIS/ASTER (MASTER) sensor (Hook et al., 2001) was flown on a King Air B200 fixed-wing aircraft at altitudes of ~2000 and ~6000 m, which gives approximate pixel sizes of 5 and 15 m, respectively. The MASTER sensor has ten TIR bands (10.15–11.45μm) with an NEΔT that ranges from 0.46 to 0.71 °C, and can scan ±43° from nadir. In another example, on Prince Edward Island (Canada), a FLIR Systems SC-3000 TIR imaging sensor mounted on a Cessna 172 fixed-wing aircraft was used to acquire images of the Trout River. These data were successfully used to detect and quantify ground water discharge to the estuary (Danielescu et al., 2009). The SC-3000 sensor has a single TIR band of 8–9μm with a NEΔT of 0.02 °C. The sensor has a fixed horizontal FOV of ±10° from nadir and a detector array of 320 × 240 pixels. In this study, the aircraft was flown at an altitude of 1000 m to give a ground sample distance (GSD, i.e. the pixel size of the imaging sensor expressed in ground units) of 1 m. Although the term GSD is often used interchangeably with pixel size, it is sometimes expressed explicitly when the ground distance represented by a pixel changes across the image, which is common for aerial imaging and oblique view geometries. In Northern Utah, a helicopter-mounted Space Instruments Firemapper 2.0 was used to collect TIR images with a 3 m GSD to identify areas of thermal refugia in the 9.6 km^2 Cutler Reservoir (Dahle 2009). The Firemapper 2.0 system has a single TIR band of 8–12μm with an NEΔT of 0.07 °C. The imaging sensor has horizontal FOV of ±22.1° from nadir and a detector array of 320 × 240 pixels.

Early work with airborne TIR imaging in riverine environments focused primarily on detecting cold water sources and longitudinal temperature patterns (Torgersen et al., 1999, 2001) with data geo-referenced to specific locations along the longitudinal extent of the river (i.e. tributary junctions, springs geo-referenced according to their distance upstream). Over the past decade, creating continuous image mosaics with specified mapping accuracy has increasingly become a requirement so that the image data can be accurately combined with other spatially explicit data layers and accurately geo-referenced field data. Although some of the early TIR images were from ground-based imaging sensors mounted on an aerial platform, more recent TIR sensor systems such as the ITRES TASI 600 (USA) push-broom hyperspectral thermal imaging sensor system are specifically designed for airborne operations. Such imaging sensors provide timing and scan information suitable for integration with airborne Global Position System (GPS)/inertial measurement units (IMU) for direct geo-referencing. Similarly, the FLIR Systems SC6000 QWIP frame based sensor (8–9.2μm) has accurate timing and triggering capability that allow direct integration with an aircraft's modern GPS which records its geographical location, and its IMU which record the aircraft's velocity, orientation, and gravitational forces. An IMU greatly simplifies the process of extracting ground control points (GCPs) used for ortho-rectification of the image data. When an IMU is not available, as is more common in older systems, these GCPs need to be extracted from other image sources, base-maps, or manual or surveyed GPS locations. The process of identifying GCPs in the image is simplified when a concurrent visible image is obtained with the TIR image, otherwise distinguishing water from the bank material can be difficult when, for example, a shaded gravel bar is colder than water and confused for a cold-water spring, or dead wood in the stream is confused for a warm water input.

TIR imaging systems have progressively increased the size of the detector arrays, allowing a broader range of options for smaller GSDs or larger ground footprints at fine and medium pixel sizes. In general, this also allows a broader range of platform options since fine-resolution pixel sizes that were once achieved from a helicopter can be designed for higher flying aircraft capable of covering greater areas in shorter amounts of time. For example, the FLIR Systems SC6000 offers multiple lens options, has a pixel array of 640×512, and an NEΔT of $0.035\,^{\circ}$C. With a $\pm 17.5^{\circ}$ FOV (25 mm lens), a 1 m pixel can be achieved with a 644 m wide ground footprint. The previously mentioned ITRES TASI 600 is a pushbroom hyperspectral thermal imaging sensor that acquires an image with 600 pixels across its track with a $\pm 20.0^{\circ}$ horizontal FOV and 32 bands within the 8–11.5μm spectral range (the advantage of multi-band thermal imaging sensors was discussed in Section 5.4.1). Finally, technological advances have made small, relatively inexpensive handheld thermal imagers appealing for mounting on airborne platforms. However, these image systems typically do not have the radiometric features or durability suitable for the airborne environment.

5.5.2.1 Helicopter versus fixed-wing aircraft

TIR imaging sensors have been mounted on both helicopter and fixed-wing airborne platforms. The selection of platform depends in a large part on the objectives of the data acquisition, the desired TIR imaging sensor spatial resolution, and the characteristics of the river.

The advantage of a fixed-wing aircraft is that the post-processing of TIR images is simplified by the relative stability of the platform, whereas a helicopter will typically have more variable flight characteristics such as altitude, yaw, roll, and pitch, which require additional processing and can introduce artifacts into the image which makes image interpretation more complex. Fixed-wing aircraft have long been used for aerial photography and other airborne remote sensing tasks. Consequently, finding a commercial charter for a fixed-wing aircraft with an existing camera port and appropriate flight characteristics, which is also located close to almost any project site, is relatively straightforward. Additionally, the instrumentation on fixed-wing aircraft is normally located inside the aircraft so that the installation, operation, and transport of the sensor is simplified when compared to operating from a helicopter. On rotary-winged aircraft, the instrumentation is most often external to the aircraft in a weatherised pod. In general, helicopters have a smaller range and higher operating costs than fixed-wing aircraft, but are more suitable for certain data-collection missions.

Fixed-wing aircraft are generally preferred in TIR image collections where flight parameters (altitude, speed) can remain relatively constant over the project area, such as for large water bodies or targeted sections of river with limited terrain relief. Helicopters are more suitable for collecting images along a sinuous corridor at very fine resolutions where the helicopter's slow speed and maneuverability are an advantage, and the acquisition of ultra-high resolution images is needed (e.g. for studying complex braided floodplains). In many cases, a fixed-wing aircraft may have to collect multiple lines of image data to capture the same areas, and may be unable to safely maneuver at the low altitudes required to capture images of the same spatial resolution.

As TIR imaging sensor systems decrease in size and cost, their application from airborne platforms is likely to increase. Smaller TIR imaging sensors can also be mounted within unmanned aerial vehicles (UAVs), including remotely controlled aircraft (e.g., Berni et al., 2009) and under balloons (N. Bergeron, Institut National de la Recherche Scientifique, Quebec, Canada, pers. comm.).

5.5.2.2 Airborne image analysis

Some of the factors and considerations of using TIR imaging sensors on airborne platforms, and their post-processing, applies also to other remote imaging sensors; thus, standard methods can be used. Airborne data also need to be calibrated. For example, as described previously in Handcock (2006), the flight team for the MASTER multispectral data collection processed the TIR data to 'radiance-at-sensor.' using onboard calibration targets and an onboard GPS to record the aircraft's location. Ground-based data can also be used to calibrate raw TIR data, as will be explored in Section 5.6.

If the objective is to create continuous image mosaics of riverine systems or large water bodies, the acquisition of overlapping images becomes especially important, as does the timeliness of image acquisition (see Section 5.6.1). The ability to capture images efficiently at fine spatial resolutions, particularly on smaller streams, is paramount to minimising the diurnal change in water temperature during the image capture. For example, a recent 48 km stretch of the Anchor River (AK, USA) was flown by helicopter in 0.6 hr at a spatial resolution of 60 cm GSD. A flight plan for the same corridor, using a fixed-wing aircraft at the same GSD, would have required a flight time of approximately 2.4 hr.

5.5.3 Satellite imaging

Space-borne TIR imaging sensors can cover a greater aerial extent than airborne TIR imaging sensors and cover a range of pixel sizes, number of bands, FOV, revisit times, and sensor sensitivities. If TIR satellite images are available for the study time, and are of a suitable pixel size compared to the thermal application (see Section 5.4.2), they can be an attractive source of broad-scale data due to their low cost, capability for regional coverage, and the potential for repeat monitoring with systematic image characteristics. While TIR satellite-based images generally cover large areas, their coverage may still not be extensive enough to fully track some water bodies. Therefore, measurements of long channel reaches may have to be compiled over several days or weeks depending on the satellite orbit.

TIR imaging sensors typically have larger pixel sizes than do visible and near infrared (VNIR) imaging sensors, with the pixel size being determined by the sensor specifications (e.g., aperture, sensitivity of the detector, and the desired NEΔT). For example, the Advanced Spaceborne Thermal Emission and Reflection (ASTER) radiometer (Kahle et al., 1991; Yamaguchi et al., 1998), mounted on NASA's Terra spacecraft, has a VNIR sensor with three bands (15 m pixel size), a shortwave infrared (SWIR) sensor with six bands (30 m pixel size), and a TIR

sensor with five bands (8.12–11.65μm) with 90 m pixel size and an NEΔT of $\leq$0.3 °C at 27 °C (Gillespie et al., 1998; Yamaguchi et al., 1998). While the pixel size of the ASTER TIR sensor is fine in terms of a satellite-based imaging sensor, it is 'medium' based on the multi-sensor criteria defined previously. The revisit time for ASTER is 16 days, and the FOV of the system is 60 km. As well as raw TOA ASTER data which are not radiometrically corrected, a number of higher-level products have been are available, including temperature, emissivity, and ground-leaving spectral radiance.

The NASA EOS MODIS sensor is also on the Terra platform (imaging in the morning), as well as a second MODIS sensor onboard Aqua (imaging in the afternoon). MODIS has ten TIR bands (6.54–14.39μm) with a pixel size of 1000 m and an estimated NEΔT of 0.05 °C at 27 °C (Barnes et al., 1998). Although MODIS has more bands than ASTER and a wider FOV (2330 km), the larger pixels limits it to observations of wide rivers. One advantage of MODIS is that images of the entire globe are acquired daily (and more frequently for higher latitudes). As well as raw TOA MODIS that have not been radiometrically corrected, a number of higher-level products are available, including temperature and emissivity.

The Landsat ETM$^+$ sensor on Landsat-7 is in the same orbit as Terra, which allows images to be acquired ~20 min apart with a 16-day revisit time. The single TIR band of Landsat ETM$^+$ (10.40–12.50μm) has inherent pixel size of 60 m (National Aeronautics and Space Administration, 1998), which is more recently available resampled to 30 m for easier comparison with other Landsat bands. The NEΔT of the Landsat ETM$^+$ TIR band is 0.22 °C at 7 °C (Barsi et al., 2003), and the FOV of the system is 185 km. The TIR sensor on Landsat was historically not always switched on to acquire data; however, the large number of images in the Landsat archive, the ongoing acquisitions, and the recent availability of free data, makes Landsat TIR an attractive option if it meets an application's specifications.

5.5.4 Key points

• The choice of TIR imaging sensor and whether the platform is ground-, air-, or satellite-based will depend on many factors, including the size of the area that has to be covered, how frequently data are required, cost and sensor availability, and whether the accuracy requirements of the application require a TIR imaging sensor with on-board calibration sources.

• Ground-based TIR imaging sensors are convenient but can only view the water from specific locations along the stream. Observation angles need to be chosen carefully to reduce the effects of scattering from bank objects.

• Airborne TIR imaging sensors provide spatially extensive images with fine pixel-sizes suitable for narrow streams and rivers, but acquisition can be costly, processing complex, and images are generally acquired over narrow swath widths.

• The pixel size and ground footprint of the TIR image is a function of the altitude above the water body at which the sensor collects data, the sensor characteristics, and the optical FOV of the sensing system.

• A balance must be found between the pixel size of the TIR sensor and the altitude at which airborne images are collected. Too high an altitude will result in more problems with mixed pixels, while too low an altitude will increase the expense and time required to acquire images over a given area.

• Advantages of fixed-wing aircraft include the relative stability of the platform that simplifies post-processing of TIR images, and the lower operating costs.

• Advantages of rotary wing aircraft include the greater flexibility of the platform for tracking sinuous riverine features and for flying at the low altitudes required for capturing very-fine resolution imagery. Such advantages require additional processing, and can introduce artifacts into the image which may make image interpretation more complicated, but software advances will likely reduce these disadvantages in time.

• UAVs hold great potential for future collection of TIR imagery, however, the lack of defined operating rules and licensing procedures for most countries severely limits their usage.

• Space-borne TIR imaging sensors provide spatially extensive images for low cost over large areas, but pixel-sizes are usually coarser than airborne data and are commonly too coarse to resolve the river channel, except for the widest rivers. However their low cost and capability for regional coverage at multiple points in time make satellite-based TIR imaging sensors attractive if the pixel size is suitable for the river size.

• The typically narrow swath widths of airborne TIR images make it more likely that overlapping scan lines must be collected and processed to create a mosaic of the river.

• Common to all sensor platforms is the issue of calibration. Some TIR imaging sensor systems have on-board calibration sources, while other systems record only relative values and must be calibrated.

5.6 Validating TIR measurements of rivers

5.6.1 Timeliness of data

Water temperature in streams and rivers changes throughout the day as the sun elevation angle and heating changes. Collecting validation data for TIR observations requires that water temperature measured is close enough to the time of the TIR image collection that the temperature has not changed significantly. This can be relatively easy to do in the case of ground-based images but is much more complex for observations from aircraft or satellite. In all situations, the collection of simultaneous measurements can be difficult due to the requirement for large numbers of personnel. Two questions that must be addressed pertain to the time of day of image capture and the duration of the collection period, both of which depend on the application. For example, to detect warm groundwater springs during winter, TIR images should be obtained when the river is the coldest.

If multiple images are to be compiled into a mosaic to provide coverage for a larger area, TIR data should be acquired when water temperature is most stable. River temperature is typically most stable in the early afternoon when air temperature is also relatively stable. However, the thermal inertia of the river water provides additional stability not found in air temperature, increasing the time over which river conditions should reflect the thermal conditions during TIR image acquisition. To study this question more closely, we have provided observations acquired at an interval of 15 min from four sites in the Pacific Northwest (USA) (Figure 5.9). Two sites are on a large river (Green River, annual flow 45.6 $m^3 s^{-1}$), whereas the other two are on smaller streams (Big Soos Creek, annual flow 3.5 $m^3 s^{-1}$; and a tributary to Covington Creek, which is a tributary to Big Soos Creek). Two sites are relatively open with limited riparian cover, whereas the other two have dense riparian cover (the tributary to Covington Creek is almost completely shaded). All four sites show a strong diurnal cycle, with minimum temperature occurring just as the sun rises above the horizon, and maximum temperature occurring by late morning or early afternoon. Changes in temperature are typically slowest around these extremes, so that early morning and early afternoon have the widest sampling windows with the least change in temperature. The width of a sampling window is highly dependent on how much of a change in temperature is acceptable. If observations

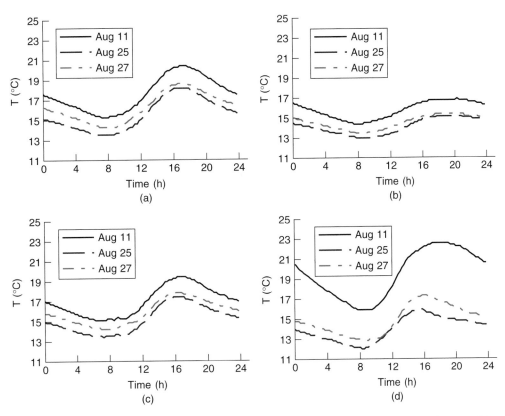

Figure 5.9 Change in temperature over time over a day for four locations: (a) a large river with an open riparian zone, (b) a small stream with an open riparian zone, (c) a large river with forested riparian zone, and (d) a small stream with forested riparian zone.

should not vary by more than ±1 °C from the TIR image, a window of more than 4 hr is possible, but to constrain measurements to within ±0.5 °C a window of less than 2 hr may be better.

5.6.2 Sampling site selection

The number of sampling sites is dictated by the (1) length of river to be measured, (2) its accessibility, and (3) the location of temperature extremes within the river. River length to be measured will be controlled by the method used for TIR image collection. For example, a single TIR image frame collected from a bridge, helicopter, or balloon may simply require field technicians to walk the length of the river reach, collecting representative temperature measurements. However, multiple aircraft images or even a single satellite image may cover 10s to 100s of kilometers of river channel, requiring in-stream sampling at many locations with multiple field teams. If recent TIR images are available, they can be used to select

validation sites, and hand-held TIR sensors can be used to survey a region to determine optimum locations for deploying thermometers for longer-term monitoring.

Where digital spatial datasets of roads and river channels are available, an initial set of sampling locations can be devised by identifying locations where rivers and roads cross or come very close to each other. Otherwise local maps can be used for a similar search. In some cases it may be possible to determine if sites are inaccessible from available maps, air photos, and other spatial data layers within a GIS or Google Earth (Google Inc., 2010). For example, such data sources can be used to identify the presence of restricted lands, steep slopes, or channels that are not visible from above through dense riparian canopies. However, in many cases determining accessibility will require a survey of the locations to avoid inappropriate sample sites.

When identifying sampling locations, care should be taken to avoid locally cool or warm locations. These can be caused by the influence of tributaries, back-water effects (where a dam or other obstruction results in changed temperature upstream), and groundwater seeps, which can change the bulk T_k of the main river channel locally. Fine-scale examples can be seen in Figures 5.1–5.3 in which seeps, springs, and tributaries affect water temperature at spatial scales of a few to 100s of m. The mixing of two very large tributaries, such as the Mississippi and Ohio rivers (USA) (Figure 5.7) can affect water temperature for many kilometers. In measurements made on the Columbia River (USA) using a hand-held thermometer, thermal eddies were revealed at the edge of the wide river, suggesting that more frequent measurements (~ 1 min^{-1}) might be necessary when the river is not well-mixed or in the presence of obstructions (Handcock et al., 2006).

5.6.3 Thermal stratification and mixing

As suggested by Wunderlich (1969), the effects of thermal stratification on data collection can be minimised by collecting measurements in shallow, well-mixed parts of the river because sites that are exposed to solar radiation, slow moving currents, or substantial cold-water inflows from seeps or springs may experience substantial thermal stratification (Nielson et al., 1994; Matthews and Berg, 1997; Torgersen et al., 2001). Thermal stratification is possible during sunny conditions or slow river flows when solar heating of the surface layer is not compensated by vertical mixing. For example, thermal stratification was observed by Torgersen et al. (2001) in some areas of the river that were not well mixed. Nielsen et al. (1994) found significant vertical stratification for flows of less than 1 ms^{-2} in the Middle Fork Eel River in northern California. They found that at depth, water could be as much as 7 °C cooler than at the surface. These examples highlight that TIR images only measure the water's surface, and that there are complex mixing processes deeper in the water column. For a good discussion of this issue and the physics of mixing, see Torgersen et al. (2001).

Wind and the breaking of waves can also disturb the skin layer and leave the surface well mixed, as has been seen in studies of ocean waves (Jessup et al., 1997), so that the measured T_r approaches the T_k of the water at the depth of mixing. Hook et al. (2003) investigated the difference between T_k in the surface layer and the T_r in the skin layer of lakes using four monitoring stations permanently moored on Lake Tahoe, California–Nevada (USA). They found a difference between T_k and T_r that varied over the diurnal cycle, which they attributed to solar heating and lower wind speeds in the morning. Water in rivers is more turbulent than in lakes and oceans, so beyond the potential for wind mixing, river flow around and over rocks, woody debris and other obstructions induces additional mixing and will break the skin layer. Because of this additional mixing, temperature derived from TIR images for turbulent rivers should be more representative of T_k than data for more static circumstances.

5.6.4 Measuring representative temperature

The use of hand-held, ground-based TIR thermometers, or small imaging radiometers, provides the most direct validation of remotely determined temperature. However, obtaining accurate temperature from such devices can be difficult. Because TIR radiation is emitted by all objects at all temperatures, the potential for contamination of readings from a radiant thermometer is high. Contact sensors, which include analog and digital sensors, are more robust and accurate, but they measure T_k at depth, thereby reducing their utility for comparison with T_r. T_k is the environmentally and ecologically important temperature but is not a direct comparison with what the remote TIR sensor detects. For example, in oceanographic applications where SST is of interest, a regression is usually applied between in situ T_k measured from buoys and ships and T_r from TIR sensors (Robinson et al., 1984) to convert the skin temperature measurement of T_k into T_k from deeper in the water column.

This regression approach is especially important for SST measurements since bulk T_k measurements have traditionally been collected at depths of up to 1 m. Such measurements can lead to values of T_k that are 0.1 to 1.5 K warmer than near-surface measurements of T_k. Capturing representative river temperature is potentially more difficult than in the ocean as water depths can change, there can be exchanges of water between the water and the streambed, and there can be spatial variability of temperature along and across the channel. For that reason, the collection of representative temperature of streams and rivers must be handled with care, both for ground-truth evaluation of remote sensing products and for capturing the ecologically sensitive values of T_k.

Precision of thermometers should be at least as good as that expected from the TIR measurements. For example, if the TIR sensor provides temperature with an NEΔT of 0.5 °C, a bulb thermometer should have lines at least every 1.0 °C so that temperature can be measured and recorded with a precision of 0.5 °C.

5.6.4.1 Radiant temperature measurement

There are many types of TIR thermometers available on the market ranging from under $100 to several thousand dollars (USA). These thermometers use similar technology to the satellite- and aircraft-based TIR imaging sensors, to detect the radiation emitted in the TIR spectrum; however, they are typically designed to measure the temperature of a single location, not to provide spatially extensive images, and are sensitive to the entire TIR spectral range rather than a narrow spectral range. Increased price for TIR imaging sensors will generally provide better optics, aiming capabilities and insulation. The latter is important for obtaining accurate temperature, as it reduces the sensitivity of the device to environmental conditions such as rapid heating or cooling of the internal components of the thermometer, which will change the amount of unintended TIR radiation reaching the sensor. Experiments with an inexpensive TIR thermometer (e.g., a TempTestr-IR from Oakton, 2005) showed that even short-term exposure to direct solar radiation in a climate-controlled building could substantially alter temperature readings from the thermometer (by several degrees Celsius). Increasing the insulation around the TIR thermometer and limiting its direct exposure to extreme conditions (such as direct solar radiation) can reduce the uncertainty in the measurement.

Care must also be taken that the TIR radiant thermometer is measuring only water temperature. These devices have a distance to target/spot size ratio that will indicate how wide the area being sensed is at different distances from target. Inexpensive sensors will typically have a lower ratio (e.g., 6:1, or $9.5°$), in which case the spot diameter increases by 1 cm for every 6 cm greater distance of the target. The target area for less expensive devices will expand more rapidly with distance than for a more expensive device (e.g., with a ratio of 75:1, or $0.76°$). The ratio should be smaller if measurements are made closer to the target in order to reduce the likelihood of interfering objects (e.g., rocks, logs) being included in the measurement.

These TIR radiant thermometers require that a single representative ε be specified, to ensure that the output temperature is appropriate for the material being measured. Higher temperature accuracy can be obtained by using accurate emissivities, but for spectrally variable targets, an accuracy limit is reached because the target is a selective radiator and not, in fact, a graybody. For example, for distilled water at 300 K the emissivities between 10 and 14 μm range from 0.965

to 0.993, but using a single average emissivity of 0.985 will result in calculated temperatures of $29.00–28.69\,°C$ (~0.3 K error). Ultimately, the change in temperature resulting from the emissivity correction is likely to be small relative to the other sources of error when using inexpensive devices.

5.6.4.2 Kinetic temperature measurement

There are a variety of technologies used for the measurement of T_k. These can be divided into analog and digital technologies and hand-held and self-contained temperature-logging units. Analog thermometers, most notably bulb thermometers, have traditionally been less expensive and therefore easier to supply to many field staff taking measurements. Digital thermometers may be easier to read and are potentially more robust. Self-contained data loggers are still the most expensive option, but they can be used to monitor water temperature for extended periods of time and in locations where regular access may be difficult. A combination of devices can provide an expanded network of observations in conjunction with an image acquisition as long as all devices are calibrated and care is taken in selecting measurement locations.

Thermometers will generally be used to measure water temperature near the water surface since the measured temperature will need to be read from the device and recorded. Because inserting the thermometer into the water will break any skin layer, such measurements will always be of the bulk water, not the skin itself. Measurements in the top 10 cm of water are most appropriate for comparison with calibrated TIR images, although the collection of simultaneous measurements can be difficult due to personnel restrictions. Temperature measured by data loggers mounted to the streambed tends to be lower than that measured using near-surface thermometers and TIR sensors. As with measurements of SST, this temperature difference is likely the result of thermal stratification (see Section 5.6.3 for more details).

In the next sections, we apply the knowledge described in this theoretical background to two examples of using TIR remote sensing to monitor water temperature.

5.6.5 Key points

• When absolute water temperature is required, it is necessary to collect validation data. If only relative temperature differences are required, validation data requirements can be reduced or eliminated.

• The number of sampling sites is dictated by the (1) length of river to be measured, (2) its accessibility, and (3) the location of temperature extremes within the river. Care should be taken to avoid locally cool or warm locations, as validation is best made where temperatures represent the bulk of the region being observed.

• Validation data can be collected from automatic gauges, or from surveys with radiant or kinetic thermometers with or without a data logger. A combination of devices can provide an expanded network of observations in conjunction with an image acquisition as long as all devices are calibrated and care is taken in selecting measurement locations.

○ Measurements using hand-held kinetic thermometers to measure T_k (i.e., analog, digital, or data logger) will generally be near the water surface, and because inserting the thermometer into the water will break any skin layer, such measurements will always be of the bulk water, not the skin itself.

○ Analog thermometers are usually less expensive, but digital thermometers may be easier to read and are potentially more robust. Both can be used to rapidly expand a sampling network with trained volunteers for a single day of observations.

○ Self-contained data loggers are the most expensive option, but they can be used to monitor water temperature deeper in the water column, for extended periods of time, and in locations where regular access may be difficult.

○ Radiant thermometers can be used to survey a region for appropriate sampling locations. They are, however, generally sensitive to the entire TIR spectral range, and require that a single representative ε be specified. Such thermometers are also generally sensitive to their own temperature so care should be taken when using them in the field.

• Measured T_r represents the 'skin' layer at the water surface. Depending on the amount of mixing in the river, it may not be representative of T_k further down in the water column.

• The effects of thermal stratification on data collection can be minimised by collecting measurements in shallow, well-mixed parts of the river because sites that are exposed to solar radiation, slow moving currents, or substantial cold-water inflows from seeps or springs may experience substantial thermal stratification.

• Water temperature should be measured close enough to the time of the TIR image collection that the temperature has not changed significantly (see Section 5.6.1).

• Precision of thermometers should be at least as good as that expected from the TIR measurements.

5.7 Example 1: Illustrating the necessity of matching the spatial resolution of the TIR imaging device to river width using multi-scale observations of water temperature in the Pacific Northwest (USA)

In this section, we show examples of TIR remote sensing of water across a range of stream widths and pixel sizes, to illustrate the necessity of matching the specifications of the TIR imaging sensor to the river characteristics such as the channel width. The combination of the spatial resolution of TIR images and river width both affect the accuracy and uncertainty of recovered in-stream T_r (Handcock et al., 2006). The accuracy is of great importance in that any detectable spatial pattern in temperature could be evidence for identifying thermal features such as springs, seeping cold water or subsurface flow. To illustrate the consistency between T_r and T_k fully, longitudinal profiles of radiant temperature from downstream to headwater stream reaches can be plotted in order to compare with the *in situ* gage observations. The mean difference and the standard error (deviation) of temperature are important metrics for evaluating the accuracy of temperature extraction.

As an example of these concepts, we have provided a group of images for reaches of the Green River, Washington (USA), as remotely sensed from TIR imaging sensors with pixel sizes of 5 m (MASTER), 15 m (MASTER), 60 m (Landsat 7 ETM[+]) and 90 m (ASTER) respectively (Figure 5.10a). Although Green River was clearly visible under MASTER and Landsat sensors, this river reach was obscured in the 90 m pixel size of the ASTER image. In our analysis, data were extracted along centre-lines of the river, in order to remove geo-referencing errors and possible along-stream mixed pixels. The results showed that in the 5 m MASTER image, the Green River along-stream radiant temperature had a standard deviation of 0.7 °C and the mean difference between radiant temperature and kinetic temperature was +1.9 °C. The fact that the 0.7 °C variability was close to the NEΔT for the MASTER sensor (0.46–0.71 °C; Hook et al., 2001) indicated no obvious influence from warmer bank temperature. In contrast, the standard deviation was 1.6 °C and the mean difference was +2.1 °C for the 15 m pixels

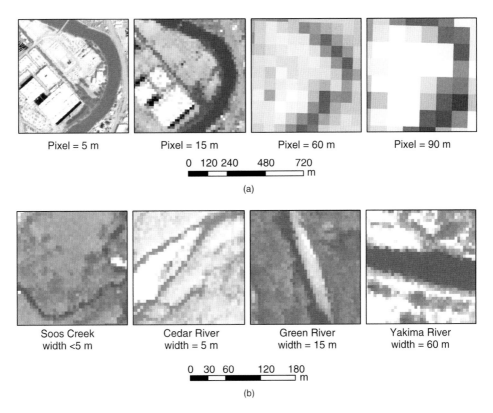

Pixel = 5 m   Pixel = 15 m   Pixel = 60 m   Pixel = 90 m

0 120 240   480   720
⊐ m

(a)

Soos Creek   Cedar River   Green River   Yakima River
width <5 m   width = 5 m   width = 15 m   width = 60 m

0 30 60   120   180
⊐ m

(b)

Figure 5.10 Examples of 1 km² TIR image subsets (a) from different TIR sensors for a single location on the Green River Washington (USA) at different pixel sizes to illustrate the effect of pixel size on what can be seen, and (b) with a set 5 m pixel size from the MASTER sensor for a range of stream widths from streams in Washington (USA) observed at different pixel sizes to show the interaction between stream width and pixel size. For these un-calibrated images, darker shades indicate lower temperatures, and lighter shades higher temperatures.

of the MASTER data. This increased variability can be explained by the effects from warmer bank temperature because the along-stream temperature variability pattern was still similar to that of the 5 m MASTER image in some locations. In comparison, for the 90 m ASTER image, the mean difference increased to 4.6 °C, with the standard deviation decreasing to 4.6 °C, which is much larger than the NEΔT of ASTER sensor (≤0.3 °C at 27 °C; Gillespie et al., 1998; Yamaguchi et al., 1998). Consequently, the Green River was fully resolved by the 5 m pixel size of the MASTER sensor, partly resolved by the 15 m pixels of the MASTER, and not resolved by the 90 m pixels of the ASTER sensor to obtain radiant temperature. See Cherkauer et al. (2005) for a detailed description and discussion.

A similar problem can be illustrated by evaluating different width reaches at a single TIR imaging sensor

resolution, such as the 5 m resolution MASTER images used in Figure 5.10b. Here, Soos Creek with a width of less than 5 m, had mixed at the bank, and the water was never fully resolved. Water pixels could occasionally be fully resolved in the Cedar River whose average width was equivalent to the resolution of the TIR image of about 5 m, whereas determining water temperature was easier for both the Green River with widths equal to approximately three pixels, and the Yakima River with greater than three pixels. As described previously, narrow stream widths will result in the problem of mixed pixels along the stream. Generally, at least three pure-water pixels are needed in order for an accurate temperature extraction (assuming that there are no sub-pixel obstructions within the stream), and as the number of pixels across the stream decreases, the accuracies decrease and uncertainties increase (Handcock et al., 2006).

5.8 Example 2: Thermal heterogeneity in river floodplains used to assess habitat diversity

In this example, we expand our scope from looking at water in narrow stream and river reaches using multi-scale data, to show an example of thermal heterogeneity in the river floodplain as an indicator of habitat diversity. River floodplains are transitional areas that extend from the edge of permanent water bodies to the edge of uplands. In their natural state, they are among the most complex, dynamic, and diverse ecosystems globally, characterised by interacting flow, thermal, and sediment pulses that provide a complex 'template' to which organisms are adapted and by which ecosystem processes are controlled (Naiman et al., 2005; Stanford et al., 2005; Tockner et al., 2010). Although the changes in the composition and the configuration of habitat types have been well documented, little is known about thermal patch dynamics at the landscape scale (cf. Cardenas et al., 2008; Smikrud et al., 2008). Thermal patch dynamics are expected to control the distribution of aquatic and terrestrial organisms as well as of animals that exhibit complex life cycles (e.g., aquatic insects, amphibians). Furthermore, information on thermal heterogeneity is required to 'scale-up' ecosystem processes from the patch to the entire ecosystem.

In a recent study, Tonolla et al. (2010) applied ground-based TIR images to quantify surface temperature patterns at 12–15 minute intervals over 24 h cycles in near-natural Alpine river floodplains (Roseg, Tagliamento River; Figure 5.11). Each habitat type exhibited a distinct thermal signature creating a complex thermal mosaic. The diel temperature pulse and maximum daily temperature were the main thermal components that differentiated the various aquatic and terrestrial habitat types. In both river floodplains, exposed gravel sediments exhibited the highest diel pulse (up to 23 °C) while in aquatic habitats the pulse was as low as 11 °C. At the floodplain scale, thermal heterogeneity was low during night-time but strongly increased during day time, thereby creating a complex shifting mosaic of thermal patches (Figure 5.11). However, TIR images only record T_r at the water surface. Within the top 29 cm of the unsaturated gravel sediments, thermal heterogeneity was as high as across the entire floodplain at the surface (Tonolla et al., 2010). This strong vertical gradient should be considered when calculating temperature-dependent ecosystem processes.

This study emphasised that remotely sensed TIR images provide a unique opportunity to simultaneously map surface temperature of aquatic and terrestrial ecosystems at a high spatio-temporal resolution, a capability not possible using non-imaging ground-based methods (e.g., Arscott et al., 2001; Kaushal et al., 2010). However, a major challenge is to link thermal patch dynamics with ecological processes. Indermaur et al. (2009a, b), for example, demonstrated that home range placement of amphibians in the Tagliamento floodplain depends on the thermal properties of the individual habitat types (e.g., large wood deposits that provide thermal refugia), as well as on the spatial configuration of these habitats. Furthermore, there is clear evidence that the diel temperature pulses are ecologically more relevant than the average daily temperature. Microbial activity, for example, immediately reacts to short-term alterations in temperature leading to rapid alterations in ecosystem respiration when temperature changes. Therefore, ignoring local-scale and short-term thermal dynamics may lead to false conclusions about environmental change impacts on ecosystems. Concurrently, the effects of global warming can be attenuated by manipulating specific habitat characteristics and processes such as vegetation cover and the exchange between subsurface and surface water.

The use of airborne vehicles mounted with TIR imaging sensors can be extended to map the riparian areas, for example, mapping floodplains at different flow conditions and studying the distribution and density of terrestrial mammals such as deer or wild boar during flood events (e.g., Naugle et al., 1996). Furthermore, the TIR technique can be used in combination with other sensors such as LIDAR to quantify the three-dimensional heterogeneity of river floodplains.

5.9 Summary

In this chapter, we showed how TIR measurements can be used for observing water temperature in riverine landscapes for practical applications. We explored the theoretical basis of TIR observations of water temperature, data sources, the processing steps necessary to obtain accurate estimates of temperature in riverine environments from TIR data, and the validation of such temperature estimates. We also provided some multi-scale examples of the application of TIR data in riverine ecology and management. At the end of each section, and in Table 5.1, we have summarised some of the key points for managers using TIR data to monitor water temperature of stream and rivers. We hope that the

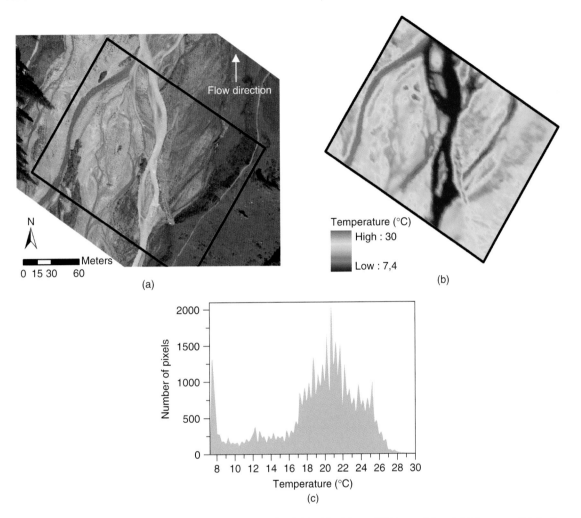

Figure 5.11 Geo-referenced natural-color image of the Roseg floodplain (a), airborne TIR image (b), and (c) frequency distribution radiant temperatures in TIR image (°C) at 15:00 (23 August 2004) (D. Tonolla, personal communication).

practical approach we took will be useful for the application of TIR remote sensing and can be successfully used in thermal monitoring of riverine landscapes.

Acknowledgements

Although the research described in the article has been funded in part by the United States Environmental Protection Agency's STAR program through grant R827675-01-0, it has not been subjected to EPA review, and does not necessarily reflect the views of the Agency, and no official endorsement should be inferred. Emissivity data reproduced from the MODIS Emissivity Library. This data set was collected by Dr. Zhengming Wan's Group at ICESS (Institute for Computational Earth System Science) located on the campus of UCSB (University of California, Santa Barbara). Any use of trade, product, or firm names is for descriptive purposes only and does not imply endorsement by the United States Government. We acknowledge the extensive contributions of M. Boyd (formerly with the Oregon Department of Environmental Quality, Portland, Oregon (USA), and currently with Watershed Sciences, Inc., Portland, Oregon, USA) for his development of river-temperature modeling techniques that utilised airborne TIR images for calibration and led to airborne TIR data collection throughout thousands of kilometers of rivers and streams in Oregon and the Pacific Northwest (USA). We also thank Diego Tonolla, and H. Piegay and two anonymous reviewers for constructive comments.

5.10 Table of abbreviations

ASTER	NASA EOS Advanced Spaceborne Thermal Emission and Reflection radiometer
DN	Digital Number
ε	Emissivity
EPA	The United States Environmental Protection Agency
FAA	Federal Aviation Administration (USA)
FOV	Field of view
FLIR	Forward looking infrared
GPS	Global positioning system
GSD	Ground sampling distance
IFOV	Instantaneous field of view
IMU	Inertial measurement unit
L_p	Atmospheric path spectral radiance
L_g	Spectral radiance emitted at the ground-surface
L_s	Spectral radiance measured at the sensor
Landsat ETM$^+$	Sensor on the Landsat-7 spacecraft
W	Spectral radiant emittance
MASTER	MODIS/ASTER airborne simulator
MODIS	NASA EOS sensor on the Terra and Aqua spacecrafts
NEΔT	Noise-equivalent temperature change
NIR	Near infrared part of the electromagnetic spectrum
ρ	Reflectivity
SST	Sea-surface temperature
SWIR	Shortwave Infrared
τ	Atmospheric transmissivity
T_k	Kinetic water temperature
T_r	Radiant water temperature
TIR	Thermal Infrared
TOA	Top of atmosphere
UAV	Unmanned Aerial Vehicle
VNIR	Visible and near infrared parts of the electromagnetic spectrum
λ	Wavelength

References

Adler-Golden, S.M., Matthew, M.W., Bernstein, L.S., Levine, R.Y., Berk, A., Richtsmeier, S.C., Acharya, P.K., Anderson, G.P., Felde, G., Gardner, J., Hike, M., Jeong, L.S., Pukall, B., Mello, J., Ratkowski, A., and Burke, H.-H. (1999). Atmospheric correction for shortwave spectral imagery based on MODTRAN4. *SPIE Proc. Imaging Spectrometry*, 3753, 61–69.

Anding, D. and Kauth, R. (1970). Estimation of Sea Surface Temperature from space. *Remote Sensing of Environment*, 1, 217–220.

Arrigoni, A.S., Poole, G.C., Mertes, L.A.K., O'Daniel, S.J., Woessner, W.W., and Thomas, S.A. (2008). Buffered, lagged, or cooled? Disentangling hyporheic influences on temperature cycles in stream channels. *Water Resources Research*, 44 (W09418).

Arscott, D.B., Tockner, K., and Ward, J.V. (2001). Thermal heterogeneity along a braided floodplain river (Tagliamento River, northeastern Italy). *Canadian Journal of Fisheries and Aquatic Sciences*, 58, 2359–2373.

Atwell, B.H., MacDonald, R.B., and Bartolucci, L.A. (1971). Thermal Mapping of Streams from airborne Radiometric Scanning. *Water Resources Bulletin*, 7 (2), 228–243.

Baird, O.E. and Krueger, C.C. (2003). Behavioral thermoregulation of brook and rainbow trout: Comparison of summer habitat use in an Adirondack River, New York. *Transactions of the American Fisheries Society*, 132, 1194–1206.

Baldridge, A.M., Hook, S.J., Grove, C.I., and G. Rivera, G. (2009). The ASTER Spectral Library Version 2.0. *Remote Sensing of Environment*, 113 (4), 711–715.

Barnes, W.L., Pagano, T.S., and Salomonson, V.V. (1998). Prelaunch characteristics of the Moderate Resolution Imaging Spectroradiometer (MODIS) on EOS-AM1. *IEEE Transactions on Geoscience and Remote Sensing*, 36 (4), 1088–1100.

Bartholomew, M.J., Kahle, A.B., and Hoover, G. (1989). Infrared spectroscopy (2, 3-20 µm) for the geological interpretation of remotely-sensed multispectral thermal infrared data. *International Journal of Remote Sensing*, 10 (3), 529–544.

Belknap, W. and Naiman, R.J. (1998). A GIS and TIR procedure to detect and map wall-base channels in western Washington. *Journal of Environmental Management*, 52, 147–160.

Berk, A., Bernstein, L.S., and Robertson, D.C. (1989). MODTRAN: a moderate resolution model for LOWTRAN7. GL-TR-90-0122. Technical Report. Geophysics Directorate, Phillips Laboratory, Hanscom, AFB, MA. 44 pp.

Berni, J.A.J., Zarco-Tejada, P.J., Suarez, L., Fereres, and E. (2009). Thermal and Narrow-band Multispectral Remote Sensing for Vegetation Monitoring from an Unmanned Aerial Vehicle. *IEEE Transactions on Geoscience and Remote Sensing*, 47, 722–738.

Beschta, R.L., Bilby, R.E., Brown, G.W., Holtby, L.B., and Hofstra, T.D. (1987). Stream temperature and aquatic habitat: Fisheries and forestry interactions. In Streamside management: Forestry and fishery interactions. Salo, E.O. and Cundy,

T.W. (eds). University of Washington, Institute of Forest Resources: Seattle, USA: 191–232.

Bolgrien, D.W. and Brooks, A.S. (1992). Analysis of thermal features of Lake-Michigan from AVHRR satellite images. *Journal of Great Lakes Research*, 18 (2), 259–266.

Boyd, M. and Kasper, B. (2003). Analytical methods for dynamic open channel heat and mass transfer: Methodology for Heat Source Model Version 7.0. http://www.deq.state.or.us/wq/TMDLs/tools.htm (Last accessed 25th July 2010)

Brown, G.W. and Krieger, J.T. (1970). Effects of clear-cutting on stream temperature. *Water Resources Research*, 6 (4), 1133–1139.

Burkholder, B.K., Grant, G.E., Haggerty, R., Khangaonkar, T., and Wampler, P.J. (2008). Influence of hyporheic flow and geomorphology on temperature of a large, gravel-bed river, Clackamas River, Oregon. *Hydrological Processes*, 22, 941–953.

Cardenas, M.B., Harvey, J.W., Packman, A.I., and Scott, D.T. (2008). Ground-based thermography of fluvial systems at low and high discharge reveals complex thermal heterogeneity driven by flow variation and bioroughness. *Hydrological Processes*, 22, 980–986.

Cardenas, M.B., Neale, C.M.U., Jaworowski, C., and Heasler, H. (2011). High-resolution mapping of river-hydrothermal water mixing: Yellowstone National Park. *International Journal of Remote Sensing*, 32, 2765–2777.

Chen, Y.D., Carsel, R.F., McCutcheon, S.C., and Nutter, W.L. (1998). Stream temperature simulation of forested riparian areas: I. watershed-scale model development. *Journal of Environmental Engineering*, 124 (4), 304–315.

Chen, C.Q., Shi, P., and Mao, Q.W. (2003). Application of remote sensing techniques for monitoring the thermal pollution of cooling-water discharge from nuclear power plant. *Journal of Environmental Science and Health Part A*, 38 (8), 1659–1668.

Cherkauer, K.A., Burges, S.J., Handcock, R.N., Kay, J.E., Kampf, S.K., and Gillespie, A.R. (2005). Assessing satellite-based thermal-infrared remote-sensing for monitoring Pacific Northwest river temperatures. *Journal of the American Water Resources Association*, October. Paper No. 03161.

Cristea, N.C. and Burges, S.J. (2009). Use of Thermal Infrared Imagery to Complement Monitoring and Modeling of Spatial Stream Temperatures. *Journal of Hydrologic Engineering*, Oct 2009, 1080:1090.

Cuenca, J. and Sobrino, J.A. (2004). Experimental measurements for studying angular and spectral variation of thermal infrared emissivity. *Applied Optics*, 43 (23), 4598–4602.

de Jong, S.M., and van der Meer, F.D. (Eds.) (2004). Remote sensing image analysis: including the spatial domain, Dordrecht: London, p. 359.

Dahle, S. (2009). Predicting the growth potential of a shallow, warm-water sport fishery: A spatially explicit bioenergetics approach. MSc Thesis, Utah State University, pp. 49.

Danielescu, S., MacQuarrie, K.T.B., and Faux, R.N. (2009). The integration of thermal infrared imaging, discharge measurements and numerical simulation to quantify the relative contributions of freshwater inflows to small estuaries in Atlantic Canada. *Hydrological Processes*, 23 (20), 2847–2859.

Dudhia, J. (1993). A non-hydrostatic version of the Penn State-NCAR mesoscale model: validation tests and simulation of an Atlantic cyclone and cold front. *Monthly Weather Review*, 121, 1493–1513.

Dunckel, A.E., Cardenas, M.B., Sawyer, A.H., and Bennett, P.C. (2009). High-resolution in-situ thermal imaging of microbial mats at El Tatio Geyser, Chile shows coupling between community color and temperature. *Geophysical Research Letters*, 36 (L23403), 5 pp, doi:10.1029/2009GL041366.

Ebersole, J.L., Liss, W.J., and Frissell, C.A. (2001). Relationship between stream temperature, thermal refugia and rainbow trout Oncorhynchus mykiss abundance in arid-land streams in the northwestern United States. *Ecology of Freshwater Fish*, 10, 1–10.

Emery, W.J. and Yu Y. (1997). Satellite sea surface temperature patterns. *International Journal of Remote Sensing*, 18 (2), 323–334.

Environmental Protection Agency (2003). EPA Region 10 guidance for Pacific Northwest state and tribal temperature water quality standards. Region 10 Office of Water, Seattle, WA. EPA 910-B-03-002. 57 pp.

Faux, R.N. and McIntosh, B.A. (2000). Stream temperature assessment. *Conservation in Practice*, 1, 38–39.

Faux, R.N., Lachowsky, H., Maus, P., Torgersen, C.E., and Boyd, M.S. (2001). New approaches for monitoring stream temperature: Airborne thermal infrared remote sensing. Remote Sensing Applications Laboratory, USDA Forest Service, Salt Lake City, Utah.

Gillespie, A.R. (1992). Spectral mixture analysis of multispectral thermal infrared images. *Remote Sensing of Environment*, 42, 137–145.

Gillespie, A.R., Kahle, A.B., and Palluconi, F.D. (1984). Mapping alluvial fans in Death Valley, California, using multichannel thermal infrared images. *Geophysical. Research. Letters*, 11 (11), 1153–1156.

Gillespie, A.R., Rokugawa, S., Matsunaga, T., Cothern, J.S., Hook, S., and Kahle, A.B. (1998). A temperature and emissivity separation algorithm for ASTER images. *IEEE Transactions on Geoscience and Remote Sensing*, 36 (4), 1113–1126.

Google Inc. (2010). Google Earth (Version 5.0), http://www.google.com/earth/index.html (Last accessed 25th July 2010).

Gu, Y., Zhang, Y., and Zhang, J. (2008). Integration of spatial-spectral information for resolution enhancement in hyperspectral images. IEEE *Transactions on Geoscience and Remote Sensing*, 46 (5), 1347–1358. DOI: 10.1109/TGRS.2008.917270.

Gustafson, W.T., Handcock, R.N., Gillespie, A.R., and Tonooka, H. (2003). Image sharpening method to recover stream temperatures from ASTER images. Proceedings SPIE International Society for Optical Engineering. *Remote Sensing for Environmental Monitoring*, 4886 (72).

Handcock, R.N., Cherkauer, K.A., Kay, J.E., Gillespie, A., Burges, S.J., and Booth, D.B. (2002). Spatial Variability in Radiant Stream Temperatures Estimated from Thermal Infrared Images. *Eos Transactions AGU*, 83 (47), Fall Meet. Suppl., Abstract H72 E-0896, 2002.

Handcock, R.N., Gillespie, A., Cherkauer, K.A., Kay, J.E., Burges, S.J., and Kampf, S.K. (2006). Accuracy and uncertainty of thermal-infrared remote sensing of stream temperatures at multiple spatial scales. *Remote Sensing of Environment*, 100, 427–440.

Hick, P.T., and Carlton, M.D.W. (1991). Practical applications of airborne multispectral scanner data for forest, agriculture, and environmental monitoring. Pages 60–61 in 24th International Symposium on Remote Sensing of Environment, Rio de Janeiro, Brazil.

High, B., Peery, C.A., and Bennett, D.H. (2006). Temporary staging of Columbia River summer steelhead in coolwater areas and its effect on migration rates. *Transactions of the American Fisheries Society*, 135, 519–528.

Hook, S.J., Myers, J.J., Thome, K.J., Fitzgerald, M., and Kahle, A.B. (2001). The MODIS/ASTER airborne simulator (MASTER) – a new instrument for earth science studies. *Remote Sensing of Environment*, 76 (1), 93–102.

Hook, S.J., Prata, F.J., Alley, R.E., Abtahi, A., Richards, R.C., Schladow, S.G., and Palmarsson, S.O. (2003). Retrieval of lake bulk and skin temperatures using Along-Track Scanning Radiometer (ATSR-2) data: A case study using Lake Tahoe, California. *Journal of Atmospheric and Oceanic Technology*, 20 (4), 534–548.

Indermaur, L., Gehring, M., Wehrle, W., Tockner, K., and Näf-Dänzer, B. (2009a). Behavior-based scale definitions for determining individual space use: requirement of two amphibians. *The American Naturalist*, 173, 60–71.

Indermaur, L., Winzeler, T., Schmidt, B.R., Tockner, K., and Schaub, M. (2009b). Differential resource selection within shared habitat types across spatial scales in sympatric toads. *Ecology*, 90, 3430–3444.

Ishiyama, T., Tsuchiya, K., and Sugihara, S. (1995). Influence of look-angle on the water-surface temperature observed with an IR radiometer, *Advances in Space Research*, 17 (1), 43–46.

Jessup, A.T., Zappa, C.J., Loewen, M.R., and Hesany, V. (1997). Infrared remote sensing of breaking waves. *Nature*, 385 (6611), 52–55.

Kay, J.E., Kampf, S.K., Handcock, R., Cherkauer, K., Gillespie, A.R., and Burges, S.J. (2005). Accuracy of lake and stream temperatures estimated from thermal infrared imagery. *Journal of the American Water Resources Association*, October. Paper No. 04102.

Kahle, A.B., Palluconi, F.D., Hook, S.J., Realmuto, V.J., and Bothwell, G. (1991). The Advanced Spaceborne Thermal Emission and Reflectance Radiometer (ASTER). *International Journal of Imaging Systems Technology*, 3, 144–156.

Kaushal, S.S., Likens, G.E., Jaworski, N.A., Pace, M.L., Sides, A.M., Seekell, D., Belt, K.T., Secor, D.H., and Wingate, R.L. (2010). Rising stream and river temperatures in the United States. Frontiers in Ecology and the Environment (e-View). doi:10.1890/090037.

Kilpatrick, K.A., Podesta, G.P., and Evans, R. (2001). Overview of the NOAA/NASA advanced very high resolution radiometer Pathfinder algorithm for sea surface temperature and associated matchup database. *Journal of Geophysical Research-Oceans*, 106 (C5), 9179–9197.

Kotchenova, S.Y., Vermote, E.F., Matarrese, R., and Klemm, Jr., F.J. (2006). Validation of a vector version of the 6S radiative transfer code for atmospheric correction of satellite data. Part I: Path Radiance. *Applied Optics*, 45, 6762–6774.

LeDrew, E.F. and Franklin, S.E. (1985). The use of thermal infrared imagery in surface current analysis of a small lake. *Photogrammetric Engineering and Remote Sensing*, 51, 565–573.

Lillesand, T.M., Kiefer, R.W., and Chipman, J.W. (2008). Remote sensing and image interpretation, 6th Ed.; John Wiley and Sons: Hoboken, USA, p. 768.

Loheide, S.P. and Gorelick, S.M. (2006). Quantifying stream-aquifer interactions through the analysis of remotely sensed thermographic profiles and *in situ* temperature histories. *Environmental Science and Technology*, 40, 3336–3341.

Madej, M.A., Currens, C., Ozaki, V., Yee, J., and Anderson, D.G. (2006). Assessing possible thermal rearing restrictions for juvenile coho salmon (Oncorhynchus kisutch) through thermal infrared imaging and in-stream monitoring, Redwood Creek, California. *Canadian Journal of Fisheries and Aquatic Sciences*, 63, 1384–1396.

Masuda, K., Takashima, T., and Takayama, Y. (1988). Emissivity of pure and sea waters for the model sea-surface in the infrared window regions. *Remote Sensing of Environment*, 24 (2), 313–329.

Matthews, K.R. and Berg, N.H. (1997). Rainbow trout responses to water temperature and dissolved oxygen stress in two southern California stream pools. *Journal of Fish Biology*, 50 (1), 50–67.

Mather, P.M. (2004). Computer Processing of Remotely Sensed Images (3rd Ed.). John Wiley, Chichester, U.K.

McCullough, D.A., Bartholow, J.M., Jager, H.I., Beschta, R.L., Cheslak, E.F., Deas, M.L., Ebersole, J.L., Foott, J.S., Johnson, S.L., Marine, K.R., Mesa, M.G., Petersen, J.H., Souchon, Y., Tiffan, K.F., and Wurtsbaugh, W.A. (2009). Research in thermal biology: Burning questions for coldwater stream fishes. *Reviews in Fisheries Science*, 17, 90–115.

Mertes, L.A.K., Dekker, A.G., Brakenridge, G.R., Birkett, C.M., and Letourneau, G. (2004). Rivers and lakes. In Remote sensing for natural resource management and environmental monitoring. Manual of remote sensing. Volume 4, Ustin, S.L. (ed.) John Wiley & Sons, Inc.: Hoboken, New Jersey, pp. 345–400.

MODIS Emissivity Library. (2010). http://www.icess.ucsb.edu/modis/EMIS/html/em.html (Last accessed 25th July 2010).

Naiman, R.J., Decamps, H., and McClain, ME. (2005). Riparia. Elsevier/Academic Press. San Diego, USA.

National Aeronautics and Space Administration, (1998). Landsat-7 science data users handbook. Greenbelt, Maryland, NASA Goddard Space Flight Center, electronic version, http://landsathandbook.gsfc.nasa.gov/handbook/handbook_toc.html (Last accessed 25th July 2010).

Naugle, D.E., Jenks, J.A., and Kernohan B.J. (1996). Use of thermal infrared sensing to estimate density of white-tailed deer. *Wildlife Society Bulletin*, 24, 37–43.

Nielsen, J.L., Lisle, T.E., and Ozaki, V.L. (1994). Thermally stratified pools and their use by steelhead in northern California streams. *Transactions of the American Fisheries Society*, 123, 613–626.

Oakton (2005). http://www.4oakton.com/ (last accessed 25th July 2010).

Ontar Corporation (1998). PcModWin v 3.7, Village Way, North Andover, MA, 01845, USA.

Oregon Department of Environmental Quality (2006). Willamette Basin total maximum daily load (TMDL). Appendix C: Temperature. Oregon Department of Environmental Quality, Portland, OR, USA.

Oregon Department of Environmental Quality (2010a) http://www.deq.state.or.us/wq/tmdls/granderonde.htm (last accessed 25th July 2010).

Oregon Department of Environmental Quality (2010b) http://www.deq.state.or.us/wq/tmdls/docs/granderondebasin/upgronde/appxa.pdf (Last accessed 25th July 2010).

Parkinson, C.L. (2003). Aqua: An earth-observing satellite mission to examine water and other climate variables. *IEEE Transactions on Geoscience and Remote Sensing*, 41 (2), 173–183.

Poole, G.C. and Berman, C.H. (2001). An ecological perspective on in-stream temperature: Natural heat dynamics and mechanisms of human-caused thermal degradation. *Environmental Management*, 27, 787–802.

Prakash, A. (2000). Thermal remote sensing: Concepts, issues and applications. XIXth ISPRS Congress, July 16–23, Amsterdam, The Netherlands.

Rayne, S. and Henderson, G.S. (2004). Airborne thermal infrared remote sensing of stream and riparian temperatures in the Nicola River watershed, British Columbia, Canada. *Journal of Environmental Hydrology*, 14, 1–11.

Robinson, I.S., Wells, N.C., and Charnock, H. (1984). The sea surface thermal boundary layer and its relevance to the measurement of sea surface temperature by airborne and spaceborne radiometers. *International Journal of Remote Sensing*, 5, 19–45.

Roganalski, A. and Chrzanowski, K. (2002). Infrared devices and techniques. *Opto-Electronics Review*, 10 (2), 111–136.

Salisbury, J.W. and D'Aria, D.M. (1992). Emissivity of terrestrial materials in the 8-14 μm atmospheric window. *Remote Sensing of Environment*, 42, 83–106.

Sentlinger, G.I., Hook, S.J., and Laval, B. (2008). Sub-pixel water temperature estimation from thermal-infrared imagery using vectorized lake features. *Remote Sensing of Environment*, 112, 1678–1688.

Smikrud, K.M., Prakash, A., and Nicholos, J. (2008). Decision-based fusion for improved fluvial landscape classification using digital aerial photographs and forward looking infrared images. *Photogrammetric Engineering and Remote Sensing*, 74, 903–911.

Stanford, J.A., Lorang, M.S., and Hauer, F.R. (2005). The shifting habitat mosaic of river ecosystems. *Verhandlungen der Internationalen Vereinigung für Theoretische und Angewandte Limnologie*, 29, 123–136.

Tockner, K. (2006). Using ecological indicators to evaluate rehabilitation projects. EAWAG News 61e, Swiss Federal Institute for Environmental Science and Technology, Duebendorf, Switzerland.

Tockner, K., Lorang, M.S., and Stanford, J.A. (2010). River flood plains are model ecosystems to test general hydrogeomorphic and ecological concepts. *River Research and Applications*, 26, 76–86.

Torgersen, C.E., Price, D.M., Li, and H.W., and McIntosh, B.A. (1999). Multiscale thermal refugia and stream habitat associations of chinook salmon in northeastern Oregon. *Ecological Applications*, 9, 301–319.

Torgersen, C.E., Faux, R.N., McIntosh, B.A., Poage, N.J., and Norton, D.J. (2001). Airborne thermal remote sensing for water temperature assessment in rivers and streams. *Remote Sensing of Environment*, 76, 386–398.

Tonolla, D., Acuna, V., Uehlinger, U., Frank, T., and Tockner, K. (2010). Thermal heterogeneity in river floodplains. *Ecosystems*, 13, 727–740.

Washington Department of Ecology (1998). Washington State Water Quality Assessment Section, Report 305(b). Olympia, WA. 56 p.

Wunderlich, W O. (1969). The fully-mixed stream temperature regime. ASCE Specialty Conference, Utah State University, Utah, August 20–23, 1969. 35 p.

Yamaguchi, Y., Kahle, A.B., Pniel, M., Tsu, H., and Kawakami, T. (1998). Overview of Advanced Spaceborne Thermal Emission and Reflection Radiometer (ASTER). *IEEE Transactions on Geoscience and Remote Sensing*, 36 (4), 1062–1071.

6

The Use of Radar Imagery in Riverine Flood Inundation Studies

Guy J-P. Schumann[1], Paul. D. Bates[1], Giuliano Di Baldassarre[2] and David C. Mason[3]

[1]School of Geographical Sciences, University of Bristol, UK
[2]Department of Hydroinformatics and Knowledge Management, UNESCO-IHE, Delft, The Netherlands
[3]Environmental Systems Science Centre, University of Reading, UK

6.1 Introduction

Flooding is a major hazard in both rural and urban areas worldwide, and has occurred regularly in the world in recent times. Between 1998 and 2004, Europe suffered over 100 major damaging floods, including the catastrophic floods along the Danube and Elbe rivers in summer 2002. Severe floods in 2005 further reinforced the need for concerted action. Since 1998 floods in Europe have caused some 700 deaths, the displacement of about half a million people and at least €25 billion in insured economic losses (EC, 2011). In the UK, there was extensive flooding due to extreme rainfall in the north and west of England in the summer of 2007 that caused a number of deaths and damage of over £3 billion. An increasingly urbanised global population and the interdependence of complex urban infrastructure means that our vulnerability to such events is increasing.

Within the more recent scientific literature, dating back ten years, it is recognised that remote sensing can support flood monitoring, modelling and management (Smith, 1997; Bates et al., 1997). In particular, satellites and aircraft carrying Synthetic Aperture Radar (SAR) sensors are valuable as radar wavelengths can penetrate cloud cover and obtain land cover information both night and day (Woodhouse, 2006).

However, given the strong inverse relationship between spatial resolution and revisit time for satellites, monitoring floods from space in near real-time or operationally is currently only possible through either low resolution (about 100 m ground pixel size) SAR imagery or satellite constellations. For instance, revisit times for SAR imagery of ~100m spatial resolution (usually termed wide swath mode) are in the order of three days and the data can be obtained within 24 h at low cost. Hence, this type of space-borne data can be used for monitoring major floods on medium-to-large rivers. For basin areas down to around 10 000 km² flood waves usually take several days to transit through the catchment river network and there is thus a reasonable chance of floodplain inundation coinciding with a satellite overpass. In smaller basins with shorter flood wave travel times the probability of imaging a flood decreases proportionately and acquisitions become increasingly opportunistic such that even

Fluvial Remote Sensing for Science and Management, First Edition. Edited by Patrice E. Carbonneau and Hervé Piégay.
© 2012 John Wiley & Sons, Ltd. Published 2012 by John Wiley & Sons, Ltd.

wide swath systems could not be relied on for operational monitoring. For finer resolution SAR systems the same issue occurs, but here revisit times can be up to 35 days and so one can only be guaranteed to capture flooding imagery in the very largest river basins such as the Amazon which have a monomodal annual flood pulse that lasts for several months. In the majority of river basins therefore the chances of imaging a flood with a high resolution SAR system becomes vanishingly small.

Moreover, for many applications river reaches tend to be studied at a much smaller scale however and therefore do actually require much finer spatial resolutions. Very fine resolution (<5 m) imagery and other types of remotely sensed data become a prerequisite when monitoring and modelling urban areas where most assets at risk of flooding are located and where the width of individual streets often determines the ability to model or monitor flood inundation patterns accurately. Here constellations of multiple fine resolution SAR systems present a possible solution. COSMO-SkyMed for instance can get a 3 m image sequence with a time from request to acquisition of the first image of 26–50 hours, then subsequent images at 12-hour intervals. With a system in place such as the ESA GPOD FAIRE system (http://gpod.eo.esa.int) that allows SAR images to be available to the user three hours after acquisition, rapid delivery of fine resolution image and information is technically feasible and might be a common form of dissemination in the near future.

This chapter aims to discuss the potential for, and uncertainties of, SAR imagery to monitor floods and support hydraulic modelling over a variety of different scales. The chapter first describes the basic principles of microwave remote sensing with a view to flood inundation studies. Then, the potential of SAR image data to monitor and map flood inundation is outlined and discussed. Finally, the support these data can offer to flood inundation modelling is reported via a number of rural and urban case studies. In all these sections, the uncertainty of the data is discussed and the need to move from deterministic binary information (e.g. wet/dry maps) to fuzzy data types (e.g. fusing different interpretations of a single image or multiple flood images) is highlighted.

6.2 Microwave imaging of water and flooded land surfaces

The variables which both scientists and practitioners involved with flood risk management would like to measure or estimate during a flood event, and hopefully over different spatial and temporal scales, might include discharge, flow velocity and direction, water volume and level, flooded area and flood edge. Remote sensing can provide information about most of these with varying degrees of accuracy, however discharge and flow velocities can only be obtained indirectly through integration with a hydrodynamic model or gauging networks. Also, water volume and in some cases water level estimation requires the use of a topographic data set.

Using the microwave region (1 mm–1 m or 300 GHz–300 MHz) of the electromagnetic spectrum has some advantages over the visible and infrared. Commonly used radar bands (Table 6.1) for monitoring flood inundation processes include L, C, X and Ka with C being the most widespread and L being generally preferred to map flooding beneath vegetation given its longer wavelength. The information that is obtained with microwave systems is complementary to that retrieved with systems used in other regions of the spectrum as microwave interactions are governed by different physical parameters. The amount of microwave energy of a particular wavelength scattered off an object or feature is affected by its size, shape, texture and water content (Woodhouse, 2006). Further advantages of microwaves are that they can penetrate clouds and measurements can be recorded at any time without relying on background illumination. The following sections outline the use of passive radiometry, active radar systems (SAR) and radar interferometry for

Table 6.1 Most commonly used parts of the microwave region of the electromagnetic spectrum (after IEEE).

Wavelength (cm)	Bands (Standard Radar Nomenclature)	Frequency Range (GHz)
30–100	P (UHF*)	0.3–1
15–30	L	1–2
8–15	S	2–4
4–8	C	4–8
2.5–4	X	8–12
1.7–2.5	Ku**	12–18
1.1–1.7	K	18–27
0.75–1.1	Ka**	27–40
0.4–0.75	V	40–75
0.27–0.4	W	75–110
<0.27	mm	110–300

*P band used to be called UHF (ultra high frequency) band
**Ku and Ka stand for K 'under' (from the German word 'unter') and K 'above', respectively

imaging of water and flooded land surfaces and retrieval of meaningful information about flood parameters.

6.2.1 Passive radiometry

Passive microwave radiometers measure emitted thermal radiation (i.e. brightness temperature) in the microwave region as opposed to active systems such as SAR that illuminate the target area and record the return signal (or backscatter). The use of passive microwave systems over land surfaces is difficult given the large angular beams of such systems (Rees, 2001) resulting in spatial resolutions as large as 20 to 100 km. Interpretation of the wide range of materials with many different emissivities (emissivity is defined as the efficiency of an object to radiate energy at a certain frequency compared to a blackbody which is a perfect absorber and emitter of all electromagnetic radiation) within the beam ground footprint is thus rendered nearly impossible.

Nevertheless, as such instruments are sensitive to changes in the dielectric constant, very large areas of water can be detected; water has a very high dielectric constant whereas most dry materials have a very low dielectric constant (see section 6.2.2). Both Sippel et al. (1998) and Jin et al. (1999) show that areal flood estimates can be obtained from changes in brightness temperature captured by passive radiometers. Sippel et al. (1998) used flood areas derived from brightness temperature changes in the Amazon River floodplain to reliably construct an area-stage relation ($R^2 = 0.87$) from altimetry- and field-based water stages. However, the uncertainty in flood area estimation can be large. Papa et al. (2006) highlight considerable disagreement in flood area estimation between passive radiometry, altimetry and climate research datasets in the order of 0.3 to $0.7 \times 10^6 \, km^2$. Another problem with area-stage relationships is that a large number of images showing different flood extents needs to be available to construct reliable and extrapolative functions (Smith et al., 1997).

6.2.2 Synthetic Aperture Radar

Although at the detailed level the operation of a synthetic aperture radar (SAR) system is quite complex, it suffices to say that SAR is an advanced form of a side-looking radar system (Figure 6.1) with a manageable size antenna that employs the Doppler shift principle to synthetically obtain finer spatial resolution than is possible with conventional beam-scanning mechanisms. While in motion, the system repeatedly illuminates a target scene with pulses of microwaves and coherently detects the

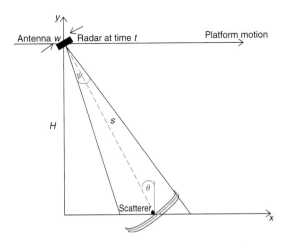

Figure 6.1 Simple geometry for considering SAR imaging. At time t, the radar is at (vt, H, 0) and a scatterer has the fixed coordinates (x, 0, 0). The figure illustrates a side looking geometry where the slant range (line-of-sight distance) s equals $H/\cos\theta$ and the radar beam of angular width ψ is given by λ/w. At different stages in t, while the sensor moves forward, the relative position of the scatterer will change and therefore produce a Doppler shift in the return signal. As both the amplitude and phase of the coherent signal are stored, any desired value of x can be resolved by extracting the component of the return signal that has the right variation of frequency with time (Rees, 2001).

waveforms received successively at the different antenna positions. These distinctive coherent signal variations are stored and then post-processed together to resolve fine resolution image elements.

Surface roughness, which determines the angular distribution of surface scattering, is considered the main factor affecting SAR backscattering, whereas the dielectric properties of an illuminated target significantly control the penetration depth and therefore the intensity of the signal. The average relative dielectric constant or relative permittivity describes the response of a material to the electric field of the signal (Ulaby et al., 1986, Raney, 1998). In the microwave domain, most natural dry materials have a low value of the dielectric constant (dimensionless, 3–8) whereas metals and materials with increased water content have much higher values, e.g. water has a value of approximately 80. In principle, increasing moisture content leads to a stronger reflectivity and an enhanced surface scattering. In contrast, higher permittivity increases diffuse volume scattering, which appears within dielectrically discontinuous media such as vegetation. For mapping purposes this means that, for pixels going through a wetting phase or experiencing a gradual increase in soil moisture, i.e. from a dry to an inundated state, SAR backscatter gradually increases until water ponds on the surface which, if covered by little or no vegetation, generally causes a sharp drop in the intensity of the return signal as a result of the signal being reflected away from the SAR antenna due to the specular reflection from a smooth water surface.

For mapping flood area and extent, optical imagery has been successfully used in numerous cases (see Marcus and Fonstad, 2008 for a detailed review). However, systematic application of visible or thermal satellite imagery to flood mapping is often hampered by persistent cloud cover during floods, particularly in small to medium-sized catchments where floods often recede before weather conditions improve. For thermal and visible image acquisition the only solution is to use recording systems on aircrafts flying below cloud cover. Although this requires extensive planning and is often regarded as a rather costly option, an effective system in place could consistently and very rapidly deliver high quality aerial photos or other types of imagery of a flood. For instance, an aircraft equipped with an infrared and visible camera as well as a LiDAR and hyperspectral imaging (CASI) system is operated by the Environment Agency of England and Wales (http://www.geomatics-group.co.uk) and has a call-out

time of only two hours, which is hardly possible to match with any satellite system.

Moreover, the inability to map flooding beneath vegetation canopies limits the applicability of these sensors. SAR, particularly in L-band, has been shown to be a promising tool for detection of flooding beneath vegetation canopy (Hess et al., 1990) as the double-bounce reflections between smooth water surfaces and vegetation trunks or branches create a bright appearance on the image.

In terms of microwave (radar) remote sensing SAR imagery seems to be at the moment the only reliable source of information for monitoring floods from space on rivers generally <1 km in width; for larger rivers, direct water level measurements with centimetric accuracy may be obtained from radar altimetry missions such as the Jason and Topex/Poseidon missions or the future NASA/CNES SWOT mission (for a detailed review see Alsdorf et al., 2007). Over the last few decades, spaceborne SAR systems (Figure 6.2) have increasingly been used for flood mapping. While past and current medium-resolution SAR satellite and space shuttle radar missions have a proven track-record for large-scale flood mapping in X- (SIR-C/X-SAR, SRTM), C- (ERS-1/2 AMI, ENVISAT ASAR, RADARSAT-1/2, SIR-C/XSAR),

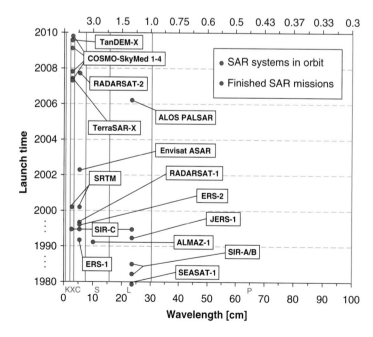

Figure 6.2 Launch of civil spaceborne SAR missions since 1978. Wavelengths, corresponding frequencies and commonly used waveband letters (in green) are also shown (Modified from Martinis, 2010 after Lillesand et al., 2004). For planned missions, see Table 6.5.

and L-band (SEASAT-1, JERS-1, ALOS PALSAR, SIR-A/B/C/X-SAR), their potential for deriving flood parameters in complex and small-scaled scenarios is clearly limited. Since 2007, the successful launch of polar orbiting platforms including TerraSAR-X, RADARSAT-2 (fine beam), a constellation of four COSMO-SkyMed satellites and more recently TanDEM-X marks a new generation of fine resolution SAR systems suitable for flood monitoring purposes. These satellites provide data up to 1 m pixel spacing, thus permitting information retrieval about detailed hydrological parameters from space in a near-real time mode. These recent efforts are evidence for the ever growing consensus among space agencies to strengthen support for such high resolution SAR imagery.

Before reporting notable case studies on flood monitoring with SAR imagery and integration with flood inundation models, the next section provides a brief outline of several studies that have employed SAR interferometry, which involves mapping change from coherent SAR image pairs, to investigate water level variations on river floodplains and wetlands over space and time.

6.2.3 SAR interferometry

Interferometry, of which Bamler et al. (1998) provide a thorough technical review, requires two SAR images from slightly different viewing geometries. Co-registration of the two images to a sub-pixel accuracy and subtraction of the complex phase (time delay) and amplitude (intensity) for each SAR image pixel allows changes in surface topography or displacements to be mapped. The value of the resulting interferometric phase at each pixel varies between $-\pi$ and $+\pi$ and is primarily a function of the distance between the radar antenna positions (or baseline) during acquisition, topographic relief, surface displacement, and the degree of correlation between the individual scattering elements that comprise each pixel location, i.e. coherence (Alsdorf et al., 2002). Moreover, it is worth noting that displacements are determined solely from interferometric phase (measured to half a wavelength) but require a reliable reference to a dataset where no motion has occurred (i.e. reference to 'permanent scatterers' (Perissin, 2006). Measurement of surface elevation is also controlled by baseline effects, which are very difficult to determine accurately. Thus, large uncertainties and inaccuracies may be introduced to interferometric SAR (or InSAR) derived digital elevation models

(DEMs), which can be in the order of (many) metres (see SRTM DEM validation studies by e.g. Farr et al. (2007), Rodriguez et al. (2005), Sun et al. (2003) and Bamler et al. (2003)). Furthermore, any unwanted variation in delay of the returned signal (e.g. caused by temporal and spatial variations in atmospheric water vapour content) can lead to significant height and displacement errors (Smith, 2002). For the global DEM produced from elevation data acquired during the Shuttle Radar Topography Mission (SRTM-DEM) at 3 arc second ($\sim$90 m) spatial resolution, average global height accuracies vary between 5 and 9 m (Farr et al., 2007). This vertical error has been shown to be correlated with topographic relief with large errors and data voids over high-relief terrain while in the low-relief sites errors are smaller but still affected by hilly terrain (Falorni et al., 2005).

Despite these limitations, topography has been mapped successfully with InSAR technology. For instance, an InSAR DSM of the UK was produced using repeat pass InSAR techniques applied to ERS-2 satellite data in the LandMap project (Muller, 2000). This has a height standard deviation of ±11m and a spatial resolution of 25 m.

The main airborne InSAR is the InterMap STAR-3i. This is a single-pass across-track X-band SAR interferometer on board a Learjet 36, with the two antennae separated by a baseline of 1 m. In the NextMap Britain project in 2002/3, an accurate high resolution DSM of the whole of Britain was built up containing over 8 billion elevation points (Mason et al., 2011). This meant that for the first time there was a national height database with height accuracies better than ±1m and spatial resolutions of 5 m (10 m) in urban(rural) areas (www.intermap.com). Using in-house software, Intermap is able to filter the digital surface model (DSM) to strip away features such as trees and buildings to generate a bare-earth digital terrain model (DTM).

For changes in water level retrieval with InSAR technology, the specular reflection of smooth open water that causes most of the return signal to be reflected away from the antenna and the roughening of the surface (by e.g. wind or wavelength properties) result in complete loss of temporal coherence between SAR images acquired at different times, rendering interferometric retrieval difficult if not impossible (Alsdorf et al., 2002). However, for inundated floodplains where there is emergent vegetation Alsdorf et al. (2000, 2001) show that it is possible to obtain reliable interferometric phase signatures of water stage changes (at centimetre scale) from the double bounced return signal of the repeat-pass L-HH-band Shuttle Imaging Radar (SIR-C). L-band penetrates the vegetation canopy and follows a double bounce path that includes the water and tree trunk surfaces, with both amplitude and phase coherence stronger than surrounding non-flooded terrain, permitting determination of the interferometric phase. Alsdorf (2003) also used these characteristics and found that decreases in water levels were correlated with increased flow-path distances between main channel and floodplain water bodies that could be modelled in a GIS. This correlation function allowed changes in water storage to be mapped over time.

Most of the studies on water level changes with InSAR technology successfully mapped relative changes; however it is often information on absolute water elevations that hydrologists and water resources managers need and so the static nature of InSAR observations reveals only limited information describing the dynamic nature of (wetland) water flow. In an attempt to overcome some of this limitation, Hong et al. (2010) have used highly coherent interferometric phases obtained with short time difference between two SAR acquisitions. Their technique transforms relative wetland InSAR observations to an absolute frame using calibration to gauge stations and generates both detailed maps of water levels and water level time series for 50 m pixels with an RMSE of 6–7 cm including 3–4 cm InSAR water level detection error. Such products could prove very useful to understand wetland surface flow patterns and allow efficient management of wetlands (Hong et al., 2010).

6.3 The use of SAR imagery to map and monitor river flooding

As a response to the summer 2007 floods, the UK Government set up the Pitt Commission to consider what lessons could be learned from those flood events (Pitt, 2008) Among its many recommendations, the Commission highlighted the need to have real-time or near real-time flood visualisation tools available to enable emergency responders to react to and manage fast-moving events, and to target their limited resources at the highest-priority areas. It was felt that a simple GIS that could be effectively updated with timings, level and extent of flooding during a flood event would be a useful system to keep the emergency services informed. In a similar context but outside near real-time management, the European Floods Directive (2007/60/EC, European Commission, 2007) aims to reduce and manage the risks that floods

pose to human lives, the environment and economic activity. The Directive requires Member States to first carry out a preliminary assessment by 2011 to identify the river basins and associated coastal areas at risk of flooding. For such zones they would then need to draw up flood risk maps by 2013 and establish flood risk management plans focused on prevention, protection and preparedness by 2015.

In addition, a near real-time flood extent could be used in conjunction with a hydraulic model of river flood flow to help predict future flood extent. The flood waterline from the image could be intersected with a LiDAR DEM to obtain water surface elevations along the waterline, and these could be assimilated into the model run, correcting the water surface elevations predicted by the model where necessary (see for example Neal et al., 2009). This would help to keep the model 'on track' so that the model's prediction of future flood extent could be viewed with more confidence. A near real-time flood detection algorithm, developed on a blueprint of those discussed in Section 6.3.1.1, giving a synoptic overview of the extent of flooding in both urban and rural areas, and capable of working during night-time and day-time even if cloud was present, could thus be a useful tool for operational flood relief management and flood forecasting.

6.3.1 Mapping river flood inundation from space

Before engaging with any process involving satellite remote sensing and flood management it is crucial to consider end user requirements and the appropriate timeline as well as the spatial resolution of the delivered products. Figure 6.3 shows the requirement in terms of spatial resolution and turnaround time for specific flood management deliverables. For instance, flood mapping for emergency management can be done at any spatial resolution really but should be made available to the end user within 48 hours whereas for insurance assessment quite the opposite situation might apply; spatial resolutions finer than 10 m are generally required but timeliness might be less important.

6.3.1.1 Operational flood detection

Near-real time flood detection is of course desirable for many obvious reasons and research efforts are continuously invested in developing algorithms that operate in near-real time (e.g. the ESA GPOD FAIRE system). By placing data analysis and decision making capabilities onboard a spacecraft, the time it takes to detect and react

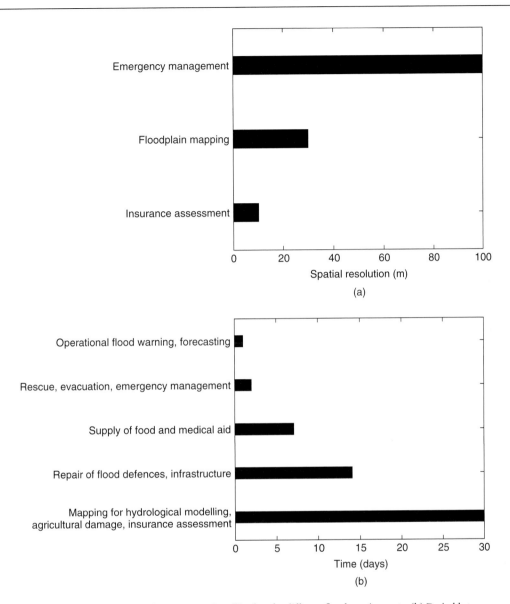

Figure 6.3 (a) Maximum acceptable resolution range of satellite data for different flood requirements. (b) Desirable turnaround time for acquisition, processing and dissemination of satellite data for flood requirements (Modified from Blyth, 1997).

to a flood event would be reduced to a few hours (Ip et al., 2006). The objective of such an experiment would be to demonstrate the ability of an autonomously controlled spacecraft to detect and react to dynamic events such as floods while optimising use of limited downlink bandwidth and maximising science return. However, to date, algorithms that enable an automatic delineation of flooded areas have not been developed to a level where they could be used operationally; although it is undeniable

that they are an essential component of any SAR based monitoring service (Matgen, 2011). What is needed at this stage is the development of new concepts for an efficient and standardised SAR based monitoring of floods. In this context, Matgen et al. (2011) propose a hybrid methodology, which combines radiometric thresholding and region growing as an approach enabling the automatic, objective and reliable flood extent extraction from SAR images. First results indicate that the proposed method may outperform manual approaches if no training data are available even if the parameters associated with these methods are determined in a non-optimal way. The results further demonstrate the algorithm's potential for accurately processing data from different SAR sensors.

Towards a similar aim, Pulvirenti et al. (2011a) propose a SAR image processing algorithm which maps flood areas and is to be inserted in the operational flood management system of the Italian Civil Protection authorities. The algorithm can be used in an almost automatic mode or in an interactive mode, depending on the user's needs. Using also simple hydraulic and contextual information, the approach is primarily based on fuzzy logic and uses three different electromagnetic scattering models to retrieve

SAR backscatter information allowing the user to also deal with more complex situations such as emergent flooded vegetation. The algorithm is designed to work with radar data at L, C, and X frequency bands and also employs ancillary data, such as a land cover map and a digital elevation model. The flood mapping procedure was tested on a flood that occurred in Albania on January 2010 using multi-temporal COSMO-SkyMed very high resolution X-band SAR data (Figure 6.4).

For fully automated unsupervised flood detection in large very fine resolution single polarised SAR data sets (~1–5 m, such as TerraSAR-X scenes), Martinis et al. (2009, see also Martinis et al., 2011) propose a split based automatic thresholding and classification

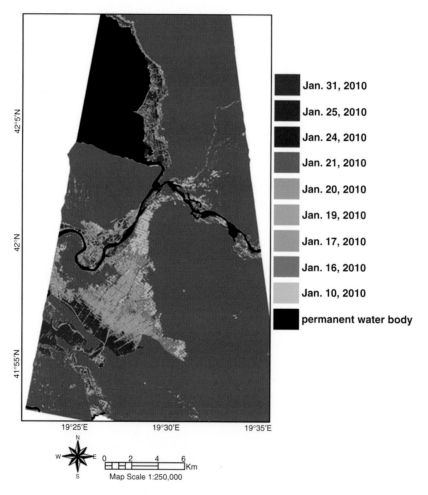

Figure 6.4 Multi-temporal map showing the time evolution of the Shkodër flood in Albania. It has been produced by applying the algorithm proposed by Pulvirenti et al. (2011a) to nine COSMO-SkyMed images with acquisition dates reported in the legend. White: no data; gray: non-flooded. Reproduced from Pulvirenti et al. (2011a) An algorithm for operational flood mapping from Synthetic Aperture Radar (SAR) data using fuzzy logic. Nat. Hazards Earth Syst. Sci., 11, 529–540.

refinement process. The approach has already been successfully applied in other operational rapid mapping activities, such as the 'Rapid Mapping' service at DLR's centre for satellite based crisis information (ZKI, http://www.zki.dlr.de). Furthermore, high resolution topographic information may be successfully combined with this multi-scale segmentation to enhance the mapping performance in areas of flooded vegetation and man-made objects as well as to remove misclassified non-water areas – however, this is largely restricted to rural areas (see Section 6.4.2 for near real-time detection in urban areas).

6.3.1.2 Large area mapping with low resolution SAR

As noted earlier, given the strong inverse relationship between spatial resolution and revisit time, routine monitoring of floods from space in near-real time seems currently only possible through low resolution (about 100 m pixel size) SAR imagery (Di Baldassarre et al., 2011) or satellite constellations. The fact that coarser resolution SAR data can be used successfully to delineate flood area and edges on larger floodplain inundations (inundation widths >500 m) has been demonstrated, for example, by Blyth (1997); Kussul et al. (2008) within a grid computing system, and many times by the International Disaster Charter (http://www.disasterscharter.org/). A large number of research studies have recently demonstrated the usefulness of coarse resolution flood images for supporting flood modelling and management (Table 6.2).

6.3.1.3 Classification accuracies of flood parameters

Many SAR image-processing techniques exist to more or less successfully derive a flood area or extent (e.g. Aplin et al., 1999), including simple visual interpretation (Macintosh, 1995; Oberstadler et al., 1997; Brivio et al., 2002), image histogram thresholding (e.g. Brivio et al., 2002; Matgen et al., 2004; Schumann et al., 2005), automatic classification algorithms (e.g. Hess et al., 1995; Bonn et al., 2005), image texture algorithms (Schumann et al., 2005), multi-temporal change detection methods (e.g. Calabresi et al., 1995; Laugier et al., 1997), of which extensive reviews are provided in Liu et al. (2004) and Lu et al. (2004). Complex auto-logistic regression

Table 6.2 Recent research on the use of low resolution SAR imagery to support flood studies.

Aim of Study	Flood Parameters Extracted	Extraction Method Used	Location	Source
Near-real time verification of a 1D hydrodynamic model	Flood area	Histogram threshold method	River Po, Italy	Di Baldassarre et al., 2009a
Near-real time verification of a 1D hydrodynamic model	Flood area and flood edge heights Water surface profiles	Histogram threshold method fused with SRTM topography	River Po, Italy	Schumann et al., 2010a
Event-specific flood risk mapping	Flood area	Multiple methods listed in Table 3	River Dee, NE Wales, UK	Schumann and Di Baldassarre, 2010
Calibration of a 1D hydrodynamic model	Flood edge heights	Automated region growing fused with SRTM topography	River Severn, SW UK	Schumann et al., 2010b
Selecting an appropriate hydrodynamic model structure	Flood area	Histogram threshold method	River Po, Italy	Prestininzi et al., 2011
Operational automated flood detection	Flood area	Automated histogram threshold method combined with region growing algorithm	River Severn, SW UK Red River basin, US and Canada	Matgen et al., 2011

(Atkinson, 2000) and principal component analysis (Matgen et al., 2006) may also be applied. Image statistics-based active contour models have been used by Bates et al. (1997), Horritt et al. (1999), de Roo et al. (1999), Horritt et al. (2001) and Schumann (2005).

Classification accuracies of flooded areas (most of the time defined as a ratio of the total area of interest where classification errors are omitted) vary considerably and only in rare cases do classification accuracies exceed 90%. Table 6.3 summarises strengths and limitations of the most widely used image processing methods, most of which are available in any standard commercial remote sensing software package. From this list it is apparent that no single method can be considered appropriate for all images, nor do all methods perform equally well for a particular type of image or flood event (Schumann et al., 2009a).

Apart from flood area and extent, if an accurate and high enough resolution digital elevation model (DEM, generated from e.g. LiDAR) is available and provided a flood boundary can be adequately extracted from SAR imagery, it is possible to also map water levels and flood depths for a given event. Various techniques for this are listed in Table 6.4 and as early as the 1980s there have been several attempts to successfully derive water stages or heights from remote sensing datasets. In general, accuracies of the resulting water stages or heights increase with the complexity of the method and the resolution of the datasets used.

6.3.1.4 Including error and uncertainty analysis

In recent years, there has been an increasing interest in environmental science to assess uncertainty in observations and models (Beven, 2006). Error and uncertainty analysis is a process which should be inherent in all studies involving remotely sensed data as these are prone to much uncertainty, the sources and magnitudes of which are relatively easy to identify and quantify. According to Atkinson and Foody (2002), characterising the sources of uncertainty and improving the uncertainty information can lead to a reduction of uncertainty in remote sensing products which in turn increases not only accuracy but also credibility. Yet, in the case of SAR remote sensing and flood mapping, the observed flood area or flood edge data used are commonly treated as deterministic when in reality these data are subject to considerable uncertainty.

In one of the first attempts to deal with uncertainty in flood mapping from SAR, Schumann et al. (2009a) demonstrate that fusing a large number of deterministic binary wet/dry maps of flood area extracted from two simultaneous SAR acquisitions of a flood on the River Dee (NE Wales, UK) using five widely used image processing algorithms leads to a multi-algorithm ensemble map that contains more meaningful information than any of the single binary wet/dry maps alone. The authors then went on to show the value of this map for calibration of a flood inundation model. More details on this aspect are given later in Section 6.4.1.

In a similar context, Schumann et al. (2010a) used uncertainty information in SAR derived flood edges as well as in coarse resolution topographic heights from the NASA SRTM (Shuttle Radar Topography Mission) DEM to generate water surface gradients for large rivers reliable enough to reject erroneous flood model simulations or even to distinguish between flood models of different complexities thereby helping select the most appropriate model structure for a given flood event (Prestininzi et al., 2011).

A general prerequisite to estimate and eventually reduce uncertainties or use them in a meaningful way is to first identify and understand the sources. Therefore, the next section describes the most common flood detection error sources in more detail.

6.3.2 Sources of flood and water detection errors

As stated previously, surface roughness is the main factor affecting radar backscattering whereas the dielectric properties control the intensity of the signal. However, there are a relatively large number of other factors affecting image 'quality' as alluded to in the next paragraph.

Image classification or interpretation errors (i.e. dry areas mapped as flooded and vice versa) may arise from a variety of sources and adversely affect algorithm performance as illustrated in Table 6.4. Error sources may include inappropriate image processing algorithm, altered backscatter characteristics, unsuitable wavelength and/or polarisations, unsuccessful multiplicative noise (i.e. speckle) filtering, remaining geometric distortions, and inaccurate image geo-coding. Horritt et al. (2001) state that wind roughening and the effects of protruding vegetation, both of which may produce significant pulse returns, complicate the imaging of the water surface. Moreover, due to the corner reflection principle (i.e. where the structure of rectangular surfaces, e.g. buildings, is such that the wave is returned to the SAR antenna and thus causes complete sensor saturation resulting in white image pixels (Rees, 2001)) in combination with the relatively coarse resolution of many SAR systems means

Table 6.3 Advantages and disadvantages of commonly used image processing techniques to obtain flood area from SAR images (After Di Baldassarre et al., 2011).

Image Processing Method	Strength	Limitation	Level of Complexity	Computational Efficiency	Level of Automation	Consist-ency*
Visual interpretation	Easy to perform in case of a skilled and experienced operator with knowledge of flood processes	Very subjective; Difficult to implement over many images; May be difficult for images that show complex flood paths;	Low to high (may have varying degrees of complex-ity)	Relatively low	Hardly possible	>0.9
Histogram thresholding	Easy and quick to apply; Objective method;	No flexibility; Optimized threshold might not be the most appropriate; Works only well if image is relatively little distorted	Very low	Very high	Full	0.8
Texture based	Takes account of the SAR textural variation; Based on statistics; Mimics human interpretation as it takes account of tonal differences;	Difficult to choose correct window size and appropriate texture measure; After application still requires threshold value to obtain flood area classification	Moderate	Moderate	Full	0.6**
Active contour modelling/ Region growing	Image statistics based; Usually provides good classification results; Easy to define seed region (e.g. on the river channel); If integrated with land elevation constraints (see Mason et al., 2007) results are improved by mimicking inundation processes	Requires several parameters to fine-tune; Slow on large image domains; Difficult to choose correct tolerance criterion; May miss separated patches of dry or flooded land (particularly in case of active contour modelling)	Moderate to high	Moderate (depending very much on domain size)	Relatively high	0.7

*Refers to consistency of binary classification between different SAR images, after Schumann et al. (2009a);
**Average of different texture measures

Table 6.4 Reported water level retrieval techniques based on remotely sensed flood extent and DEM fusion and their accuracies (After Di Baldassarre et al., 2011).

Method	Error (RMSE)	Validation Data	Source
Landsat TM-derived flood extent superimposed on topographic contours for volume estimation	±21 percent	Field data	Gupta and Banerji (1985)
ERS SAR flood extent overlain on topographic contours	0.5 m–2 m	Field data	Oberstadler et al. (1997)
ERS SAR flood extent overlain on topographic contours	up to 2 m	Model outputs	Brakenridge et al. (1998)
Inter-tidal area water from multiple ERS images superimposed onto simulated water heights	mean error of 0.2–0.3 m	Field data	Mason et al. (2001)
Flooded vegetation maps from combined airborne Land C-SAR integrated with LiDAR vegetation height map	around 0.1 m	Field data	Horritt et al. (2003)
Integration of high-resolution elevation data with event wrack lines	<0.2 m	Model outputs	Lane et al. (2003)
Fusion of RADARSAT-1 SAR flood edges with LiDAR	correlation coefficient of 0.9	TELEMAC-2D model outputs	Mason et al. (2003)
Complex fusion of flood aerial photography and field based water stages from various floodplain structures	0.23 m	Mean between maximum and minimum estimation	Raclot (2006)
Fusion of ENVISAT ASAR flood edges with LiDAR and interpolation modelling	0.4–0.7 m	Field data	Matgen et al. (2007)
Fusion of ENVISAT ASAR flood edges with LiDAR and regression modelling	<0.2 m	Field data	Schumann et al. (2007)
Fusion of ENVISAT ASAR flood edges with LiDAR/topographic contours/SRTM and regression modelling	<0.35 m/0.7 m/1.07 m	1D model outputs	Schumann et al. (2008)
Fusion of hydraulically sensitive flood zones from ENVISAT ASAR imagery and LiDAR	0.3 m; 0.5 m uncertainty	Mean between maximum and minimum estimation	Hostache et al. (2009)
Fusion of ERS SAR flood edges from active contour modelling with LiDAR	mean error of up to 0.5 m	Aerial photography	Mason et al. (2009)
Fusion of TerraSAR-X flood edge with LiDAR (rural area)	error of 1.17 m (note that relative changes in levels from TerraSAR-X and aerial photography were mapped with an accuracy of 0.35 m compared to gauge data)	One field gauge	Zwenzner and Voigt (2009)
Fusion of uncertain ENVISAT ASAR WSM flood edges with SRTM heights	error in median estimate of 0.8 m	LiDAR derived water levels	Schumann et al. (2010a)

that to date it has been impossible to extract flooding information for urban areas, which for obvious reasons would be desirable when using remote sensing for flood management. The magnitude of the deteriorating effects is a function of wavelength, radar look angle and polarisation (i.e. the direction of oscillations of electromagnetic waves). Henry et al. (2006) compare different polarisations (VV, HV and HH) for flood mapping purposes and conclude that HH is most efficient in distinguishing flooded areas. De Roo et al. (1999) identify the geometric correction and geo-coding (or ortho-rectification) of a SAR image as the most difficult and time-consuming step in the entire image processing chain. It is worth noting here that although this might be true for most historic imagery, geo-positioning of a SAR image can now be achieved at a subpixel accuracy (∼1 m) for very fine spatial resolution using knowledge of the orbit and atmospheric delay modelling (Breit et al., 2010).

In summary, the major sources of image distortions which are largely responsible for difficulties and errors in detecting flood areas in SAR imagery may be classified as either system-related distortions or distortions related to the characteristics of the flooded land surface. System-related distortions are caused by a combination of (i) terrain geometry effects such as layover, foreshortening and shadowing (Figure 6.5), (ii) speckle, which

is an interference effect resulting in noisy images; and (iii) inappropriate wavelength polarisations. Layover and shadow effects are irreversible, whereas foreshortening can be rectified with topographic data sets. The deteriorating effect of speckle can be attenuated using different methods such as multi-look processing or filtering in the spatial, frequency, and time domains; however this is at the cost of spatial resolution.

Successful flood detection may also be hindered by open water surfaces that are roughened by waves, moderate to strong winds or intense rainfall. Flooded vegetation is another feature that can hamper interpretation and detection of flood area, however, the use of L- instead of the more widely used C-band can partially overcome this problem due to the longer wavelengths penetrating the vegetation canopy and reaching the flooded surface (see e.g. Hess et al., 1990). Figure 6.6a illustrates most of the significant interactions between the radar signal and the wet and dry land.

Perhaps the most important limitation of SAR imagery with respect to flood inundation studies is inside urban areas where layover and shadow areas persist in flooded streets between adjacent buildings, as well as specular reflection from water surfaces and double bounce effects between roads and buildings in flooded and non-flooded conditions (Figure 6.6b, Mason et al. 2010).

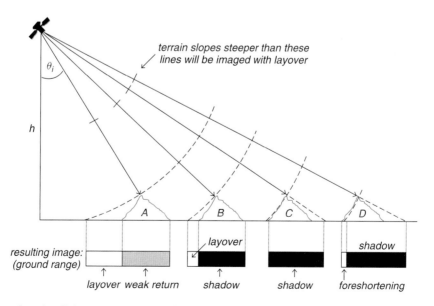

Figure 6.5 Effects of terrain relief on SAR images (Modified from Lillesand et al. (2004)).

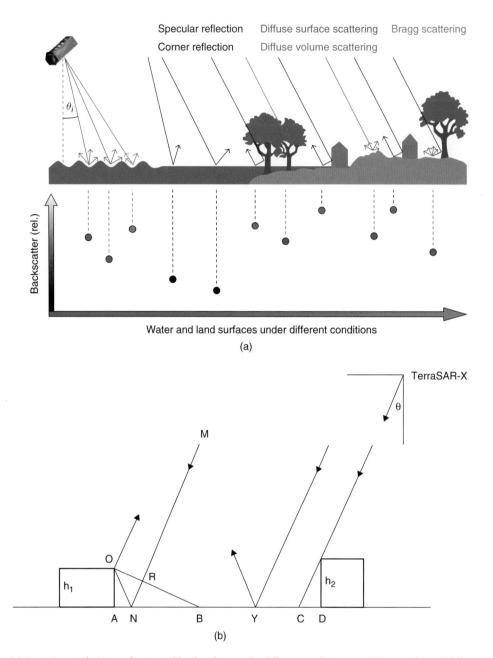

Figure 6.6 (a) Scattering mechanisms of water and land surfaces under different conditions as well as specular and diffuse components of surface scattered radiation as a function of incidence angle and surface roughness (Martinis, 2010). (b) (AB) Layover and (CD) shadow regions in a (AD) flooded street between adjacent buildings of height h1 and h2 (θ = incidence angle) (Mason et al., 2010).

6.3.3 Integration with flood inundation modelling

The integration of remotely sensed flood parameters with flood inundation models requires a profound understanding of the many factors underlying both the remote sensing and the flood modelling part. For this reason, relatively few studies have looked at this complex interplay. This chapter has so far illustrated that remote sensing can provide information on both flood extent and area as well as water level data which can be used in model calibration and evaluation, and in building and understanding model structures.

Obtaining accurate flood extents and centimetre-scale accuracies in water level estimation from remote sensing that are within expected accuracies of model predictions enables the modeler to evaluate and also improve uncertain flood inundation predictions (Schumann et al., 2009b). Acknowledging and examining the extent of uncertainty in both data and model should be generally accepted as a key element in flood risk management exercises (Pappenberger and Beven, 2006), especially when integrating uncertain spatially distributed observation data such as those derived from SAR remote sensing with uncertain model structures. This integration can be performed indirectly, through employing more traditional model-data comparison techniques (see e.g. Pappenberger et al., 2007 and Hunter et al., 2005), or directly via assimilation into flood models (see e.g. Neal et al., 2009; Matgen et al., 2010 and Giustarini et al., 2011).

Combining satellite data with hydrodynamic modelling has now become established as a powerful approach, the robustness of which needs, however, to be examined further, in particular for flood forecasting. Schumann et al. (2009b), who provide a detailed review on recent progress in integration of remote sensing-derived flood parameters and hydraulic models, argue that what is certainly to be gained from this development is that fundamental research issues in terms of both model evaluation and remote sensing data processing techniques will be addressed in one way or another. In a very similar sense, Cazenave et al. (2004) argue that scientists have much to gain from current and future satellite observations and missions to provide (global) hydrological data sets that could be used to evaluate process models. Moreover, it is expected that satellite measurements combined with models that allow direct integration of such data would

provide the basis for assimilation of remotely sensed data (Alsdorf et al., 2005), for instance, in operational flood forecasting systems.

In order to outline more clearly the techniques applied and values associated with integrating SAR-derived flood parameters with flood inundation models, the next section describes by means of illustrative case studies how SAR image data have recently been employed to support flood modelling.

6.4 Case study examples

The following three case studies are taken from the recent scientific literature and demonstrate the value of low and high resolution SAR imagery to support flood inundation modelling. The case studies are complementary to the many elements discussed in this chapter and are intended to give the reader an appreciation of recent research activities in flood mapping and uncertainty, flood mapping specifically applied to a rural-urban interface, and flood mapping integrated with hydraulic modelling. More specifically, the first case describes a rural application where uncertainty in image data was used to improve flood model parameter identifiability and also to map associated flood risk. The second case reports a first application of a very high resolution TerraSAR-X image inside an urban setting, and finally, the third example outlines the use of multi-temporal SAR images to map floodplain wetting and drying patterns to help understand hydrodynamic model limitations.

6.4.1 Fuzziness in SAR flood detection to increase confidence in flood model simulations

As noted earlier in this chapter, SAR-derived inundation maps are typically treated as deterministic (i.e. binary wet/dry classification) maps. However, these flood extent maps are unavoidably affected by many sources of inaccuracies. This example refers to a rural case study, a river reach of the Lower Dee (UK), and aims at discussing the uncertainties in SAR-derived flood area maps and how these might be used to improve flood model calibration. This test site is of particular interest because, during a flood in December 2006, both coarse resolution (ENVISAT ASAR WSM) and medium resolution (ERS-2 SAR) satellite imagery were acquired simultaneously

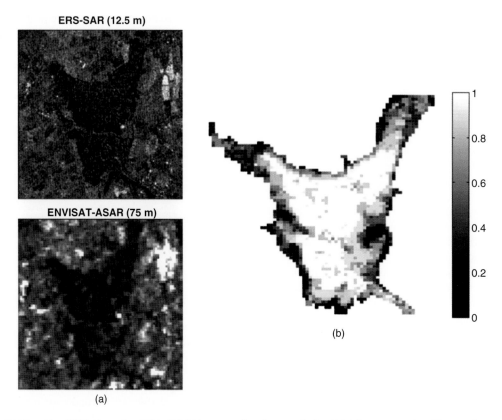

Figure 6.7 River Dee, UK in December 2006: A) SAR imagery of medium resolution (ERS-SAR, top row) and low resolution (ENVISAT-ASAR, lower row); B) uncertain flood inundation map generated by combining ten flood extent maps derived from the two imagery applying five different image processing techniques (Schumann et al., 2009a).

(Figure 6.7). This is quite a unique set of SAR images as in terms of the inundation process, the true flood extent at both acquisitions was the same. Given that all other significant acquisition parameters (e.g. frequency, polarisation and incidence angle) were the same for both images all apparent difference and therefore uncertainty in flood extent mapping was attributed to differences in spatial resolution and flood area extraction algorithms (Schumann and Di Baldassarre, 2010). Schumann et al. (2009a) processed these two images to derive a variety of flood extent maps by using five different image processing procedures: visual interpretation, histogram threshold, active contour modelling, image texture variance and Euclidean distance. This resulted in ten different flood extent maps, for which significant differences were observed. Fusing these different flood maps using equal weighting, Schumann et al. (2009a) produced a fuzzy map (Figure 6.7) that expresses for each pixel the possibility

of being inundated according to the multi-algorithm ensemble generated.

An investigation into the value of such a map for flood model evaluation revealed that accounting for the uncertainty in extracting flooded area increases the information content and leads to a more identifiable flood model parameter set than any of the conventional binary flood maps. Di Baldassarre et al. (2009b) presented a technique to produce an uncertain flood inundation map from LISFLOOD-FP (Bates and De Roo, 2000) model simulations conditioned on the fuzzy SAR flood map. This calibration exercise clearly demonstrated the necessity to move from traditional, deterministic binary (wet/dry) maps to fuzzy or probabilistic maps of flood extent. Furthermore, it was argued that accounting for uncertainties in flooded area observations considerably increases confidence in flood model results and ultimately flood inundation predictions.

In another study, Schumann and Di Baldassarre (2010) illustrated how this fuzzy flood map may be a useful tool for event-specific flood risk mapping. In this case, the possibility of inundation of each SAR pixel may be expressed as a flood hazard level and combined with a pixel based vulnerability index that is associated with a given land cover or land use class.

6.4.2 Near real-time flood detection in urban and rural areas using high resolution space-borne SAR images

The vast majority of a flooded area may be rural rather than urban, but it is very important to detect the urban flooding because of the increased risks and costs associated with it. Flood extent can be detected in rural floods using SARs such as ERS and ASAR, but these have too low a resolution (25 m) to detect flooded streets in urban areas. However, a number of SARs with spatial resolutions as fine as 1 m have recently been launched that are potentially capable of detecting urban flooding. They include TerraSAR-X, RADARSAT-2, and the four COSMO-SkyMed satellites.

As outlined in Section 6.3.1.1, an automatic near real-time flood detection algorithm using single-polarisation TerraSAR-X data has been implemented by Martinis et al. (2009, 2011). This searches for water as regions of low SAR backscatter using a region-growing iterated segmentation/classification approach, and is very effective at detecting rural floods, but would require modification to work in urban areas containing radar shadow and layover.

A semi-automatic algorithm for the detection of flood-water in urban areas using TerraSAR-X has also been developed by Mason et al. (2010). It uses a SAR simulator (Speck et al., 2007) in conjunction with LiDAR data to estimate regions of the image in which water would not be visible due to shadow or layover caused by buildings and taller vegetation. Ground will be in radar shadow if it is hidden from the radar by an adjacent intervening building. The shadowed area will appear dark, and may be misclassified as water even if it is dry. In contrast, an area of flooded ground in front of the wall of a building viewed in the range direction may be allocated to the same range bin as the wall, causing layover which generally results in a strong return, and a possible misclassification of flooded ground as un-flooded. The algorithm is aimed at detecting flood extents for validating an urban flood inundation model in an offline situation, and requires user interaction at a number of stages.

The objective of the work by Mason et al. (2012, in press) outlined here was to build on a number of aspects of the existing algorithms to develop an automatic near real-time algorithm for flood detection in urban and rural areas. The algorithm assumes that high resolution LiDAR data are available for at least the urban regions in the scene, so that a SAR simulator may be run in conjunction with the LiDAR data to generate maps of radar shadow and layover in urban areas. It is therefore limited to urban regions of the globe that have been mapped using LiDAR. However, in the UK most major urban areas in flood-plains have now been mapped, and the same is true for many urban areas in other developed countries.

The algorithm first detects flooding in the rural areas. It is well-known in image processing that an improved classification can be achieved by segmenting an image into regions of homogeneity and then classifying them, rather than by classifying each pixel independently using a per-pixel classifier. Following Martinis et al. (2009, 2011), the image is segmented into homogeneous regions using the multi-resolution segmentation algorithm of the eCognition Developer software (Definiens AG, 2009). These regions can then be classified on the basis of their backscatter, texture, shape and contextual features. Classification is performed by assigning all segmented regions with mean SAR backscatter less than a threshold to the 'flood' class. To determine the threshold, training regions for 'flood' are automatically selected from regions giving no return in the LiDAR data (water), and for 'non-flood' from un-shadowed areas well above the flood level. The classification step classifies the majority of the flooded rural area correctly. The initial segmentation is refined using a variety of rules e.g. flood regions having LiDAR mean heights significantly above the local flood height are reclassified as non-flood.

A simpler region-growing technique is used in the urban areas, guided by knowledge of the local waterline heights in adjacent rural areas. A set of seed pixels having backscatter less than the threshold, and heights less than or similar to the adjacent rural waterline heights, is identified. Seed pixels are clustered together provided that they are close to other seeds. Regions of shadow and layover are masked out in the processing.

The algorithm was developed using the data set acquired for the 1-in-150 year flood of the rivers Severn and Avon at Tewkesbury, UK, in July 2007. This resulted in substantial flooding of urban and rural areas, about 1500 homes in Tewkesbury being inundated. A 3 m resolution TerraSAR-X image was acquired just after the

Figure 6.8 TerraSAR-X sub-image of the July 2007 flood in the urban areas of Tewkesbury (2.6 × 2 km, dark areas = water) (© DLR 2007).

flood peak (Figure 6.8). LiDAR data provided a DEM in the urban area of Tewkesbury.

The flood extent estimated by TerraSAR-X in the urban and rural areas was validated using the flood extent estimated from aerial photos (Mason et al., 2012, in press). In the urban area, 75% of the urban water pixels visible to TerraSAR-X were correctly detected, though this percentage reduced somewhat if the urban flood extent visible in the aerial photos and detected by TerraSAR-X was considered, because flooded pixels in the shadow/layover areas not visible to TerraSAR-X then had to be taken into account. Better flood detection accuracy was achieved in rural areas, with almost 90% of water pixels being correctly detected by TerraSAR-X.

As well as algorithm development, it is also necessary to consider operational impediments to the production of a near real-time flood extent. In order to ensure that a SAR image was obtained in near real-time, it would be necessary to minimise the time delay between an overpass and the production of the resulting SAR flood extent. Pre-processing operations such as calculation of radar shadow and layover could be carried out in parallel with tasking the satellite to acquire the image. Taking as example TerraSAR-X imaging of the UK, download of the raw SAR data to the ground station, followed by near real-time processing to a multi-look image, geo-registering this automatically, and delineating the flood extent, could in theory be carried out in four to five hours after overpass. The time to extract the flood extent would be small compared to the time to carry out the SAR processing. However, as far as is known, operational systems for near real-time supply of geo-located high

resolution SAR data to users (analogous to ESA-ESRIN's FAIRE system for ASAR/ERS-2 data, which delivers processed geo-registered images to users in about three hours (Cossu et al., 2009)) have still to be developed.

Figure 6.9 shows a possible multi-scale visualisation of the flood extents in the rural and urban areas, with flooding shown as blue in the rural area and yellow in the urban area. Regions coloured brown in the urban area are areas of shadow/layover that are below the waterline height

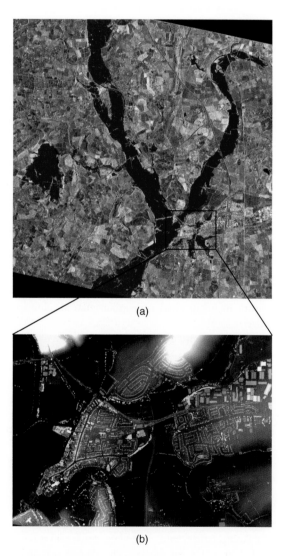

(a)

(b)

Figure 6.9 Possible multi-scale visualisation of flood extents in (a) rural (blue = predicted flood), and (b) urban areas (yellow = predictedflood, brown = shadow/layover areas that may be flooded (after Mason et al., 2012, in press)).

threshold, and therefore may or may not be flooded, as effectively they cannot be imaged by TerraSAR-X.

6.4.3 Multi-temporal SAR images to inform about floodplain dynamics

Although most studies integrating remote sensing derived flood parameters and flood inundation models have been very successful in one way or the other, they are mostly restricted to rural areas and single image data, and only very few have attempted to observe and map flood dynamics. The most notable efforts to date were undertaken by Bates et al. (2006) on a predominantly rural reach of the River Severn in west-central England, and more recently by Schumann et al. (2011) in the town of Tewkesbury and also by Pulvirenti et al. (2011b) for a river reach in northern Italy.

On a 16 km reach of the River Severn, west-central England, Bates et al. (2006) use airborne synthetic aperture radar to map river flood inundation synoptically at fine spatial resolution (1.2 m). Images were obtained at four times through a large flood event between 8 and 17 November 2000 and processed using a statistical active contour algorithm (Horritt, 1999) to yield the flood shoreline at each time. Intersection of these data with a high vertical accuracy survey of floodplain topography obtained from airborne laser altimetry permitted the calculation of dynamic changes in inundated area, total reach storage and rates of reach dewatering. In addition, comparison of the data to gauged flow rates, the measured floodplain topography and map data giving the location of embankments and drainage channels on the floodplain yielded new insights into the factors controlling the development of inundation patterns at a variety of scales.

Finally, the data were used to assess the performance of a simple two-dimensional flood inundation model, LISFLOOD-FP (Bates and de Roo, 2000), and allowed, for the first time, to validate the dynamic performance of the model (Figure 6.10). This process is shown to give new information into structural weaknesses of the model and suggests possible future developments, including the incorporation of a better description of floodplain hydrological processes in the hydraulic model to represent more accurately the dewatering of the floodplain.

Like the case study reported above, remote sensing studies of flooding dynamics to date have focused almost exclusively on rural reaches. Urban areas, where most assets are located and where flood risk is thus very high, have been given relatively little attention, mainly due to

lack of adequate data distributed over space and time. As a consequence it is in urban areas where observations of flood dynamics are most needed. This is addressed in a recent study by Schumann et al. (2011) where a unique space-borne radar data set consisting of five images with three additional aerial photographs over one single event hydrograph of the summer 2007 flooding of the town of Tewkesbury (England, UK) has been assembled.

Previous observations of urban flooding have used single image and ground wrack mark data and have therefore been unable to adequately chart the propagation and recession of flood waves through complex urban topography. By using a combination of spaceborne radar and aerial imagery Schumann et al. (2011) were able to show that remotely sensed imagery, particularly from the new TerraSAR-X radar, can reproduce dynamics adequately and support flood modelling in urban areas. They illustrated that image data from different remote sensing platforms reveal sufficient information to distinguish between models with varying spatial resolution, particularly toward the end of the recession phase of the event (Figure 6.11).

Findings also suggest that TerraSAR-X is able to compete with aerial photography accuracies and can point to structural model errors by revealing important hydraulic process characteristics which might be missing in models; however, as Schumann et al. (2011) note, this has only been demonstrated for a particular urban setting and different results should be anticipated for different types of urban area.

Nevertheless, from this work it can be concluded that SAR imagery from as far as several hundred kilometres from the Earth's surface possesses the potential to deliver important information about floodplain dynamics that can be used to identify and help build suitable models, even in urban environments.

A study by Pulvirenti et al. (2011b) using COSMO-SkyMed images confirms the ability of metric resolution SAR to map flooding with an unprecedented precision. They successfully mapped flood dynamics over a period of five days from a set of very fine resolution (1 × 1 m) X-band SAR images for a flood event in northern Italy in 2009. An unsupervised clustering algorithm was developed to segment the images into homogeneous, non-noisy regions. To associate the segments to different stages of the flood evolution, some distinctive multi-temporal backscattering trends, common to many regions, have been singled out. The authors noted that changes in vegetation and scattering mechanisms contribute most to the radar signal as the water level changes and that these

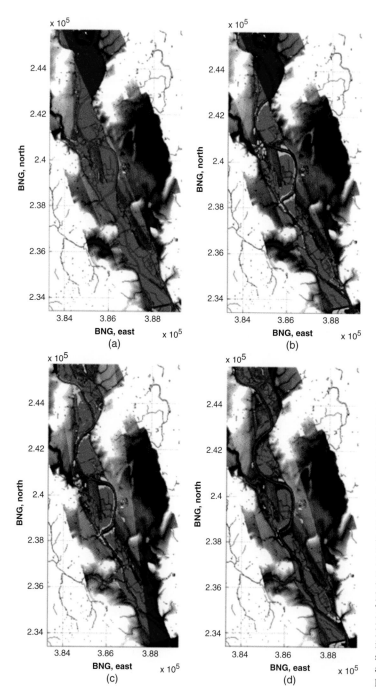

Figure 6.10 Comparison of inundation extent predicted using LISFLOOD-FP with that predicted using the ASAR imagery: (a) 8 November 2000; (b) 14 November 2000; (c) 15 November 2000; (d) 17 November 2000. Light blue indicates areas predicted as flooded using both LISFLOOD-FP and the ASAR; dark blue represents areas predicted as flooded by the model but lying outside the limit of the ASAR swath; red indicates areas predicted as inundated in LISFLOOD-FP but not in the ASAR; yellow indicates areas predicted as inundated in the ASAR but not LISFLOOD-FP; the underlying grey scale represents the DEM used in the model, with the lighter band showing the extent and location of the ASAR swath for each overflight. The blue vectors show the drainage network. Reprinted from Journal of Hydrology, 328, Bates, P.D. et al., Reach scale floodplain inundation dynamics observed using airborne synthetic aperture radar imagery: Data analysis and modelling, 306–318, Copyright 2006, with permission from Elsevier.

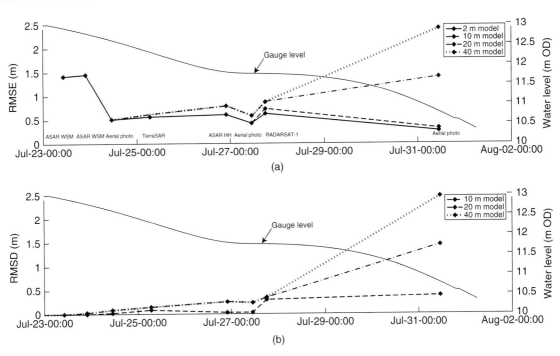

Figure 6.11 Comparison of water level dynamics from remotely sensed imagery and hydraulic models of different resolutions for the July 2007 event in the town of Tewkesbury. (a) RMSE (root mean squared error) of the remote sensing-derived water level dynamics when assessed with models of different grid resolutions. (b) RMSD (root mean squared deviation) of modelled dynamics when taking the 2 m model results as a reference. A stage hydrograph measured at a nearby bridge is shown as a solid line. Reproduced from *Remote Sensing of Environment*, Volume 115, Issue 10, Schumann et al., The accuracy of sequential aerial photography and SAR data for observing urban flood dynamics, a case study of the UK summer 2007 floods, pp. 2536–2546. Copyright 2011, with permission from Elsevier.

must be correctly taken into account to reproduce the measurements.

Pulvirenti et al. (2011b) conclude that more case studies are required to confirm the suitability of such methodology and to establish the requirements of an operational flood management system.

6.5 Summary and outlook

This chapter has described the utility of radar imagery in riverine flood inundation studies. While mainly focusing on active synthetic aperture radar (SAR) imagery, the potential of passive radiometry for acquiring estimates of inundation area over very large rivers or wetlands has been briefly outlined. With the ability to acquire data during nearly all meteorological conditions, day and night, and its capability to provide information about the extent of open water bodies, SARs present an alternative to optical imagery, aerial photography, and hydraulic model simulations for mapping flood extents over large areas and

thus may facilitate effective flood disaster management (Schumann et al., 2007). This was illustrated in various sections in this chapter by providing a critical account of theories, methods and recent advances in the use of radar imagery for flood hazard mapping and risk management.

Important to note is that, until very recently, a more widespread application of SAR remote sensing has been hampered mainly by a lack of adequate spatial and temporal resolutions. However, with the recent launch of satellites carrying very fine resolution SAR (typically <5 m) such as TerraSAR-X, RADARSAT 2 and the COSMO-SkyMed satellites, this situation is expected to change dramatically. As illustrated in Section 6.4 of this chapter, considerable research efforts are currently being made to exploit the full potential that such high resolution image data could offer to support flood hazard modelling and risk management, even in urban areas and with multiple images.

Recent and planned radar satellite missions dedicated to hydrology and flood inundation studies (Table 6.5)

Table 6.5 Recent and planned satellite missions dedicated to hydrology using radar technology.

Mission/ Satellite	Launch Year	Band (GHz)	Revisit Time at Equator (Days)	Spatial Resolution (m)	Primary Use for Flood Hydrology
ALOS	2006	L (1.3)	46	10–100	Flood mapping Soil moisture at small scale
TerraSAR-X (TerraSAR-X 2 planned)*	2007	X (9.6)	11	1–18	High resolution flood mapping
RADARSAT-2 (RADARSAT constellation mission – RCM – planned)**	2007	C (5.3)	24	3–100	High resolution flood mapping Soil moisture at small scale
COSMO-SkyMed	2007	X (9.6)	16 (4 to <1, with 4 satellites in constellation)	1–100	High resolution flood mapping
SMOS	2009	L (1.4) (passive)	2-3	30–50 km	Large scale flood mapping Soil moisture at large scale
Cryosat-2	2010	Ku (13.575) Interferometric altimeter	369	~250 m imaging	Water level measurement Topography
TanDEM-X	2010	X (9.6) (Interferometry)	11 (in tandem orbit with TerraSAR-X)	3 m	Water level dynamics Topography
Sentinel-1	2013	C (5.3)	12 (6, with 2 satellites in constellation)	5–100	High resolution flood mapping Soil moisture at small scale
TanDEM-L (potential)	2015	L (1–2) (Interferometry)	8	5–100	Water level dynamics Topography
SWOT	2019	Ka (35) Interferometric altimeter	22	<100 m imaging	Water level measurement Topography

*Source: http://www.infoterra.de
**Source: http://www.eoportal.org/

will clearly cause a shift from a data-poor to a data-rich environment in which scientists may need to rethink the way they handle data and develop models. For instance, Bates (2004) foresaw that such a data-rich environment gives a potential for model redesign to take advantage of new information, for research into optimal ways to assimilate data into hydraulic models, for investigations of parameter scaling behaviour, for studies of the physical meaning of grid-scale effective parameters in models of different dimensionality or discretisation, and for an exploration of the ways in which new data sources may reduce uncertainty in model predictions.

References

Alsdorf, D.E., Melack, J.M., Dunne, T., Mertes, L.A.K., Hess, L.L., and Smith, L.C. (2000). Interferometric radar measurements of water level changes on the Amazon floodplain, *Lett. Nat.*, 404, 174–177.

Alsdorf, D.E., Smith, L.C., and Melack, J.M. (2001). Amazon floodplain water level changes measured with interferometric SIR-C radar. *IEEE Transactions on Geoscience and Remote Sensing*, 39(2), 423–431.

Alsdorf, D.E. (2002). Interferometric SAR observations of water level changes: potential targets for future repeat-pass AIR-SAR missions. In AIRSAR Earth Science and Application Workshop, Pasadena, California, March 4–6 2002. JPL, CIT, NASA.

Alsdorf, D.E. (2003). Water storage of the central Amazon floodplain measured with GIS and remote sensing imagery. *Annals of the Association of American Geographers*, 93(1), 55–66.

Alsdorf, D., Rodriguez, E., Lettenmaier, D., and Famiglietti, J. (2005). WatER: The Water Elevation Recovery satellite mission – Response to the National Research Council Decadal Survey request for information, technical report, Ohio State Univ., Columbus. (Available at http://bprc.osu.edu/water/publications/WatER_NRC_RFI.pdf)

Alsdorf, D., Rodriguez, E., and Lettenmaier, D.P. (2007). Measuring surface water from space, *Rev. Geophys.*, 45, RG2002, doi:10.1029/2006RG000197.

Aplin, P., Atkinson, P.M., Tatnall, A.R., Cutler, M.E., and Sargent, I.M. (1999). SAR for flood monitoring and assessment, paper presented at Remote Sensing Society 1999: Earth Observation: From Data to Information, Remote Sens. Soc., Nottingham, UK.

Atkinson, P.M. (2000). Autologistic regression for flood zonation using SAR imagery, paper presented at 26th Annual Conference, Remote Sens. Soc., Leicester, UK.

Atkinson, P.M. and Foody, G.M. (2002). Uncertainty in remote sensing and GIS: Fundamentals. In G. Foody and P.M. Atkinson (eds), *Uncertainty in remote sensing and GIS*. Wiley, Chichester, UK, pp. 1–18.

Bamler, R. and Hartl, P. (1998). Synthetic aperture radar interferometry, *Inverse Problems*, 14, R1–R54.

Bamler, R., Eineder, M., Kampes, B., Runge, H., and Adam, N. (2003). SRTM and beyond: current situation and new developments in spaceborne InSAR. In ISPRS/EARSeL Workshop, Hannover, Germany, October 6–8 2003. ISPRS.

Bates, P.D., Horritt, M.S., Smith, C.N., and Mason, D.C. (1997). Integrating remote sensing observations of flood hydrology and hydraulic modelling, *Hydrol. Processes*, 11, 1777–1795.

Bates, P.D. and De Roo, A.P.J. (2000). A simple raster-based model for floodplain inundation. *Journal of Hydrology*, 236, 54–77.

Bates, P.D. (2004). Invited commentary: Remote sensing and flood inundation modelling. *Hydrological Processes*, 18, 2593–2597.

Bates, P.D., Wilson, M.D., Horritt, M.S., Mason, D., Holden, N., and Currie, A. (2006). Reach scale floodplain inundation dynamics observed using airborne synthetic aperture radar imagery: Data analysis and modelling, *J. Hydrol.*, 328, 306–318.

Beven, K. (2006). A manifesto for the equifinality thesis, *J. Hydrol.*, 320, 18–36.

Blyth, K. (1997). Floodnet: A telenetwork for acquisition, processing and dissemination of Earth observation data for monitoring and emergency management of floods, *Hydrol. Processes*, 11, 1359–1375.

Bonn, F. and Dixon, R. (2005). Monitoring flood extent and forecasting excess runoff risk with RADARSAT-1 data, *Nat. Hazards*, 35, 377–393.

Brakenridge, G.R., Tracy, B.T., and Knox, J.C. (1998). Orbital SAR remote sensing of a river flood wave, *Int. J. Remote Sens.*, 19(7), 1439–1445.

Breit, H., Fritz, T., Balss, U., Lachaise, M., Niedermeier, A., and Vonavka, M. (2010). TerraSAR-X SAR processing and products, *IEEE. Trans. Geoscience Rem. Sens.*, 48(2), 727–740.

Brivio, P.A., Colombo, R., Maggi, M., and Tomasoni, R. (2002). Integration of remote sensing data and GIS for accurate mapping of flooded areas, *Int. J. Remote Sens.*, 23(3), 429–441.

Calabresi, G. (1995). The use of ERS data for flood monitoring: An overall assessment, paper presented at 2nd ERS Applications Workshop, Eur. Space Agency, London.

Cazenave, A., Milly, P., Douville, H., Beneveniste, J., Lettenmaier, D., and Kosuth, P. (2004). International workshop examines the role of space techniques to measure spatio-temporal change in terrestrial waters, *Eos Trans. AGU*, 85(6), 8.

Cossu, R., Schoepfer, E., Bally, Ph., Fusco, L. (2009). Near real-time SAR based processing to support flood monitoring, *J. Real-Time Image Processing*, 4(3), 205–218, 2009.

Definiens A.G. (2009). Definiens Developer 8 User Guide, Document Version 1.2.0, Munich, Germany.

De Roo, A.P.J., Van Der Knijff, J., Horritt, M., Schmuck, G., and De Jong, S. (1999). Assessing flood damages of the 1997 Oder flood and the 1995 Meuse flood, paper presented at 2nd International ITC Symposium on Operationalization of Remote Sensing, Int. Inst. for Geo-Inf. Sci. and Earth Obs., Enschede, Netherlands.

Di Baldassarre, G., Schumann, G.J.-P. and Bates, P.D (2009a). Near real time satellite imagery to support and verify timely flood modelling. *Hydrological Processes: Scientific Briefing*, 23 (5), 799–803.

Di Baldassarre, G., Schumann, G.J.-P., and Bates, P.D. (2009b). A technique for the calibration of hydraulic models using uncertain satellite observations of flood extent. *Journal of Hydrology*, 367, 276–282.

Di Baldassarre, G., Schumann, G.J.-P., Brandimarte, L., and Bates, P.D. (2011). Timely low resolution SAR imagery to support floodplain modelling: a case study review. *Surveys in Geophysics*, 32, 255–269.

European Commission, EC (2011). A New European Floods Directive. Available: http://ec.europa.eu/environment/water/flood_risk/

Falorni, G., Teles, V., Vivoni, E.R., Bras, R.L., and Amaratunga, K.S. (2005). Analysis and characterization of the vertical accuracy of digital elevation models from the Shuttle Radar Topography Mission, *J. Geophys. Res.*, 110, F02005, doi:10.1029/2003JF000113.

Farr, T.G., Rosen, P.A., Caro, E., Crippen, R., Duren, R., Hensley, S., Kobrick, M., Paller, M., Rodriguez, E., Roth, L., Seal, D., Shaffer, S., Shimada, J., Umland, J., Werner, M., Oskin, M., Burbank, D., and Alsdorf, D. (2007). The Shuttle Radar Topography Mission, Rev. *Geophys.*, 45, RG2004, doi:10.1029/2005RG000183.

Giustarini, L., Matgen, P., Hostache, R., Montanari, M., Plaza, D., Pauwels, V.R.N., De Lannoy, G.J.M., De Keyser, R., Pfister, L., Hoffmann, L., and Savenije, H.H.G. (2011). Assimilating SAR-derived water level data into a hydraulic model: a case study. *Hydrol. Earth Syst. Sci.*, 15, 2349–2365.

Gupta, R.P. and Banerji, S. (1985). Monitoring of reservoir volume using LANDSAT data, *J. Hydrol.*, 77, 159–170.

Henry, J.B., Chastanet, P., Fellah, K., and Desnos, Y.L. (2006). ENVISAT multi-polarised ASAR data for flood mapping, *Int. J. Remote Sens.*, 27(10), 1921–1929.

Hess, L.L., Melack, J.M., and Simonett, D.S. (1990). Radar detection of flooding beneath the forest canopy: A review, *Int. J. Remote Sens.*, 11, 1313–1325.

Hess, L.L., Melack, J.M., Filoso, S., and Wang, Y. (1995). Delineation of inundated area and vegetation along the Amazon floodplain with the SIR-C synthetic aperture radar, IEEE Trans. *Geosci. Remote Sens.*, 33(4), 896–904.

Hong, S, Wdowinski, H.S., Kim, S.W, and Won, J.S. (2010). Multi-temporal monitoring of wetland water levels in the Florida Everglades using interferometric synthetic aperture radar (InSAR). *Remote Sensing of Environment*, 114, 2436–2447.

Horritt, M.S. (1999). A statistical active contour model for SAR image segmentation, *Image Vision Comput.*, 17(3), 213–224.

Horritt, M.S., Mason, D.C., and Luckman, A.J. (2001). Flood boundary delineation from synthetic aperture radar imagery using a statistical active contour model, *Int. J. Remote Sens.*, 22(13), 2489–2507.

Horritt, M.S., Mason, D.C., Cobby, D.M., Davenport, I.J., and Bates, P.D. (2003). Waterline mapping in flooded vegetation from airborne SAR imagery, *Remote Sens. Environ.*, 85(3), 271–281.

Hostache, R., Matgen, P., Schumann, G.J.-P., Puech, C., Hoffmann, L., and Pfister, L. (2009). Water level estimation and reduction of hydraulic model calibration uncertainties using satellite SAR images of floods, *IEEE Trans. Geosci. Remote Sens.*, 47, 431–441.

Hunter, N.M., Bates, P.D., Horritt, M.S., De Roo, P.J., and Werner, M. (2005). Utility of different data types for flood inundation models within a GLUE framework, *Hydrol. Earth Syst. Sci.*, 9, 412–430.

Ip, F., Dohm, J.M., Baker, V.R., Doggett, T., Davies, A.G., Castaño, R., Chien, S., Cichy, B., Greeley, R., Sherwood, R., Tran, D., and Rabideau, G. (2006). Flood detection and monitoring with the Autonomous Sciencecraft Experiment onboard EO-1. *Remote Sensing of Environment*, 101, 463–481.

Jin, Y-Q. (1999). A flooding index and its regional threshold value for monitoring floods in China from SSM/I data, *Int. J. Remote Sens.*, 20(5), 1025–1030.

Kussul, N., Shelestov, A., and Skakun, S. (2008). Grid system for flood extent extraction from satellite images. *Earth Sci Inform*, 1, 105–117.

Lane, S.N., James, T.D., Pritchard, H., and Saunders, M. (2003). Photogrammetric and laser altimetric reconstruction of water levels for extreme flood event analysis, *Photogramm. Record*, 18(104), 293–307.

Laugier, O., Fellah, K., Tholey, N., Meyer, C., and De Fraipont, P. (1997). High temporal detection and monitoring of flood zone dynamic using ERS data around catastrophic natural events: The 1993 and 1994 Camargue flood events, paper presented at 3rd ERS Symposium on Space at the Service of Our Environment, Eur. Space Agency, Florence, Italy.

Lillesand, T.M., Kiefer, R.W., and Chipman, J.W. (2004). *Remote sensing and image interpretation* – Fifth edition, John Wiley & Sons, New York, USA.

Liu, Y., Nishiyama, S., and Yano, T. (2004). Analysis of four change detection algorithms in bi-temporal space with a case study, *Int. J. Remote Sens.*, 25(11), 2121–2139.

Lu, D., Mausel, P., Brondizio, E., and Moran, E. (2004). Change detection techniques, *Int. J. Remote Sens.*, 25(12), 2365–2407.

MacIntosh, H. and Profeti, G. (1995). The use of ERS SAR data to manage flood emergencies at the smaller scale, paper presented at 2nd ERS Applications Workshop, Eur. Space Agency, London.

Marcus, W.A. and Fonstad, M.A. (2008). Optical remote mapping of rivers at sub-meter resolutions and watershed extents, *Earth Surf. Processes Landforms*, 33, 4–24.

Martinis, S., Twele A., and Voigt, S. (2009). Towards operational near real-time flood detection using a split-based automatic thresholding procedure on high resolution TerraSAR-X data, *Natural Hazards and Earth System Sciences*, 9, 303–314.

Martinis, S. (2010). Automatic near real-time flood detection in high resolution X-band synthetic aperture radar satellite data using context-based classification on irregular graphs, PhD Thesis, Faculty of Geosciences, Ludwig-Maximilians University, Munich, Germany.

Martinis, S., Twele A., and Voigt, S. (2011). Unsupervised extraction of flood-induced backscatter changes in SAR data using Markov image modeling on irregular graphs, *IEEE Trans. Geoscience Rem. Sens.*, 49(1), 251–263.

Mason, D.C., Davenport, I.J., Flather, R.A., Gurney, C., Robinson, G.J., and Smith, J.A. (2001). A sensitivity analysis of the waterline method of constructing a digital elevation model for intertidal areas in ERS SAR scene of Eastern England. *Estuarine, Coastal and Shelf Science* 53, 759–778.

Mason, D.C., Cobby, D.M., Horritt, M.S., and Bates, P.D. (2003). Floodplain friction parameterization in two-dimensional river flood models using vegetation heights derived from airborne scanning laser altimetry, *Hydrol. Processes*, 17, 1711–1732.

Mason, D.C., Bates, P.D., and Dall'Amico, J.T. (2009). Calibration of uncertain flood inundation models using remotely sensed water levels, *J. Hydrol.*, 368, 224–236.

Mason, D.C., Speck, R., Devereux, B., Schumann, G.J.-P., Neal, J.C., and Bates, P.D. (2010). Flood detection in urban areas using TerraSAR-X. IEEE. *Trans. Geoscience Rem. Sens.*, 48(2), 882–894.

Mason, D.C., Schumann, G.J.-P. and Bates, P.D. (2011). Data utilization in flood inundation modelling. In: Pender, G. and Faulkner, H. (2011). *Flood risk science and management*, Wiley-Blackwell, Chichester, UK, pp. 211–233.

Mason, D.C., Davenport, I.J., Neal, J.C., Schumann, G.J.-P., and Bates, P.D. (2012). Near real-time flood detection in urban and rural areas using high resolution synthetic aperture radar images, IEEE. *Trans. Geoscience Rem. Sens.*, in press, 50(8), DOI:10.1109/TGRS.2011.2178030.

Matgen, P., Henry, J.B., Pappenberger, F., De Fraipont, P., Hoffmann, L., and Pfister, L. (2004). Uncertainty in calibrating flood propagation models with flood boundaries observed from synthetic aperture radar imagery, paper presented at 20th ISPRS Congress, Int. Soc. for Photogramm. and Remote Sens., Istanbul.

Matgen, P., El Idrissi, A., Henry, J.-B., Tholey, N., Hoffmann, L., De Fraipont, P., and Pfister, L. (2006). Patterns of remotely sensed floodplain saturation and its use in runoff predictions, *Hydrol. Processes*, 20(8), 1805–1825.

Matgen, P., Schumann, G.J.-P., Henry, J., Hoffmann, L., and Pfister, L. (2007). Integration of SAR-derived inundation areas, high precision topographic data and a river flow model toward real-time flood management, *Int. J. Appl. Earth Obs. Geoinf.*, 9(3), 247–263.

Matgen, P., Hostache, R., Schumann, G.J.-P., Pfister, L., Hoffmann, L., and Savenije, H.H.G. (2011). Towards an automated SAR-based flood monitoring system: Lessons learned from two case studies. *Physics and Chemistry of the Earth*, 36, 241–252.

Matgen, P., Montanari, M., Hostache, R., Pfister, L., Hoffmann, L., Plaza, D., Pauwels, V.R.N., De Lannoy, G.J.M., De Keyser, R., and Savenije, H.H.G. (2010). Towards the sequential assimilation of SAR-derived water stages into hydraulic models using the Particle Filter: proof of concept, *Hydrol. Earth Syst. Sci.*, 14, 1773–1785.

Muller, J-P. (2000). The LandMap Project for the automated creation and validation of multiple resolution orthorectified satellite image products and a 1" DEM of the British Isles from ERS tandem SAR interferometry. In: Proceedings of the 26th Annual Conference of the Remote Sensing Society, 12–14 September 2000, University of Leicester.

Neal, J., Schumann, G.J.-P., Bates, P.D., Buytaert, W., Matgen, P., and Pappenberger, F. (2009). An assimilation approach to discharge estimation from space, *Hydrol. Processes*, 23, 3641–3649.

Oberstadler, R., Hönsch, H., and Huth, D. (1997). Assessment of the mapping capabilities of ERS-1 SAR data for flood mapping: A case study in Germany, *Hydrol. Processes*, 10, 1415–1425.

Papa, F., Prigent, C., Rossow, W.B., Legresy, B., and Remy, F. (2006). Inundated wetland dynamics over boreal regions from remote sensing: The use of TOPEX-Poseidon dual-frequency radar altimeter observations, *Int. J. Remote Sens.*, 27(21), 4847–4866.

Pappenberger, F. and Beven, K.J. (2006). Ignorance is bliss: Or seven reasons not to use uncertainty analysis, *Water Resour. Res.*, 42, W05302, doi:10.1029/2005WR004820.

Pappenberger, F., Frodsham, K., Beven, K., Romanowicz, R., and Matgen, P. (2007). Fuzzy set approach to calibrating distributed flood inundation models using remote sensing observations, *Hydrol. Earth Syst. Sci.*, 11, 739–752.

Perissin, D. and Rocca, F. (2006). High-accuracy urban dem using permanent scatterers. IEEE Transactions on Geoscience and Remote Sensing, 44, 3338–3347.

Raney, R.K. (1998). Radar fundamentals: Technical perspective. In Henderson, F.M. and Lewis, A.J. (eds.) *Manual of remote sensing: Principles and applications of imaging radar* – Third edition, John Wiley & Sons, New York, USA.

Pitt, M. (2008). Learning lessons from the 2007 floods. U.K. Cabinet Office Report, June 2008. Available: http://archive.cabinetoffice.gov.uk/pittreview/thepittreview.html.

Prestininzi, P., Di Baldassarre, G., Schumann, G.J.-P., and Bates, P.D. (2011). The use of low-resolution satellite imagery to guide the selection of the appropriate hydraulic model structure. *Advances in Water Resources*, 34, 38–46.

Pulvirenti, L., Pierdicca, N., Chini, M., and Guerriero, L. (2011a). An algorithm for operational flood mapping from Synthetic Aperture Radar (SAR) data using fuzzy logic. *Nat. Hazards Earth Syst. Sci.*, 11, 529–540.

Pulvirenti, L., Chini, M., Pierdicca, N., Guerriero, L., and Ferrazzoli, P. (2011b). Flood monitoring using multi-temporal COSMO-SkyMed data: Image segmentation and signature interpretation. *Remote Sensing of Environment*, 115, 990–1002.

Raclot, D. and Puech, C. (2003). What does AI contribute to hydrology? Aerial photos and flood levels, *Appl. Artif. Intel.*, 17, 71–86.

Rees, W.G. (2001). *Physical Principles of Remote Sensing*, Cambridge Univ. Press, Cambridge, UK.

Rodriguez, E., Morris, C.S., Belz, J.E., Chapin, E.C., Martin, J.M., Daffer, W., and Hensley, S. (2005). An assessment of SRTM topographic products. Technical report, Pasadena, California.

Schumann, G.J.-P., Henry, J.B., Hoffmann, L., Pfister, L., Pappenberger, F., and Matgen, P. (2005). Demonstrating the high potential of remote sensing in hydraulic modelling and flood risk management, paper presented at Annual Conference of the Remote Sensing and Photogrammetry Society with the NERC Earth Observation Conference, Remote Sens. and Photogramm. Soc., Portsmouth, UK.

Schumann, G.J.-P., Matgen, P., Pappenberger, F., Hostache, R., Puech, C., Hoffmann, L., and Pfister, L. (2007). High-resolution 3D flood information from radar for effective flood hazard management, *IEEE Trans. Geosci. Remote Sens.*, 45, 1715–1725.

Schumann, G.J.-P., Matgen, P., Cutler, M.E.J., Black, A., Hoffmann, L., and Pfister, L. (2008). Comparison of remotely sensed water stages from lidar, topographic contours and SRTM, *ISPRS J. Photogramm. Remote Sens.*, 63, 283–296.

Schumann, G.J.-P., Di Baldassarre, G., and Bates, P.D. (2009a). The utility of space-borne radar to render flood inundation maps based on multi-algorithm ensembles. *IEEE Transactions on Geoscience and Remote Sensing*, 47, 2801–2806.

Schumann, G.J.-P., Bates, P.D., Horritt, M.S., Matgen, P., and Pappenberger, F. (2009b). Progress in integration of remote sensing–derived flood extent and stage data and hydraulic models, Reviews of Geophysics, 47, RG4001, doi:10.1029/2008RG000274.

Schumann, G.J.-P., Di Baldassarre, G., Alsdorf, D., and Bates, P.D. (2010a). Near real-time flood wave approximation on large rivers from space: application to the River Po, Northern Italy. Water Resources Research, 46, doi: 10.1029/2008WR007672.

Schumann, G.J.-P. and Di Baldassarre, G. (2010). The direct use of radar imagery for event-specific flood risk mapping. *Remote Sensing Letters*, 1, 75–84.

Schumann, G.J.-P., Neal, J.C., and Bates, P.D. (2010b). Global scale simulation of floodplain inundation with low resolution space-borne data, Proceedings of the IAHS Remote Sensing and Hydrology 2010 Symposium, 27–30 September 2010, Jackson Hole, WY, USA.

Schumann, G.J.-P., Neal, J.C., Mason, D.C., and Bates, P.D. (2011). The accuracy of sequential aerial photography and SAR data for observing urban flood dynamics, a case study of the UK summer 2007 floods. *Remote Sensing of Environment*, in press.

Sippel, S.J., Hamilton, S.K., Melack, J.M., and Novo, E.M.M. (1998). Passive microwave observations of inundation area and the area/stage relation in the Amazon River floodplain, *Int. J. Remote Sens.*, 19(16), 3055–3074.

Smith, L.C. (1997). Satellite remote sensing of river inundation area, stage, and discharge: A review, *Hydrol. Processes*, 11, 1427–1439.

Smith, L.C. (2002). Emerging applications of Interferometric Synthetic Aperture Radar (InSAR) in geomorphology and hydrology. *Annals of the Association of American Geographers*, 92(3), 385–398.

Speck, R, Turchi, P., and Süß, H. (2007). An end-to-end simulator for high-resolution spaceborne SAR systems, Proc. SPIE Defense and Security. 6568.

Sun, G., Ranson, K.J., Kharuk, V.I., and Kovacs, K. (2003). Validation of surface height from shuttle radar topography mission using shuttle laser altimetry, *Remote Sensing of Environment*, 88, 401–411.

Ulaby, F.T., Moore, R.K., and Fung, A.K. (1986). *Microwave Remote Sensing: Active and Passive*. Vol. III – From theory to applications –. Artech House, Dedham, Massachusetts, USA.

Woodhouse, I.H. (2006). *Introduction to microwave remote sensing*. CRC Press, Boca Raton, Florida, USA.

Zwenzner, H. and Voigt, S. (2009). Improved estimation of flood parameters by combining space based SAR data with very high resolution digital elevation data. *Hydrol. Earth Syst. Sci.*, 13, 567–576.

7

Airborne LiDAR Methods Applied to Riverine Environments

Jean-Stéphane Bailly[1], Paul J. Kinzel[2], Tristan Allouis[3], Denis Feurer[4] and Yann Le Coarer[5]

[1]AgroParisTech, UMR TETIS and UMR LISAH, Montpellier, France
[2]USGS, Geomorphology and Sediment Transport Laboratory, Golden, CO, USA
[3]Irstea, UMR TETIS, Montpellier, France
[4]IRD, UMR LISAH, Montpellier, France
[5]Irstea, Unité de Recherche, HYAX, Aix-en-Provence, France

7.1 Introduction: LiDAR definition and history

LiDAR is an acronym for Light Detection And Ranging. Other names, such as ALSM (Airborne Laser Swath Mapping), LADAR (LAser Detection And Ranging), altimetry laser, or laser scanner, are also used, though less frequently. LiDAR is an active remote sensing technique. The principles of LiDAR are quite similar to RADAR, but instead of using radio wavelengths (1–10 cm), it operates in the infrared, visible, or ultraviolet wavelengths of the electromagnetic spectrum (250 nm up to 11 µm) (Weitkamp, 2005, p.4). LiDAR systems include hardware and software for analysing the properties of laser pulses reflected from multiple surfaces back to the system by a remote target. In a mono-static configuration, a LiDAR system is composed of a laser transmitter, a telescope, photo-detectors, and a computer.

LiDAR has its origins in the atmospheric sciences. The early development of LiDAR instruments included ground-based light pulse systems developed by the two French meteorologists Barthelémy and Bureau in 1935, to measure the elevation of clouds (Flamant, 2005, p.865).

The American scientist Hulburt used these systems in 1937 to measure air density profiles in the upper atmosphere (Weitkamp, 2005, p.3). The first successful ground-based ranging LiDAR system was developed by Smullin and Fiocco (1962), who used it to measure the distance between the Earth and the Moon. The first airborne LiDAR for earth surface observation was developed in 1965 as a 'bathymeter' to locate submarines (Ott, 1965). Today, LiDARs can be placed in four categories: elastic scattering LiDAR, used in ranging applications; inelastic or fluorescence LiDAR for chemical, composition or concentration measurements; Doppler LiDAR for speed measurements; and differential-absorption LiDARs for gaseous species detection (Flamant, 2005, p. 871).

In this chapter, airborne LiDAR systems and experiments conducted in riverine environments are presented, specifically for bathymetry. In the first section, the theoretical background of airborne LiDAR is presented. Examples of LiDAR system capabilities and practical survey information are provided in the following two sections. In the fourth section, the techniques used to derive information from LiDAR signals within the scope of riverine environments are explained. Two river survey case studies are presented in the fifth section.

Fluvial Remote Sensing for Science and Management, First Edition. Edited by Patrice E. Carbonneau and Hervé Piégay.
© 2012 John Wiley & Sons, Ltd. Published 2012 by John Wiley & Sons, Ltd.

7.2 Ranging airborne LiDAR physics

7.2.1 LiDAR for emergent terrestrial surfaces

When considering emergent terrestrial surfaces, both the ground and vegetation have a high reflectance ratio in the near-infrared range (NIR), 700 nm–1,400 nm (Caloz and Collet, 1992). Vegetation also has a high transmission ratio in that range, which facilitates the travel of an NIR laser to the ground through the vegetation. As a result, LiDAR, using an NIR wavelength called topographic LiDAR, is able to simultaneously gather information on both the ground and vegetation.

The most commonly used infrared wavelength is 1,064 nm, which is the characteristic wavelength of frequency-doubled Nd:YAG (neodymium-doped yttrium aluminum garnet; $Nd:Y_3Al_5O_{12}$) diode-pumped, crystalline lasers. Nd:YAG lasers are preferred as they induce greater power efficiency per transmitted pulse that promotes a high signal-to-noise ratio to ensure a full waveform. However, other lasers are also used in topographic LiDAR (Jutzi and Stilla, 2006), such as erbium fiber amplified lasers, which provide a 1,550-nm wavelength with more powerful emitted pulses (Samson and Torruellas, 2005).

7.2.1.1 Physical equations

LiDAR for terrestrial surface measurements emits pseudo-Gaussian laser pulses and receives the sum of the pulses' energies reflected by each target the laser beam reaches. Consequently, the received waveform is the sum of the response functions of the different targets, convolved by the transmitted signal and by the receiver impulse function.

The received power $P(t)$ can be written as (Wagner et al., 2006) (equation 7.1):

$$P(t) = \sum_{i=1}^{N} \frac{S_{fov}}{\pi^2 R_i^4 \beta_t^2} P_T(t) * \sigma_i'(t) * \Gamma(t) \qquad (7.1)$$

where $\Gamma(t)$ is the receiver impulse function, $\sigma'i(t)$ is the back-scatter cross-section that combines the target parameters, $P_T(t)$ is the transmitted signal, S_{fov} is the area of the receiver optics, β_t is the transmitter beam width, R_i is the target range, and N is the number of targets.

7.2.1.2 Signal registration: echoes and full waveform

Once the back-scattered energy is collected by the LiDAR's telescope, two types of signal processing are possible (Figure 7.1a). The first is a real-time process in which the received signal extracts the range of targets. Such systems, called multi-echo or multi-pulse LiDAR, detect significant echoes with a threshold method (Thiel and Wehr, 2004). The second possible process is the digitisation and recording of the entire received waveform. This kind of system is called full waveform LiDAR. A full waveform topographic LiDAR typically records the waveform at a 1-GHz frequency, which is equivalent to one sample per nanosecond (resolution of 15 cm in air). The target ranges are then extracted by post-processing the recorded waveform. A fit of Gaussian functions on the waveform is traditionally performed, assuming that the signal refection from the target is Gaussian (Hofton et al., 2000). Such a process enhances the 15-cm vertical resolution, allowing the system to fit a Gaussian function between two samples and to provide more ranging measurements than multi-pulse systems (Chauve et al., 2009). Furthermore, full waveform systems can provide the echo amplitude and width that is useful for characterising the type of reflective surface (Reitberger et al., 2008).

The ranging measurements are combined with a global positioning system (GPS) and inertial measurement unit (IMU) of the aircraft to place the target's position within a global reference system. Finally, a dense (X,Y,Z) point cloud coverage is obtained. The points represent the back-scatter both from objects (buildings, vegetation) and from bare terrain, and they provide information about the vertical distribution of targets if the targets are sufficiently sparse. Indeed, in the case of back-scatter from a building, the laser will be completely reflected from the building surface, but in the case of sparse vegetation (see Figure 7.1a), the laser pulse will be reflected three times.

The first echo will give the maximum elevation of the vegetated surface while the last echo will provide the elevation of the bare Earth. Figure 7.1b shows that the last returns are better able to provide bare-Earth elevations under sparse object cover and to provide information on the internal structure of the target. Nevertheless, the first returns better describe the top of the surface, which corresponds to vegetation canopies or roofs.

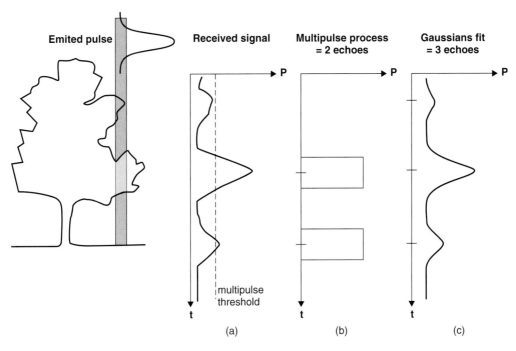

Figure 7.1a The emitted pulse on the scene, the received signal waveform (a), the result of the multi-pulse process (b) and the result of Gaussian fitting process (c) for target range extraction.

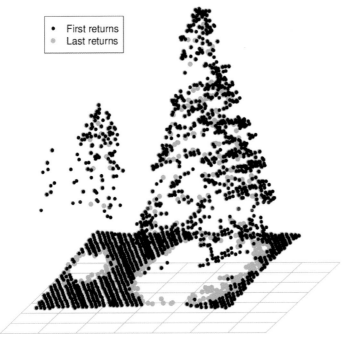

Figure 7.1b 3D point clouds of the first return echoes (black) and last return echoes (grey) on two trees and a flat, bare terrain. © Cemagref.

7.2.2 LiDAR for aquatic surfaces

LiDAR sensors designed for surveys over aquatic areas are called bathymetric hydrographic LiDARs, or ALB for Airborne LiDAR Bathymeters. These LiDARs were primarily developed to survey across the land-water boundary in near-coastal areas, but since 2005, there has been an increasing use of these systems in riverine environments.

7.2.2.1 Physical equations

Compared to LiDARs used for terrestrial surfaces, bathymetric LiDARs use a green wavelength (532 nm) to penetrate the water surface with minimal optical attenuation. The relevant portion of the bathymetric LiDAR signal corresponds to a back-scattered waveform, which usually contains two peaks, typically indicating water surface and bottom reflections. A schematic description of this system is shown in Figure 7.2a

When considering only the green signal, the foundation equations of the collected signal power P by the receptor for an emitted pulse can be stated as a sum of four dependent components (Guenther, 1985; Jurand et al., 1989), as in Equation (7.1):

$$P = P_s + P_{bsc} + P_b + P_{bg} + P_n \qquad (7.2)$$

where P_s denotes the power returned by the water surface, and $P_{bsc}(d)$ denotes the power returned by the water column at a given depth d. Consequently, the total water column distribution P_{bsc} is computed through Equation (7.1):

$$P_{bsc} = \int P_{bsc}(d).dd \qquad (7.3)$$

P_b denotes the power returned from the bottom; P_{bg} is the background power coming from sunlight or backscatter in the air; and P_N is the power from system noise. More detailed formulations can be found in Tulldahl and Steinvall (2004). A number of authors expound on these complex physical equations and describe each of the interdependent components (e.g., Guenther, 1985; Jurand et al., 1989; Tulldahl and Steinvall, 1999, 2004; Zege et al., 2004).

The first condition controlling the shape of each component is the typically Gaussian form of the transmitted pulse $P_T(t)$ itself. Indeed, the LiDAR waveform and component forms result from the convolution product of $P_T(t)$, with the range of target backscatter properties for targets encountered by the laser corresponding to time t.

Optical properties such as scattering and attenuation increase the diameter of the pulse and bias towards the

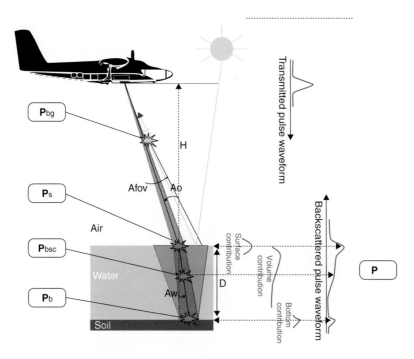

Figure 7.2a A scheme of the bathymetric LiDAR system (green laser) and main components convolving to the LiDAR waveform. Courtesy of Airborne Hydrography AB.

Table 7.1 River parameters controlling green laser waveforms.

River Property	Waveform Impact	Impact on LiDAR Measurement Accuracy
Water velocity	• Decrease backscatter signal power	• weak
Water surface roughness	• Increase or decrease the surface backscatter signal power • (with an 'optimum')	• increase water-surface altimetry accuracy • decrease very shallow water depth-measurement capacities
Water surface slope	• decrease surface backscatter signal power • decrease bottom backscatter signal power	• enlarge LiDAR measure resolution • decrease very shallow water depth-measurement capacities
Water turbidity (from sediment, dissolve organic matter or phytoplankton)	• decrease bottom backscatter signal power • increase water column backscatter signal power	• decrease deep water depth-measurement capacities • decrease bathymetry accuracy
River bottom material	• Increase bottom backscatter signal power with reflective materials	• increase bathymetry accuracy in case of reflective materials
River bottom roughness	• decrease bottom backscatter signal power	• decrease very shallow water depth-measurement capacities
River bottom slope	• decrease bottom backscatter signal power	• decrease very shallow water depth-measurement capacities
Presence of algae	• Increase water column backscatter signal power	• decrease bathymetry accuracy

undercutting region which compromises depth accuracy. Measurement biases can be magnified by physical and biological properties inherent to riverine environments. Some of the main physical and biological properties of rivers controlling the collected signal power over time, i.e., waveform $P(t)$, are listed in Table 7.1.

The waveforms are the raw data of bathymetric LiDAR, from which derivative variables can be estimated (see Section 7.4). For instance, information such as water depth is derived from the surface peak (t_1) and bottom (t_2) positioning in a waveform through (in the case of a nadir laser beam):

$$d = \frac{c_2(t_2 - t_1)}{2} \text{ where } c_2 \text{ is light speed in water}$$
$$(2.25 \; 10^8 \, \text{m.s}^{-1}).$$

To perform simulations of bathymetric waveforms and derivative information, simulation based on these equations have been implemented (Zege et al., 2004; Lesaignoux et al., 2007). These tools permit the exploration of LiDAR signals in various aquatic conditions and system design configurations. These tools also assist in evaluating the performance and accuracy of various signal processing algorithms to retrieve water parameters. For example, Zege et al. (2004) developed the software AOLS (Airborne Oceanic Lidar Simulator) and verified retrieval techniques using analytical inversion of the Raman lidar equation from a set of Raman waveform simulations. Another example can be found in Lesaignoux et al. (2007): the use of a huge set of simulated green waveforms as examples shown in Figure 7.2b, allowed the assessment of the performance and limitations for very shallow-water bathymetry using a bi-Gaussian fitting process on waveforms.

7.2.2.2 Multispectral systems

As shown in section 7.2.2.1, a unique characteristic of bathymetric LiDARs is the use of different wavelengths in transmission or reception. These different wavelengths are usually 1,064 and 532 nm in transmission, and in the SHOALS (Scanning Hydrographic Operational Airborne LiDAR Survey) LiDAR system (Irish and Lillycrop, 1999), the Raman wavelength of 634 nm is also analysed in the

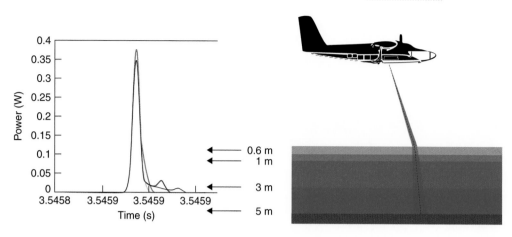

Figure 7.2b Examples of a simulated 1-ns sampled green waveform for 60-cm (light grey), 1-m (grey), and 2-m clear water depth (thin black line).

receiver. The Raman signal comes from inelastic interactions between the green laser and the excitation energy from the O:H bonds of water molecules directly below the air-water boundary. Green laser-induced fluorescence results as the bonds excite to a higher quantum state to be backscattered while emitting the red wavelength. Both the 1,064- and 532-nm wavelengths for emission result from the diffraction process of a single Nd:YAG 1,064-nm laser pulse (Zege et al., 2004).

The use of both green and infrared wavelengths scanned collinearly during a LiDAR survey provides a redundancy and cross-validation of water-surface positioning, therefore it can help estimate water-bed positioning and depth more accurately (Wozencraft and Millar, 2005). Indeed, light celerity is different in air and water, the time budget of laser beams inside these two media have to be precisely determined as it highly conditions the ranging measures, i.e., altimetric positions.

Even if green LiDAR can be used for terrestrial surveying over riverbanks, the higher transmission of infrared radiation in vegetation permits a better description of the terrain under the vegetation.

In contrast to false infrared returns, if the green return is noticeably weaker than the volume backscatter return it may provide assistance when surface detections are incorrect due to land reflection or the presence of unexpected targets occur, such as birds. Raman returns are accurate regardless of wind speed and standing waves, while wind speed weakens infrared surface returns. Therefore, infrared and green Raman wavelengths should be coupled together to reliably predict water surface (Gunther et al.,

1999; Allouis et al., 2010). In extremely shallow environments a more detailed interpretation of the Raman signal may provide information about the water column depth, water temperature, and composition, e.g. chlorophyll (Burikov et al., 2004, Peeri and Philpot, 2007).

7.3 System parameters and capabilities: examples

Up to 2009, four LiDAR systems have been used for large bathymetry surveys: the SHOALS system (Canada), the LADS system (Australia) (http://www.navy.gov.au/Laser_Airborne_Depth_Sounder), the HawkEye system (Sweden) and the EAARL system (US). As representative systems, the two latter systems are detailed below.

7.3.1 Large footprint system: HawkEye II

HawkEye, originally designed by Saab Dynamics (http://www.airbornehydro.com/hawkeyeii), is an airborne combined bathymetric and topographic LiDAR system that is capable of surveying emergent and submerged topography simultaneously using near-infrared and green lasers. At present, the HawkEye system is developed and manufactured by Airborne Hydrography AB, a company formed from the original Saab group. The HawkEye II sensor simultaneously collects bathymetric measurements at 4 Khz and topographic measurements at 64 kHz. The scanner pattern is generated by the two axes of a servo-controlled scanner mirror, which provides a constant incidence angle that is typically 20

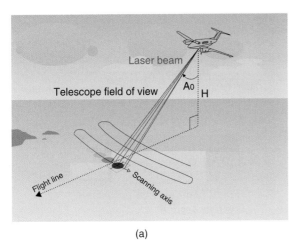
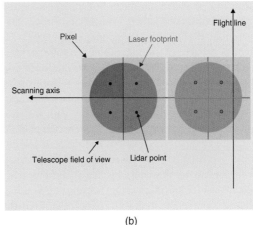

(a)   (b)

Figure 7.3 Illustration of the HawkEye survey principles: a) scanning system with constant incidence angle (a); b) 2D spatial pattern and resolution of measurements. (a) Courtesy of Airborne Hydrography AB and Blom Areofilms.

degrees to the water surface (Figure 7.3), thus facilitating air/water interface refraction correction effects when calculating water-surface and bottom positions. The 5-mJ green laser pulse has a 5-ns width, which allows the laser beam to penetrate water up to 3 Secchi depths (i.e., 50 m) in very clear waters. A unique characteristic of the Hawkeye LiDAR is its division of the telescopic field of view into four quarters, which allows each laser beam footprint (15 mrad for aperture angle) to be divided into four pixels (Figure 7.3). The spatial resolution of the data is then multiplied by 2. Also, the images are collected simultaneously during the survey through a 2 to 16-Mpixel digital camera at 1 Hz. These images facilitate the interpretation of data when experiencing spurious responses due to unexpected targets.

Mounted on an aircraft flying at a speed of 75 meters per second with a maximum swath width of 330 meters, the system can survey up to 350 square km per day. HawkEye typically delivers data (point cloud of topography, water bottom, and water surface) with a density ranging from 0.08 to 0.35 points per square meter for bathymetry and from 1 to 4 points per square meter for topography. The vertical RMSE (root mean square error) of the data is typically less than 25 cm for bathymetry and 15 cm for topography.

7.3.2 Narrow footprint system: EAARL

The Experimental Advanced Airborne Research LiDAR (EAARL) system was designed and built by C. Wayne Wright while working as an instrumentation scientist

for the National Aeronautics and Space Administration (NASA) in Wallops Island, Virginia, USA. The EAARL was initially used to survey coral reefs (Brock et al., 2004) and coastal areas (Nayegandhi et al., 2006; Sallenger et al., 2004; Nayegandhi et al., 2009), but it also found use along rivers and streams (Kinzel et al., 2007; McKean et al., 2008). The system offers a radically different design that sets it apart from other bathymetric LIDARs, as the green (532-nm) laser has a relatively low power (70 μJ/pulse compared with the 5 mJ of other bathymetric LiDARs), which enables eye-safe operation at a much narrower field of view than other systems (Wright and Brock, 2002; Feygels et al., 2003). The area illuminated by the EAARL at a typical surveying altitude (300 m) is approximately 0.20 m, in contrast to a spot size greater than 1 m of other bathymetric systems. The EAARL also uses a much narrower pulse width (1.2 ns compared to up to 7 ns in other systems). While these design parameters offer some advantages for clear, shallow-water environments, they also reduce the maximum surveying depth to less than 25 m (Wright and Brock, 2002).

The EAARL laser transmitter is a Continuum EPO-5000 doubled YAG laser that is currently flown on a Pilatus PC-6 aircraft at a nominal altitude of 300 m. At this altitude, a scanning mirror directs a laser swath that measures approximately 240 m across with a laser sample every 2 m × 2 m along the center and 2 m × 4 m along the edge of the swath. The position of the EAARL laser samples are determined using two precision dual-frequency kinematic carrier phase GPS receivers and an integrated miniature inertial measurement unit. These instruments provide a sub-meter horizontal geolocation for each EAARL laser sample (Wright and Brock, 2002; Nayegandhi et al., 2006; Nayegandhi et al., 2009). The system is capable of storing the entire time series of backscattered laser intensity or waveforms reflected from targets using an array of four high-speed waveform digitisers connected to four sub-nanosecond photo-detectors. The photo-detectors vary in sensitivity, with the most sensitive channel receiving 90% of the reflected photons, the middle channel receiving 9%, and the least sensitive channel receiving 0.9%. A fourth channel is also available for Raman or infrared backscatter. This range in the detector's sensitivities is important because targets along a flight path can vary in physical and optical characteristics, influencing the intensity and structure of the reflected laser intensity.

7.3.3 Airborne LiDAR capacities for fluvial monitoring: a synthesis

Table 7.2 summarises the use of Airborne LiDAR (bathymetric or topographic), using a narrow footprint (NFP) or large footprint (LFP) system for fluvial monitoring, which can be found in the international literature up to mid-2010.

7.4 LiDAR survey design for rivers

7.4.1 Flight planning and optimising system design

Planning a LiDAR survey over rivers for bathymetry or topography requires the use of GPS ground-control

Table 7.2 Fluvial properties surveyed from airborne LiDAR.

Fluvial Property	Surveyed River Area (km²) or Length (km)	LiDAR Type	LiDAR Density	Reference
River bottom elevation	Gardon, FR (2 km)	LFP Bathymetric	high	Bailly et al., 2010
Stream habitat	Bear Valley Creek, USA (10 km)	NFP Bathymetric	high	McKean et al., 2008
Channel morphological features	Bear Valley Creek, USA (10 km)	NFP Bathymetric	high	McKean et al., 2008
	Yakima River, USA (10 km)	LFP Bathymetric	high	Millar et al., 2005
Riparian area delineation	? (10 km)	NFP Bathymetric	high	McKean et al., 2008
Bathymetry	Platte River, USA (16 km)	NFP Bathymetric	high	Kinzel et al., 2007
	Yakima and Trinity Rivers, USA	LFP Bathymetric	high	Hilldale and Raff, 2008
Channel, levee delineation, colluvial hillslope and fan deposits	Dijle and Amblève catchments, FR (1,900 km²)	NFP Topographic	low	Notebaert et al., 2009
Sediment movements	Colorado River, USA (4 km)	LFP Bathymetric	high	Millar et al., 2008
Floodplain hydraulic friction	River Severn, UK (6*6 km²)	NFP Topographic	low	Mason et al., 2003
Riverbed vegetation	Saint-Laurent estuary, CA (1*3 km²)	LFP Bathymetric	high	Collin et al., 2010
Riparian vegetation discrimination	Garonne, FR (2 km²)	NFP Topographic	high	Antonarakis et al., 2008

stations, typically within a 10-km radius, to obtain accurate aircraft geodetic positioning (see next section). In bathymetric surveys, additional information may be required.

As the typical flight height is 250–500 m over the surface and flight lines must be straight, airplane pilots need to know the topography around the river to avoid risks when turning back between flight lines. This consideration leads to problems in mountainous areas and canyons, with rivers having high longitudinal slopes and energy. Here, LiDAR may be advantageous because of the difficulties encountered with sonar surveys. A helicopter platform can be used to reduce these risks in canyons (Millar 2008).

As bathymetric LiDAR systems often use equalisers or signal amplification for photons backscattered by the water bottom, a pre-calibration of this electronic hardware may be required when water properties change (Josset, 2009). This pre-calibration can be done by knowing before flying the dominant optical properties of the waters and river bottoms (albedo, Secchi depths, etc.) or with onboard calibration performed while flying for a short period over an area with constant water depth. This latter case, however, can rarely be accomplished for rivers except by using a helicopter in a static position.

All bathymetric LiDAR systems are class IV lasers, i.e., eye-safe in accordance with the international standard IEC 60825-1. However, as they can emit high-energy laser pulses in the visible domain when flying at low altitudes,

it is important to inform the public that they should not look up at the plane with lenses. Even with a reduced background noise signal, nocturnal surveys are not eye-safe because the human eye normally collects about 12 times more light at night than during the day.

The usual curvilinear shape of rivers makes bathymetric LiDAR surveys less efficient economically than more classical terrestrial LiDAR surveys due to the numerous turns that are necessary between flight lines (see Figure 7.4). This factor can increase the flight time up to five times and, consequently, the LiDAR use time. Once again, the use of a helicopter platform can reduce this difficulty.

Timing errors from large angle beam propagation can be minimised when scanner incident angles do not exceed 15 to 20 degrees, whereas smaller angles may over saturate the laser footprint and distort return waveforms. Semi-circular, rectangular or elliptical swath patterns scanned in-line or cross-track from the aircraft render uniform pulse spacing. It is good practice to overlap flight coverage and swath patterns to maximise data-point collection and lessen the potential for data gaps in the survey. Repeated verification of scanner, beam angle, and time calculations during the survey assure optimal data accuracy and survey integrity (Guenther, 2007).

Another important system design consideration for riverine environments is the footprint size related to laser beam divergence angle. Large footprint (LFP) systems prevent the proper representation of high-slope areas on the river bed, returning the elevation of the shallowest

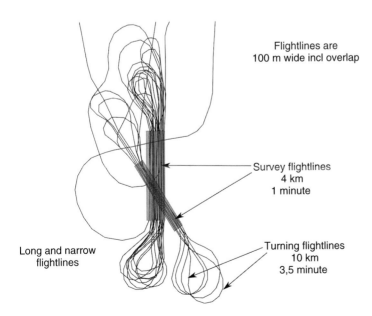

Flightlines are
100 m wide incl overlap

Survey flightlines
4 km
1 minute

Long and narrow
flightlines

Turning flightlines
10 km
3,5 minute

Figure 7.4 Aircraft flight lines on a simple shape example survey (light grey boxes in the middle) with the HawkEye LiDAR system; Courtesy of Airborne Hydrography AB and Blom Areofilms.

feature within the footprint and locating that elevation at the centre of the footprint. Although this may not create large errors when comparing measured and modeled water-surface elevations, it adds bias on the river section with over estimation of bottom elevation. Using these sections for hydraulic modeling can affect our ability to predict velocity conditions, critical to riverine habitats. Consequently, there is also a benefit to more accurately defining meso-scale topographic features of the river bed for volume comparisons related to morphological changes. This can be achieved by using a narrow footprint (NFP) system, such as the EAARL system, or as the four X-Y-Z points obtained from a single footprint with the HawkEye (Figure 7.3).

7.4.2 Geodetic positioning

Ranging LiDAR-delivered data are usually X,Y,Z points. These geodetic positions result from a geometrical chain that includes several processes from optical and electronic time delays up to aircraft attitude and location.

An accurate distance measurement between the LiDAR system and the target is needed, which results from a time waveform or pulse transform as explained in Section 7.2.1.1. The calibration of the timing system is to sub-nanosecond accuracy because a 1-ns error in time results in an altimetrical error of about 15 cm.

The LiDAR system positioning in time as well as laser beam angle must also be precisely determined. This is done by first using Kinematic GPS (KGPS) onboard, with the on-the-fly (OTF) technique, carrier-phase ambiguity resolution and correction with a near-ground base DGPS station. This step prevents erroneous initialisations, and it provides highly precise (sub-decimeter accuracy) horizontal and vertical positions for aircraft with respect to the WGS-84 ellipsoid. Additionally, the aircraft's attitude and, thus, the laser beam angles are collected in time through an inertial measurement unit (IMU) for pitch, roll, and yaw registration. As Guenther (1985) has stated, 'at a 400-m altitude and with a nominal 20-degree nadir angle, a system angle error of 0.05 degrees (<1 mrad), which equates to a nadir angle error of 0.10 degrees, would yield a 25-cm error in the vertical height of the aircraft'. If 0.01 degrees is the acceptable limit for angle accuracy, ' this can only be accomplished by applying an inverse algorithm to flight data collected occasionally for the purpose of angle calibration' (Guenther et al., 1999). Moreover, as GPS and IMU collect data at a 1-Hz frequency, they need to be properly interpolated in time to the laser frequency.

Consequently, a better altimetric (decimetric) and planimetric (metric) accuracy is obtained in LiDAR data than in photogrammetry data (Baltsavias 1999).

7.5 River characterisation from LiDAR signals

7.5.1 Altimetry and topography

7.5.1.1 Emergent terrain topography

The last returns, or echos, of an emitted laser pulse do not necessarily reflect from the ground, especially if the target is dense. In addition, some negative outliers originating from multipath reflection may be positioned below the ground surface (Kobler et al., 2006). Consequently, it is necessary to mathematically separate the bare-earth reflection from other reflections to produce digital elevation models (DEM). As shown in Figure 7.5a, the last echoes are first extracted from the entire point cloud, and they are filtered to keep only bare-earth echoes. These latter points are interpolated to produce a raster DEM.

Several filtering techniques have been developed that can be summarised in four groups (Sithole and Vosselman, 2004). The first group, called 'slope-based', compares the slope between two points to a given threshold. One of the two points is then classified as an object if the slope exceeds the threshold or as bare-earth if it does not. The second group, called 'block-minimum', directly produces a raster DEM by selecting the minimal height value of points in each raster cell. The third group, called 'surface-based', constructs a surface from a minimum number of selected points. The surface is iteratively made more complex by the addition of successive points selected by certain distance criteria. The main method belonging to this group, called 'iterative TIN', is the most commonly used method for producing a DEM from LiDAR data (Axelsson, 2000). The last group, called 'clustering/segmentation', studies the positions of point clusters by comparing them to their neighbours to detect objects with noticeable edges and elevations. A raster DEM is then computed using interpolation techniques (kriging, inverse distance weighting), with the splines give the best results (Brovelli et al., 2004).

In general, all of these filters produce satisfactory results for low complexity landscapes (low slope, sparse vegetation, small buildings). However, landscapes with bare-earth discontinuities, such as mountainous areas or steep riverbanks still pose problems for filtering algorithms.

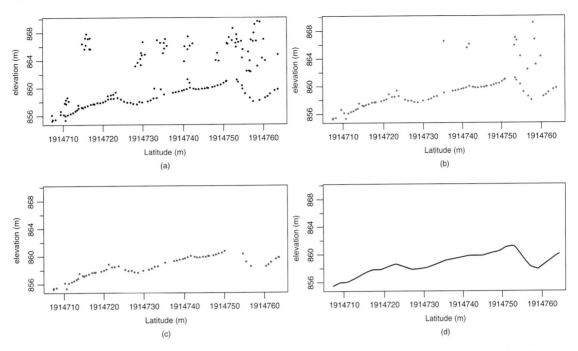

Figure 7.5a A generic method for producing DEM illustrated along a latitudinal profile: a) row points cloud, b) selection of last echoes, c) extraction of terrain points, d) terrain points interpolation to produce a raster DEM.

7.5.1.2 Vegetation structure and hydraulic friction

Compared to optical techniques, LiDAR remote sensing has the ability to provide information about the vertical distribution of quite sparse targets, making LiDAR an innovative technique for studying vegetation. The best strategy for extracting information about vegetation from LiDAR data is to perform a spatial analysis of the point cloud. Using this approach, Antonarakis et al. (2008) recognised the land-cover type (planted forest, natural forest, type of soil) of three river meanders based on the spatial distribution and reflectance of the LiDAR returns. They determined the type of forest with an accuracy between 83% and 98%. Nevertheless, because of the discrete sampling of the LiDAR and due to occlusions that can occur under a return, a spatial analysis of the point cloud can pose some problems. Consequently, interpolated rasters are traditionally used to derive vegetation parameters. With this approach, a canopy height model (CHM) is computed as the difference between the digital surface model (DSM), which comes from the interpolation of the first returns, and the digital elevation model (DEM) (Figure 7.5b). The CHM, therefore, represents the vegetation surface that is distinct from the ground surface.

CHM is widely used to identify the location, height, and crown width of trees (Popescu and Wynne, 2004). In addition, Geerling et al. (2007) combined CHM, other LiDAR-derived rasters representing vegetation structure (minimum, mean, median, standard deviation, and range heights), and spectrographic image rasters to classify and map the plant community of the vegetation in a floodplain. If the classification successfully recognises bush (87% accuracy) and forest (100% accuracy) from LiDAR rasters alone, the combination with spectrographic image rasters allows the recognition of bare and pioneer communities, grass and herbaceous vegetation, and herbaceous and low woody vegetation with a 57% to 64% accuracy.

Because LiDAR is capable of collecting information under sparse targets and provides data at a high spatial density, the floodplain vegetation structure can be mapped efficiently. Consequently, LiDAR has renewed the use of remote sensing in hydraulic modeling, particularly in the estimation of floodplain friction parameters from vegetation height statistics (Asselman et al., 2002; Cobby et al., 2001; Cobby et al., 2003; Mason et al., 2003) or from fusion between high spatial resolution imagery and LiDAR data (Forzieri et al., 2010).

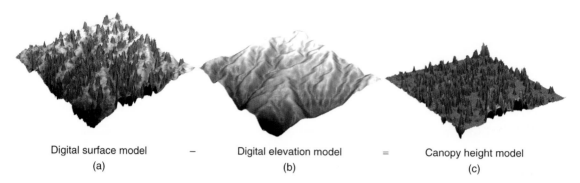

Digital surface model   –   Digital elevation model   =   Canopy height model
(a)   (b)   (c)

Figure 7.5b Computation of the Canopy Height Model (CHM, a) as the difference between the Digital Surface Model (DSM, b) and the Digital Elevation Model (DEM, c). Courtesy of Airborne Hydrography AB.

Similarly, the canopy heights or positions of isolated trees can be extracted from LiDAR data and used in hydraulic models (Straatsma and Baptist, 2008).

7.5.2 Prospective estimations

7.5.2.1 Water column properties

As a photon of light is transmitted through water, it may interact with water molecules, dissolved substances, or suspended particles. As a result, the photon may be absorbed, backscattered (elastic LiDAR), or re-emitted as a photon at a slightly altered wavelength (inelastic LiDAR) (Peeri and Philpot 2007).

In the case of elastic LiDAR, as explained in Sections 7.2.2 and 7.4.1, waveforms of green laser pulses penetrating the water surface show decreasing power between the water surface and bottom peaks due to the exponential decay of transmission within the water column. This decreasing waveform power (P) due to water volume backscattering can be simplified as the following equation:

$$P(d) = B\exp^{-2kt} \qquad (7.4)$$

where t denotes time, and it can be transformed in depth (d) by:

$$d = c_2 t \qquad (7.5)$$

where k denotes the attenuation coefficient (approximation of the absorption coefficient), c_2 denotes the light celerity in water and B is proportional to the volume backscatter coefficient (Billard et al., 1986, p.2081). As a consequence, the slope of this decreasing part of the waveform is theoretically a measure of k, and a relationship to turbidity can be established. Recently, this approach

has only been applied in coastal areas (Billard et al., 1986; Churnside, 2008; Shamanaev, 2007). For example, Billard et al. (1986) used the LADS system in a deep area near the Australian coast. The signal used for analysis ranged from approximately 1 m up to 33 m in depth to remove the water surface and bottom peaks. By regression and by using the Kalman filter, they estimated the B and k parameters. Then, the turbidity profiling across the survey was considered homogeneous in the vertical column with an estimated 15% relative error. However, in all of these studies, estimations are only possible for water depths greater than about five meters, which make them generally unsuitable for riverine environments.

In the case of inelastic LiDAR, the relation between water-column characteristics to Raman waveforms using the ranging SHOALS system has been successfully explored by Peeri and Philpot (2007). We can also mention in passing the use of more specific fluorosensors on oceanic waters for phytoplankton pigment mapping (Hoge et al., 1983; Babichenko et al., 1993).

7.5.2.2 River bottom features

The interaction of LiDAR signals with river bottoms has the future potential to provide information on the nature of the substratum and the presence of immersed objects or vegetation. To date, the authors are not aware of any published studies (Irish and White, 1998) concerning river bottom feature extraction from bathymetric LiDAR data. The most recent developments of this technique have primarily concerned coastal areas (Wang and Philpot, 2007, Collin et al., 2011a, Collin et al., 2011b), but these developments foreshadow what could be possible in riverine environments. First, the enhancement and interpretation of a LiDAR-derived DEM allows the extraction of features of specific interest, such as seabed morphology (Finkl, 2005) or immersed objects (West and Lillycrop, 1999). These last authors reported a 100% probability for detecting a 2-m immersed cube with a sampling density of 3 m x 3 m after Guenther et al. (1996). Full waveform signals would also allow the detection of mid-water objects as long as their energy return is not mixed with volume scattering or weakened by too large of a beam dispersion at great depths.

Temporal studies based on a DEM comparison showed the possible detection of moving sandbars (Irish and White, 1998) or artificial objects such as a disposal mound under three meters of water (Irish and Lillycrop, 1999). Full waveform information can be used to derive not only DEMs but also bottom roughness maps, focusing on signal variability (Brock et al., 2004).

Finally, both amplitude and full waveform information are processed to classify the seabed into different bottom types (Wang and Philpot, 2007; Kuus et al., 2008; Collin et al., 2010, Collin et al., 2011b). In these last studies, the unmixing between contributions of depth, local bottom shape, and acquisition configuration was solved through statistical analysis of the full waveforms datasets, which allows the classification of bottom types into either maps describing vegetation cover or maps of seabed habitats.

7.6 LiDAR experiments on rivers: accuracies, limitations

7.6.1 LiDAR for river morphology description: the Gardon River case study

The Gardon River is a Rhône River tributary located in the South of France. The Gardon flows from an elevation of 1,450 m to 30 m over a 120-km length. It is a gravel-bed river within a sub-humid Mediterranean hydrological context, i.e., with strong rain events in autumn that alter the river morphology.

A 1.5-km long reach located in the mid-reaches of the Gardon River was surveyed in March 2007 with the HawkEye II system mounted on an Aerocommander 690A airplane flying at a height of 250 m. It took 30 minutes to fly over this reach using three parallel and overlapping strips, each having a width of about 110 m. The LiDAR waveform data were processed using the coastal survey studio suite, which provided X,Y,Z LiDAR points on the wetted riverbed, the water surface, and the riverbanks or vegetation canopy (Bailly et al., 2010). Below, only the wetted riverbed points are considered.

The surveyed Gardon River section contained two riffle-pool sequences and two different types of river bottom. Upstream, there was a rough bedrock made of marls, and downstream there was a smooth gravel bed composed of silica gravels. Both sub-sections had the same bottom albedo because periphyton covered the entire reach. At the time of the survey, the water had a Secchi depth of about 0.8 m, and the water velocity had a maximum of 1.5 m.s^{-1} in riffles. The water depths ranged up to 4.45 m in pools with a mean depth of 0.64 m, and 43% of its area had a depth less than 0.4 m. The mean longitudinal slope of the reach was 0.12%. There was little riparian vegetation on the right riverbank.

Concurrently with the LiDAR flight, tacheometry was used to survey the polylines marking the riverbanks, and dual frequency GPS was used to survey the river bottom. A total of 5,443 topographic points of about 2-cm accuracy for Z position inside the wetted riverbed were collected and referenced. In depths beyond those that can be waded, a tagline mounted on a prism onboard a boat, combined to a tacheometer to accurately shoot the water surface position, were used. Higher positioning errors can occur in these higher depths (dispersion error of ∼5 cm for one standard deviation on Z). A depth was attributed to each of the reference points. The spatial sampling scheme of the topographic points consisted of profiles across the river (Figure 7.6a-A).

Approximately 50,000 Lidar X,Y,Z points on the river bottom were obtained (depicted in colour as the Z values in Figure 7.6a-B). This set of points gave a mean density of one point per 0.9 m^2. Computing the accuracy of data having a spatial location (interpolation problem) and resolution (scaling problem) different from the reference data is not direct. For that reason, a method was developed to compute LiDAR riverbed elevation data bias and accuracy (random error) in comparison with the reference and data limitations, i.e., the minimum and maximum detectable depth (Bailly et al., 2010). This method was based on geostatistics using anisotropic, ordinary block-kriging (Atkinson and Tate, 2000) within a curvilinear channel-fitted coordinate reference system similar to the one proposed by Legleiter and Kyriakidis (2006). This technique permits upscaling as well as the interpolation of reference data, and it takes into account the uncertainties in the results (the so-called weighted errors).

Regarding the LiDAR accuracy, the computations for river bottom elevation exhibited a bias depending on water depth (Figure 7.6b-A). Added to that bias, a random error with a 0.32-m standard deviation was found

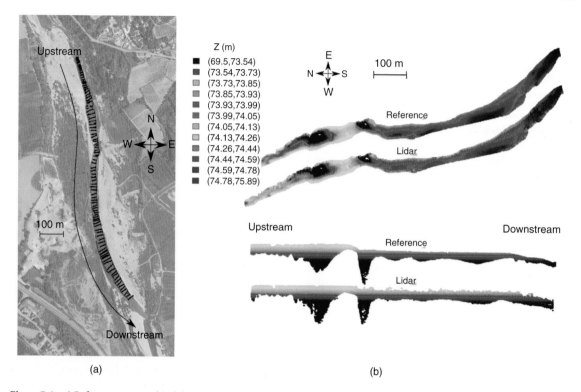

Figure 7.6a a) Reference topographical data sampling scheme over the Gardon River reach superimposed on an red, green, blue (RGB ortho image ©IGN. b) 1- up: reference altimetric image of the riverbed (top) compared to the LiDAR (down) in rotated geographical space with colours depending on elevation (Z); and 2- down: altimetric images along the river's upstream-downstream curvilinear main axis.

regardless of the depth. The LiDAR bias, due to the inherent multi-scattering within the water column, can be empirically corrected using some control points (Guenther, 1985, p. 283). However, measuring a water depth less than 32 cm (28% of the Gardon each area) is unrealistic when considering the signal-to-noise ratio. In this case study, we observed that depths up to 3.9 meters (about 5 times the Secchi depth) had been reached. However, 32 cm appears to be a sufficient accuracy within a 'line-thickness' related to riverbed roughness because the granular distribution of the gravel produces highly variable micro-topography (roughness), and altimetrical comparisons highly sensitive to small planimetric displacement.

Moreover, this experiment shows that LiDAR provides a continuous topographic surface from the underwater riverbed to the riparian areas, as seen in Figure 7.6b-B. LiDAR shows continuous profiles, thus significantly improving the overall surface model (Hilldale and Raff, 2008). The landforms of wetted riverbed (pools, rifles) are well reproduced (McKean et al., 2008). The obtained accuracy is definitely suitable for mesoscale hydroecological studies of rivers and for mesohabitat mapping (Borsanyi et al., 2004), but the accuracy may be more critical for numerical hydraulic studies on small spatial extents (Casas et al., 2006).

7.6.2 LiDAR and hydraulics: the Platte River experiment

Conventional LiDAR surveys in emergent areas provide spatially explicit topographic data sets that have been incorporated into hydraulic models for computing flow over floodplains (Tayefi et al., 2007) or in conjunction with passive optical data collected in the wetted portion of a river for modeling in-channel flows (Hicks et al., 2006). The simultaneous measurement of emergent and submerged channel topography with a single sensor (EAARL) was attempted in 2002 and 2005 along a 160-km reach of the Platte River (Kinzel et al., 2007). The Platte River is a wide (100 to 300 m), shallow (<1 m), braided,

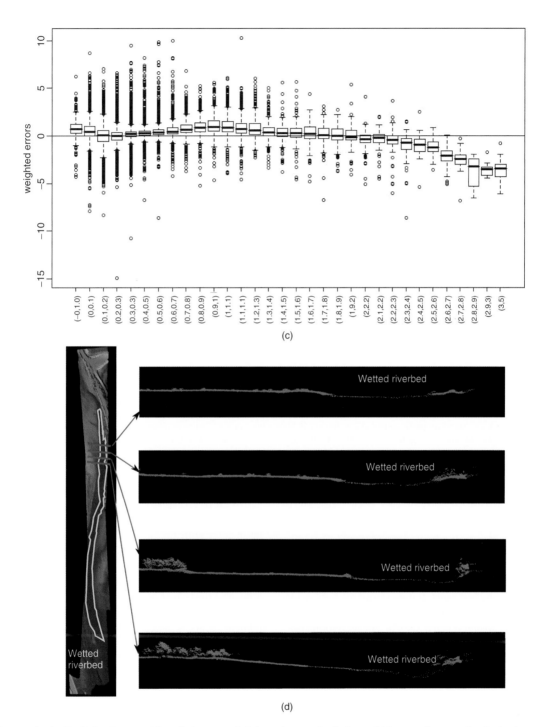

(c)

(d)

Figure 7.6b a) LiDAR river bottom altimetrical error statistic box-plots per depth classes. b) images of topography obtained continuously inside and outside the wetted riverbed with the corresponding profiles of the initial LiDAR point cloud (courtesy of Airborne Hydrography cd): colours are related to topography.

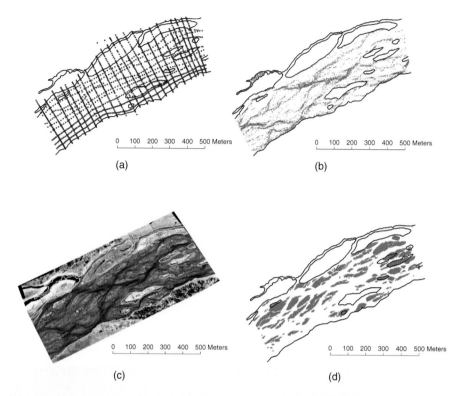

Figure 7.7a a) Topographic survey points collected in the study reach. b) The location of waveforms identified by the ALPS shallow-water bathymetry algorithm in 2002 along the Rowe Sanctuary study reach. c) Georeferenced aerial natural-colour imagery taken on March 26, 2002. For reference and clarity the black lines in Figures a, b, and d (representing the edges of the banks and islands) are shown in yellow on Figure c. d) Location of emergent sandbars defined using aerial thermal infrared imaging. Flow is from lower left to upper right.

sand-bedded channel in central Nebraska, USA. The river channel and riparian corridor have been transformed by diminution of flows from upstream water-resource development.

On March 26, 2002 ten passes were made with the EAARL over a 1-km reach of the Platte River within the Rowe Sanctuary. Simultaneously, measurements of channel topography and water-surface elevations were made by wading in the channel using a survey-grade global positioning system (Figure 7.7a-A). The waveforms collected from the EAARL in the reach, both from emergent and submerged topography, were first processed with the USGS Airborne LiDAR Processing Software (ALPS) using a first-return algorithm (Bonisteel et al., 2009). This algorithm operated on all the waveforms regardless of their shape and associated target type. The waveforms were then processed using a shallow-water bathymetric algorithm. Fewer points resulted from bathymetric processing because of the processing criteria used for discrimination.

Figure 7.7a-B shows the location of the waveforms identified by the ALPS bathymetry algorithm. The points follow the braided channels. An aerial natural colour photograph taken at the same time as the EAARL survey is shown for comparison (Figure 7.7a-C). The locations of the bathymetric waveforms correspond well to the darker portions of the aerial photograph indicating locations where water influences the transmission of sunlight to the camera.

The following procedure was used to examine the influence of using EAARL-collected channel topography over conventional measurements in a multidimensional computational hydraulic model. The topographic points collected with the conventional survey were used with EAARL points collected along the emergent floodplain and islands to map the topography of the river reach. These points were used as input to MD_SWMS, a software interface created by the United States Geological Survey for multidimensional hydraulic models (McDonald et al., 2006). The depth-averaged 2-dimensional hydraulic

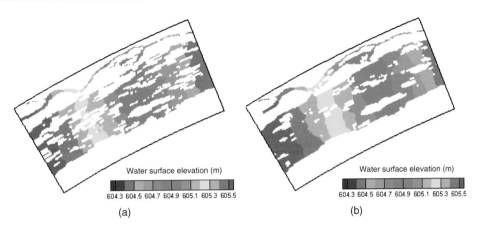

Water surface elevation (m)

604.3 604.5 604.7 604.9 605.1 605.3 605.5

(a)

Water surface elevation (m)

604.3 604.5 604.7 604.9 605.1 605.3 605.5

(b)

Figure 7.7b Map of inundation and water-surface elevation in the study reach predicted by FASTMECH model using in-channel topography collected with a) a ground survey b) in-channel topography collected with EAARL sampled at the same locations as the ground survey.

model FaSTMECH (Flow and Sediment Transport with Morphologic Evolution of Channels) was applied to predict depth, velocity, and water-surface elevation. FaST-MECH solves the vertically and Reynolds averaged Navier Stokes equations on curvilinear, orthogonal coordinate system. A 5-m by 5-m curvilinear orthogonal numerical grid was created using the grid generation tools in MD_SWMS. The topographic points were mapped to the grid using a template method which searches in a bin of specified dimensions around each grid node for the topographic point that comes closest to it. If a single point is located in the search bin, the node is given that elevation value. If multiple points are found, the node is given an elevation value that is the inverse distance weighted average of those points. If no point is located, the search bin is expanded in size until a point or points are found. The boundary conditions for the model were specified: a discharge ($19 \, m^3/s$) and a downstream water-surface elevation. The downstream water-surface elevation was obtained from the ground survey as were longitudinal measurements of water-surface elevation collected throughout the reach. A single-valued drag coefficient was used to parameterise hydraulic roughness in the reach. The drag coefficient was adjusted to provide the best fit between the predicted and measured water-surface elevations through the reach (root mean square error 0.045 m). Aerial thermography was used to identify the locations of emergent sandbars at the modeled streamflow (Figure 7.7a-D). The predicted water-surface elevation in the reach using the ground survey derived measurements of in-channel topography and the FaSTMECH model is shown in Figure 7.7b-A. There was good correspondence

between the prediction of unwetted area in the channel (Figure 7.7b-A) and the locations of emergent sandbars (Figure 7.7a-D).

To ensure any vertical bias in the LiDAR survey was identified and corrected for, the ground-based GPS measurements made in 2002 on the emergent sand targets, less complex with regard to laser backscatter than submerged targets, were compared with EAARL points processed with the first-return algorithm and located less than 0.50 m away from a surveyed ground-truth point. The GPS ellipsoid heights were subtracted from the corresponding EAARL ellipsoid heights and a mean error of 0.18 m was identified between the data sets. Because the measured EAARL heights were less than the corresponding GPS measurements, 0.18 m was added to each EAARL point to remove the bias associated with the mean error. The shallow-water bathymetry algorithm was then applied to the data set to identify those waveforms, and the same bias adjustment was applied.

A second topographic dataset was created for modelling that included EAARL-derived floodplain and island points and EAARL in-channel points. The EAARL in-channel points were selected from a dataset that included a combination of EAARL points processed with the bathymetry algorithm and EAARL points processed with the first-return algorithm. Points processed with both algorithms were required to map the emergent sandbars and submerged areas within the channel. We removed the first-return points that fell inside a 2-m radius of each EAARL bathymetric point. This was done to minimise the influence of the more spatially dense first-return points on the neighbouring sparser bathymetric points but also to preserve the first-return points that were needed to map the emergent sandbars in the reach. To ensure that the spatial densities of the points were equivalent for modelling, we interpolated the EAARL data to a surface with kriging and selected points from that surface that were coincident to the ground survey.

The FaSTMECH model was run with the EAARL dataset using the same boundary conditions and drag coefficient used for the topographic dataset collected on the ground. The predicted water-surface elevation for this simulation is shown in Figure 7.7b-B. While the simulation also predicts emergent sandbars in the reach, the root mean square error comparing the predicted water-surface elevation to the water-surface elevations measured on the ground was higher than from the ground survey, 0.097 m. The predicted water-surface elevations were also higher than the measured elevations by an average of 0.09 m. A possible reason for this is that the EAARL system measured the deeper portions of the channel higher than their actual elevation (Kinzel et al., 2007). This was confirmed by a comparison of the ground survey and the coincident LiDAR which points indicated a mean error of -0.1 m. The higher elevation LiDAR points had the effect of reducing the conveyance through the reach and predicting the water-surface elevation higher than what was observed.

7.7 Conclusion and perspectives: the future for airborne LiDAR on rivers

In the future, hardware and software developments will enhance the resolution, accuracy, and types of data products derived from riverine airborne LiDAR surveys. First, the use of polarised LiDAR will probably enhance the capacities of extremely shallow water bathymetry (Mitchell et al., 2010). With regard to spatial coverage and resolution, the pulse repetition frequency of current bathymetric LiDARs is much smaller than airborne topographic mapping LiDARs. Using the EAARL sensor as an example, an increase in the pulse repetition frequency from 5,000 Hz to 30,000 Hz is planned (C. Wayne Wright, USGS personal communication, 2010). The laser power will also be increased from 70 micro joules to 700 micro joules. To remain eye safe, the laser energy will be divided over three laser spots, each of the spots will remain at the current (~ 20 cm) size. The effect will be to triple the number of laser samples in each raster or swath. This greater rate of sampling and laser power per sample could be advantageous in fluvial settings where multiple flight lines have been used to increase point density and spatial resolution.

Techniques that improve ranging accuracy to shallow-water riverine targets are also needed. Extracting information from bathymetric waveforms may be improved by combining theoretical models of waveform shape with advanced pattern-recognition software routines. Increased incorporation of passive optical sensors with LiDAR systems is also beneficial. Synergistic integrated processing of hyperspectral imagery and LiDAR has been identified as a growing trend (Crane et al., 2004). This involves using one data set to facilitate the processing of another. These possibilities are especially tantalising with respect to riverine surveys. If used with bathymetric waveforms, hyperspectral data may be a valuable complementary data set with the potential of refining shallow depth estimation. Additionally, hyperspectral data could be leveraged to provide information about the edges and composition of riverine features. These emerging datasets will provide managers a continuous representation of the river channel and floodplain which can inform decisions on a variety of issues including: connectivity of in-channel and riparian habitat, flood risk, land use, fate and transport of pollutants, and navigational concerns. These data would also provide input to models and simulations which could be used by stakeholders to evaluate or illustrate the consequences of proposed management actions. The future development of remote sensing technologies and LiDAR for mapping and monitoring river systems will continue to be stimulated by competing demands for water and with societal concern for the inventory and preservation of aquatic resources.

However, the accuracy and spatial resolution of satellite LiDAR data on continental waters is far better than radar data (Braun et al., 2004; Baghdadi et al., 2011) and considering the progress and increasing use of spatial laser technology, we can speculate that future satellite LiDAR missions would probably help in accurate monitoring of river system dynamics.

Any use of trade, firm, or product names is for descriptive purposes only and does not imply endorsement by the U.S. Government.

References

Allouis, T., Bailly, J., Pastol, Y., and LeRoux, C. (2010). 'Comparison of LiDAR waveform processing methods for very shallow water bathymetry using Raman, near-infrared and green sign', *Earth Surface Processes and Landforms* **35**(6), 640–650.

Antonarakis, A.S., Richards, K.S., and Brasington, J. (2008). 'Object-based land cover classification using airborne LiDAR', *Remote Sensing of Environment* **112**(6), 2988–2998.

Asselman, N., Middlekoop, H., Ritzen, M., and Straatsma, M. (2002). 'Assessment of the hydraulic roughness of river flood plains using laser altimetry', *in* FJ Dyer, MC Thoms, and JM Olley, ed., 'Structure, function and management implications of fluvial sedimentary systems', IAHS Publication vol. 276, International Commission On Continental Erosion and Unesco, Alice Springs, pp. 381–388.

Atkinson, P.M. and Tate, N.J. (2000). 'Spatial scale problems and geostatistical solutions: A review', *The Professional Geographer* **52**(4), 607–623.

Axelsson, P., (2000). 'DEM generation from laser scanner data using adaptive TIN models'. International Archive of Photogrammetry and Remote Sensing XXXIII (Part B4), pp. 110–117.

Babichenko, S., Poryvkina, I., Arikese, V., Kaitala, S., and Kuosa, H. (1993). 'Remote sensing of phytoplankton using laser-induced fluorescence', *Remote Sensing of Environment* **45**, 43–50.

Baghdadi, N., Lemarquand, N., Abdallah, H., and Bailly, J.S. (2011). 'The relevance of GLAS-ICESat elevation data for the monitoring of river networks', *Remote Sensing* **3**, 708–720.

Bailly, J.S., LeCoarer, Y., Languille, P., Stigermark, C., and Allouis, T. (2010). 'Geostatistical estimation of bathymetric LiDAR errors on rivers', *Earth Surface Processes and Landforms* **35**, 1199–1210.

Baltsavias, E.P. (1999). 'A comparison between photogrammetry and laser scanning', *ISPRS Journal of Photogrammetry and Remote Sensing*, 54(2-3), 83–94.

Billard, B., Abbot, R.H., and Penny, M.F. (1986). 'Airborne estimation of sea turbidity parameters from the WRELADS laser airborne depth sounder', *Appl. Opt.* **25**(13), 2080–2088.

Bonisteel, J.M., Nayegandhi, A., Wright, C.W., Brock, J.C., and Nagle, D.B., 2009, 'Experimental Advanced Airborne Research Lidar (EAARL) Data Processing Manual': U.S. Geological Survey Open-File Report, 2009-1078, 38p.

Borsanyi, P., Alfredsen, K., Harby, A., Ugedal, O., and Kraxner, C. (2004). 'Meso-scale habitat classification method for production modelling of Atlantic salmon in Norway', *Hydroécologie Appliquée* **14**, 119–138.

Braun, A., Cheng, K., Csatho, B., and Shum, C. (2004). 'ICESat Laser Altimetry in the Great Lakes', *in* 'Proceedings of the 60th Annual Meeting of the Institue of Navigation', 409–416.

Brock, J., Wright, C., Clayton, T., and Nayegandhi, A. (2004). 'LIDAR optical rugosity of coral reefs in Biscayne National Park, Florida', *Coral Reefs* **23**, 48–59.

Brovelli, M.A., Longoni, U.M., and Cannata, M. (2004). 'LIDAR data filtering and DTM interpolation within GRASS', *Transactions in GIS*, **8**(2), 155–174.

Burikov, S.A., Churina, I.V., Dolenko SA, Dolenko, T.A., Fadee, V.V. (2004). 'New approaches to determination of temperature and salinity of seawater by laser Raman spectroscopy'. *EARSeL eProceedings*, **3**(3), 298–305.

Caloz, R. and Collet, C. (2001). *Précis de télédétection: vol.3: traitements numériques d'images de télédétection*, Presses de l'Université du Québec, Sainte-Foy, CAN.

Casas, A., Benito, G., Thorndycraft, V., and Rico, M. (2006). 'The topographic data source of digital terrain models as a key element in the accuracy of hydraulic flood modelling', *Earth Surface Processes and Landforms* **31**, 444–456.

Chauve, A., Vega, C., Bretar, F., Durrieu, S., Allouis, T., Pierrot-Deseilligny, M., and Puech, W. (2009). 'Processing full-waveform lidar data in an alpine coniferous forest: assessing terrain and tree height quality', *International Journal of Remote Sensing*, **30**(19), 5211–5228.

Churnside, J.H. (2008). 'Polarization effects on oceanographic lidar', *Optics express* **16**(2), 1196–1207.

Cobby, D., Mason, D., and Davenport, I. (2001). 'Image processing of airborne scanning laser altimetry data for improved river flood modelling', *ISPRS Journal of Photogrammetry and Remote Sensing* **56**(2), 121–138.

Cobby, D., Mason, D., Horritt, M., and Bates, P. (2003) 'Two-dimensional hydraulic flood modelling using a finite-element mesh decomposed according to vegetation and topographic features derived from airborne scanning laser altimetry', *Hydrological Processes*, **17**, 1979–2000.

Collin, A., Long, B., Archambault, P. (2010). 'Salt-marsh characterization, zonation assessment and mapping through a dual-wavelength LiDAR', *Remote Sensing of Environment*, 114 (3), 520–530.

Collin, A., Long, B., and Archambault, P. (2011). 'Benthic classifications using bathymetric LIDAR waveforms and integration of local spatial statistics and textural features', *Journal of Coastal Research* **62**, 86–98.

Collin, A., Archambault, P., and Long, B. (2011). 'Predicting species diversity of benthic communities within turbid nearshore using full-waveform bathymetric LiDAR and machine learners', *PLoS ONE* **6**(6).

Crane, M., Clayton, T., Raabe, E., Stoker, J., Handley, L., Bawden, G., Morgan, K., and Queija, V. (2004). 'Report of the U.S. Geological Survey LiDAR Workshop Sponsored by the Land Remote Sensing Program and held in St. Petersburg, FL, November 2002', U.S. Geological Survey Open-File Report, 2004-1456, 72 p.

Feygels, V., Wright, C., Kopilevich, Y., and Surkov, A. (2003). 'Narrow-field-of-view bathymetrical lidar: theory and field test', *in* 'Proc. Ocean Remote Sensing and Imaging II; SPIE', 1–11.

Finkl, C. (2005). 'Interpretation of seabed geomorphology based on spatial analysis of high-density airborne laser bathymetry', *Journal of Coastal Research* **21**(3), 501–514.

Flamant, P.H. (2005). 'Atmospheric and meteorological LiDAR: from pioneers to space applications', *C.R. Physique* **6**, 864–875.

Forzieri, G., Moser, G., Vivoni, E.R., Castelli, F., and Canovaro, F. (2010). 'Riparian vegetation mapping for hydraulic roughness estimation using very high resolution remote sensing data fusion', *Journal of Hydraulic Engineering*, **136**, 855–867.

Geerling, G.W., Labrador-Garcia, M., Clevers, J.G.P.W., Ragas, A.M.J., and Smits, A.J.M. (2007). 'Classification of floodplain vegetation by data fusion of spectral (CASI) and LiDAR data', *International Journal of Remote Sensing* **28**(19), 4263–4284.

Guenther, G. (1985). 'Airborne laser hydrography: system design and performance factors', Technical report, NOAA professional paper series.

Guenther, G.C., Thomas, R.W.L., and LaRocque, P.E. (1996). 'Design considerations for achieving high accuracy with the SHOALS bathymetric lidar system', *in* 'Proc. SPIE', pp. 54–71.

Guenther, G.C., Cunningham, A.G., LaRocque, P.E., and Reid, D.J. (1999). 'Meeting the accuracy challenge in airborne lidar bathymetry', *EARSel*.

Guenther, G. (2007). 'Digital Elevation Model Technologies and Applications: The DEM Users Manual', American Society for Photogrammetry and Remote Sensing, Chapter 8: 'Airborne lidar bathymetry', pp. 253–320.

Hicks, D., Shankar, U., Duncan, M., Rebuffé, M., and Aberle, J. (2006). *Braided Rivers: Process, Deposits, Ecology and Management*, Blackwell, Use of remote-sensing with two-dimensional hydraulic models to assess impacts of hydro-operations on a large, braided, gravel-bed river: Waitaki River, New Zealand, pp. 311–326.

Hilldale, R.C. and Raff, D. (2008). 'Assessing the ability of airborne LiDAR to map river bathymetry', *Earth Surface Processes and Landforms* **33**(5), 773–783.

Hofton, M.A., Minster, J.B., and Blair, J.B. (2000). 'Decomposition of laser altimeter waveforms', *IEEE Transactions on Geoscience and Remote Sensing* **38**, 1989–1996.

Hoge, F.E. and Swift, R.N. (1983). 'Airborne dual laser excitation and mapping of phytoplankton photopigments in a Gulf Stream Warm Core Ring', *Applied Optics* **22**, 2272–2281.

Irish, J. and White, T. (1998). 'Coastal engineering applications of high-resolution lidar bathymetry', *Coastal Engineering* **35**(1-2), 47–71.

Irish, J.L. and Lillycrop, W.J. (1999). 'Scanning laser mapping of the coastal zone: the SHOALS system', *ISPRS Journal of Photogrammetry and Remote Sensing* **54**, 123–129.

Josset, D. (2009). 'Etude du couplage radar-lidar sur plates-formes spatiales et aéroportées: Application à l'étude des nuages, des aérosols et de leurs interactions', PhD thesis, Université Pierre et Marie Curie - Paris VI.

Jurand, T., Kam, M., and Fischl, R. (1989). 'Mathematical model of echoes in laser-based aerial bathymetric surveying', *in* 'Proceedings of the 281h Conference on Decision and Control, Tampa, Florida, December 1989', pp. 3–15.

Jutzi, B. and Stilla, U. (2006). 'Characteristics of the measurement unit of a full-waveform laser system', *Revue Francaise de Photogrammetrie et de Teledetection*, **182**, 17–22.

Kinzel, P.J., Wright, C.W., Nelson, J.M., and Burman, A.R. (2007). 'Evaluation of an experimental LiDAR for surveying a shallow, braided, sand-bedded river', *Journal of Hydraulic Engineering* **133**(7), 838–842.

Kobler A., Pfeifer N., Ogrinc P., Todorovski L., Oštir K., and Džeroski S. (2006). 'Using redundancy in aerial LIDAR point cloud to generate DTM in steep forested relief'. Proceedings of the Workshop on 3D Remote Sensing in Forestry, Vienna, Austria. T. Koukal, W. Schneider (Eds.)

Kuus, P., Clarke, J.H., and Brucker, S. (2008). 'SHOALS 3000 Surveying Above Dense Fields of Aquatic Vegetation Quantifying and Identifying Bottom Tracking Issues', *in* 'CHC 2008, Victoria B.C., May 5–8, 2008'.

Legleiter, C.J. and Kyriakidis, P. (2006). 'Forward and inverse transformations between Cartesian and channel-fitted coordinate Systems for meandering rivers', *Mathematical Geology* **38**(38), 927–957.

Lesaignoux, A., Bailly, J., Allouis, T., and Feurer, D. (2007). 'Épaisseur d'eau minimale mesurable en rivière sur fronts d'ondes lidar simulés', *Revue Francaise de Photogrammetrie et de Teledetection* **186**, 48–53.

Mason, D., Cobby, D., Horritt, M., and Bates, P. (2003). 'Floodplain friction parameterization in two-dimensional river flood models using vegetation heights derived from airborne scanning laser altimetry', *Hydrological Processes* **17**(9), 1711–1732.

McDonald, R., Nelson, J., and Bennett, J. (2006). 'Multidimensional surface-water modeling system user's guide', *U.S. Geological Survey Techniques and Methods*, 6-B2, 156 p.

McKean, J., Isaak, D., and Wright, C. (2008). 'Geomorphic controls on salmon nesting patterns described by a new, narrow-beam terrestrial-aquatic lidar', *Frontiers in Ecology and the Environment* **6**(3), 125–130.

Millar, D. (2005). 'Using airborne LIDAR bathymetry to map shallow water river environments' Coastal GeoTools 2005, Myrtle beach, SC.

Millar, D. (2008). 'Using airborne LIDAR bathymetry to map shallow river environments: a successful pilot on the Colorado River' Geophysical Research Abstracts,'.

Mitchell, S., Thayer, J.P., and Hayman, M. (2010). 'Polarization lidar for shallow water depth measurement', *Appl. Opt.* **49**(36), 6995–7000.

Nayegandhi, A., Brock, J.C., and Wright, C.W. (2009). 'Small-footprint, waveform-resolving lidar estimation of submerged and sub-canopy topography in coastal environments', *International Journal of Remote Sensing* **30**(4), 861–878.

Nayegandhi, A., Brock, J.C., Wright, C.W., and O'Connell, M.J. (2006). 'Evaluating a small footprint, waveform-resolving LiDAR Over coastal vegetation communities', *Photogrammetric Engineering & Remote Sensing* **72**(12), 1407–1417.

Notebaert, B., Verstraeten, G., Govers, G., and Poesen, J. (2009). 'Qualitative and quantitative applications of LiDAR imagery in fluvial geomorphology', *Earth Surface Processes and Landforms* **34**, 217–231.

Ott, L. (1965). 'Underwater ranging measurements using blue-green laser'(NADC-AE-6519), Technical report, Naval Air Development Center, Warminster, PA.

Peeri, S. and Philpot, W. (2007). 'Increasing the existence of very shallow-water LIDAR measurements using the red-channel waveforms', *IEEE Transactions on Geoscience and Remote Sensing* **45**(5), 1217–1223.

Popescu, S.C. and Wynne, R.H. (2004). 'Seeing the trees in the forest: Using lidar and multispectral data fusion with local filtering and variable window size for estimating tree height', *Photogrammetric Engineering and Remote Sensing* **70**(5), 589–604.

Reitberger, J., Krzystek, P., and Stilla, U. (2008). 'Analysis of full waveform LIDAR data for the classification of deciduous and coniferous trees', *International Journal of Remote Sensing* **29**(5), 1407–1431.

Sallenger, A., Wright, C., Guy, K., and Morgan, K. (2004). 'Assessing storm-induced damage and dune erosion using airborne lidar: Examples from Hurricane Isabel', *Shore and Beach* **72**, 3–7.

Samson, B. and Torruellas, W. (2005). 'High Peak Power Fiber Amplifiers Operating at Eye-Safe Wavelengths', Technical report, NASA Tech Briefs.

Shamanaev, V. (2007). 'Estimation of lidar return signal levels in two-beam hydrooptical laser sensing', *Russian Physics Journal* **50**, 1178–1182.

Sithole, G. and Vosselman, G. (2004). 'Experimental comparison of filter algorithms for bare-Earth extraction from airborne laser scanning point clouds', *ISPRS Journal of Photogrammetry and Remote Sensing* **59**(1-2), 85–101.

Smullin, L. and Fiocco, G. (1962). 'Optical echoes from the Moon', *Nature* **194**, 1267.

Straatsma, M.W. and Baptist, M.J. (2008). 'Floodplain roughness parameterization using airborne laser scanning and spectral remote sensing', *Remote Sensing of Environment* **112**(3), 1062–1080.

Tayefi, V., Lane, S., Hardy, R., and Yu, D. (2007). 'A comparison of one- and two-dimensional approaches to modeling flood inundation over complex upland floodplains', *Hydrological Processes* **21**, 3190–3202.

Thiel, K.H. and Wehr, A. (2004). 'Performance capabilities of laser scanners: an overview and measurement principle analysis', *in* 'Proceedings of the ISPRS working group 8/2, Laser-Scanners for Forest and Landscape Assessment. International Archives of Photogrammetry, Remote Sensing and Spatial Sensing, pp. 14–18. ISPRS, Freiburg'.

Tulldahl, H.M. and Steinvall, K.O. (1999). 'Analytical waveform generation from small objects in LiDAR bathymetry', *Applied Optics* **38**, 1021–1039.

Tulldahl, H.M. and Steinvall, K.O. (2004). 'Simulation of sea surface wave influence on small target detection with airborne laser depth sounding', *Applied Optics* **12**, 2462–2483.

Wagner, W., Ullrich, A., Ducic, V., Melzer, T., and Studnicka, N. (2006). 'Gaussian decomposition and calibration of a novel small-footprint full-waveform digitising airborne laser scanner', *ISPRS Journal of Photogrammetry and Remote Sensing* **60**(2), 100–112.

Wang, C.K. and Philpot, W.D. (2007). 'Using airborne bathymetric lidar to detect bottom type variation in shallow waters', *Remote Sensing of Environment* **106**(1), 123–135.

Weitkamp, C., ed. (2005). *LiDAR range-resolved optical remote sensing of the atmosphere*, New York: Springer.

West, G. and Lillycrop, W. (1999). 'Feature Detection and Classification with Airborne LiDAR - Practical Experience', *in* 'Proc. Shallow Survey 99, October 18-20, Sydney, Australia.'

Wozencraft, J.M., Millar, D. (2005). 'Airborne lidar and integrated technologies for coastal mapping and charting', *Marine Technology Society Journal*, **39**(3), 27–35.

Wright, C.W., and Brock, J.C., 2002, 'EAARL: A lidar for mapping shallow coral reefs and other coastal environments', *in* Proceedings of the Seventh International Conference on Remote Sensing for Marine and Coastal Environments, Miami, FL, 20-22 May 2002, 8 p.

Zege, E., Katsev, I., Prikhach, A., and Malinka, A. (2004). 'Elastic and raman lidar sounding of coastal waters. Theory, computer simulation, inversion possibilities', EARSeL eProceedings, **3**(2), 248–260.

8

Hyperspatial Imagery in Riverine Environments

Patrice E. Carbonneau[1], Hervé Piégay[2], Jérôme Lejot[3], Robert Dunford[4] and Kristell Michel[2]

[1]Department of Geography, Durham University, Science site, Durham, UK
[2]University of Lyon, CNRS, France
[3]University of Lyon, CNRS, France
[4]Environmental Change Institute, Oxford University Centre for the Environment, Oxford, UK

8.1 Introduction: The Hyperspatial Perspective

The bibliometric analysis in Chapter 1 of the present volume quite clearly shows that to date, the bulk of publications in the area of fluvial remote sensing (FRS) are very strongly grounded in classic remote sensing methods, approaches and perspectives. We find a large volume of papers examining large scale features such as river estuaries (e.g., Hedger et al., 2007; Chen et al., 2009) and vegetation mapping (in the riparian zone) is also very common (e.g., Akasheh et al., 2008; Bertoldi et al., 2011). Published work which uses FRS to examine smaller scale parameters, such as woody debris and hydraulic variables, which are usually of interest to fluvial geomorphologists and lotic ecologists are much less common and generally quite recent. Marcus and Fonstad (2008) and Marcus and Fonstad (2010) review and discuss the progress and fundamental changes occurring in FRS. One of the crucial elements of this progress is the ability to collect image data at resolutions often below 10 cm, and as low as 1.5 cm, from airborne platforms. This imagery has the power to resolve landscape features such as individual clasts and individual trees. Larger in-channel features such as bedforms, even if smaller than 1 meter, can now be densely sampled. For example, Figure 8.1 shows a series of images of a river reach with large exposed gravel bars. These images were acquired from airborne and, for comparison, satellite platforms. Figure 8.1a was acquired from an Unmanned Aerial Vehicle and has a spatial resolution of 13 cm. In this image and in the zoomed subset, we can clearly see small morphological units and individual shrubs. Figure 8.1b was acquired from a fixed wing, traditional aircraft (seen in Figure 1.1c of Chapter 1 in this volume). It has a spatial resolution of 50 cm. Due to the lower resolution, many small scale features are lost but we can still appraise the morphological change that has occurred in the two-year timespan between photos 8.1a and 8.1b. Figure 8.1c shows a Quickbird, pansharpened, image with a simulated resolution of 60 cm. Even at such resolutions, and despite the fact that the red, green and blue bands in Quickbird do not exactly correspond to those in RGB photography, we can identify features which are similar to those in Figure 8.1a which was taken four month earlier. In contrast, Figures 8.1d to 8.1f all show images with resolutions in excess of 1m. Clearly in these cases we can see that the level of feature representation drops

---

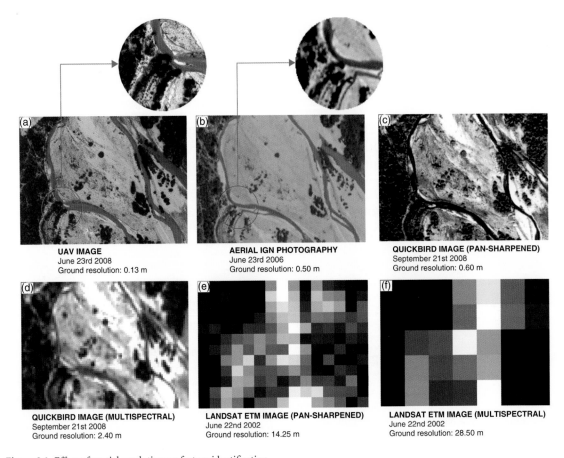

Figure 8.1 Effect of spatial resolution on feature identification.

rapidly as spatial resolution increases from 2.4m to 28.5m. Figure 8.1 clearly demonstrates that as spatial resolutions drop below 1m and approach the 10 cm range, individual landscape features which are relevant to river sciences become visible. As a further example, Figure 8.2 shows another series of images with resolutions ranging from 1.5 cm to 16 cm. In these images, individual trees, aquatic plants, boulders, cobbles, sedimentary plumes and other morphological features all become clearly resolved to the point where quantitative measurements on their size and number can reliably be made directly from the imagery. Clearly, such imagery with centimetric scale resolutions has a strong potential as a data acquisition method in the broad field of river sciences and management.

In other fields of remote sensing, the term 'hyperspatial' is currently emerging as a descriptor of such imagery capable of resolving small, individual, spatial landscape features. For example, Greenberg et al. (2009), working in the area of forestry remote sensing, define hyperspatial

imagery as having pixels smaller than single trees. Furthermore, Rango et al. (2009) give a quantitative definition of hyperspatial imagery as imagery with sub-decimetric spatial resolutions. While the selection of fixed boundaries applicable to the term 'hyperspatial resolutions' is somewhat subjective, we argue that the sub-decimetric threshold identified by Rango et al. (2009) is sensible in a fluvial context. Therefore we propose that the FRS community adopt the term 'hyperspatial imagery' when referring to sub-decimetric imagery. This new generation of image products offers a powerful new perspective to river scientists and managers. In particular the ability to sample small features such as illustrated in Figure 8.2 has the potential to change the way traditional reach scale field studies are conducted. Existing knowledge and the published literature in the fields of fluvial geomorphology and lotic ecology are dominated by small to medium scale studies with the main data acquisition methods based on labour intensive fieldwork (Kondolf and Piégay, 2003).

(a) Meander bend with small tributary
(Luc-en-diois / Drôme-alt.: 210 m, res: 10 cm)

(b) Oxbow lake section with power lines
(Malourdie / Rhône-alt.: 178 m, res: 8.5 cm)

(c) Limestone outcrops
(Varambon / Ain-alt.: 335 m, res: 16 cm)

(d) Sedimentary plume
(Bellegarde / Ain-alt.: 141 m, res: 6.7 cm)

(e) Regenerating forest
(Bellegarde / Ain-alt.: 258 m, res: 12.3 cm)

(f) Woody debris / grain size structure
(Gévrieux / Ain-alt.: 36 m, res: 1.7 cm)

(g) Floating aquatic vegetation
(Planet / Ain-alt.: 31 m, res: 1.5 cm)

(h) Submerged aquatic vegetation
(Brotalet / Rhône-alt.: 160 m, res: 7.7 cm)

Figure 8.2 Further examples of the features which become resolved in hyperspatial imagery (alt. – altitude of flight; res. – spatial resolution).

These studies have focused on a variety of topic areas including channel morphology, grain size and aquatic habitats (Thorne, 1998; Gordon et al., 2004; Ramakrishna et al., 2011). However, the limitations of traditional field measurements are well known. Areas of interest are often considerably larger than can be exhaustively sampled by field survey, continuous spatial sampling is generally ideal but rarely feasible at high resolutions and large extents. Consequently, the scale and resolution of field-based studies has been constrained by purely methodological limitations. As argued by Fausch et al. (2002) this limitation has hampered progress in river sciences since natural processes often operate on longer scales both in space and time. Hyperspatial imagery could make a crucial contribution in this area by allowing for sampling schemes and experimental designs that can continuously sample at sub-decimetric resolutions over reach scales or even over catchment scales. River monitoring studies could also benefit from hyperspatial imagery. From a purely scientific perspective, many current key questions concern the evolution of the hydrographic network and the ecology it supports. Experimentally, the study of these questions requires a network of sampling sites with paired biological and physical information that acquire data at a sampling frequency capable of capturing change at timescales which match the geomorphic and ecological processes in operation. Such datasets would also be invaluable in a policy context to river managers and decision-makers who need feedback on the efficiency of any restorations actions undertaken. Owing to the same methodological constraints mentioned above, these idealised datasets have been nearly impossible to collect. Hyperspatial imagery is not in itself a solution to the need for image data sets with a higher temporal resolution. However, in parallel to the development of hyperspatial imagery, a number of lightweight and easy to deploy imaging platforms have become readily available. As a result, the acquisition of hyperspatial imagery at monthly or even daily temporal resolutions is now logistically and economically possible for river reaches of hectametric or kilometric scales.

In this chapter, we will first summarise the elements of technical progress in terms of acquisition platform and camera hardware which are making hyperspatial imagery a feasible option for scientists and managers. The chapter then discusses the specific problems and technical issues that arise when using hyperspatial image data. Finally, we present case studies which use hyperspatial imagery as a core data acquisition methodology.

8.2 Hyperspatial image acquisition

When considering an experimental or management design based on imagery in standard colour format (RGB), three key ideal image parameters should first be identified based on the scientific and/or management requirements of the project:
- spatial resolution,
- spatial extent (i.e. size of the study site),
- temporal resolution.

The successful acquisition of imagery possessing any given combination of these parameters will be determined by the following factors:
- acquisition platform:
  - ground-tethered devices (e.g. blimps and kites),
  - Unmanned Aerial Vehicles or Systems (UAV or UAS),
  - manned, Ultra Light Aerial Vehicles (ULAV),
  - general aviation aircraft (e.g. helicopters and Cesna),
  - presence/absence of infrastructure to enable platform launch,
- acquisition sensor:
  - spatial resolution,
  - in-flight performance,
- logistic and economic factors:
  - total cost,
  - acquisition conditions and flight execution,
  - post-processing requirements (staff and software).

8.2.1 Platform considerations

The four platform types listed above are those which, to date, allow for hyperspatial image acquisition. Each of these platforms has a range of strong and weak points. Table 8.1 provides an overview and brief literature and the following sections discuss the characteristics of the various systems.

8.2.2 Ground-tethered devices

Ground tethered aerial devices such as kites and blimps are ideal for applications where it is necessary to monitor a small area, at either a fixed scale or at multiple scales. Blimps and kites with sizes from 1m to 2m are generally capable of lifting loads of 1–3 kg depending on the exact design. This allows for full size SLR (Single-Lens Reflex) digital cameras to be carried on specially designed mounts that maintain the camera in a vertical orientation. The camera is generally controlled by a radio-remote or

Table 8.1 Examples of low altitude remote sensing ('Spec' denotes spectral-based application; 'DEM' denotes use for Digital Elevation Model generation; 'MP' denotes millions of pixels).

Reference	Platform Type	Sensor	Area (ha) or Length (km)	Pixel size(cm)	Altitude (m)	Load (kg)	Application Field
Aber et al., 2002	Kite	Olympus stylus epic, film: RGB / Canon EOS RebelX, film: NIR / Canon Powershot (2MP)	0.5–1 ha	2.5–10	50–150	>1	Forestry; Agriculture; Others
Smith et al., 2009;	Kite	Nixon D70 (6MP) RGB	>1 ha	4	50-150	>1.5	Topography
Wundram and Loffler, 2008	Kite	Nixon Coolpix (4MP) RGB	1-10 ha	~10	150	'low'	Vegetation monitoring
Ries and Marzolff, 2003	Blimp	Film: RGB/NIR	<10 ha	Varies	<350	<25	Monitoring gully erosion
Jensen et al., 2007	Blimp	Kodak DC3000 (1MP) RGB / NIR with Hoya R72 filter	2 ha	25	400	<1	Agriculture
Guichard et al., 2000	Blimp	Active NIR transmitter/receiver	13 ha	2	>100	<6	Intertidal zone monitoring
Rango et al., 2009	UAV	Aiptek Pencam SD (2MP) RGB	?	~5	10-100	<28	Rangeland monitoring
Dunford et al., 2009	UAV	Canon Powershot G5 (5MP) RGB / Canon EOS 5D (12 MP) RGB	~174 ha	3.2–21.8	~150	4	Floodplain forestry
Lejot et al., 2007	UAV	Canon Powershot G5 (5MP) RGB / Canon EOS 500 N Reflex	0.2km – 4.7 km	3.6–14	76 - 290	4	Topography and bathymetry
Hervouet et al., 2011	UAV & ULAV	Canon Powershot G5 (5MP) RGB / Canon EOS 5D (12 MP) RGB / Sony DSLR-A350	<43.6	3-15	c.150	4+	Understanding vegetation colonisation.
Carbonneau et al., 2004;	Helicopter	Undisclosed (6MP) RGB	80 km	3 cm	c.150	N.A.	Grain size mapping
Dugdale et al., 2010	Aircraft	Canon EOS-1DS M. 2 (16MP) RGB	Not specified	3 cm	c. 450	N.A.	Grain size mapping
Carbonneau et al., 2011	Aircraft	Canon EOS-1DS M. 2 (16MP) RGB	16 km	3 cm	c.450	N.A.	Fluvial Geomorphology

sometimes by timed release. These systems are primarily used in stationary mode but they can be mobile at the local scale. Most often these tethered devices are small and work at relatively low altitudes (<350 m) to focus on study areas if less than 10 ha. However, potential users should check local airspace regulations since some aviation authorities prohibit high altitude kite flying. For example, in the UK, airspace regulations allow all tethered devices to operate freely at altitudes up to 60 metres. Any operations above this altitude require special permission (Chubey et al., 2006). The almost toy-like nature and raw materials involved mean that in principle these devices can often be produced relatively cost-effectively making use of light, low-cost cameras. The examples in Table 8.1 show typical performance parameters for this platform type. We can see that blimps and kites have been used in applications ranging from topography mapping to vegetation dynamics. When flying at low altitudes, resolutions are in the hyperspatial range. However, we

can see that in this case, the areas covered are relatively small and thus this platform is clearly not suited to large scale sampling schemes. Interested readers are referred to Aber et al. (2010) for a detailed treatment of small format aerial photography using tethered platforms and the flying platforms mentioned in the next section.

8.2.2.1 Unmanned Aerial Vehicles (UAVs) and Ultra-Light Aerial Vehicles (ULAV)

In recent years, small aircraft, both unmanned (UAVs) and manned (ULAVs), have begun to offer very interesting alternatives to traditional platforms (e.g. Figure 8.3). Unmanned Aerial Vehicles (UAVs), synonymously called Unmanned Aerial Systems (UAS), provide a low-cost opportunity to deliver hyperspatial imagery with a rapid deployment time that allows the user to collect imagery at monthly, weekly or even daily timescales. In contrast with tethered systems, they are most often used to

Powered Paraglider

Mass (kg): 25 + 0.7 kg/L (petrol)
Payload (kg): 165 kg incl. pilote
Cruise speed (km/h): 40
Endurance (hour): variable
Size: span 11.63 m
Max Flight Height (m): < 6000

Micro UAV SmartOne

Mass (kg): 1.1 incl. battery and camera
Cruise speed (km/h): 50
Endurance: 35 min - 1 hour
Size: span 1.2 m
Max Flight Height: about 200 m

Close Range UAV Pixy

Mass (kg): 5.6
Payload (kg): 4
Cruise speed (km/h): 60
Endurance (hour): 1
Size (cm): 84 x 63 x 53 - canopy 2.95 m
Max Flight Height (m): 500

Helicopter Robinson R22

Mass (kg): max. 621
Payload (kg): 248
Cruise speed (km/h): 177
Endurance: approx. 2 hours
Size (m): 2.7 x 8.8 x 1.9
Max Flight Height (m): 4267

(a) (b)
(c) (d)

Figure 8.3 Examples of recently developed platforms.

cover larger areas up to maximum areas approaching a square kilometre. The first aerial photography from a UAV was taken in 1955 for a military application (Newcome, 2004). However, UAVs are currently undergoing a tremendous expansion. These systems generally employ airframe technology developed in the field of model-making and available for several decades. However, much of their new-found popularity can be attributed to recent technological developments enabling automated flight and controlled image acquisition. Thanks to the miniaturisation of GPS technology which can measure the aircraft position in 3D space and to the development of small electronic Inertial Measurement Units (IMU) capable of measuring aircraft orientation in 3D space, these systems can now generally assist the user in manual piloting and even fly autonomously. As a result, many systems are now capable of pre-programmed flight paths with automated, controlled, image acquisition. When combined to recent developments in small format, cheap, digital cameras intended for the wider consumer market, UAV platforms now offer scientists and managers an opportunity for 'do-it-yourself' remote sensing hyperspatial data acquisition.

For example, the entries in Table 8.1 relating to UAV work illustrate the typical performance of these systems. After proper training and adequate experience with the system, pilots can expect to cover areas which exceed 100 ha (= 0.1 km^2). Flying altitudes are low which results in image resolutions well within the hyperspatial range. An important caveat which is not mentioned in Table 8.1 and which will be encountered by new UAV users is the need to obey air traffic regulations. UAVs are increasingly recognised in air traffic regulations and potential users should examine these regulations with care as they place constraints on UAV operations. In Europe the general principle is that UAV operations are possible in unpopulated areas if the pilot keeps the UAV within line of sight (approx. 500 m) and flies at low altitudes. For example, in the UK, small UAVs with weights below 7 kg can operate freely outside of urban areas provided that the pilot maintains visual contact at all times and that the flight altitude is below 123 metres (400 ft) (Defourny et al., 2006). Consequently, UAV surveys are best targeted to smaller study areas at the reach or small river scale. This new, exciting, technology is not yet suited to catchment scale data acquisition.

In a similar manner, piloted, Ultra-Light Aerial Vehicles (ULAVs) are an excellent means to provide data over slightly larger study areas. They do this, however, at the cost of some of the flexibility and rapid-response capability of their unmanned counterparts. ULAVs are larger, more expensive, and require significantly more skill, experience (as well as a fearless pilot!) than is required for UAV work. As a result, it is common for ULAV work to be tasked from existing private companies rather than river managers or research teams having access to their own ULAV. Nevertheless, ULAV services are offered by more specialised companies, at more affordable costs and with more flexible re-flight opportunities than traditional aerial photographic surveys. There is again a great variation within ULAV platforms including a variety of light paragliders, small planes and small helicopters. The difference in size between these and the generally smaller UAVs provides additional advantages in terms of greater carrying capacity for larger and more complex sensors. Peer-reviewed publications making use of ULAV-acquired data are still quite rare. However, in Table 8.1, we see that Hervouet et al. (2011) report spatial resolutions ranging from 3–15 cm and study areas approaching 0.5 km^2.

8.2.2.2 Traditional aerial photographic surveys

In addition to these new technologies, traditional aerial photography also offers hyperspatial imagery and remains an important data acquisition platform. Conventional planes and helicopters are able to cover very large areas and traditional aircraft still remain the only sensible option for study areas which significantly exceed 1 km^2. Furthermore, full sized aircraft can lift heavy loads thus allowing for a variety of image capture devices to be used including optical, multispectral RADAR and LiDAR devices. However, their significant disadvantage is cost and logistics, especially in cases where repeat imagery is required. Whilst the cost of airborne surveys is dropping significantly (Carbonneau et al., 2011), repeat imagery remains costly and logistically complex. The evidence of this is that, to our knowledge, there is no published work in the peer-reviewed literature which relies on repeated hyperspatial imagery acquired at catchment-scales. Furthermore, since scientists and managers very rarely have access to aircraft, this platform is rarely scrambled within hours or days of a trigger event such as a flood. The examples in Table 8.1 illustrate the typical performances of this traditional platform. Typically, full sized aircraft can deliver hyperspatial imagery when flying below 500 m. Even at such low altitudes, it is quite feasible to sample long river reaches or several 10s of kilometres. Furthermore, unpublished data now exists which covers several hundreds of kilometres at hyperspatial resolutions (N.E. Bergeron 2012, *personal communication*).

8.2.3 Camera considerations

A hyperspatial imaging system is a combination of the platform and the sensor. It is vital to consider both simultaneously to ensure the data provision needs of the intended application are met. This entails considering three further issues: (i) camera type; (ii) spatial resolution and (iii) motion blur.

8.2.3.1 Camera type

The camera hardware used for image acquisition is progressively moving away from traditional and expensive metric (i.e. scientific) cameras towards lower cost digital cameras designed for professional photographers and low cost small format cameras designed for the mass consumer market. Furthermore, in the context of remote sensing, the film-based camera is now firmly on the path to extinction. Given the rapid technical development in the area of digital photography, professional grade cameras can now offer many advantages in terms of cost and ease of use. Often perceived as the most important characteristic of a digital camera, sensor resolution is the first criterion in camera selection. High-end digital cameras are now routinely capable of delivering imagery in excess of 15 million pixels (MP). In the examples illustrating this chapter, all the platforms were equipped with commercial cameras, mainly digital cameras with resolution varying from 5 to 14 million of pixels (e.g., Canon EOS 500D/12.8 MP (reflex), Canon Powershot G5/5 MP (compact); Canon PowerShot G9/12.1 MP (compact), Fujifilm FinePix S3Pro/12 MP (reflec); Sony DSLR-A350/14.2 MP (reflex)).

8.2.3.2 Spatial resolution

The resulting ground resolution for a given sensor used at a given flying height can be determined with a few simple geometric equations. Figure 8.4 gives a schematic representation of an airborne camera in the process of acquiring an image. Beams of light from the ground target converge at the focal point of the lens and are then projected onto the imaging sensor of the camera. The distance from the convergence point to the imaging sensor is known as the focal length. Geometrically, the two triangles in Figure 8.4 are similar and we can state:

$$d/D = f/H \qquad (8.1)$$

Where d is the width of the imaging sensor, D is the ground footprint of the image, f is the lens focal length and H is the flying height. The final ground resolution of

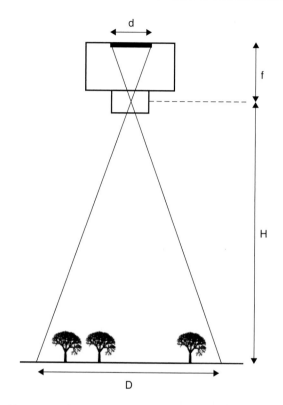

Figure 8.4 Schematic drawing showing the different parameters for calculating survey width (D) and spatial resolution: flight height (H), width of the imaging sensor (d) and lens focal length (f).

the image can then be determined as the ratio of the D and the number of pixels in an imaging sensor width. As an example, Figure 8.5a presents four curves giving image resolution as a function of flying height for hypothetical SLR-type image sensors of 10, 15, 20, and 25 megapixels and a focal length of 50 mm. Another crucial question, also captured in Equation (8.1) is the ground foot of the image (D). In Figure 8.5b, we have given D as a function of flying height for the four sensors shown in Figure 8.5a. Figure 8.5 clearly shows that high end digital cameras with resolutions above 15 megapixels are well suited to imaging small to medium rivers with resolutions below 5 cm and with a sufficient width to span a single channel. The basic rule of thumb is simply that platforms flying below 500 m and equipped with cameras in excess of 10 MP (with standard 35–50 mm lenses) will be capable of acquiring hyperspatial imagery. These criteria are now easily met when readily available camera technology is mounted on existing platforms.

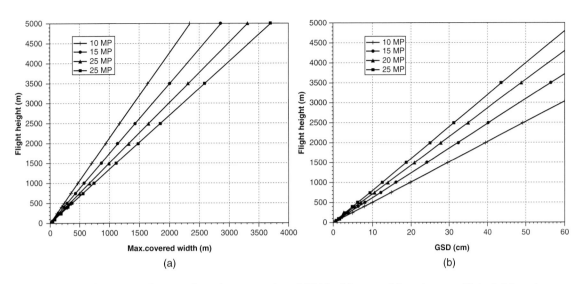

Figure 8.5 Expected camera performance for various sensor sizes. a) Width of the ground footprint versus Flying height and sensor size. b) Spatial resolution (ground sampling distance, GSD) versus flying height and sensor size for hypothetic SLR-type image sensors of 10, 15, 20 and 25 Megapixels and focal length of 50 mm.

8.2.3.3 Motion blur

Camera resolution should not be the only consideration in the planning of a hyperspatial image survey. Even if resolution and ground footprint are often the primary concern of end users, high resolution does not guarantee high quality. One factor that is often overlooked in the determination of flying and image acquisition conditions is motion blur which is caused by camera motion during the time of exposure. The motion blur present in an image can be calculated with the following formula:

$$B = V^*T^*H^*f^*10^{-3}{}^*p^*10^{-6} \qquad (8.2)$$

Where B is the motion blur in pixels, V is the velocity of the aircraft in m/s, T is the exposure time in seconds, H is the altitude above ground level in meters, f is the focal length in millimeters and p is the size of 1 pixel on the sensor in microns. Basically, this equation calculates the number of pixels that a single point on the ground will traverse during the time of exposure. Motion blur is especially relevant to hyperspatial image acquisition since the objective of the highest possible resolution requires flights at minimal altitudes which, according to Equation (8.2), increases the motion blur. As a typical example, consider the case of an aircraft travelling at 40 m/s and an altitude of 500 m. The aircraft has a 12 megapixel SLR digital camera with a 50 mm lens. Each pixel in the sensor has a width of 9 microns. In excellent lighting conditions,

typical exposures can be as short as 1/1000 s thus yielding a blur of 0.4 pixels. In poor lighting conditions, exposure times can often be as high as 1/250 s in which case the pixel blur would reach a very visible 1.76 pixels.

Motion blur can be a key driver of image quality and can significantly offset any potential gains in resolution. Purchasing a camera with a higher megapixel count is not guaranteed to produce high quality imagery with an improved resolution and potential users should think carefully before purchasing a camera on the sole basis of the imaging sensor size (in pixels). Digital SLR cameras usually employ sensors with a size of 23 mm × 15 mm. This is about two thirds the size of traditional film. Whilst camera manufacturers have dramatically increased the pixel count for their sensors, the size of the sensor has remained similar. Given that the sensor area remains similar, more light will be needed in order to ensure that each pixel on the imaging sensor receives a sufficient sample of photons. The result is a need for increased exposure times which increases motion blur. According to Equation (8.2), the flying speed of the aircraft could be reduced to offset the increased exposure. However, significant reduction in flying speeds may not be possible since most aircraft must maintain a minimum speed in order to remain airborne. Furthermore, reducing aircraft speed makes the flight less stable and the associated small-scale chaotic motion further reduces the quality of the images. Another option would be to increase the flying height. This would obviously cancel any gains in resolution. However, it would produce images with larger ground footprints which reduce the final size of the image database. As the final target ground resolution increases, motion blur will become increasingly important. One option would be to use a helicopter capable of stationary flying. Another significant cause of motion blur is low light associated with cloudy weather. The experience of the authors quite clearly indicates that in the case of professional grade SLR digital cameras with pixel counts above 12–15 megapixels, image blur will be significant unless weather conditions are perfect and there is abundant light. Furthermore, other authors (Dugdale, *personal communication*) have found that the type and altitude of the cloud cover can also have a significant impact on image quality with high cloud allowing for better quality imagery than low cloud.

Smaller format cameras such as those produced for the mass consumer market are also susceptible to motion blur. One key difference between small format and SLR cameras is the much reduced diameter and quality of the lens system. Furthermore these cameras generally employ smaller imaging sensors. As a result, these cameras need brighter light conditions and are quite sensitive to motion blur and not suited for use on board full sized aircraft. However, these cameras have found a niche in UAV and ULAVs applications. The payload of UAVs currently available is often restricted to a few hundred grams which is insufficient to carry full SLR cameras which generally weigh in excess of 1 kg. Fortunately, small UAVs can fly at very low speeds (often below 10 m/s). This therefore allows for a reduction in motion blur which makes these cameras a viable option.

Advanced imaging systems often use what is termed 'forward motion compensation' in order to reduce motion blur. The basic principle is simply to move the camera in the opposite direction to flight during the split second of image exposure. This therefore results in a reduction in the effective ground speed of the camera and therefore leads to a reduction of motion blur (Pacey and Fricker, 2005). Such systems have been widely used in larger format aerial cameras since the 1980s (e.g. in the classic Leica WildRC20 camera). Generally speaking, forward motion compensation requires additional weight. This is not an issue for general aviation aircraft but remains problematic for small UAS platforms.

8.2.4 Logistics and costs

The diverse range of platforms discussed above implies a range of logistic complications. Tethered platforms and small UAVs have similar logistic constraints to that of traditional fieldwork albeit with a requirement for good weather which may be difficult to fulfil in certain climates and in the immediate aftermath of floods. However, once purchased their operational costs are generally quite low since they can be operated by any staff with sufficient training. In the case of traditional aircraft platforms which must be contracted out to external companies, the logistic challenge is often the rapid deployment of field crews which are required to collect ground data in order to calibrate and validate the subsequent data analyses procedures (e.g. see Chapter 9). The associated costs are generally determined by the length (or area) of the survey. Figure 8.6 shows some sample costs for outsourced airborne data acquisition. Some common satellite data has been added for the purpose of comparison. However, it should be noted that the costs in Figure 8.6 do not take into account the mobilisation or deployment costs. These costs therefore assume that the aircraft is based near the study site. If this is not the case, the cost of moving the aircraft in position closer to the study site will always be directly passed on to the client.

8.3 Issues, potential problems and plausible solutions

Several technical problems are associated with hyperspatial image datasets. Many of the relatively new platforms discussed above are still experimental and as a result, acquisition is not yet perfectly controlled. Indeed, aircraft orientation and elevation are not perfectly constant during the acquisitions thus resulting in variations in the ground footprint of the image and the spatial resolution of the image. This complicates the already challenging issue of georeferencing the imagery. Moreover, since we are often using consumer cameras not specifically designed for aerial use, image quality can often be sub-optimal. Typically, the authors' experience has shown that image brightness can vary significantly even when ambient light conditions have not varied. Another common issue is shading caused by bankside vegetation. This causes significant problems in subsequent analysis methods such as depth mapping and even basic classification. In fact the process of classification is significantly more difficult with hyperspatial imagery. Finally, even if all these hurdles are progressively overcome with better equipment and more advanced post-processing solutions, hyperspatial image datasets are usually very large and often number in the thousands of images. Image management and data

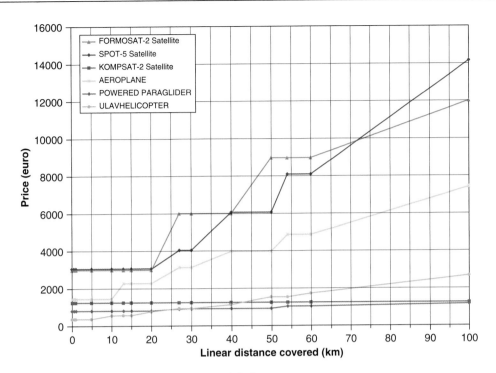

Figure 8.6 Cost estimates versus linear distance for a range of platforms.

mining are becoming increasingly important if the large volume of data contained in hyperspatial image datasets is to be effectively transformed into knowledge.

8.3.1 Georeferencing

As stated previously, hyperspatial images will tend to have relatively small ground footprints. Therefore, if acquired for long river reaches or even entire rivers, the resulting hyperspatial image database will contain hundreds or even thousands of images which are raw, without any real coordinates and geographical orientations. Georeferencing, also called image registration, can be defined as the process whereby an image raster is mapped to real world coordinates. The objective of this process is therefore to derive a transformation relationship which will associate each image pixel with a position in a given real world map coordinate system. The process can function in two dimensions or in three dimensions. In two dimensional georeferencing, the implicit assumption is that the underlying topography is flat and therefore that the entire image was acquired at a constant scale and resolution. In three dimensions, previous knowledge of the underlying topography can allow for the production of an 'orthorectified

image' where variations in scale following variations of topographic elevation are corrected and where the entire image is reprojected in order to achieve rigorously constant scale and resolution throughout. We can list three main approaches to image georeferencing: A field-based approach which requires ground control points (GCPs) of known coordinates, an algebraic approach where the position and attitude (i.e. orientation) of the camera at the time of exposure is accurately known and an image registration approach where a pre-existing, georeferenced image of the same area is used to establish the position of a new image. In practice, these three methods of georeferencing described here do not operate independently. Multiple methods can be combined in order to improve results. However, for the purpose of clarity, we will discuss the three approaches independently.

Georeferencing a large volume of data is not straightforward. In the case of the field-based approach, the standard procedure is to collect an absolute minimum 2 GCPs per image in the case of 2D georeferencing. However, in order to minimise errors, 3 to 6 GCPs per image are generally recommended. In the case of orthorectification, a digital elevation model will be required in addition to the GCPs. The georeferencing process also requires that each GCP be visible in the imagery. In general, this is accomplished by placing survey targets in the field area prior to image acquisition. Alternatively, uniquely identifiable features in the landscape such as man-made structures, boulders or fallen trees can be used as control targets. This option is especially attractive in the case of hyperspatial imagery where individual landscape features are easily recognisable. These targets, either natural or man-made, then need to be surveyed to high accuracy and therefore professional geodetic GPS equipment is usually preferred. Once the field data is collected, GIS or remote sensing software must be used in order to georeference the imagery. This usually involves the need for a user to manually identify each GCP in each image. Clearly, such a manual approach would be extremely labour intensive in the case of large hyperspatial image databases.

Alternatively, the imagery can be georeferenced without ground control points if the external orientation (i.e. horizontal position, altitude above ground level and orientation) of the camera is known at the time of exposure. If we assume level ground, simple geometry can be used to calculate the position of the image. However, this approach is highly sensitive to errors in the external orientation. Whilst modern differential GPS systems can readily give high accuracy position data at the time of image acquisition, the orientation is more problematic. One solution is the use of Inertial Navigation Systems (INS). These systems use gyroscopes and accelerometers in order to track changes in orientation and position. They

were developed mainly for military navigation applications and are capable of accurately logging position and orientation. Until recently, these were too bulky and prohibitively expensive for routine use in image surveys. However, recent developments in electronics have led to a range of electronic INS products which are commercially available to the public. These devices can cost as little as 200 \$US and weigh as little as 6 grams. Whilst not commonly in use, government mapping agencies and private sector companies are increasingly taking advantage of the new availability of INS sensors. However, INS systems are not the only option for the in-flight determination of aircraft (and camera) orientation. Several UAV models now incorporate infrared sensors to facilitate flight and navigation. These sensors generally detect the horizon line as the mid-point between the sky (low infrared signal) and the ground (high infrared signal). By having infrared sensors in multiple orientations, the aircraft can estimate its orientation. This information can then be used to automatically mosaic (i.e. stitch) the imagery and produce a georeferenced product without the recourse to ground control. For example, Figure 8.7 presents a georeferenced mosaic of a large erosional feature which was formed by a mass failure in the wake of a flood near Durham, UK. The figure is comprised of 72 images acquired at an altitude of 125 m with a small UAV manufactured by SmartPlanes AB. in Sweden (see Figure 8.3b). The raw imagery was acquired with a small format 7 megapixel Canon IXUS camera and had a resolution of 5 cm per pixel. This UAV

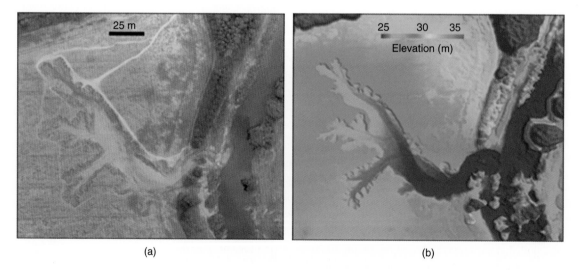

(a)   (b)

Figure 8.7 Example of mosaics (a) and digital elevations (b) generated from hyperspatial imagery acquired from a UAV platform.

records external orientation parameters for each image. It uses an onboard GPS to acquire position and a series of infrared sensors to acquire orientation. This data is crucial to the production of georeferenced mosaics.

We now consider a third approach to image geo-referencing: image co-registration. This approach offers another potential alternative to traditional field based georeferencing methods and can be defined as the process by which multiple images are aligned into a single coordinate system. For most areas of the world, government agencies or private sector companies have accumulated vast databases of low to medium resolution imagery either from high altitude airborne or satellite platforms. This imagery is often available to the public and is usually georeferenced with reasonable accuracy. Therefore, it becomes possible to use pattern matching approaches in order to identify features in both an un-georeferenced hyperspatial image (called the sensed image) and a pre-existing georeferenced image (called the reference image). Once the sensed image has been localised within the reference image, the georeferencing data from the reference image can be used to calculate the position of the sensed image. Carbonneau et al. (2010) present an automated approach which applies this image co-registration strategy in order to georeference large databases of hyperspatial imagery. This work uses normalised cross correlation algorithms to match the pixel intensities within a search window or the entire sensed image to a specific location in the reference image. Normalised cross correlation has now become an established matching algorithm (e.g. (Wang et al., 2007) which is implemented in commercial software such as MATLAB). This algorithm computes a correlation coefficient γ from -1 to 1 between a smaller template image, t, and a larger image f according to equation (8.3) (Carbonneau et al., 2010):

$$\gamma(u,v) = \frac{\sum_{xy}\left[f(x,y) - \bar{f}_{uv}\right]\left[t(x-u,y-v) - \bar{t}\right]}{\sqrt{\sum_{xy}\left[f(x,y) - \bar{f}_{uv}\right]^2 \sum_{xy}\left[t(x-u,y-v) - \bar{t}\right]^2}}$$

(8.3)

Where x and y are the pixel coordinates in the reference image f, u and v are the pixel coordinates in the sensed image t, is the mean of the sensed image and is the mean of the reference image.

The method described by Carbonneau et al. (2010) relbreakies on the metric or sub-metric resolution satellite or airborne data and has produced encouraging results. Figure 8.8 shows an example of georeferenced sensed images (with a white frame added) overlain on base image for the Ste-Marguerite river. Hyperspatial imagery with a resolution of 3 cm was automatically georeferenced to airborne imagery with a resolution of 10 cm. The initial impression is that the white framed sensed images overlay extremely well with the underlying reference image. Closer inspection at the border of the sensed images reveals that this overlay is not perfect but nonetheless the agreement is very close. Carbonneau et al. (2010) cite an RMS error of 1.7 m. Multiple sources of error affect these results and weighing the relative importance of these errors is not straightforward. One factor that was observed as important is the time elapsed between the acquisition of the sensed and reference images along with any potential changes in the scenery between the

(a) (b)

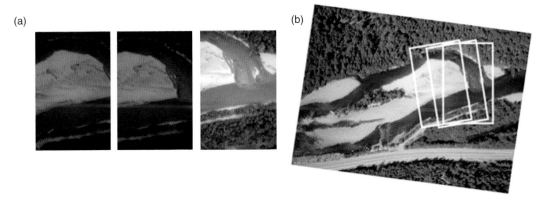

Figure 8.8 Example of automated georeferencing. a) Three hyperspatial images, unreferenced acquired from a helicopter platform at a spatial resolution of 3 cm. b) Hyperspatial image with 10 cm resolution with white frames showing the resulting position as calculated from an automated georeferencing procedure.

two acquisitions. Pattern matching approaches will obviously be sensitive to such changes in the scenery. For example, seasonal changes in vegetation and/or water colour can, theoretically, have a significant impact on such approaches. Additionally, it was found that rapid changes in flow regime can significantly change river imagery and negatively impact automated georeferencing attempts. For example, in addition to the airborne data shown in Figure 8.8, Carbonneau et al. (2010) also tested the use of Quickbird imagery acquired three days after the sensed imagery. In such a short time span, the surrounding vegetation and infrastructure remained virtually the same. However, during this period, river discharge decreased significantly and consequently, dry, exposed, gravel bars in the Quickbird reference image occupy a significantly larger area. In extreme cases, some very shallow riffles became exposed thus causing an important change in the rivers' appearance. This effect was at least in part responsible for an increase of the RMS error to 5.5 m.

8.3.2 Radiometric normalisation

The large hyperspatial image databases which can cause georeferencing issues often suffer from non-uniform illumination conditions. When hundreds or even thousands of images are acquired, maintaining a uniform level of illumination for the entire dataset is nearly impossible at the time of acquisition. First, over the timespan required to collect the imagery, weather conditions can change. Second, if camera settings are not adjusted properly, differences in exposure and/or aperture can lead to very significant changes in brightness within the dataset, even for images acquired within 1 second of each other. This factor is further complicated by the use of small commercial cameras designed for the mass market. These camera designs often automate image acquisition parameters beyond user control. Unfortunately, since they were not designed for aerial use, the automation parameters are often sub-optimal in the case of airborne acquisition. The resulting lack of uniformity in the radiometric levels of the imagery can have a detrimental impact on the data quality of the information derived from processes such as depth and grain size mapping (see Chapter 9). Whilst there are many standardised approaches to correct for illumination effects developed mainly for satellite imagery (Dai et al., 2010), such approaches typically give poor results for river imagery owing to the complexity and spatial variability of riverine environments. With few solutions present in the current literature, we propose

a new approach to normalise brightness values in fluvial hyperspatial image datasets. This method is based on a similar principle to the automated georeferencing described above. This georeferencing procedure relied on the presence of a reference image with pre-existing georeferencing data. In the radiometric correction procedure we propose here, the reference image is used to provide baseline radiometric values which can be used to normalise a set of sensed images.

The procedure is conceptually simple. First, both sensed and reference images must be classified in order to group pixels into vegetation, channel and dry classes. Second, the sensed images must be georeferenced in order to establish their position within the reference image. Third, for each sensed image, pixel histograms are compared separately for each class. Finally, by using histogram matching, the histograms from the reference image are used as target histograms and thus the histograms for the sensed image classes are adjusted to those of the reference image. To perform the matching, a process is realised not on the RGB channel but LAB mode (L: luminance (%), AB: colour range green/red and blue yellow). Given that the reference image is a single image, all the sensed images will be adjusted to the radiometry of the reference image. Figures 8.9 and 8.10 give some sample results for the same data shown in Figure 8.8. The figure shows that this simple method has successfully normalised the brightness values in the three images which now appear much more continuous.

8.3.3 Shadow correction

A similar approach can also be applied to the correction of illumination problems within a single image. In particular, shadows within the channel can pose a significant problem. One crucial application of river imagery which is gaining in importance is the mapping of channel bathymetry from image data. Depth can be extracted from images of clear water rivers either through photogrammetric analysis (Beal et al., 1997; Lane, 2000; Butler et al., 2002; Ballester et al., 2003) or, more commonly, finding a relationship between imaged river brightness and water depth (Lyon et al., 1992; Garguet Duport et al., 1995; Winterbottom and Gilvear, 1997; Gilvear, 2004; Legleiter and Goodchild, 2005; Carbonneau et al., 2006; Lejot et al., 2007). As such, shadows are one of the largest obstacles to the applicability image based bathymetry as they radically change the relationship between image radiance and channel depth. This shadow problem is most acute during times of the year and/or day when the sun angle

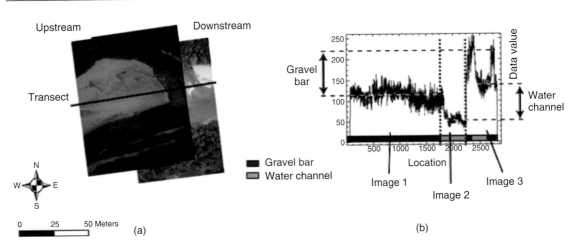

Figure 8.9 Radiometric variability in 3 consecutive images, Ste-Marguerite River, Quebec, Canada. a) Georeferenced images with one image noticeably brighter. b) Radiometric profile for the red band along the transect in a). The water in Image 3 is clearly brighter than in Image 2 without any significant change in real depth.

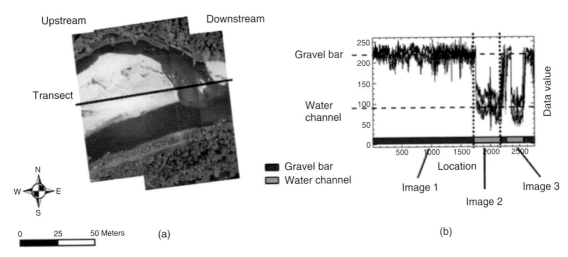

Figure 8.10 Radiometric normalisation procedure applied to the data in Figure 8.9, Ste-Marguerite River, Quebec, Canada. a) Georeferenced imagery. b) Radiometric profile along the same transect showing improved correspondence between the brightness levels in all 3 images.

is low and larger portions of the channel are affected by shading. Consequently, any attempts to quantify the depth of a channel section based on its radiometry using established methods (e.g., Winterbottom and Gilvear, 1997; Conyers and Fonstad, 2005; Carbonneau et al., 2006) will result in a sharp depth discontinuity since the darkened, shaded, areas will yield falsely exaggerated depths. Furthermore, higher-level analyses such as the extraction of particle sizes in rivers (Carbonneau et al., 2004), stream hydraulics (Lorang et al., 2005), and stream habitats (Wright et al., 2000; Marcus et al., 2003; Legleiter

and Goodchild, 2005; Hedger et al., 2006) would be as much or more influenced by the effects of shadows as is depth, so a basic approach to dealing with shadows is crucial in river studies. Therefore, a transferable method designed to remove the effects of shadows would make the widespread application of image based bathymetry far simpler and more accurate.

Once again there are no methods published in the literature which could alleviate the shadow problem. Given the effectiveness of histogram equalisation in normalising brightness values for entire images, it was also investigated as a potential solution for shadow removal. However, instead of referring to a reference image, we propose the use of adjacent un-shaded areas in order to establish normalisation targets. Conceptually this approach is very similar to the normalised procedure described above. First, image classification is used to delineate shaded and non-shaded areas. Second, pixel histograms are extracted for shaded and un-shaded portions of the river bed and third, histogram equalisation is used to transform the shaded area into a simulated unshaded area.

Figure 8.11 shows an example from the Dartmouth River in Quebec, Canada. In this example, we have chosen an image where a bridge spans the river and casts a shadow across the entire channel width. The shadow removal procedure discussed above will be used to restitute the brightness values in the shaded area. In order to assess the success of the procedure, we have defined two cross sections A and B. Cross section A is immediately outside the shaded area and cross section B is 3 meters

downstream of cross section A but well within the shaded area. Given the close proximity of the two cross sections, we would expect that their brightness profiles will be very similar in unshaded conditions. In Figure 8.11, we can see that prior to shadow removal, the shaded and unshaded cross sections have significantly different brightness values. The exception is the blue band where shading seems to have had very little effect. However, once shadow removal was performed, the corrected cross section displays a cross sectional profile which becomes very similar to the unshaded profile. This is a promising result which indicates that these targeted histogram matching approaches have the potential of removing shadows and of restituting channel bathymetry within these areas.

However, the main conceptual problem with this approach is the underlying assumption that the image histogram for the bright, unshaded areas is a good estimate of required histogram in the corrected areas.

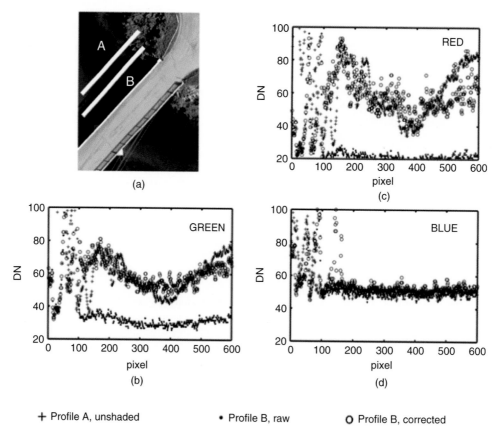

+ Profile A, unshaded   • Profile B, raw   O Profile B, corrected

Figure 8.11 Shadow correction procedure. a) Image where a bridge creates a shadow spanning the full channel. Effect of the correction procedure (b) on the red band, (c) on the green band, and (d) on the blue band.

This therefore assumes that the optical properties of the shaded area are only modified by the presence of shading and that no physical properties of the river in terms of bottom reflectance, depth or turbidity have changed. Such spatial autocorrelation in river properties can be reasonably assumed over short distances but discontinuities may exist such as random boulders, man-made bank stabilisation and localised incisions. We could easily conceive of a case where the shaded area covers a pool at the outside of a meander bend while the unshaded area is limited to shallow water. In such a case, the algorithm would use a shallow water histogram to correct a shaded pool histogram and the results would inevitably be poor. However, despite these 'what if?' considerations, in the case of the Dartmouth river images used in Figure 8.11, the assumption held and the results are promising. It should be remembered that in the absence of any corrections, the depth estimates in shaded areas have very large errors. Therefore any correction method, even if imperfect represents an improvement. However, it should be remembered that the image in Figure 8.11 is relatively small and the channel change within each footprint was generally small. In cases where channel change can be important within a single image footprint, the assumption that the unshaded area is a good approximation for the shaded area may not hold and it could become necessary to select unshaded pixels only in close proximity to shaded areas in order to establish target histograms. However, the ultimate, fundamental, limitation of this method lies with the radiometric resolution of the sensor and the compression of the histogram into few radiometric values. Figure 8.12 explicitly examines the effect of shadows on the image histogram for the red band. In Figure 8.12a, if we examine the histogram in the shaded area, the width of this histogram at ±1 standard deviation is 4.6 DN. In the illuminated areas (Figure 8.12b), the width is 27.8 DN. This demonstrates that the effect of shading compresses the histogram to the point where a very limited range of DN values are available. In such conditions, the digitisation of the analogue emitted radiation into a very limited number of integer DN values leads to a significant loss of information. The only potential improvement here is to improve the radiometric resolution of the sensor (see Chapter 1 for a definition of radiometric resolution). Unfortunately, this goes against the current trend in river remote sensing instrumentation. Current commercial development efforts all seem very focused on improving sensor resolutions and there seems to be little impetus in the mainstream imaging industry to produce high radiometric resolution cameras on a mass scale thus lowering the cost of these devices. This suggests that fluvial remote sensing is approaching the point where customised, fit-for-purpose, instrumentation will become a crucial step if substantive scientific progress is to be maintained.

8.3.4 Image classification

Image classification is arguably the most fundamental of all remote sensing analysis methods. This basic process whereby the features in an image are first segmented into distinct groups and then classified into a land-use type (e.g. river, exposed bar and vegetation) is generally the first step in any analysis of remotely sensed data. The basic algorithms for this process all rely on the statistical distribution of pixel brightness values in the available bands in order to segment the image into groups of pixels having similar spectral characteristics (Lillesand et al.,

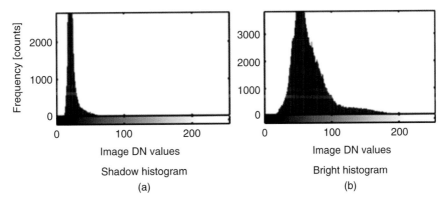

Figure 8.12 Effects of shading on the image histogram. a) Shaded area histogram, b) Bright, unshaded, histogram for a morphologically similar area.

2008). However, when dealing with RGB hyperspatial imagery, this basic assumption that a given feature in the landscape (or riverscape) will have similar spectral (colour) characteristics begins to break down. When the spatial resolution increases, the inherent internal heterogeneity of features becomes increasingly visible. For example, in the case of vegetation, instead of seeing a continuous expanse of green, we see green interspersed with dark shadows and light wood colours. As a result, classic approaches developed for traditional satellite data such as Landsat do not perform well when applied to hyperspectral imagery (Townshend et al., 2000). Consequently, the classification of hyperspatial imagery requires more recent approaches such as object-orientated analysis. This type of analysis uses principles of pattern recognition in order to somewhat mimic the human ability to identify features in an image. Readers are referred to Gao (2009) for an in-depth yet accessible discussion of object-based classification. Recent published works show that the object-based approach delivers better results when compared to the traditional pixel-based approaches (e.g., Sugumaran and Harken, 2005). For example, Figure 8.13 illustrates the potential improvements in classification accuracy when object-based classification is applied to

vegetation classification in forested riparian zones. Here the classifier is attempting to identify dead tree branches which present a complex structure. The object-based approach performs well in the first two cases where the dead branches occupy a significant area. In addition to riparian zones, object-based classification has been shown to perform well in urban areas (Jacquin et al., 2008), vegetation mapping (Hall-Beyer et al., 2003; Chubey et al., 2006), spatial changes (Defourny et al., 2006). The benefits of object-based classification applied to FRS will also be illustrated in Section 8.4 of the present chapter.

8.3.5 Data mining and processing

Even if the issues discussed above can be resolved and a final hyperspatial image dataset is compiled which is georeferenced and radiometrically correct, there still remains the issue of managing and analysing large image databases. For example, the results in Carbonneau et al. (2004) and Carbonneau et al. (2005) are based on a hyperspatial image dataset having in excess of 5000 images. The work of Hervouet et al. (2011) is based on a set of 3750 photos and the work of Carbonneau et al. (2011) is based on 193 hyperspatial images. The authors' experience has shown quite clearly that dealing with datasets

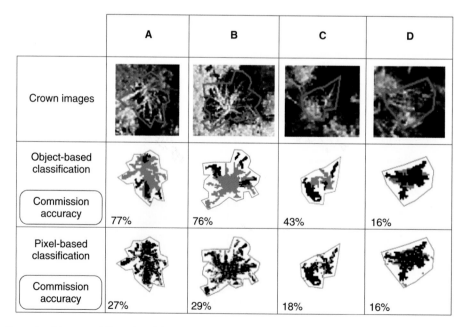

Figure 8.13 Example of the potential benefits of the application of Object-based classification. Detection of dead branches performed for 4 individual tree crowns (A, B, C, D). Comparison between object-based classification and pixel-based classification (in green: good detection, in black: wrong detection, in red: omission). The red line is the limit of the crown (e.g., the study area). The percent value is the proportion of good detection. From Dunford et al. (2009).

in excess of 100 images with standard remote sensing approaches, which usually function on a per-image basis, is extremely labour intensive. Furthermore, crucial spatial relationships which connect features in separate images are easily lost or obfuscated due to the complexity of the dataset. Therefore, enhanced data management is generally required when dealing with hyperspatial image datasets. The first option is to make better use of the image metadata which is commonly available. Information such as resolution (spatial and spectral), georeference, ground footprint, can be encoded in the metadata and thus allow for an increasing level of automation in terms of management and analysis. For example, Figure 8.14 shows an example of image footprints plotted spatially along with the spatial resolution which is plotted in a colour scale.

In this case we are dealing with a small number of images and the added metadata is easily managed. However, in the case of the large image databases discussed above, the enhanced use of metadata is insufficient since standard GIS packages are simply not designed to deal with such large databases.

A few researchers are therefore modifying or creating GIS software that is suited to the study of fluvial environments with remotely sensed data. For example, Thorp et al. (2010) discuss an integrated GIS approach which uses multiple data sources in order to characterise the hydrogeomorphology of an entire catchment and quantify the success and viability of rehabilitation efforts. McKean et al. (2009) use a freely available toolkit for Arc GIS called the River Bathymetry toolkit. This toolkit

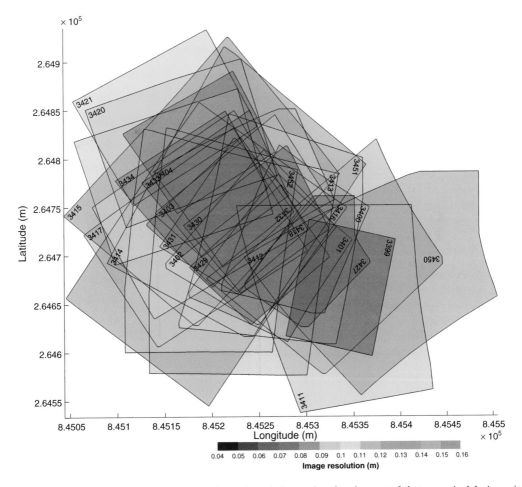

Figure 8.14 Example of advanced image management. Image boundaries overlays for a large set of photos acquired during a single campaign along the Drôme River, in 2005.

is an excellent example of data mining used in fluvial remote sensing. The toolkit allows users of bathymetric LiDAR (see Chapter 7) acquired by the USGSto organise and querry the large data volumes which are typical of LiDAR datasets. Carbonneau et al. (2011) present results obtained with a newly designed system: the Fluvial Information System (FIS), which integrates a suite of cutting edge procedures designed to analyse hyperspatial imagery and manage the results in a spatially explicit framework which is tailored to river environments. The aim of the FIS is to apply the riverscape concept discussed by several authors such as Fausch et al. (2002), Ward et al. (2002) and Wiens (2002) in order to provide a continuous, spatially explicit analysis of rivers which includes key variables such as width, depth, particle size, and elevation. This data therefore allowed Carbonneau et al. (2011) to derive hydraulic information (velocity, stream power, Froude number and shear stress) in order to provide parameters for aquatic habitat characterisation and modelling over the entire length of a small Scottish river.

8.4 From data acquisition to fluvial form and process understanding

The last part of this chapter is focused on different examples which illustrate the use of hyperspatial imagery in practical applications. The examples are grouped under two separate headings: (i) Feature detection with hyperspatial imagery, (ii) Repeat monitoring with hyperspatial imagery.

8.4.1 Feature detection with hyperspatial imagery

In this section we present two contrasting examples which illustrate the opportunities presented by hyperspatial imagery. The first example focuses on understanding the dieback of cottonwood tree crowns observed along the river Drôme, France. The second example uses hyperspatial imagery in order to provide a baseline geometry of secondary channels in a restoration project along the Rhône River, France.

In the Drôme case study, a specific procedure was developed in order to assess the health of cottonwood units based on UAV, ULAV, airborne and satellite images. The study area was a 5 km reach downstream of Luc-en-Diois. This reach drains a sub-catchment of 225 km². The reach has both single thread channels with average widths of c.10m and larger braided reaches with widths in excess of 200 m. The riparian corridor has similar width to the

channel and is mainly covered by willow and cottonwood patches. Field reconnaissance showed significant crown dieback of cottonwoods which, it was hypothesised, was a link to water table depletion following a period of active mining during the 1980s (Lejot et al., 2011).

Hyperspatial imagery was first acquired for the entire reach with a 'Pixy' UAV (see Figure 8.3c). The Pixy UAV is a flying paramotor developed by the French IRD (Institut de Recherche pour le Développement). It has a paraglider-type nylon tubular wing from which the aluminium chassis is suspended. It has a light gas-powered engine (7 kg) which allows for a maximum payload 4 kg. It is launched like a plane and lands by gliding. Take-offs require a flat open surface (asphalt or short grass) of at least 40 m in length and 20 m in width. Since this very rarely occurs in the vicinity of rivers, a takeoff track usually needs to be built. The pixy operates at speeds between 15 and 35 km/h. This low velocity allows high quality image acquisition with minimal motion blur. Maximum flight altitude is 800 m and maximum flight duration is approximately 1 hour. The pixy is also equipped with an onboard GPS (Garmin II+) which logs flights tracks. Altitude, speed, and trajectory are all controlled by radio-remote with live feedback sent by the aircraft to a field laptop PC. Flight parameters (altitude, position, and velocity) are recorded and used in order to generate metadata. The pixy UAV has been shown capable of collecting data over wide spatial scales with both Lejot et al. (2007) and Hervouet et al. (2011) collecting high resolution imagery for river corridors of 5 km in length. However, this range was not surveyed from a single launch point and several such points were distributed along the reach.

This UAV dataset was supplemented with ULAV data acquired from a manned paraglider (see Figure 8.3a). Compared to the pixy, the paraglider has the advantages of flying higher, up to 1500–1800 m thanks to a gas powered engine and a 30 m² soft fabric wing. It requires from 50 m to 400 m for take-off and can land in even shorter spaces as small as 25m. Furthermore, this landing area can be a relatively rough surface such as a gravel bar. With a 20 litre tank of fuel, the paraglider can remain airborne for 2–3 hours. From this position, the pilot uses a camera equipped with a spirit level and manually takes photographs. This imagery is usually at a higher altitude to the UAV imagery and therefore provides a set of context images of slightly lower resolution but of greater ground footprints and total areas without the need to land and re-launch.

In the data analysis phase, a continuous processing of the hyperspatial imagery was found to be overly labour intensive and, as a result, a multi resolution sampling strategy was developed. The process was as follows: (i) mature cottonwood forest stands were mapped using a Quickbird image isolating deciduous from conifers, (ii) forest establishment age was assessed for these stands using a time-series of 50 cm aerial photographs taken between 1948 and the present (Figure 8.15a and b), (iii) from this map mature cottonwood patches (35–45 years) were selected, (iv) from these mature cottonwood stands dead branches were detected using available hyperspatial images (Figure 8.13). As the hyperspatial data provided a sufficient spatial resolution that dead wood branches could be individually identified with the naked eye an image processing methodology was developed that used not only the spectral properties of the dead wood, but also textural and structural properties discernible from the images (see Dunford et al., 2009). Figure 8.15c shows a plot of dead branch density versus plot elevation. A positive and significant correlation is observed thus supporting the initial hypothesis. In this example, hyperspatial imagery greatly reduced the field effort needed to characterise the dieback of cottonwoods and provided highly useful additional textural and structural parameters to assist the remote characterisation of dead branches.

Our second example comes from restoration work on the Malourdie reach of the Rhône River in France. This site is a former channel which was dredged in 2003 in order to recreate water ponds with low velocity on the margins of the fast flowing main channel in order to provide habitat for fish spawning and rearing and to improve local biodiversity (Figure 8.16). In such a context, monitoring is essential in order to evaluate the intervention's success and to estimate its life span and evolution. Hyperspatial ULAV campaigns were done in April 2006 and March 2007 supported with field surveys providing ground control and water depth measurements (119 in 2006 and 59 in 2007; Figure 8.16a). By applying regression models linking the water depth measurements to the radiometric values from the hyperspatial imagery, predictions of water depth could be generated (Figure 8.16b). This provided bathymetric map covering the full restored reach covered by the imagery for each of the years surveyed (Figure 8.16b – April 2006). The quality of two models ranged from $0.71 \leq R^2 \leq 0.89$. Residual errors of models have been also estimated, ranging from 9 to 12.5 cm, and mapped showing a homogeneous spatial distribution of residuals error. Sedimentation is also monitored at the site every two years. The initial year following the restoration, the sedimentation rate was significant but this has

considerably reduced through time and from this data a trend has been extrapolated (Figure 8.16c). This model is used then to predict bathymetry at a range of times (Figure 8.16d).

8.4.2 Repeated surveys through time

A further advantage of flexible platforms is their capability to perform repeated hyperspatial surveys allowing seasonal changes and inter-annual changes to be evaluated. Multidate approaches bring with them a number of constraints in term of acquisition which must be considered by operators when designing campaigns. One of the key constraints is non-uniform illumination during different survey dates which can lead to difficulties when comparing images from different epochs. Causes for these differences include atmospheric condition, season and time of day. In addition channel spectral properties may also vary as a result of changes in either the channel bed (e.g., biofilm development, spatial variation in bedload remobilisation), surface conditions (e.g., algae development, laminar versus turbulent flow conditions) or water column conditions (variations in discharge and suspended sediment concentration). All these variations become factors that complicate analyses methods which are based on the spectral properties of the image such as bathymetric mapping or vegetation phenology classification.

Here we discuss a multidate study of the Mollon reach of the Ain river, France. The site was surveyed 10 times during 2010 mainly during the months of April to September. The range of discharge during the six month acquisition period was 20 to 80 $m^3.s^{-1}$ with a mean annual discharge of 120 $m^3.s^{-1}$. From these repeated surveys during the spring/summer period, it is possible to map the vegetation patches and monitor monthly changes involving patterns of growth and senescence (Figure 8.17). It is also possible to map the colonisation of the former channel by floating vegetation between June and July 2010 (Figure 8.17, pink square in the upper right corner and Figure 8.18) and the early spring colonisation of herbaceous vegetation on the gravel bar and its senescence at the beginning of the summer. The use of image data from different dates allows for the classification to distinguish different vegetation patches – thus separating high, low and sparse riparian vegetation patches. However, as the georeferencing of each photo is not perfect, the images do not overlay perfectly. Therefore the boundaries of the different land-use types are fuzzy, and may often require a filter to mitigate for these spatial errors.

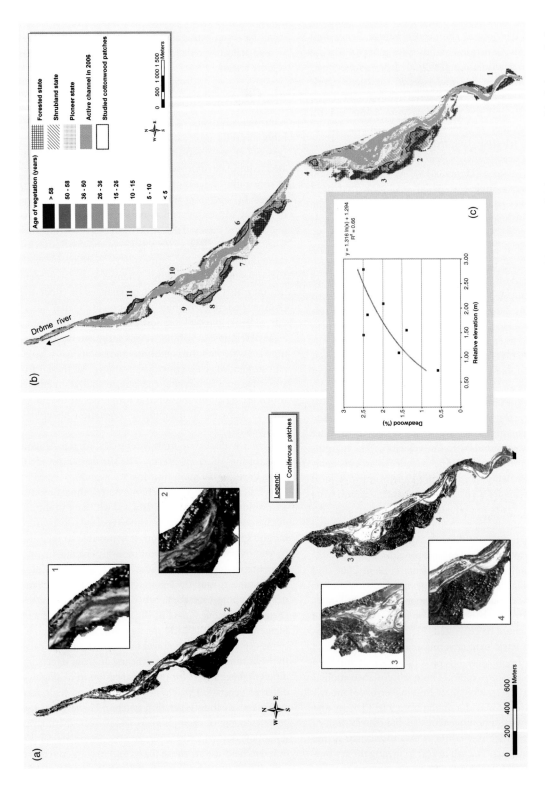

Figure 8.15 a–b Mapping of mature unit of cottonwood from Quickbird image to detect decidous from conifers and from national airborne 50 cm archived photo to isolate popular units established 36 to 58 years ago. c) Deadwood coverage versus ground elevation from sampling plots selected within the previous detected patches.

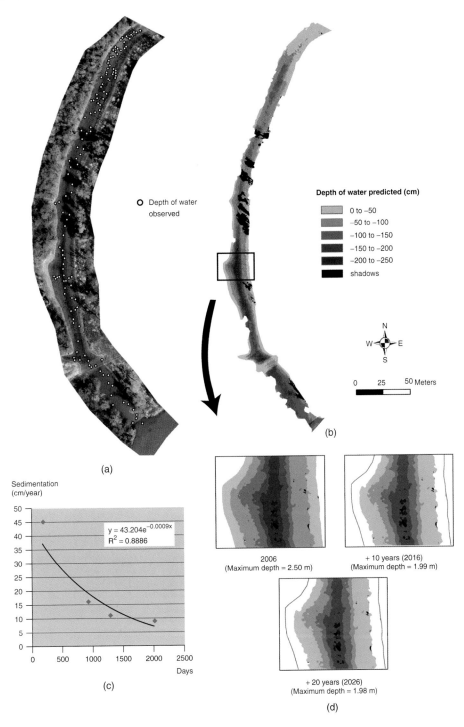

(a)

(b)

(c)

(d)

Figure 8.16 Characterisation of the 2006 bathymetry of the restored former channel of the Malourdie, on the upper Rhône, France and simulation of its evolution : a) Image acquired from a UAV survey, b) Bathymetric map provided by the statistical relationships linking observed depth and calculated depth from radiometric values, c) Prediction of temporal evolution of the sedimentation rate model calibrated from field measures (red dots) d) Application of the model in (c) on the observed bathymetry (b) for the prediction of bathymetric states in 2016, 2026 and 2036.

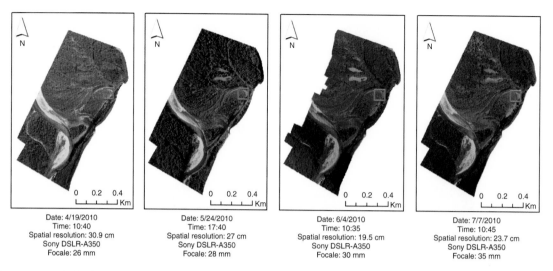

Date: 4/19/2010
Time: 10:40
Spatial resolution: 30.9 cm
Sony DSLR-A350
Focale: 26 mm

Date: 5/24/2010
Time: 17:40
Spatial resolution: 27 cm
Sony DSLR-A350
Focale: 28 mm

Date: 6/4/2010
Time: 10:35
Spatial resolution: 19.5 cm
Sony DSLR-A350
Focale: 30 mm

Date: 7/7/2010
Time: 10:45
Spatial resolution: 23.7 cm
Sony DSLR-A350
Focale: 35 mm

Figure 8.17 Multidate hyperspatial imagery used to monitor the evolution of a riparian buffer zone. The photos were taken between April 2010 and July 2010, on the site of Mollon (Ain River, France) with a powered paraglider.

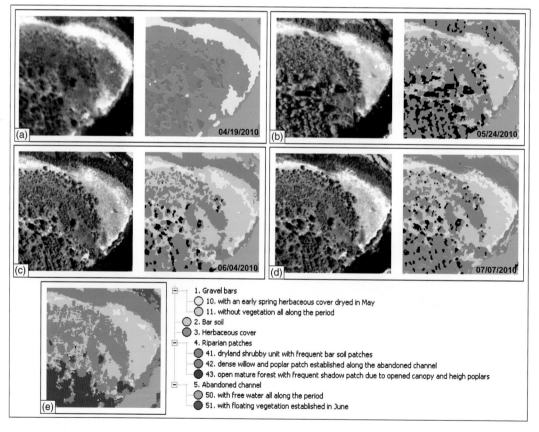

Figure 8.18 Vegetation maps (e) resulting from the multi-date remote sensing analysis (a to d) using oriented-object procedure showing main vegetation patches and their seasonal changes.

Our second example focuses on river restoration. For river restoration, the ability to monitor a restoration before and after it is completed is crucial in order to evaluate the success of the restoration and to monitor the evolution of the restored system. This procedure has been applied in an EU LIFE program focused on sediment reintroduction of the Ain River, France. To evaluate the channel bathymetry of these efforts, several surveys were done using high resolution images acquired by the Pixy UAV. The photos (Figure 8.19a) show one of the three sites before, immediately after and a few months after 18,800 m³ of sediment was introduced. Bathymetric models were produced for July 2005 and June 2006 in order to map the evolution of wetted channel topography (Figure 8.19b). Estimates of residual error of the bathymetric models were realised from field measurements. Estimated elevation errors ranged from 12.5 to 21 cm. These errors were produced by

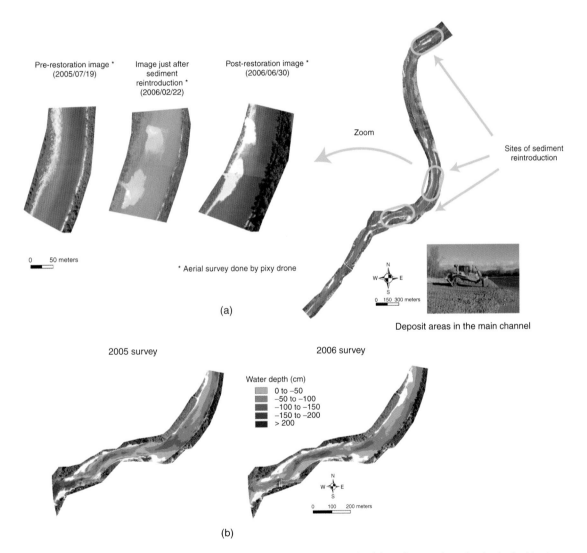

Figure 8.19 Monitoring restoration efforts with hyperspatial image data. Example of the sediment reintroduction in the Ain river channel, France. a) Images acquired at different dates showing the site before introduction, after introduction and after the first flood. b) Bathymetric maps following field and airborne surveys. c) map showing the different stages needed to assess morphological changes from bathymetric comparisons.

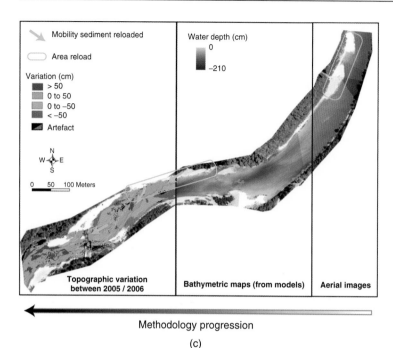

Mobility sediment reloaded

Area reload

Variation (cm)
- > 50
- 0 to 50
- 0 to –50
- < –50
- Artefact

Water depth (cm)
0
–210

N
W–E
S

0   50   100 Meters

Topographic variation
between 2005 / 2006

Bathymetric maps (from models)

Aerial images

Methodology progression

(c)

Figure 8.19 (*continued*).

(i) non-uniform illumination conditions on the image database used and (ii) the presence of different substrate types on the monitoring site (submerged aquatic vegetation, limestone outcrops and different structures of grain size inside the water channel) which led to errors in bathymetric models. The results clearly show significant morphological change. We observed a migration of sediment from the reintroduction area and an accumulation on a downstream reach mobilised by several annual floods between the two surveys. A sediment wave 380 meters long was detected on the downstream reach from the reintroduction site. The resulting sediment budget presents an excess of sediment of 7007 m^3 (Lejot et al., 2011). Such an approach therefore provides detailed quantitative information to river managers in order to demonstrate the success of a restoration strategy. In this case, rarely accessible data on sediment local transport as a result of gravel addition has been made available. Furthermore the hyperspatial data has proved invaluable for fish ecologists who are surveying the effects of such measures on fish distribution and fish habitat.

8.5 Conclusion

Hyperspatial imagery has the potential to dramatically change the way in which small rivers are managed and studied. In this chapter, we have attempted to deliver a very pragmatic overview of the new data acquisition options that are now available to river scientists and managers. We hope to have shown that hyperspatial image acquisition methods are now well within reach of most projects. With the proliferation of image platforms and affordable cameras, even low-budget endeavours should be able to acquire hyperspatial imagery over small reaches whilst at moderate budgets, catchment scale hyperspatial data acquisition is now well within the reach of river scientists and managers. Furthermore, many newly available UAV and ULAV platforms allow for a 'do-it-yourself' approach which is ideally suited to repeat image collection. This potential improvement in the temporal resolution of hyperspatial images is likely to be a key factor in their expanding usage since it allows for monitoring at monthly or even daily timescales. Such monitoring methods with both high temporal and spatial resolutions are crucially needed in river sciences and monitoring.

However, it should be noted that whilst hyperspatial imagery can now allow for automated measurements of many important habitat parameters such as grain size, depth and riparian vegetation types, the need for field-based sampling persists. Indeed, the analysis methods cited in this chapter all have their limitations and readers are greatly cautioned from treating this new data source as a silver bullet capable of meeting all our sampling needs. A very obvious and simple reminder of this fact is the inapplicability of these methods to small channels with significant overarching vegetation (i.e. tunnelling). If we cannot see the channel from the air, even hyperspatial imagery will be ineffective as a sampling tool. Another crucial area which still requires fieldwork is the sampling of biota. The recent progress in river sciences has left us in a situation where our ability to map abiotic parameters and physical habitat greatly exceeds our ability to map lotic biota. This discrepancy will clearly need to be addressed. However in the near future, hyperspatial imagery can be a powerful addition to the river sciences 'toolbox' when used in conjunction with other data sources and field-based sampling.

Acknowledgements

The authors thank all the persons who very kindly participated in the Drone Pixy project applied to the fluvial corridor, all students and colleagues who helped during the pixy campaigns, especially Thierry Fournier and Marie-Laure Trémélo and the ZABR which funded and supported this effort. We also thank Rodolphe Montagnon the pilot of the power paraglider, A. Hervouet, M. Gagnage, B. MacVicar, James, T.D., and J. Riquier, for kindly providing data and figures for this chapter. Furthermore, we would like to thank Dr Mark Fonstad for discussions and assistance in the resolution of the shadow removal problem.

References

Aber, J.S., Aber, S.W., and Firooza, P. 2002. Unmanned small-format aerial photography from kites for acquiring large-scale, high-resolution, multiview-angle imagery. *Mid-Term symposium in Conjunction with Pecora 15/Land Satellite Information IV Conference, 10-15*, Denver, USA.

Aber, J.S., Marzolff, I., and Ries, J.B. 2010. *Small-format aerial photography principles, techniques and geoscience applications*, Elsevier. 266 pp.

Akasheh, O.Z., Neale, C.M.U., and Jayanthi, H. 2008. Detailed mapping of riparian vegetation in the middle Rio Grande River using high resolution multi-spectral airborne remote sensing. *Journal of Arid Environments*, **72**(9), 1734–1744.

Ballester, M.V.R., Victoria, D.D., Krusche, A.V., Coburn, R., Victoria, R.L., Richey, J.E., Logsdon, M.G., Mayorga, E., and Matricardi, E. 2003. A remote sensing/GIS-based physical template to understand the biogeochemistry of the Ji-Parana

river basin (Western Amazonia). *Remote Sensing of Environment*, **87**(4), 429–445.

Beal, R.C., Kudryavtsev, V.N., Thompson, D.R., Grodsky, S.A., Tilley, D.G., Dulov, V.A., and Gaber, H.C. 1997. The influence of the marine atmospheric boundary layer on ERS 1 synthetic aperture radar imagery of the Gulf Stream. *Journal of Geophysical Research-Oceans*, **102**(C3), 5799–5814.

Bertoldi, W., Gurnell, A.M., and Drake, N.A. 2011. The topographic signature of vegetation development along a braided river: Results of a combined analysis of airborne lidar, color air photographs, and ground measurements. *Water Resources Research*, **47**. 10.1029/2010WR010319

Butler, J.B., Lane, S.N., Chandler, J.H., and Porfiri, E. 2002. Through-water close range digital photogrammetry in flume and field environments. *Photogrammetric Record*, **17**(99), 419–439.

Carbonneau, P.E., Bergeron, N., and Lane, S.N. 2005. Automated grain size measurements from airborne remote sensing for long profile measurements of fluvial grain sizes. *Water Resources Research*, **41**(11). DOI:10.1029/2005WR003994

Carbonneau, P.E., Dugdale, S.J., and Clough, S. 2010. An automated georeferencing tool for watershed scale fluvial remote sensing. *River Research and Applications*, **26**(5), 650–658.

Carbonneau, P.E., Fonstad, M.A., Marcus, W.A., and Dugdale, S.J. 2011. Making riverscapes real. *Geomorphology*. DOI:10.1016/j.geomorph.2010.09.030

Carbonneau, P.E., Lane, S.N., and Bergeron, N. 2006. Feature based image processing methods applied to bathymetric measurements from airborne remote sensing in fluvial environments. *Earth Surface Processes and Landforms*, **31**(11), 1413–1423.

Carbonneau, P.E., Lane, S.N., and Bergeron, N.E. 2004. Catchment-scale mapping of surface grain size in gravel bed rivers using airborne digital imagery. *Water Resources Research*, **40**(7). 10.1029/2003WR002759

Chen, S.S., Fang, L.G., Zhang, L.X., and Huang, W.R. 2009. Remote sensing of turbidity in seawater intrusion reaches of Pearl River Estuary – A case study in Modaomen water way, China. *Estuar Coast Shelf S*, **82**(1), 119–127. 10.1016/j.ecss.2009.01.003

Chubey, M.S., Franklin, S.E., and Wulder, M.A. 2006. Object-based analysis of Ikonos-2 imagery for extraction of forest inventory parameters. *Photogrammetric Engineering and Remote Sensing*, **72**(4), 383–394.

Conyers, M.M. and Fonstad, M.A. 2005. The unusual channel resistance of the Texas Hill Country and its effect on flood flow predictions. *Physical Geography*, **26**(5), 379–395.

Dai, M., He, B., Huang, W., Liu, Q., Chen, H., and Xu, L. 2010. Sources and accumulation of organic carbon in the Pearl River Estuary surface sediment as indicated by elemental, stable carbon isotopic, and carbohydrate compositions. *Biogeosciences*, **7**(10), 3343–3362.

Defourny, P., Desclee, B., and Bogaert, P. 2006. Forest change detection by statistical object-based method. *Remote Sensing of Environment*, **102**(1-2), 1–11. 10.1016/j.rse.2006.01.013

Dugdale, S.J., Carbonneau, P.E., and Campbell, D. 2010. Aerial photosieving of exposed gravel bars for the rapid calibration of airborne grain size maps. *Earth Surface Processes and Landforms*, **35**(6), 627–639.

Dunford, R., Michel, K., Gagnage, M., Piégay, H., and Trémélo, M.L. 2009. Potential and constraints of Unmanned Aerial Vehicle technology for the characterization of Mediterranean riparian forest. *International Journal of Remote Sensing*, **30**(19), 4915–4935.

Fausch, K.D., Torgersen, C.E., Baxter, C.V., and Li, H.W. 2002. Landscapes to riverscapes: Bridging the gap between research and conservation of stream fishes. *Bioscience*, **52**(6), 483–498.

Gao, J. 2009. *Digital analysis of Remotely Sensed Imagery*, McGraw-Hill. 645

Garguet-Duport, B., Girel, J., and Pautou, G. 1995. Contribution of the ERS-1 data in the remote sensing analysis of floodplain landscape structure. *Comptes Rendus De L Academie Des Sciences Serie Iii-Sciences De La Vie-Life Sciences*, **318**(12), 1253–1259.

Gilvear, D.J. 2004. Patterns of channel adjustment to impoundment of the upper River Spey, Scotland (1942-2000). *River Research and Applications*, **20**(2), 151–165. 10.1002/rra.741

Gordon, D.J., McMahon, T.A., Finlayson, B.L., Gippel, C.J., and Nathan, R.J. 2004. *Stream Hydrology – An Introduction for Ecologists*, J. Wiley and Sons. 526 pp.

Greenberg, J.A., Dobrowski, S.Z., and Vanderbilt, V.C. 2009. Limitations on maximum tree density using hyperspatial remote sensing and environmental gradient analysis. *Remote Sensing of Environment*, **113**(1), 94–101. 10.1016/j.rse.2008.08.014

Guichard, F., Bourget, E., and Agnard, J.P. 2000. High-resolution remote sensing of intertidal ecosystems: A low-cost technique to link scale-dependent patterns and processes. *Limnology and Oceanography*, **45**(2), 328–338.

Hall-Beyer, M., Flanders, D., and Pereverzoff, J. 2003. Preliminary evaluation of eCognition object-based software for cut block delineation and feature extraction. *Canadian Journal of Remote Sensing*, **29**(4), 441–452.

Hedger, R.D., Dodson, J.J., Bourque, J.F., Bergeron, N.E., and Carbonneau, P.E. 2006. Improving models of juvenile Atlantic salmon habitat use through high resolution remote sensing. *Ecological Modelling*, **197**(3-4), 505–511.

Hedger, R.D., Malthus, T.J., Folkard, A.M., and Atkinson, P.M. 2007. Spatial dynamics of estuarine water surface temperature from airborne remote sensing. *Estuar. Coast. Shelf. Sci.*, **71**(3-4), 608–615. 10.1016/j.ecss.2006.09.009

Hervouet, A., Dunford, R., Piégay, H., Belletti, B., and Trémélo, M.L. 2011. Analysis of post-flood recruitment patterns in braided-channel rivers at multiple scales based on an image series collected by Unmanned Aerial Vehicles, Ultra-light Aerial Vehicles, and satellites. *Giscience & Remote Sensing*, **48**(1), 50–73.

Jacquin, A., Misakova, L., and Gay, M. 2008. A hybrid object-based classification approach for mapping urban sprawl in periurban environment. *Landscape Urban Plan*, **84**(2), 152–165. 10.1016/j.landurbplan.2007.07.006

Jensen, T., Apan, A., Young, F., and Zeller, L. 2007. Detecting the attributes of a wheat crop using digital imagery acquired from a low-altitude platform. *Computers and Electronics in Agriculture*, **59**(1-2), 66–77.

Kondolf, G.M. and Piégay, H. 2003. *Tools in fluvial geomorphology*, J. Wiley and Sons. 384 pp.

Lane, S.N. 2000. The measurement of river channel morphology using digital photogrammetry. *Photogrammetric Record*, **16**(96), 937–957.

Legleiter, C.J. and Goodchild, M.F. 2005. Alternative representations of in-stream habitat: classification using remote sensing, hydraulic modeling, and fuzzy logic. *International Journal of Geographical Information Science*, **19**(1), 29–50.

Lejot, J., Delacourt, C., Piégay, H., Fournier, T., Trémélo, M.L., and Allemand, P. 2007. Very high spatial resolution imagery for channel bathymetry and topography from an unmanned mapping controlled platform. *Earth Surface Processes and Landforms*, **32**(11), 1705–1725. 10.1002/esp.1595

Lejot, J., Piégay, H., Hunter, P.D., Moulin, B., and Gagnage, M. 2011. Characterisation of alluvial plains by remote sensing: case studies and current stakes. *Géomorphologie, relief, processus et environment*, **2**, 157–172.

Lillesand, T.M., Keifer, R.W., and Chipman, J. 2008. *Remote Sensing and Image Interpretation* 6th Ed., John Wiley and Sons. 804 pp.

Lorang, M.S., Whited, D.C., Hauer, F.R., Kimball, J.S., and Stanford, J.A. 2005. Using airborne multispectral imagery to evaluate geomorphic work across floodplains of gravel-bed rivers. *Ecological Applications*, **15**(4), 1209–1222.

Lyon, J.G., Lunetta, R.S., and Williams, D.C. 1992. Airborne Multispectral Scanner Data for Evaluating Bottom Sediment Types and Water Depths of the St Marys River, Michigan. *Photogrammetric Engineering and Remote Sensing*, **58**(7), 951–956.

Marcus, W.A. and Fonstad, M.A. 2008. Optical remote mapping of rivers at sub-meter resolutions and watershed extents. *Earth Surface Processes and Landforms*, **33**(1), 4–24.

Marcus, W.A. and Fonstad, M.A. 2010. Remote sensing of rivers: the emergence of a subdiscipline in the river sciences. *Earth Surface Processes and Landforms*, **35**(15), 1867–1872.

Marcus, W.A., Legleiter, C.J., Aspinall, R.J., Boardman, J.W., and Crabtree, R.L. 2003. High spatial resolution hyperspectral mapping of in-stream habitats, depths, and woody debris in mountain streams. *Geomorphology*, **55**(1-4), 363–380.

McKean, J., Nagel, D., Tonina, D., Bailey, P., Wright, C.W., Bohn, C., and Nayegandhi, A. 2009. Remote Sensing of Channels and Riparian Zones with a Narrow-Beam Aquatic-Terrestrial LIDAR. *Remote Sensing*, **1**(4), 1065–1096.

Newcome, L.R. 2004. *Unmanned Aviation: A Brief History of Unmanned Aerial Vehicles*, AIAA. 166 pp.

Pacey, R. and Fricker, P. 2005. Forward Motion Compensation (FMC) – Is it the same in the digital imaging world?

Photogrammetric Engineering and Remote Sensing, **71**(11), 1241–1242.

Ramakrishna, C., Muralikrishna, C., and Rao, D.M. 2011. Remote sensing and geoprocessing use for water resources monitoring of the Granulite Dome and river basin from Eastern Ghats, India. *Asian Journal of Chemistry*, **23**(8), 3559–3562.

Rango, A., Laliberte, A., Herrick, J.E., Winters, C., Havstad, K., Steele, C., and Browning, D. 2009. Unmanned aerial vehicle-based remote sensing for rangeland assessment, monitoring, and management. *Journal of Applied Remote Sensing*, **3**. 10.1117/1.3216822

Ries, J.B. and Marzolff, I. 2003. Monitoring of gully erosion in the Central Ebro Basin by large-scale aerial photography taken from a remotely controlled blimp. *Catena*, **50**(2-4), 309–328.

Smith, M.J., Chandler, J., and Rose, J. 2009. High spatial resolution data acquisition for the geosciences: kite aerial photography. *Earth Surface Processes and Landforms*, **34**(1), 155–161.

Sugumaran, R. and Harken, J. 2005. Classification of Iowa wetlands using an airborne hyperspectral image: a comparison of the spectral angle mapper classifier and an object-oriented approach. *Canadian Journal of Remote Sensing*, **31**(2), 167–174.

Thorne, C.R. 1998. *Stream Reconnaissance Handbook*, Wiley-Blackwell. 142 pp.

Thorp, J.H., Flotemersch, J.E., Delong, M.D., Casper, A.F., Thoms, M.C., Ballantyne, F., Williams, B.S., O'Neill, B.J., and Haase, C.S. 2010. Linking ecosystem services, rehabilitation, and river hydrogeomorphology. *Bioscience*, **60**(1), 67–74. 10.1525/bio.2010.60.1.11

Townshend, J.R.G., Huang, C., Kalluri, S.N.V., Defries, R.S., Liang, S., and Yang, K. 2000. Beware of per-pixel characterization of land cover. *International Journal of Remote Sensing*, **21**(4), 839–843.

Wang, H.W., Ding, J.L., Shi, Q.D., and Zhang, F. 2007. Water quality analysis of Aksu-Tarim river based on remote sensing data – art. no. 667917. *Remote Sensing and Modeling of Ecosystems for Sustainability Iv*, **6679**, 67917–67917.

Ward, J.V., Tockner, K., and Malard, F. 2002. Landscape ecology: a framework for integrating pattern and process in river corridors. *Landscape Ecology*, **17**, 35–45.

Wiens, J.A. 2002. Riverine landscapes: taking landscape ecology into the water. *Freshwater Biology*, **47**(4), 501–515.

Winterbottom, S.J. and Gilvear, D.J. 1997. Quantification of channel bed morphology in gravel-bed rivers using airborne multispectral imagery and aerial photography. *Regulated Rivers-Research & Management*, **13**(6), 489–499.

Wright, A., Marcus, W.A., and Aspinall, R. 2000. Evaluation of multispectral, fine scale digital imagery as a tool for mapping stream morphology. *Geomorphology*, **33**(1-2), 107–120.

Wundram, D. and Loffler, J. 2008. High-resolution spatial analysis of mountain landscapes using a low-altitude remote sensing approach. *International Journal of Remote Sensing*, **29**(4), 961–974.

9 Geosalar: Innovative Remote Sensing Methods for Spatially Continuous Mapping of Fluvial Habitat at Riverscape Scale

Normand Bergeron[1] and Patrice E. Carbonneau[2]

[1]Institut National de Recherche Scientifique, Centre Eau Terre et Environnement, Québec, Canada
[2]Department of Geography, Durham University, Science site, Durham, UK

9.1 Introduction

In 2002, a multidisciplinary group of researchers (biologists, geomorphologists, engineers) member of CIRSA (Centre Interuniversitaire de Recherche sur le Saumon Atlantique) initiated the Geosalar project. The aim of this research initiative funded by the GEOIDE Network Centres of Excellence Program (http://www.geoide.ulaval.ca/) was to develop and apply geomatics technology to the problem of modelling Atlantic salmon (*Salmo salar*) production in relation to fluvial habitat characteristics.

In the spirit of the riverscape approach fostered by Fausch et al. (2002), the project proposed to develop the tools needed to increase the scope (extent of studied area/resolution) and spatial continuity of fish/habitat relationship investigations. In their landmark paper, Fausch et al. (2002) demonstrated that the research community had so far failed to provide river managers with information and tools at the scale needed to efficiently conserve stream fish populations, and they suggested that this gap contributed to the constant decline of many fish populations. Indeed, while stream managers most-often face issues caused by large-scale anthropogenic disturbances of the habitat, the scientific information on which they must base their management decisions arises mainly from studies conducted over relatively short (50–500 m) and spatially discontinuous river segments (Fausch et al., 2002). Because of their small spatial extent, such studies cannot easily account for the important effects of habitat heterogeneity and spatial organisation of habitat patches on fish distribution and abundance. In his dynamic landscape model of stream fish population ecology and life history, Schlosser (1991) emphasised the role of habitat heterogeneity in providing the various types of habitat required by fish at different life stages for spawning, feeding and finding refugia from harsh environmental conditions. Because such a variety of habitat type can only be found on stream segments that are relatively long, studies conducted on a small-scale invariably fail to include in their analysis all habitat components that fish need to access in order to complete their life history.

Fausch et al. (2002) also argued that studying several short sample sections distributed along a river only provided a fragmented view of the riverscape even when based

Fluvial Remote Sensing for Science and Management, First Edition. Edited by Patrice E. Carbonneau and Hervé Piégay.
© 2012 John Wiley & Sons, Ltd. Published 2012 by John Wiley & Sons, Ltd.

on a logical statistical design. They therefore stressed the importance of describing fluvial habitats in a continuous manner in order to detect unique habitats (e.g. local cool water input) or disturbance events (e.g. barrier to movements) at specific locations that can affect the distribution and abundance of fish over the entire riverscape.

However, they acknowledged that implementing their approach was a challenge due to the lack of appropriate technology to work at the intermediate scale ($10^3 - 10^5$ m) encompassing all necessary habitats (Figure 9.1). On the one hand, traditional field-based methods offer good ground resolution of fluvial habitat variables at the micro-habitat scale but they are labour intensive and not well suited to the continuous characterisation of long river segments. On the other hand, satellite-based imagery offer a large-scale synoptic description of entire fluvial systems but their ground resolution is currently not sufficient for fine-scale habitat modelling purposes. One of the main focuses of the Geosalar project was therefore to fill the gap between these approaches by developing a new set of remote sensing methods allowing the production of high-resolution spatially continuous maps of fluvial habitat variables over long river segments. Typically, four variables are used to describe the physical habitat of riverine fishes: bed material grain size, water depth, flow velocity and water temperature. The emphasis of the Geosalar research effort was put on the quantification of bed material grain size and water depth. Post-Geosalar developments later addressed the quantification of flow velocity and water temperature. This chapter presents the innovative remote sensing methods that were developed during the Geosalar project for the quantification of river habitat variables over long spatially continuous river segments. The usefulness of these methods is then exemplified by applying them to the case of the Sainte-Marguerite River (Québec, Canada) for the analysis of Atlantic salmon juvenile and adult habitat.

9.2 Study area and data collection

The Geosalar research was conducted on the Principale branch of the Sainte-Marguerite River (SMR) ($48\,^\circ 27'$ N, $69\,^\circ 95'$ W), a gravel-cobble bed river draining a Canadian Shield catchment of approximately 1000 km^2 in the Saguenay region. Bed material is composed of well mixed igneous and metamorphic rocks. The lithological composition of this mixture is stable along the channel length and thus no spatially dependent clast colour variations can be observed. Suspended sediment load along the

channel is not altered by tributaries and therefore the suspended sediment load can be assumed as constant for the whole channel. The SMR supports Atlantic salmon and brook trout (*Salvelinus fontinalis*) populations that have been the subject of numerous ecological studies by CIRSA researchers and their students since the creation of the research group in 1995. Field work for the various components of the project was facilitated by CIRSA's research station located downstream from the confluence between the main stem and North-East (1100 km^2) branches of the SMR.

In August 2002, during the period of summer low flow, the XEOS^{TM} imaging system developed by Génivar Inc. was fitted to a helicopter and used to obtain plan view digital high resolution optical images covering the entire 80 kilometres of the main channel of the Principale branch of the SMR. A first survey conducted at a constant altitude of 155 m above ground resulted in a dataset comprising of c. 5550 standard colour images with a spatial resolution of 3 cm. A second survey conducted at an elevation of 450 m above ground generated another set of 1600 colour images with a spatial resolution of 10 cm. Image format was 3008 pixels × 1960 pixels in the standard visible bands of red, green and blue. Images were collected at 60% overlap. An onboard GPS provided the position of the centre point of each image. These surveys provided one of the first large scale hyperspatial (defined in Chapter 8) image datasets reported in the river science literature. The project also supported a number of field efforts which provided a range of data concerning the abundance and spatial distribution of Atlantic salmon at various life stages.

9.3 Grain size mapping

Bed material grain size is one of the most fundamental descriptors of salmonid habitat. Generally speaking, the freshwater stage of the salmon's life cycle requires coarse substrate (Armstrong et al., 2003). The presence of clay, silt and sand is well established as having a negative impact on the survival of eggs (Sear, 1993) and juveniles (a group name for alevins, fry and parr). Consequently the quality of the substrate is often considered as a primary indicator of the health of a salmon river. However, assessing the substrate status for an entire river has always been problematic. Certain methods, such as the River Habitat Survey (RHS) protocol developed in the UK (Raven et al., 1997), employ walking surveys and visual appraisals in order to get a semi-quantitative sampling

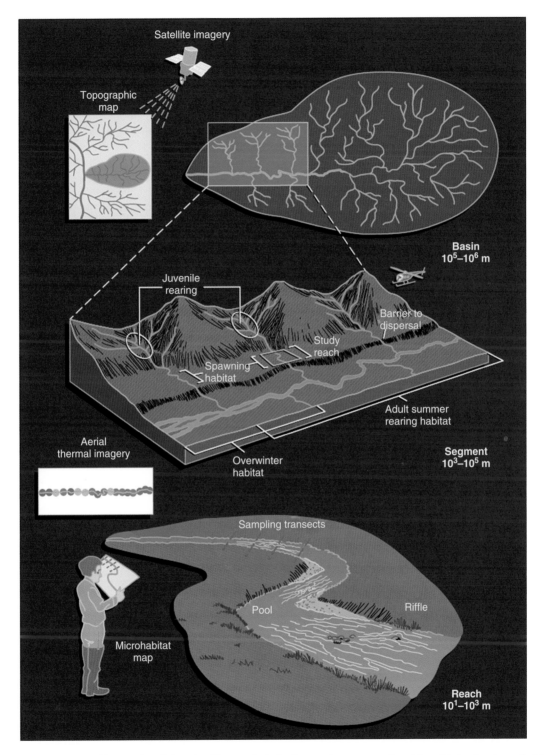

Figure 9.1 Diagram used by Fausch et al. (2002) to illustrate the lack of appropriate conceptual models and technology to work at the intermediate segment scale (10^3–10^5 m) where information is least available. Reproduced from Fausch et al. (2002), with permission from the American Institute of Biological Sciences.

of substrate quality (and other parameters) for very long reaches of a channel. However, methods such as the RHS have some key limitations. First they require more or less continuous access to the river banks. Whilst this might be possible in many European rivers which now flow through agricultural or urban landscapes, many key salmon rivers of the world, notably in Scandinavia and North America, flow through undeveloped forested areas and, as a result, complete access to the entire channel from the ground is often difficult. Another limitation with ground based approaches is that they frequently use visual appraisal methods in order to save time in the field and allow for greater distances and channel lengths to be sampled. Whilst this strategy does recognise the need to sample rivers over increasing scales as advocated by Fausch et al. (2002), the resulting data is hard to reconcile with documented habitat preferences which are collected by manual sampling of individual clasts. Furthermore, the type of qualitative data collected by visual appraisal methods is also incompatible with physical models of sediment transport which could allow the prediction and validation of habitats distribution at catchment scales. There is therefore a clear need for substrate quantification approaches which are capable of delivering accurate and quantitative measurements of substrate with little or no field effort over riverscape scales.

Such requirements can be fulfilled by remote sensing approaches and this section examines the contribution of the Geosalar project and others to the quantitative assessment of sediment size with a focus on habitat.

9.3.1 Superficial sand detection

The detrimental effects of fine sediment in fluvial gravels on salmonid habitat quality are well documented (e.g. Chapman et al., 1986). During the incubation phase, it has been shown that the presence of fine sediment within the gravel matrix reduces the survival rate of the eggs (Wu, 2000; Soulsby et al., 2001). Furthermore, Cunjak et al. (1998) suggested that during the juvenile life stage, sand deposition on the surface of the bed could block access to the large interstitial voidspaces used by juvenile salmon as shelter during overwintering. Clearly, from a purely visual perspective, it is relatively easy to detect the presence of sand on the surface of a gravel bar and therefore it can be deduced that an image of sand deposited on gravels will somehow have the information which allows for the identification of the sand patch. A method could therefore potentially be devised whereby terrestrial photographs are taken in the field during a

ground survey which could then be analysed in order to estimate sand content in a quantitative manner. The problem that must be solved is the manner in which the image information can be converted to a quantitative measurement of sand area.

In remote sensing and image analysis terms, this is a segmentation problem requiring the delineation of a feature in an image based on pixel properties. Once a feature is segmented, its type can be attributed in the process commonly called 'classification'. For example, a classic classification application is the simple identification of land-use types (e.g. forest, urban and/or agricultural) in an air photo or a satellite image. Basic segmentation and classification approaches rely on the brightness values in the image and assume that an object or land-use of any given type will most likely be of a specific colour or grey level. For example, trees are assumed to be green, water is assumed to be dark and sediment is generally light grey with a slight reddish hue depending on exact composition. However, in the case of sand identification, a close examination of ground imagery reveals that the solution may not be straightforward. Figure 9.2a shows a small image covering 20 × 20 cm of a sand patch with a few protruding gravels and cobbles. Visually, the identification of the clasts in the image is quite natural. One could therefore expect that the frequency distribution of pixels brightness values should reveal a clear bimodal distribution with one mode for the sand and a second mode for the clasts. However, the interpretation of Figure 9.2b is much more ambiguous. While this histogram does have two poorly distinguished modes, the modal tails show considerable overlap. Figure 9.2c shows the result of the application of Otsu's segmentation algorithm (Otsu, 1979) which was specifically designed for bimodal histograms. It can clearly be seen that while the clast is effectively delineated, many sand particles were also delineated. This is simply due to the fact that colour, or in this case grey level, is not the key distinctive parameter. Sand grains having a colour similar to that of the clast will be falsely identified as clasts. Given that lithology exerts a dominant control on clast colour, the presence of sand grains with the same colour as the clasts in any given river is highly likely. Carbonneau et al. (2005b) therefore hypothesised that another image property could act as a better discriminator in the classification process.

Close observation of Figure 9.2a clearly shows that sand is characterised by its mixed colour. Grains of sharply differing colour are in close spatial proximity. This leads to the suggestion that the defining feature of a patch of sand is not its average colour but the variability of

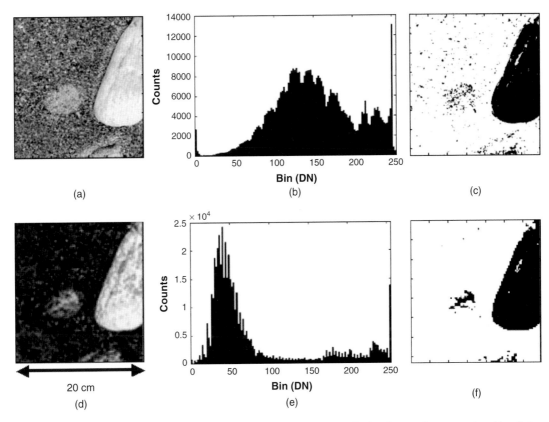

Figure 9.2 Surficial sand identification based on normal image brightness values (a, b, c) and textural entropy values (d, e, f). In a and d, pixel brightness is respectively proportional to reflected radiation and neighbourhood image texture. b and e are the histograms of images a and d respectively. c and f show the results of the binary segmentation process obtained respectively from the radiation (b) and texture (e) histograms. Reproduced from Carbonneau et al. (2005b), with permission from Wiley-Blackwell.

this colour within a local spatial neighbourhood. While a simple metric such as local variance could be used to quantify local colour variability, the concept of image texture is most often employed. Image texture is defined as: 'an attribute representing the spatial arrangement of the grey levels of the pixels in a region' (Haralick and Shapiro, 1985). This texture property therefore allows the production of a new image where the brightness of each pixel is in fact proportional to the texture of a given neighbourhood instead of being proportional to the reflected radiation. The result of this process is shown in Figure 9.2d. Here it can be noticed that the perceived brightness difference between the sand patch and the clast has been greatly amplified. As a result, the histogram in Figure 9.2e shows a much greater separation between the two modes. Consequently, the resulting segmentation shown in Figure 9.2f is now much less speckled and is a more accurate reflection of reality. After testing on a

range of 20 images, Carbonneau et al. (2005b) found that the texture based classification predicted values of sand coverage in terms of a surface % of the whole image, with an R^2 of 93% versus only 56% for traditional grey level brightness segmentation. Therefore, this approach can be effectively used during walkover surveys. All that is needed is an image with a known scale, and sand coverage can be sampled. Given that a digital picture takes seconds to acquire, this approach is very cost effective in the field. Furthermore, the approach can be combined to dubbed photosieving methods that use terrestrial images in order to measure particle sizes (Ibbeken and Schleyer, 1986; Dugdale et al., 2010). However it should be noted that this method is best used at periods of very low flow when significant areas of river bed are dry and exposed. In the case of submerged bed material, the image histogram is compressed in a manner which is similar to the shading effect discussed in Chapter 8 (see Figure 9.12 of that chapter). The compressed brightness levels will reduce image texture thereby making a clear segmentation of sand and gravels more prone to errors.

9.3.2 Airborne grain size measurements

While the methods mentioned above are an improvement over visual surveys, they retain one key limitation of traditional field based sampling from the ground: they do not allow for a continuous coverage at riverscape scales. The obvious solution is therefore to move towards the use of airborne imagery. As mentioned in Chapters 2 and 8, high resolution, hyperspatial imagery of fluvial environments carries a wealth of details. Figure 9.3 shows a hyperspatial image (spatial resolution: 3 cm) where it is relatively easy to distinguish coarse versus fine gravels (or sand) and deep versus shallow water.

This image, and others like it, leads once again to hypothesise that the information content in the image should allow for a quantitative measurement of bed material grain size and water depth. However, the reader should be reminded that even hyperspatial imagery cannot rival ground based imagery in terms of spatial resolution. In the case of Figure 9.2a, the ground resolution of the image was 0.3 mm. The 3 cm resolution of Figure 9.3 is therefore two orders of magnitude coarser by comparison. This has important implications for any grain size mapping process. In the case of Figure 9.2a, the size of each pixel is similar to the size of an individual grain of sand and

much smaller than the clasts. In Figure 9.3, the pixel size is well within the gravel range. Consequently, individual clasts cannot be delineated in the airborne hyperspatial imagery. Therefore, photosieving methods mentioned above, which rely on such particle delineations, will only function in the cases of particles in the cobble to boulder range (Dugdale et al., 2010). However, textural methods such as that used to identify sand patches do not rely on the delineation of the individual objects that create the texture pattern (sometimes called 'texels'). A key question is therefore the extent to which the texture metrics used to identify sand are in fact capable of a continuous measurement of the spatial distribution of particle sizes.

9.3.2.1 Dry gravel bar grain size mapping

Although fish live underwater, the determination of bed material size on exposed gravel bars is important since at higher flows, these zones become inundated and may make-up an important proportion of the available habitat. The quantification of bed material size from hyperspatial imagery is thus first explored for that simplest case: that of a dry exposed gravel bar with a range of materials from sands to cobbles. Figure 9.3 clearly allows identifying sandy patches in the centre of the mid-channel bar and coarser material can also be seen on either side. Carbonneau et al. (2004) hypothesised that image texture can be correlated to local grain size in the image. In order to test this hypothesis, hyperspatial imagery of the

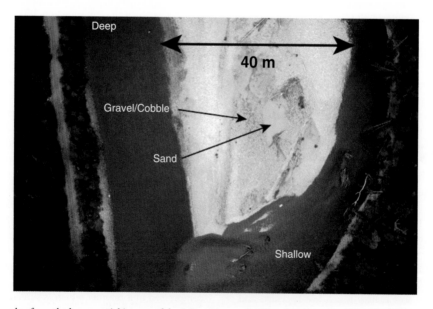

Figure 9.3 Example of one the hyperspatial images of the Sainte-Marguerite River (Québec, Canada) obtained for the Geosalar project. This 3 cm ground resolution image was taken at an altitude of 150 m above ground.

SMR was used in conjunction with field data in order to perform a direct empirical verification. With a simple field procedure, local samples of median grain size were then taken in the field. These samples were precisely positioned with a DGPS system which allowed their location in the airborne hyperspatial imagery to be identified. After experimenting with several types of image texture metrics, Carbonneau et al. (2004) opted for the two dimensional image semivariance. This is a similar measure to the texture discussed above. A flat uniform image with nearly identical brightness values will have a low semivariance as well as a low texture. Inversely, an image with a strong 'salt and pepper' aspect where pixels of very different brightness are in close spatial proximity will have a high texture.

Figure 9.4 shows the result of this hypothesis test. Figure 9.4a shows the calibration relationship which regresses the local semivariance with the grain size observed in the field.

Figure 9.4b shows the results of an independent validation where field based observations are tested against the predictions of the calibration equation. These results were quite encouraging since they established that the information in the image had the potential to yield quantitative measurements of particle size with minimal fieldwork. Whilst fieldwork is indeed required for the calibration of the relationship, this calibrated prediction of grain size can be applied to an entire image dataset in a fully automated manner thus allowing for the median grain size of every exposed surface to be measured.

9.3.2.2 Grain size mapping along the wetted area

In most areas of river science and in lotic ecology in particular, grain size measurements limited to dry exposed areas are not sufficient and particle size information for the wetted area is needed. For example, even for images taken at summer low flow (e.g. Figure 9.3), a significant fraction of the riverbed is underwater. The crucial question then becomes the extent to which the information content in this wetted but visible bed is sufficient for particle size measurements.

In Carbonneau et al. (2005a), it was demonstrated that in clear shallow flow situations, enough information is contained in the images to extract particle sizes. However, it was also demonstrated that the presence of the water degraded the precision and accuracy of the grain size mapping results. Figure 9.5 shows the results of the calibration process in submerged areas. The calibration equation in Figure 9.5a shows a strong correlation. However, the validation equation in Figure 9.5b, with a slope of 1.23, shows that larger particle sizes are over-predicted. This loss of data quality is not entirely surprising since the water interface inevitably degrades the image quality for submerged areas. Furthermore, the extent of this degradation will obviously be a function of water clarity and the exact, quantitative, relationship between water

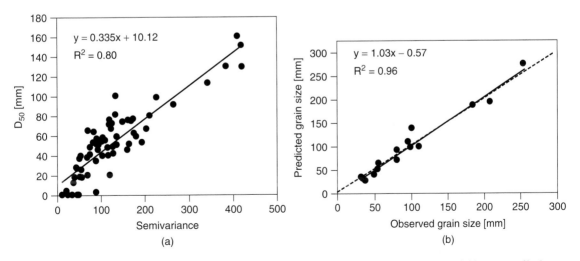

Figure 9.4 a) Calibration curve between the local semivariance of pixel brightness and the corresponding field measure of bed material size (D50) on a dry gravel bar. b) Validation curve showing the relationship between the observed and predicted grain size values. The dashed line shows the expected 1:1 relationship. From Carbonneau, P.E., Bergeron N.E., Lane, S.N. (2004), Catchment-scale mapping of surface grain size in gravel bed rivers using airborne digital imagery, Water Resources Research, 40, W07202, DOI:10.1029/2003WR002759. Copyright 2004 American Geophysical Union. Reproduced by permission of American Geophysical Union.

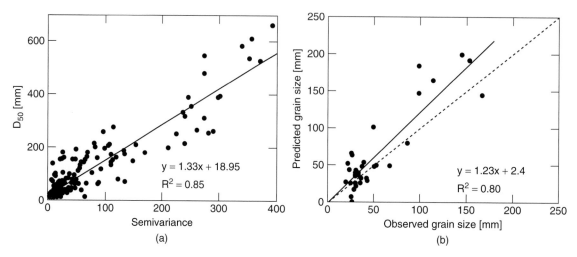

Figure 9.5 a) Calibration curve between the local semivariance of pixel brightness and the corresponding field measure of bed material size (D50) for the wetted are of the channel. b) Validation curve showing the relationship between the observed and predicted grain size values. The dashed line shows the expected 1:1 relationship. From Carbonneau, P.E., N. Bergeron, and S.N. Lane (2005), Automated grain size measurements from airborne remote sensing for long profile measurements of fluvial grain sizes, Water Resources Research, 41, W11426, DOI:10.1029/2005WR003994. Copyright 2005 American Geophysical Union. Reproduced by permission of American Geophysical Union.

clarity and the quality of image-based grain size mapping outputs is not yet known. This therefore leads to a crucial point about the application of remote sensing to particle size mapping: the measurement of particle sizes for large areas at high spatial resolutions comes at the cost of a loss of accuracy and precision for each individual measurement. If a single particle size measurement derived from the methods of Carbonneau et al. (2004) or Carbonneau et al. (2005a) is compared to those derived from ground based methods, it will inevitably be found that the quality of individual measurements are inferior. It is therefore important to appreciate the value of these airborne methods in view of the large spatial extent and high resolution of the coverage they provide.

9.3.3 Riverscape scale grain size profile and fish distribution

Automated airborne grain size mapping methods based on imagery can be extended to entire channels provided suitable imagery is available. Using the image analysis method described above on the Geosalar image data set of the Sainte-Marguerite River, Carbonneau et al. (2005a) proceeded to show an upstream profile of median grain sizes (Figure 9.6). This profile was constructed by taking the median of all available grain size measurements within each 20 m reach of the entire 80 km length of the river. The only breaks in this continuous profile are near km 18

and 80 and are associated with short periods of camera malfunctions where no images were collected.

9.3.4 Limitations of airborne grain size mapping

As mentioned previously, grain size mapping methods based on airborne imagery generate data whose quality is somewhat lesser than ground based methods. At the outset, users interested in the application of this technology must realise that these methods do not measure the size of each visible clast with millimetric accuracy. Instead, they measure the median diameter of a patch of gravel (usually 1 m^2) with precisions in the area of ±10–30 mm. Another key limitation of such image based methods is the requirement that the patch of clasts be visible in the imagery. This precludes any size measurements below the exposed surface of the gravel layer. Furthermore, in the case of the wetted perimeter, the feasibility and resulting quality of grain size mapping process is heavily dependent on the clarity of the water. Currently, only Carbonneau et al. (2005a) and Carbonneau et al. (2012) have published remotely sensed grain size values in the wetted perimeter. Therefore, it is still difficult to empirically define the required threshold of water clarity. However, as a rule of thumb, a simple visual appraisal is suggested: if the river bed cannot be visually seen from the banks or from a bridge due to high turbidity, then conventional

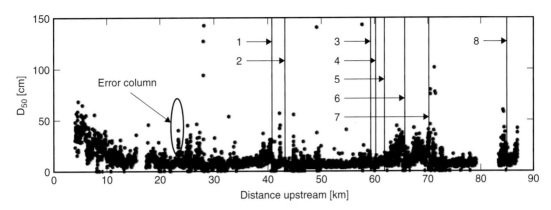

Figure 9.6 Upstream profile of airborne grain size measurements (D50) versus upstream distance along the main stem of the Ste-Marguerite River (Québec, Canada). From Carbonneau, P.E., N. Bergeron, and S.N. Lane (2005), Automated grain size measurements from airborne remote sensing for long profile measurements of fluvial grain sizes, Water Resources Research, 41, W11426, DOI:10.1029/2005WR003994. Copyright 2005 American Geophysical Union. Reproduced by permission of American Geophysical Union.

colour photography will not do any better. In terms of data acquisition, this clearly indicates that images must be acquired during periods of low rainfall, turbidity and water discharge.

Another limitation of the grain size mapping methods developed during the Geosalar project is the need for field calibration data. While the required fieldwork is generally not onerous, this does increase the final cost and limit the method to accessible rivers. However, a few authors have sought to lift the need for calibration data and there has been significant progress in this area. Dugdale et al. (2010) have developed a method whereby the grain size mapping algorithm is calibrated by direct on-screen measurements. This approach, dubbed 'Aerial Photosieving', therefore mimics traditional ground-based photosieving methods. The method relies on hyperspatial imagery. With this imagery, the user measures the b-axis (intermediate axis) of the coarse clasts which are visible in the image. This measurement is done directly on-screen and requires no fieldwork provided that the scale of the image (i.e. the pixel size) is accurately known. This use of on-screen data thus enables the grain size mapping process to be applied to areas which are inaccessible or to archival imagery. However, the method was found to result in a slight systematic overestimation of sizes and relies on the presence of coarse clasts that can be distinguished in the image (approximately larger than 2–4 pixels). Furthermore, significant progress has been made by other authors working in coastal environments. Buscombe and Masselink (2009) and Buscombe et al. (2010) have developed a method based on Fourier analysis which can derive particle sizes without any calibration

from field or on-screen data. This method has been successfully applied to coastal environments and shows much promise for river environments.

For most river sciences applications, it is suggested that the advantages offered by the airborne grain size mapping methods far outweigh their limitations. Indeed, in the Sainte-Marguerite River study case, the method allowed over 3 million bed material size measurement to be obtained over the entire 80 km-long studied river section. Clearly, such a large volume of data points would be impossible to collect with any other methods. Long awaited for by stream ecologists, such a level of habitat description can now start being incorporated into new study designs that will help improve the understanding and modelling of fish/habitat interactions.

9.3.5 Example of application of grain size maps and long profiles to salmon habitat modelling

Figure 9.7 shows an example of the unique map product that can be obtained by combining the automated airborne methods of grain size measurements for both the wet and dry portions of the channel.

Despite some loss of accuracy and precision compared to ground based methods, such an image provides a high-resolution, synoptic description of grain size over the entire image. Moreover, the process can be quickly and easily reproduced for the hundreds of images necessary to cover the entire riverscape.

As part of the Geosalar project, Hedger et al. (2006) showed how such grain size maps could be used to improve the prediction of juvenile Atlantic salmon density

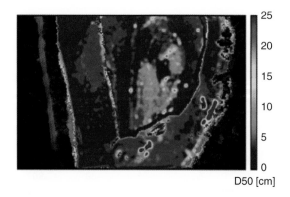

D50 [cm]

Figure 9.7 Example of a merged wet/dry grain size map obtained from the airborne method. The central portion of the channel corresponds to a dry exposed gravel bar. Lighter blue and yellow/orange colours correspond to coarser bed material.

on the Ste-Marguerite River. Using historical fry and parr density data obtained from 1997 to 2004 at 48 parcels (5 m × 20 m) distributed along the river, they derived substrate preference models using substrate size (D50) measurements obtained, 1) directly inside the parcel at time of density estimation using the traditional Wolman count method and 2) inside the larger grain size map of the image including the fishing parcel obtained using the automated airborne grain size mapping methods. They showed that, although the shape of the relationships between juvenile salmon density and D50 were similar for the two models, the relationship was stronger using mean image D50, suggesting that the habitat surrounding the location of the fishing parcel had a direct effect on fish density. Clearly, this example shows that one benefit of automated methods of grain size measurements is to allow multi-scale analysis of fish habitat relationships, a possibility that would be labour intensive using traditional ground based methods.

The grain size profile information obtained from the automated grain sizing methods allowed the identification of distinct sequences of downstream grain size fining along the Sainte-Marguerite River. Indeed, it was observed that, rather than exhibiting a single longitudinal decrease of grain size from headwater to mouth, the river could be segmented into a number of discrete sedimentary links, each characterised by a node of coarse sediment supply followed by a gradual downstream fining of substrate. The sedimentary link concept was originally developed for high mountain river environments where the supply of coarse sediment is mainly related to tributary inputs, valley-side landslides and tributary fan contacts (Rice and Church, 1998). However, using the Geosalar grain

size data set, Davey and Lapointe (2007) adapted and extended the original concept to account for sedimentary links of lower mountain landscapes of North Eastern Canada where coarse sediment inputs are often related to supply zones (rather than point sources or nodes) originating in bedrock canyon reaches or valley bottom deposits of glacial drift.

Because the downstream changes in substrate and associated slope along sedimentary links are accompanied by changes in channel morphology and hydraulics, they create a longitudinal sequence of aquatic habitat types moving from steep, fast flowing and turbulent boulder bed channels at the head of links to meandering, slow-flowing, low-gradient sand channels at the downstream end. Rice et al. (2001) were the first to demonstrate the usefulness of sedimentary links for ecological modelling by applying the concept to explain the longitudinal structuration of benthic macroinvertebrate communities. Davey and Lapointe (2007) then showed how such information on the large-scale variations of substrate size could help understand the spatial organisation of Atlantic salmon spawning habitat. Studying the location of Atlantic salmon (*Salmo salar*) spawning sites along eight sedimentary links of the Sainte-Marguerite River, they found that the centroïd of the spawning sites on each link tended to occur towards the middle to downstream end of the cobble-gravel fining sequence (Figure 9.8b). Within these zones, D50 was always in the suitable range of 40–60 mm. Higher upstream, bed material size was too coarse to allow female fish to dig their redds. Below, the absence of spawning activity was probably related to poor embryo survival associated with the high percentage of sand in riffle substrates.

Similar progress in the understanding of the large-scale pattern of juvenile salmon spatial distribution was obtained by analysing the spatial correspondence between semi-continuous measures of parr density and grain size variations of the sedimentary links of the Sainte-Marguerite River (Figure 9.8a). Clearly, peaks in parr density correspond to the heads of sedimentary links where boulder-rich reaches offer both good summer feeding habitats and abundant bed interstices providing winter shelters to juveniles.

These two examples demonstrate that the grain size information derived from high resolution images of rivers can help quickly identify 'hot spots' in the production of salmonid within the riverscape, even for rivers where only few or sparse fish or habitat information is currently available.

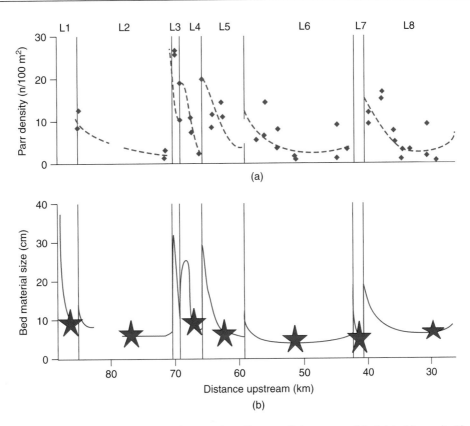

Figure 9.8 Illustration of the spatial correspondence between the sedimentary link structure of the Sainte-Marguerite River and a) the distribution of Atlantic salmon parr densities (Data extracted from Bouchard and Boisclair (2008)) b) the centroïd of spawning sites (blue stars), on each sedimentary link. Adapted from Davey and Lapointe, 2007.

9.4 Bathymetry mapping

Juvenile salmon also express a strong preference for relatively shallow flows and therefore water depth is another important habitat descriptor (Armstrong et al., 2003). However, in this case, there is a wealth of literature on the subject and much detailed literature (see Chapter 3 and (Legleiter et al., 2009; Legleiter et al., 2011). One classic approach is to establish an empirical calibration between geolocated depth measurements and image brightness values according to the method proposed by Lyzenga (1978). In this part of the Geosalar project, a Real-Time Kinematic (RTK) GPS was used to efficiently collect over 1000 geolocated depth measurements. However, this empirical approach was initially developed for larger water bodies sampled with coarse resolution imagery. As a result, its application to hyperspatial image data sets posed certain specific problems which were addressed during the

project. The key issue was found to relate to the number of images in hyperspatial data sets and to the stability of the lighting conditions. In the case of the hyperspatial image dataset for the Ste-Marguerite river, the acquisition of 5550 images over 80 kilometres meant that lighting conditions slightly changed over the duration of the flight (see the discussion on radiometric normalisation in Chapter 8). This led to a challenging condition where it was impossible to calibrate the depth mapping process for all 5550 images. For example, Figure 9.9 shows calibration and validation relationships of depth versus image brightness in the case of raw imagery where radiometric normalisation is a significant issue. These relationships, from Carbonneau et al. (2006), were created from ground points which span three separate images. An effort was made to select an area which was thought to be representative of the river. This area contained a mid-channel bar with a typical fining structure going from cobbles to sparse sandy patches. The bathymetry of the area was

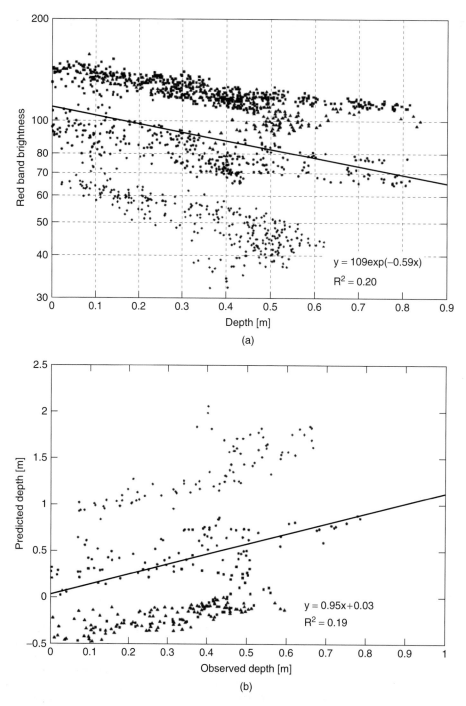

(a)

(b)

Figure 9.9 a) Calibration relationship of field-measured depth values versus image brightness in the red band with regression equation allowing for the prediction of depth from image brightness values. Three parallel bands can be seen which correspond to three slightly different levels of base illuminations in the imagery b) Validation relationship testing the predictions of the calibration equation versus additional, independent, field data. From Carbonneau, P. E., Lane, S. N., and Bergeron, N. 2006. Feature based image processing methods applied to bathymetric measurements from airborne remote sensing in fluvial environments. Earth Surface Processes and Landforms, 31(11), 1413–1423. Reproduced from Carbonneau et al. (2006), with permission from Wiley-Blackwell.

gradually varied and no significant over-arching vege-tation was present. Given the lack of uniformity in the illumination conditions, this has led to three different calibration and validation relationships which are clearly visible in the figure. However, Carbonneau et al. (2006) noted that these three relationships were parallel to each other suggesting that the optical attenuation of the light as it passes through the water column occurred at the same rate in all three images. Therefore, what was required was simply a re-adjustment of the base illumination in each image. The complete solution involved an adaptive pro-cedure whereby portions of dry exposed sediments were used to establish a wet/dry interface (Carbonneau et al., 2006). Immediately adjacent to the dry portion of this interface, the water depth could be assumed to be zero. This feature-based property allowed for an adaptive recal-ibration of the image brightness which minimised drift in the depth measurements artificially caused by changes in the illumination of the imagery. This recalibration was applicable to the whole image dataset.

Figure 9.10 shows the result of this correction proce-dure when applied to the data in Figure 9.9. The three parallel relationships have now collapsed into a single relationship which makes the process of depth mapping reliable. Figure 9.11 shows a typical result of the finalised depth mapping process when applied to an image. It can be seen that depth was set to zero at the interface of the gravel bar and then increases smoothly up to a seemingly constant, saturated, value of 1.5 m. This saturation value is another key limiting factor in such approaches. Unsur-prisingly, the success and feasibility of depth mapping from standard colour imagery depends on the visibility of the river bed in the imagery which is dependent on water clarity. It is therefore not uncommon to observe a satu-ration depth below which the riverbed is not visible and depth mapping cannot operate. In the case of Figure 9.10, this depth was found to be roughly 1.5 m. Given that water clarity conditions can be highly variable, there is no fixed value for this saturation depth. Even for a given river, changes in the suspended sediment load will lead to large changes of water clarity. Therefore, the empirical depth mapping described in Carbonneau et al. (2006) is only valid for depth data collected on the day of image acquisition.

The problem of modelling river bathymetry from image data remains a complex one. Legleiter and Roberts (2009) have used computer simulations in order to thoroughly examine the limitations and weaknesses of this approach. They conclude that one key parameter which could improve image-based bathymetry mapping is improved radiometric resolution (see Chapter 1 for a definition).

Unfortunately this seems to go against the current trends in the instrumentation used for Fluvial Remote Sens-ing (FRS). Rather than using bespoke imaging equip-ment, FRS is increasingly using standard photography equipment. The work of Legleiter and Roberts (2009) is a clear indicator that the FRS community may soon have to consider a move towards more advanced sen-sors more suited to the specific requirements of riverine environments.

9.5 Further developments in the wake of the Geosalar project

Because of the momentum it generated, the Geosalar project continued to stimulate research and development after its completion in 2008. Issues related to the man-agement and visualisation of the extraordinary volume of data generated by the new remote sensing methods were more specifically addressed. Work leading to the devel-opment of in-house airborne image acquisition systems was also conducted.

9.5.1 Integrating fluvial remote sensing methods

The Geosalar project delivered one of the largest hyper-spatial image databases currently available. Additionally, it prompted the development of some important FRS methods which were needed to analyse the images. Both inside and outside of the Geosalar project, the significant recent progress in fluvial remote sensing methods (see Marcus and Fonstad, 2010) has generally focused on sin-gle papers presenting single methods. There is currently a severe paucity of papers which analyse the riverine envi-ronment with a range of integrated hyperspatial remote sensing approaches. However, the Geosalar project pro-vided a needs-driven impetus to the development of an integrated interface which could allow users to manip-ulate and, crucially, analyse the large volume of data in the image database. In 2003–2004, a prototype Fluvial Geographic Information System (FGIS) was developed by Geosalar researchers. The goal of this system was pre-cisely the integration of the depth and grain size mapping methods along with automated channel width measure-ment done directly from the images. This early prototype successfully produced the data seen in Figure 9.7. How-ever, one of the key limitations of the FGIS prototype was georeferencing of the image data. Georeferencing can be defined as the process whereby an image raster is mapped to real world coordinates. This process is crucial in order to preserve the spatial relationships (e.g. the

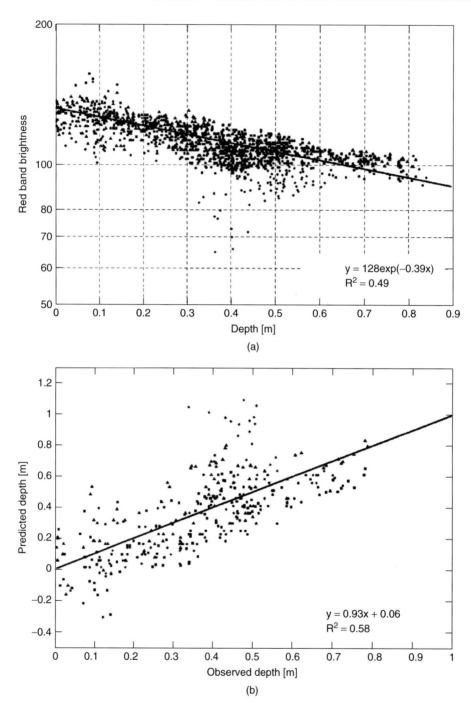

Figure 9.10 a) Final calibration curve between red band brightness and water depth after application of the correction procedure. The three distinct bands seen in Figure 9.9 have merged into 1 after the correction procedure, b) Final validation relationship testing the predictions of the calibration equation versus additional, independent, field data. From Carbonneau, P. E., Lane, S. N., and Bergeron, N. 2006. Feature based image processing methods applied to bathymetric measurements from airborne remote sensing in fluvial environments. Earth Surface Processes and Landforms, 31(11), 1413–1423. Reproduced from Carbonneau et al. (2006), with permission from Wiley-Blackwell.

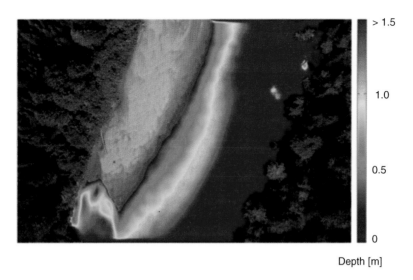

> 1.5

1.0

0.5

0

Depth [m]

Figure 9.11 Example of a bathymetric map obtained from the analysis of an hyperspatial image using the red band brightness versus water depth relationship of Figure 9.10a.

downstream distance) between objects and features in different images. In the case of the Geosalar hyperspatial image database, the coarser imagery with a spatial resolution of 10 cm was manually georeferenced over the course of the (entire) summer following image acquisition. This process rested on ground control points collected during an intensive field campaign. Unfortunately, this field effort only produced sufficient field data in order to georeference the coarser imagery (two ground control points per image). It was found that manual georeferencing of the c. 5500 3 cm images was too labour intensive to be feasible. As a result, the early FGIS functioned without explicit georeferencing information. In the FGIS, the inter-image spatial relationships were estimated based on the known overlap between successive images and the known dimensions of each pixel. In essence, the FGIS measured distances downstream by counting pixels and accounting for image overlaps whilst this process was functional, it likely induced some error because it approximated that, in the short 90 m span of a single image, the channel was straight.

Recently, the FGIS was taken beyond the prototype stage for applications in commercial settings. A new system was developed and dubbed simply the 'Fluvial Information System' (FIS). The FIS maintained the initial FGIS objective of integrating a range of remote sensing methods. However, the FIS benefited from recent developments in the field of georeferencing. As discussed in Chapter 8 of this volume, Carbonneau et al. (2010) developed an automated georeferencing approach which was much better suited to large hyperspatial image databases. Furthermore, the

FIS implemented a River Coordinate System as described by Legleiter and Kyriakidis (2006) in order to produce a unique, locally orthogonal, coordinate system which gives the position [s,n] of any point in the river as a combination of distance downstream [s] and distance across stream [n]. With this coordinate system in place, the FIS then integrates a range of remote sensing methods which are well established in the literature (e.g. Carbonneau et al., 2004; Fonstad and Marcus, 2005; Carbonneau et al., 2006; Legleiter and Kyriakidis, 2006). Carbonneau et al. (2012) successfully applied the FIS in order to demonstrate that FRS technology is now at the point where it can contribute to the broad range of river sciences. These authors showed that hyperspatial image data can be successfully acquired, managed, processed and analysed in order to produce meaningful habitat parameters at sub-metric resolutions for an entire river in the Scottish highlands. In addition to the data presented in Carbonneau et al. (2012), the FIS is also capable of producing innovative habitat visualisations as discussed below.

9.5.2 Habitat data visualisation

The availability of continuous maps for two key habitat parameters (water depth and bed material grain size) open up important new avenues in terms of habitat mapping. In line with the recommendations of Fausch et al. (2002), it is now possible to examine the spatial distribution of fish in a spatially explicit description of fluvial habitat at the riverscape scale with metric resolution. However, the representation and usage of this data poses once again new challenges. The depth map and grain size map outputs of the Geosalar project represent over 3 million measurements spread over an 80 km channel with an average width of 22 m. Visual representations of such a dataset are problematic. Since the width of the channel is less than 1% of its length, any scaled map representations will lose all the detail in the representation. For this reason, Carbonneau et al. (2012) use a new synoptic, abstract, representation of river channels where the entire channel is represented as a rectangular field where widths are presented in a normalised manner. The main goal of this visualisation strategy is to achieve a synoptic view of a given parameter which preserves both downstream and cross-stream variability. The cross section data are first reprojected to a constant width of 100 pixels. Experience has shown that for small rivers, 100 pixels preserves lateral features (albeit with a loss of scale due to the reprojection). Second, successive cross-sections are concatenated (i.e. stacked) in the horizontal direction. This transforms the river into a rectangular array where each vertical column represents one cross section and downstream distance is given horizontally. This abstract river visualisation allows for the display and viewing of the entire river in a single compact figure. In essence, this new synoptic view straightens the channel and widens the width to allow lateral patterns to become visible. While scaling and shape are lost due to the strong cross-stream exaggeration and the elimination of curvature, patterns of variability in both the cross stream and downstream direction can now be observed. Figure 9.12 gives an example. On the left panel, an 11 km stretch of the SMR is represented as a rectangle where the horizontal scale gives the distance upstream in kms and the vertical scale gives the channel widths as a normalised percentage. The colour scheme in Figure 9.12 is designed to give 'at a glance' information

on habitat suitability. Here the depth and grain size data were combined to produce a dichromatic range of colours ranging from yellow to blue. Simply put, yellow areas in Figure 9.9 represent good juvenile salmon habitat whilst dark, white and/or blue areas are unsuitable.

9.5.3 Development of in-house airborne imaging capabilities

While the airborne optical images of the Geosalar project were contracted to a private firm, the need to better control both the cost and parameters of the images stimulated researchers to develop their own airborne imaging capabilities.

At INRS, Normand Bergeron's laboratory developed a helicopter-based imaging system capable of acquiring high resolution images over hundreds of kilometres of river (Figure 9.13). The system integrates a FLIR SC660 thermal imaging camera (0.3 megapixels) and Canon EOS 550D digital SLR (18.7 megapixels) with a precise hardware triggering system in order to acquire thermal and optical imagery at 20 cm and 3 cm resolution respectively (from 300 m altitude). The two cameras are fixed to a pan-tilt unit allowing the operator to 'frame' the image from inside the helicopter, thereby aiding image acquisition quality. The system is mounted within a helipod, so that it can be used with any suitably licensed and equipped helicopter operator. The helicopter's GPS position and attitude is logged using specialised Matlab code and stored alongside the images on a laptop computer. This acquisition system allows imagery to be obtained rapidly (i.e. 40–50 river kilometres per hour), and at a fraction of the cost normally associated with similar remote sensing work.

This platform was recently successfully used to obtain a continuous coverage of more than 500 km of

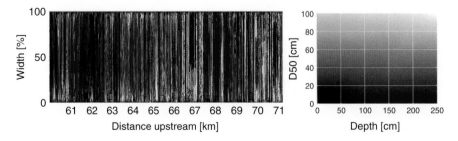

Figure 9.12 Dichromatic plot of D50 and depth. This plot represents an 11 km stretch of the St-Marguerite. Here river width is normalised from 0 to 100% and successive cross sections are stacked horizontally in order to achieve an abstract, synoptic representation of this entire reach which allows us to conserve lateral variability as well as downstream variability. The colour key on the right gives the combined values for depth and grain size at any given point.

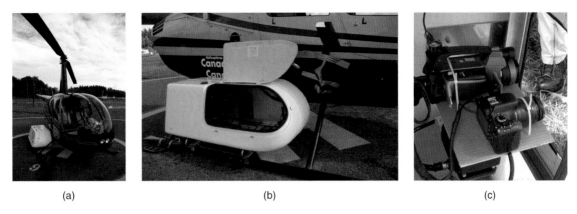

(a)   (b)   (c)

Figure 9.13 Photographs of helicopter-based image acquisition system. a) Helipod attached to helicopter. b) Interior of the helipod containing the imaging system. c) FLIR SC660 (thermal) and Canon EOS 550D mounted on pan-tilt.

(a)   (b)

Figure 9.14 a) Optical (3 cm spatial resolution) and b). Thermal infrared (TIR) (20 cm spatial resolution) imagery of the same section of the Milnikek river in the Gaspésie region (Québec, Canada). On the TIR image, the darker blue section on the right corresponds to a cold thermal anomaly created by the input of a small groundwater fed tributary.

the Matapédia River (Québec, Canada) and of all its tributaries accessible to Atlantic salmon (Figure 9.14). This work demonstrates that producing spatially continuous maps of fluvial habitat over entire riverscapes is now at hand.

9.6 Flow velocity: mapping or modelling?

While flow velocity constitutes one of the most important fluvial habitat variables, its measurement at riverscape scale probably represents the major remaining challenge in terms of habitat mapping. Although large-scale particle image velocimetry (LSPIV) can be used to obtain detailed surface flow velocity fields from shore-based video imagery of short river reaches (see Chapter 16), its application to longer river segment appears impractical. Fujita and Hino (2003) demonstrated that LSPIV

measurement of velocity fields from helicopter-based imagery is possible by stabilising the images from the moving helicopter using known coordinates of fixed points on the image. Although the possibility of repeating this procedure at several contiguous locations along the river seems feasible, no one has yet attempted to do so. Another potential avenue which shows promise is along-track interferometric SAR from high resolution satellite platforms. Romeiser et al. (2007) have demonstrated that TerraSAR-X data can be used to calculate surface velocities for large to medium rivers. The TerraSAR-X satellite is a synthetic aperture radar sensor/platform. Romeiser et al. (2007) use a Doppler effect in order to deduce surface flow velocity. Since TerraSAR-X data has a spatial resolution of 3 m, this method cannot be applied to rivers less than a few tens of meters wide.

An alternative approach which was made possible by Geosalar innovations is the direct estimation of flow velocities based on known discharge and image-based depth and grain size measurements. In traditional flow modelling methods, water depth and bed roughness are iteratively solved based on flow and continuity (i.e. conservation of mass) equations. However many researchers have considered the problem of the hydraulic roughness exerted by the bed and its relationship to flow depth and bed particle size. For example, Richards (1982) present a classic form (equation 9.1) of the 'law of the wall' where the logarithmic velocity profile is integrated over the flow depth in order to give an equation where dimensionless velocity (mean column velocity V/shear velocity u_*) is calculated as a function of water depth, H, and D_{65} (the 65th percentile of the local grain size distribution). Such equations have long been difficult to apply in flow modelling because the detailed maps of depth and grain size were simply not available.

$$\frac{V}{u_*} = 5.75 \log\left(\frac{H}{D_{65}}\right) + 6 \qquad (9.1)$$

With the grain size and depth data discussed earlier in this chapter, and with a simple continuity assumption that in a given river cross-section, the discharge will partition itself laterally in proportion to the depth and roughness, it is possible to make estimates of local velocities. Figure 9.15 shows an example of such velocity estimates for a 1 km stretch of the St-Marguerite River using the rectified projection where the river is represented as a rectangle. Figure 9.15 provides reasonable estimates of velocity which quite clearly captures the fast flowing areas associated with flow constriction and the slow flowing areas associated with an increase in cross-sectional area. Clearly such a simplified estimation of velocity will not be as precise or accurate as a fully calibrated and validated computer fluid dynamics model. However, this lower precision is offset by the potential to function over large areas and even entire rivers. Furthermore, the type of data presented in Figure 9.15 is well suited to juvenile salmonid habitat models since they can detect areas of slow flow which are well known to be preferred by juvenile salmon (Armstrong et al., 2003).

The use of precision as the standard quality metric in this case is inappropriate. An alternate validation approach can be found in geostatistical methods. In such methods, the quality of a dataset is estimated not by the likely error of a single point but by the overall properties of the dataset. For example, Carbonneau et al. (2003) used the scaling properties of photogrammetrically derived digital elevation models (DEMs) as part of the quality assessment procedure. The scaling properties of the DEM were compared to the established scaling properties of natural surfaces in order to demonstrate that the DEMs were not dominated by error (which also has clear scaling properties). A similar approach is applicable to the simplified velocity estimation method presented above. Lamouroux et al. (1995) present an alternative type of velocity prediction. These authors show that with the readily available reach scale parameters of discharge, mean roughness, mean depth and mean width, the shape of the velocity distribution can be accurately predicted. Rather than predicting single point, localised, velocities, Lamouroux et al. (1995) show that we can predict the probability distribution of velocities.

The approach of Lamouroux et al. (1995) was therefore used to validate the data presented in Figure 9.15. The required discharge data was obtained. Mean depth, mean roughness and mean width were derived from the image data and from the image processing methods described above. When applied to the model of Lamouroux et al. (1995), this resulted in a prediction envelope shown in Figure 9.16. This figure shows the prediction of Lamouroux et al. (1995), in green, for the reach in Figure 9.15 overlain on the actual calculated velocity

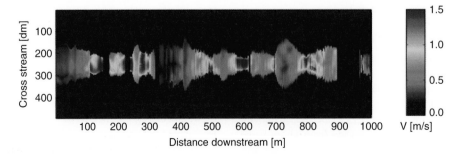

Figure 9.15 Flow velocity (V) map estimated from bed material grain size, discharge and water depth for a 1000 m reach of the Sainte-Marguerite River (Québec, Canada).

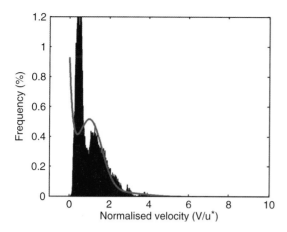

Figure 9.16 Velocity validation. This figure shows, in blue, the distribution of the Normalised velocity (V/u_*) (see Equation 9.1) for the data presented in Figure 9.15. In green, we show the velocity distributions as predicted by the methods of Lamouroux et al. (1995). Both distributions have a consistent shape.

distributions in Figure 9.15. Overall, the distributions are similar which indicates that the direct velocity estimation approach described here could potentially provide a valuable contribution to catchment scale velocity estimations.

9.7 Future work: Integrating fish exploitation of the riverscape

Now that new remote sensing methods are available to describe the riverscape at metric resolution, we suggest that the necessary next step is to provide biological information at similar scope on how fish exploit the mosaic of habitats comprising the riverscape. While catch per unit effort fish sampling techniques such as electrofishing or snorkelling can reveal population-level patterns of fish/habitat relationships, they cannot help understand the processes causing the fish movements that are creating these patterns.

We therefore suggest that more progress will be made by aiming to document how individual fish move across the riverscape throughout their life time in order to access the various habitats they need at different life stages to complete their life-cycle. The well-known distinction in fish spatial behaviour between 'stayers' and 'movers' exemplifies that given a set of habitat conditions, different fish will exploit the habitat differently (Grant and Noakes, 1987). However, it appears that the more individual fish

are tracked in natural environments, the more researchers are discovering new spatial behaviour corresponding to different tactics of habitat exploitation by fish.

For example, Bujold (2010) recently used stationary passive integrated transponder (PIT) systems to provide continuous remote monitoring of the longitudinal movement of PIT-tagged juvenile salmonids along a 2.5 km-long section of a small tributary of the Sainte-Marguerite River (Québec, Canada). This approach allowed the identification of a new type of spatial behaviour characterising fish that were called 'commuters' because of their tendency to enter the tributary at sunrise, travel upstream often as far as 2.5 km, and return to the Sainte-Marguerite River before sunset. This peculiar spatial behaviour applied to 16% of the PIT-tagged individuals, a group too large to be ignored when investigating fish/habitat relationships.

Similarly interesting results were obtained by Roy, M. (*unpublished data*, Department of geography, University of Montreal.) in a study of juvenile Atlantic salmon individual mobility using a large array of passive integrated transponder antennas buried in the bed of a natural stream. This system, described in details in Johnston et al. (2009), allowed the continuous remote monitoring of PIT-tagged fish locations at high spatial and temporal frequency, from which four types of daily behaviour were identified: stationary, sedentary (low mobility), floater (frequent movements in restricted home range) and explorer (movements across the reach). The surprising result is that most individual fish were found to exhibit all types of daily behaviour during the study period, thus challenging the traditional description of a population composed of fractions of sedentary and mobile individuals.

The studies described above are examples amongst others suggesting that how fish exploit the habitat mosaic of the riverscape is probably far from being a resolved issue. More research is thus required in order to assess the geographic path of fish moving from one habitat to the other throughout their lifetime. Such information will allow predictions to be made of how fish are affected by anthropogenic disturbances causing a modification of the available habitat or a reduction of the connectivity between the different habitats comprising the riverscape.

9.8 Conclusion

Although the riverscape approach proposed by Fausch et al. (2002) was immediately perceived by stream ecologists as the necessary next step in the analysis of fish/habitat relationships, its application remained difficult due to the lack of appropriate technology to characterise fluvial habitats at the appropriate scale and resolution. One of the main contributions of the Geosalar project was to participate in the development of new remote sensing methods that are now making the concept of riverscape 'real' (Carbonneau et al., 2012). Image analysis methods are indeed now available to produce spatially continuous high-resolution maps of important fluvial habitat variables, such as bed material size and water depth. These methods are currently applied on hyperspatial images but the constant improvement of the resolution of satellite sensors suggest that they may soon be applied to satellite imagery. Now that significant improvements were made to the quantification of fluvial environments over long river segments, it appears that the *new* necessary next step is to provide equally detailed biological data describing the exploitation by fish of the mosaic of habitats comprised within the riverscape.

Acknowledgements

The authors wish to thank the researchers who participated in all components of the Geosalar project. A special thank to Julian Dodson, PI of the project. The project was funded by the GEOIDE Network Centres of Excellence Program (http://www.geoide.ulaval.ca/) with contributions from our partners : Hydro-Québec, Génivar, Ministère des Resources Naturelles et de la Faune du Québec, Aquasalmo inc., ZEC Sainte-Marguerite, Fondation pour le Saumon du Grand Gaspé. Special thanks also to Keith Thompson and Nicholas Chrisman who were, in turn, directors of GEOIDE. The comments and suggestions of Steve Rice and Hervé Piegay on an earlier version of this chapter are gratefully acknowledged.

References

Armstrong, J.D., Kemp, P.S., Kennedy, G.J.A., Ladle, M., and Milner, N.J. 2003. Habitat requirements of Atlantic salmon and brown trout in rivers and streams. *Fisheries Research*, **62**(2), 143–170.

Bouchard, J. and Boisclair, D. 2008. The relative importance of local, lateral, and longitudinal variables on the development of habitat quality models for a river. *Canadian Journal of Fisheries and Aquatic Sciences*, **65**(1), 61–73.

Bujold, J.-N. 2010. Utilisation de la technologie des transpondeurs passifs (PIT-tags) pour l'étude du comportement spatial des salmonidés dans un tributaire de la rivière Ste-Marguerite (Saguenay, Québec). INRS-Eau, Terre et Environnement. Masters thesis.

Buscombe, D. and Masselink, G. 2009. Grain-size information from the statistical properties of digital images of sediment. *Sedimentology*, **56**(2), 421–438.

Buscombe, D., Rubin, D.M., and Warrick, J.A. 2010. A universal approximation of grain size from images of noncohesive sediment. *Journal of Geophysical Research-Earth Surface*, **115**.

Carbonneau, P.E., Bergeron, N., and Lane, S.N. 2005a. Automated grain size measurements from airborne remote sensing for long profile measurements of fluvial grain sizes. *Water Resources Research*, **41**(11). DOI:10.1029/2005WR003994.

Carbonneau, P.E., Bergeron, N.E., and Lane, S.N. 2005b. Texture-based image segmentation applied to the quantification of superficial sand in salmonid river gravels. *Earth Surface Processes and Landforms*, **30**(1), 121–127.

Carbonneau, P.E., Dugdale, S.J., and Clough, S. 2010. An Automated Georeferencing Tool for Watershed Scale Fluvial Remote Sensing. *River Research and Applications*, **26**(5), 650–658.

Carbonneau, P.E., Fonstad, M.A., Marcus, W.A., and Dugdale, S.J. 2012. Making riverscapes real. *Geomorphology*, **137**(1), 74–86. DOI:10.1016/j.geomorph.2010.09.030.

Carbonneau, P.E., Lane, S.N., and Bergeron, N. 2006. Feature based image processing methods applied to bathymetric measurements from airborne remote sensing in fluvial environments. *Earth Surface Processes and Landforms*, **31**(11), 1413–1423.

Carbonneau, P.E., Lane, S.N., and Bergeron, N.E. 2003. Cost-effective non-metric close-range digital photogrammetry and its application to a study of coarse gravel river beds. *International Journal of Remote Sensing*, **24**(14), 2837–2854.

Carbonneau, P.E., Lane, S.N., and Bergeron, N.E. 2004. Catchment-scale mapping of surface grain size in gravel bed rivers using airborne digital imagery. *Water Resources Research*, **40**(7). DOI:10.1029/2003WR002759.

Chapman, D.W., Weitkamp, D.E., Welsh, T.L., Dell, M.B., and Schadt, T.H. 1986. Effects of River Flow on the Distribution of Chinook Salmon Redds. *Transactions of the American Fisheries Society*, **115**(4), 537–547.

Cunjak, R.A., Prowse, T.D., and Parrish, D.L. 1998. Atlantic salmon (Salmo salar) in winter: "the season of parr discontent"? *Canadian Journal of Fisheries and Aquatic Sciences*, **55**, 161–180.

Davey, C. and Lapointe, M. 2007. Sedimentary links and the spatial organization of Atlantic salmon (Salmo salar) spawning habitat in a Canadian Shield river. *Geomorphology*, **83**(1–2), 82–96.

Dugdale, S.J., Carbonneau, P.E., and Campbell, D. 2010. Aerial photosieving of exposed gravel bars for the rapid calibration of airborne grain size maps. *Earth Surface Processes and Landforms*, **35**(6), 627–639.

Fausch, K.D., Torgersen, C.E., Baxter, C.V., and Li, H.W. 2002. Landscapes to riverscapes: Bridging the gap between research and conservation of stream fishes. *Bioscience*, **52**(6), 483–498.

Fonstad, M.A. and Marcus, W.A. 2005. Remote sensing of stream depths with hydraulically assisted bathymetry (HAB) models. *Geomorphology*, **72**(1–4), 320–339.

Fujita, I. and Hino, T. 2003. Unseeded and seeded PIV measurements of river flows videotaped from a helicopter. *Journal of Visualization*, **6**(3), 245–252.

Grant, J.W.A. and Noakes, D.L.G. 1987. Movers and Stayers – Foraging Tactics of Young-of-the-Year Brook Charr, Salvelinus-Fontinalis. *Journal of Animal Ecology*, **56**(3), 1001–1013.

Haralick, R.M. and Shapiro, L.G. 1985. Image Segmentation Techniques. *Computer Vision Graphics and Image Processing*, **29**(1), 100–132.

Hedger, R.D., Dodson, J.J., Bourque, J.F., Bergeron, N.E., and Carbonneau, P.E. 2006. Improving models of juvenile Atlantic salmon habitat use through high resolution remote sensing. *Ecological Modelling*, **197**(3–4), 505–511.

Ibbeken, H. and Schleyer, R. 1986. Photo-Sieving – a Method for Grain-Size Analysis of Coarse-Grained, Unconsolidated Bedding Surfaces. *Earth Surface Processes and Landforms*, **11**(1), 59–77.

Johnston, P., Berube, F., and Bergeron, N.E. 2009. Development of a flatbed passive integrated transponder antenna grid for continuous monitoring of fishes in natural streams. *Journal of Fish Biology*, **74**(7), 1651–1661.

Lamouroux, N., Souchon, Y., and Herouin, E. 1995. Predicting Velocity Frequency-Distributions in Stream Reaches. *Water Resources Research*, **31**(9), 2367–2375.

Legleiter, C.J., Kinzel, P.J., and Overstreet, B.T. 2011. Evaluating the potential for remote bathymetric mapping of a turbid, sand-bed river: 2. Application to hyperspectral image data from the Platte River. *Water Resources Research*, **47**. DOI:10.1029/2011wr010592.

Legleiter, C.J. and Kyriakidis, P.C. 2006. Forward and inverse transformations between cartesian and channel-fitted coordinate systems for meandering rivers. *Mathematical Geology*, **38**(8), 927–958.

Legleiter, C.J. and Roberts, D.A. 2009. A forward image model for passive optical remote sensing of river bathymetry. *Remote Sensing of Environment*, **113**(5), 1025–1045.

Legleiter, C.J., Roberts, D.A., and Lawrence, R.L. 2009. Spectrally based remote sensing of river bathymetry. *Earth Surface Processes and Landforms*, **34**(8), 1039–1059.

Lyzenga, D.R. 1978. Passive Remote-Sensing Techniques for Mapping Water Depth and Bottom Features. *Applied Optics*, **17**(3), 379–383.

Marcus, W.A. and Fonstad, M.A. 2010. Remote sensing of rivers: the emergence of a subdiscipline in the river sciences. *Earth Surface Processes and Landforms*, **35**(15), 1867–1872.

Otsu, N. 1979. Threshold Selection Method from Gray-Level Histograms. *Ieee Transactions on Systems Man and Cybernetics*, **9**(1), 62–66.

Raven, P.J., Fox, P., Everard, M., Holmes, N.T.H., and Dawson, F.H. 1997. "River habitat survey: A new system for classifying rivers according to their habitat quality." In: *Freshwater Quality: Defining the Indefinable?*, P.J. Boon and D.L. Howell, eds., Stationery Office Books, Edinburgh, UK, 215–234.

Rice, S., and Church, M. 1998. Grain size along two gravel-bed rivers: Statistical variation, spatial pattern and sedimentary links. *Earth Surface Processes and Landforms*, **23**(4), 345–363.

Rice, S.P., Greenwood, M.T., and Joyce, C.B. 2001. Tributaries, sediment sources, and the longitudinal organisation of macroinvertebrate fauna along river systems. *Canadian Journal of Fisheries and Aquatic Sciences*, **58**(4), 824–840.

Richards, K. (1982). *Rivers*, The Blackburn press. 361 pp.

Romeiser, R., Runge, H., Suchandt, S., Sprenger, J., Weilbeer, H., Sohrmann, A., and Stammer, D. 2007. Current Measurement in Rivers by Spaceborne Along-Track InSAR. *Ieee Transactions on Geoscience and Remote Sensing*, **45**, 4019–4030.

Schlosser, I.J. 1991. Stream Fish Ecology – a Landscape Perspective. *Bioscience*, **41**(10), 704–712.

Sear, D.A. 1993. Fine Sediment Infiltration into Gravel Spawning Beds within a Regulated River Experiencing Floods – Ecological Implications for Salmonids. *Regulated Rivers-Research & Management*, **8**(4), 373–390.

Soulsby, C., Malcolm, I.A., and Youngson, A.F. 2001. Hydrochemistry of the hyporheic zone in salmon spawning gravels: A preliminary assessment in a degraded agricultural stream. *Regulated Rivers-Research & Management*, **17**(6), 651–665.

Wu, F.C. 2000. Modeling embryo survival affected by sediment deposition into salmonid spawning gravels: Application to flushing flow prescriptions. *Water Resources Research*, **36**(6), 1595–1603.

10 Image Utilisation for the Study and Management of Riparian Vegetation: Overview and Applications

Simon Dufour[1], Etienne Muller[2], Menno Straatsma[3]
and S. Corgne[1]

[1]LETG – Rennes COSTEL, CNRS, Université Rennes 2, Place Recteur
Henri le Moal, France
[2]Université de Toulouse; CNRS, INP, UPS; EcoLab (Laboratoire Ecologie
Fonctionnelle et Environnement), France
[3]University of Twente, Enschede, The Netherlands

10.1 Introduction

Riparian vegetation is an important component of fluvial landscapes. The extent of vegetated areas within riparian systems, defined by Naiman et al., 2005, as 'transitional semi-terrestrial areas regularly influenced by freshwater, usually extending from the edges of water bodies to the edges of upland communities', is variable in space and time. It fluctuates under the control of internal processes such as biological interactions, but also due to land cover changes. Riparian vegetation affects morphological changes of the channel, flood hazard and ecological functioning (Gurnell and Gregory, 1995; Tabacchi et al., 1998; Tal et al., 2004; Corenblit et al., 2007). It is also a natural and cultural resource that provides services such as biodiversity, refuge for endangered species, recreational areas and pollution limitation (Malanson, 1993; Klimo, 2008). The recognition of the interest in such ecosystems resulted in an increased activity in managing riparian vegetation and led to scientific studies into the understanding of riparian systems. Thus, both managers and scientists need a set of tools that are able to sense, describe and monitor vegetation quality but that are also efficient in riparian corridors. High spatial complexity and temporal fluctuations complicate the transferability of these tools. The main dichotomy in these tools is between image analysis and field survey.

Over the last few decades, image analysis has undergone a huge evolution due to an increase in sources and the type of imagery (Figure 10.1). For example, very high resolution satellite images, such as IKONOS or GeoEye, combine high spatial resolution with a broad areal coverage. The emergence of new technologies gives access to new information such as forest stand structure. Moreover, image analysis has greatly progressed, resulting in an increase in computing capacity, geographic information system (GIS) techniques and advanced image analysis

Fluvial Remote Sensing for Science and Management, First Edition. Edited by Patrice E. Carbonneau and Hervé Piégay.
© 2012 John Wiley & Sons, Ltd. Published 2012 by John Wiley & Sons, Ltd.

Figure 10.1 Various examples of images, a) Commercial aerial photo of the Arve River (France, 2000), true colour, spatial resolution = 0.5 m, produced by the French *Institut Géographique National* (IGN), b) Aerial photo of the same site as image a. but in 1984, and with infrared colour spectral resolution. Source: French Institut National de l'Information Géographique et Forestière (IGN), c) Landsat ETM+ image, June 2002, Asse River (France), spatial resolution = 30 m, composition with bands 2, 3 and 4. NASA Landsat Program, USGS, Sioux Falls, d) Very high resolution picture taken from an unmanned aerial vehicle (UAV), Ain River (France), spatial resolution = 0.05 m. © CNRS, e) Oblique picture of the Ain River (France), done by helicopter in summer 1991. © CNRS, (f), (g) and (h). Three images of the same site (Laou River, Morocco) with different resolutions: f) landsat ETM+ image, June 2002, spatial resolution = 30 m, composition with bands 2, 3 and 4. NASA Landsat Program, USGS, Sioux Falls; g) Spot 5 image, August 2008, spatial resolution = 10 m, composition with bands 2, 3 and 4. Cnes 1986-2010, Distribution Spot Image and h) GeoEye image summer 2011, spatial resolution = 0.5 m (from Google Earth© site).

software. Thus, image analysis and remote sensing tools have been widely used for resource management (Belward and Valenzuela, 1991), and vegetation studies (Walsh et al., 1994; Alexander and Millington, 2000), notably in forested ecosystems (Iverson et al., 1989; Boyd and Danson, 2005) and wetlands (Lehman and Lachavanne, 1997; Ozesmi and Bauer, 2002; Adam et al., 2010). They have also been used in fluvial geomorphology (Gilvear & Bryant, 2003) and for the characterisation of riparian vegetation (Muller, 1997; Goetz, 2006; Yang, 2007).

In this chapter, we review how imagery can be used for the description and analysis of the structure and dynamics of riparian vegetation. It is neither an exhaustive overview nor a technical catalogue, but rather an application-oriented synthesis. We also discuss the limitations that can strongly reduce the efficiency of these tools; despite the vast amount of raw data collected on riparian vegetation, current methods only partly cover the information required by both researchers and river managers. Indeed, if from a technical point of view the range of the potential contribution of images of riparian vegetation studies is broad; from an applied point of view the diversity of situations encountered by managers can make it difficult to suggest universal rules. So the question is not only what we can know but what we want to know, and is remote sensing the best way to go about it? Both of these aspects are then illustrated with different applications using different media such as aerial photographs, satellite images or Light Detection and Ranging (LiDAR) data.

10.2 Image analysis in riparian vegetation studies: what can we know?

Since the 1970s, riparian studies have benefited from significant progress in remote sensing and GIS science (Table 10.1). Firstly, the diversification of platforms and instruments associated with an increase in spectral and spatial resolution obviously has benefits for vegetation

studies. Secondly, a huge development in data management has made the imagery more easily available. Progress has been seen most notably in computer science and GIS. Thus, new tools have been developed for classification such as objected-oriented procedures (Mathieu et al., 2007; Arroyo et al., 2010) or neural networks.

Imagery has been used in riparian vegetation studies with four main objectives: mapping vegetation types, mapping vegetation species, mapping historical changes, and measurement of environmental and vegetation parameters such as tree height or Leaf Area Index.

10.2.1 Mapping vegetation types and land cover

The most common use of imagery in riparian vegetation concerns vegetation type mapping. A vegetation type is defined as a plant community with relatively homogeneous floristic composition and/or physiognomy, although they may differ on other characteristics such as stand density, biomass or percentage of shade (Muller,

Table 10.1 Chronology of remote sensing tools developed for riparian vegetation studies.

Period	Selected Tools	Uses and Advances for Riparian Vegetation Studies	Examples
19th c.	Photographs	Riparian past landscape reconstitution	Grams and Schmidt, 2002
1930	Aerial photographs	Riparian past landscape reconstitution, possible to quantify area, composition and configuration	Miller et al., 1995, Marston et al., 1995, Mendonca et al., 2001, Greco and Plant, 2003
1970	Infra red sensors	Widening of the available spectra significantly increases the amount of information acquired	Girel, 1986, Otahel et al., 1994, Neale, 1997
	Satellite images (e.g. Landsat TM or Spot images)	Larger scenes, wider spectral resolution	Butera, 1983, Mertes et al., 1995
1990	Active technology (e.g. Radar, LiDAR)	New set of contextual or structural data (e.g. flood extension or tree height)	Townsend, 2001 and 2002, Mason et al., 2003, Genç et al., 2004, Straatsma and Middelkoop, 2007, Antonarakis et al., 2008b
	GIS development	Integration of ancillary data (network map, DEM), for example, to preselect area of interest	Congalton et al., 2002, Ehlers et al., 2003, Johnson and Zelt, 2005, Alber and Piégay, 2011
2000	Satellite: very high resolution images (e.g. QuickBird, IKONOS or GeoEye)	Reaching aerial photo spatial resolution with more bands and for larger scenes	Franklin and Wulder, 2002
	Object oriented classification	Combine automation and contextual information	Johansen et al., 2007b, Arroyo et al., 2010
2010	Light Aerial Remote Sensing (e.g; UAV)	Centimetric resolution, high flexibility	Corbane et al., 2006, Lejot et al., 2007, Thompson and Gregel, 2008, Dunford et al., 2009

0 2 4 Kilometers

Riparian land use/cover (1973)

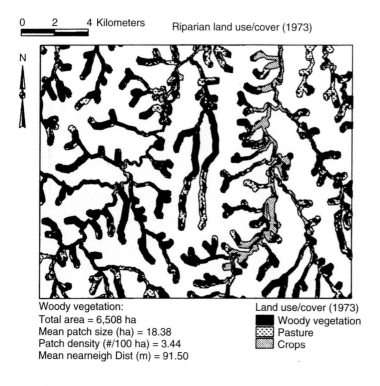

Woody vegetation:
Total area = 6,508 ha
Mean patch size (ha) = 18.38
Patch density (#/100 ha) = 3.44
Mean nearneigh Dist (m) = 91.50

Land use/cover (1973)
- ■ Woody vegetation
- ▨ Pasture
- ▨ Crops

0 2 4 Kilometers

Riparian land use/cover (1997)

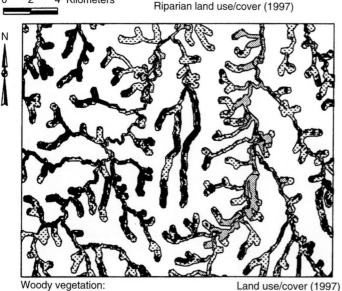

Woody vegetation:
Total area = 3,247 ha
Mean patch size (ha) = 6.09
Patch density (#/100 ha) = 5.18
Mean nearneigh Dist (m) = 104.5

Land use/cover (1997)
- ■ Woody vegetation
- ▨ Pasture
- ▨ Crops

Figure 10.2 Changes in land cover and selected landscape metrics in a portion of the Lockyer Valley catchment (south-east Queensland, Australia), maps obtained from Landsat thematic mapper (TM) digital images. Maps and indexes indicate the fragmentation of riparian vegetation over time. Reprinted from *Landscape and Urban Planning*, 59, Apan, A.A., Raine, S.R., Paterson, M.S., Mapping and analysis of changes in the riparian landscape structure of the Lockyer Valley catchment, Queensland, Australia, 43–57, Copyright 2002, with permission from Elsevier.

Table 10.2 Accuracy in riparian vegetation mapping in selected studies.

Reference	Image, Data	Accuracy	Area	Comments
	ortho-rectified digital aerial photographs	81%	Hunter region (Australia)	
Yang (2007)	Spot-4 multispectral (XS)	63%	Hunter region (Australia)	
	Landsat-7 ETM+	53%	Hunter region (Australia)	
Johansen et al. (2007b)	high resolution images (QuickBird) and an object-oriented classification	79%.	Lost Shoe Creek (Canada, scene size 50 km^2)	The accuracy was increased between 2 and 19% with texture integration.
Antonarakis et al. (2008a)	LiDAR, approach based on point elevation distribution	94–95%	Garonne River (France)	
Geerling et al. (2009).	CASI image and LiDAR data combination of spectral and structural data	74%	Waal River (Netherlands)	LiDAR data takes up most of the treatment time; this combination increases the accuracy by between 8 and 19% compared to a map derived from only one source
Akasheh et al. (2008)	high resolution multi-spectral airborne data (0.5 m)	82–92%.	Rio Grande River (USA)	
Ge et al. (2006)	aerial photographs to map *Tamarix parviflora*	71–78%	Cache Creek (USA)	As for habitat mapping, texture integration in classification increases the accuracy in species mapping

1997; Brohman and Bryant, 2005; Alencar-Silva and Maillard, 2010). From the image to the map, the common steps are: (1) data acquisition and pre-processing; (2) map legend development (i.e. which units were expected to be identified); (3) classification (i.e. to group objects together into types or classes based on shared characteristics); (4) incorporation of ancillary data, accuracy assessment (usually by field verification); and lastly (5) delineation of the geographic distribution and extent of vegetation classes (Muller, 1997; Franklin and Wulder, 2002; Brohman and Bryant, 2005).

In brief, we can distinguish two scales and associated objectives: to map the extension of riparian vegetation at a broad scale (i.e. riparian vegetation versus surrounding land cover at the river network scale) (e.g. Narumalani et al., 1997; Apan et al., 2002; Lawson et al., 2007) (Figure 10.2) or to map the riparian vegetation mosaic from local to medium scale (i.e. identify several cover types within the corridor at reach scale (Figure 10.1a and 10.1b) (e.g. Girel, 1986; Johnson et al., 1995). For a long time, broad-scale mapping has been difficult due to the fact that the available images cover limited areas (Aguiar and Ferreira, 2005). Since the 1970s,

satellite images have provided wider views and thus allow broad scale mapping (Figure 10.1c). However, the low spatial resolution of early images limited use in some very specific contexts such as large floodplains (e.g. a 2700 km^2 scene for the Amazon floodplain, Mertes et al., 1995) and so most of the river network was not included. The increase in spatial resolution since the end of the 1990s allows the combination of riparian delineation at broad scale with accurate vegetation type characterisation (Goetz et al., 2003; Derose et al, 2005; Goetz, 2006; Johansen et al., 2007b) (Table 10.2 and Figure 10.1f, g and h). Vegetation type mapping can also be done or improved by the integration of structural information such as vegetation height or crown diameter provided by LiDAR data (Farid et al., 2006; Antonarakis et al., 2008a; Geerling et al., 2009; Johansen et al., 2010a; Johansen et al., 2010b) (Table 10.2). Lastly, structural parameters extracted at tree and community scales both from aerial and terrestrial laser scanning tools provide very useful information for the quantification of floodplain roughness, for example (Straatsma and Baptist, 2008; Antonarakis et al., 2009; Antonarakis et al., 2010) or Leaf Area Index (Farid et al., 2008).

10.2.2 Mapping species and individuals

Until the 1990s infrared colour photographs were the main medium used to identify some species, such as non native species (Neale, 1997; Lonard et al., 2000; Everitt et al., 2007), and the diachronic analysis of the vegetation type was used as a surrogate to monitor species dynamic (Peinetti et al., 2002). Indeed the colour of the foliage of several species (e.g. *Salix alba*) makes it easy to delineate units dominated by such species. Improvement in spatial and spectral resolution now allows us to target the species with improved accuracy (Table 10.2 and Figure 10.1d). This has mainly been used to monitor the installation and distribution of exotic or invasive species (Rowlinson et al., 1999; Dipietro et al., 2002; Underwood et al., 2006; Hamada et al., 2007). Development of very high resolution images provides data that are well adapted to this scale. For example, Dunford et al. (2009), used colour photographs along the Drôme River (France) provided by unmanned aerial vehicles (UAV) with a spatial resolution finer than 25 cm. They were thus able to distinguish between species (mainly *Salix* sp., *Populus nigra* and *Pinus nigra*) and also able to separate deciduous and coniferous dead crowns. Hyperspectral images can also achieve such detail (Pengra et al., 2007).

10.2.3 Mapping changes and historical trajectories

Because the presence of the observer is not necessary in order to access information provided by images, they have been extensively used to reconstruct landscape trajectories and thus define past conditions of the river. In the European context a new emphasis on historical studies is being promoted by the European Water Framework Directive in order to identify reference conditions for water body functional assessment. The nature of imagery employed by researchers relates directly to the latest technology available at the start of their study periods and the timeframe studied: satellite images since the 1970s, aerial photographs since the 1930s, oblique pictures since the nineteenth century (Charlton, 2000; Grams and Schmidt, 2002), or old maps mainly since the seventeenth or eighteenth century. Aerial photographs have been widely used to map landscape configuration changes, turn-over rates for geomorphic features, and the recovery process after flooding (Miller et al., 1995; Marston et al., 1995; Mendonca et al., 2001; Sloan et al., 2001; Freidman and Lee, 2002; Peinetti et al., 2002; Greco and Plant, 2003; Ferreira et al., 2005; Geerling et al., 2006; Petit, 2006; Gonzalez

et al., 2010). The majority of studies using diachronic analysis are carried out on land cover or vegetation types.

Additional maps can be used in the reconstruction of the changes of vegetation if we look back to the 1930s and earlier (Muller et al., 2002; Gurnell et al., 2003). For instance in France, the Napoleon land registers, dating from the first half of the nineteenth century, are a good source of information because all parcels are drawn and detailed by land uses (Piégay and Salvador, 1997). Old documents can have a very fine scale, such as the 1/2372 land register drawn between 1728 and 1738 in the Kingdom of Sardinia and used by Girel and Manneville (1998). Old maps are usually easier to find for large rivers, which were the first to have flow control structures and canals constructed (see for example, in California, the California Debris Commission maps dating from the end of nineteenth century). Thus, for a 10 km reach of the Danube River in Austria, Hohensinner et al. (2004) found more than 100 maps drawn between 1714 and 1991, 41 of which were useful for their study (i.e., with scales ranging from 1/6900 to 1/25 000). The main limitations of old maps concern the characterisation of information, as land cover typology that can be vague and unevenly represented. For example, there are frequently few details for vegetation units in undeveloped areas, whereas there are more details in cultivated areas. Furthermore, older maps have greater issues with the accuracy of the image, and greater difficulty in rectification for comparison with other images.

10.2.4 Mapping other floodplain characteristics

Besides mapping vegetation, remote sensing is also used to study abiotic parameters relevant to riparian vegetation studies. For instance, morphological features within the floodplain can be identified and delineated from LiDAR data (Jones et al., 2007). Maximum stand age can be inferred from the age of the landforms obtained by a historical analysis of channel mobility in the corridor (Greco et al., 2007; Stella et al., 2012). This superposition of historical channel boundaries also allows researchers to estimate erosion rates and life spans of different landforms (Van der Nat et al., 2003; Latterel et al., 2006; Nakamura et al., 2007).

Inundation extent and some substrate characteristics can also be obtained from images. Muller and James (1994) reconstructed substrate characteristics along a reach of the Garonne River (France) from multidate Landsat Thematic Mapper (TM) images, even if they could not infer fluvial landforms from this information. Radar images can be used to very accurately map flood extent; 95% and 85% respectively for stands in leaf-off and leaf-on conditions (Townsend and Foster, 2002; Hess and Melack, 2003). Inundation conditions have also been mapped with some Moderate Resolution Imaging Spectroradiometer (MODIS) Images by Sakamoto et al. (2007) in the Mekong delta. As with LiDAR data, such information can improve riparian vegetation mapping like soil moisture estimations given by RADARSAT-1 Synthetic Aperture Radar (SAR) images (Makkeasorn et al., 2009) or by Spot images (Muller and Décamps, 2001). Lastly, hyperspectral data can give information about vegetation phenology and physiology, and thus indirectly participate in water budgeting by quantifying evapotranspiration (Ringrose, 2003; Loheide and Gorelick, 2005, Nagler et al. 2005a; Bawazir et al., 2009).

10.3 Season and scale constraints in riparian vegetation studies

10.3.1 Choosing an appropriate time window for detecting vegetation types

The spectral contrasts existing between plant species depends on the phenological stages of the local vegetation. Three phenophases, directly related to biomass changes, can be easily described for each wood category: leafing, full foliage and leaf falling (Muller, 1995). In the growing season, the effective time window for detecting such contrasts is relatively limited and explains the inappropriateness of most images (Figure 10.3). The most obvious differences are detected either in spring (leafing) or in autumn (leaf fall) when changes correspond to the modification of biomass with possible time shifts from one species to another and with corresponding increases or decreases in the near infrared bands (TM4 or Spot 3) or in the Short-wave infrared (TM5 or SPOT4). Unfortunately, in the visible bands, little information can be easily extracted due to low signal levels and poor differentiation between species. In addition, other important phenophases are not detectable (e.g. flowering, seed ripening, dispersal, and additional leafing phases).

10.3.2 Minimum detectable object size in the riparian zone

The quality of the perception of an object can be roughly estimated by simply considering the proportion of pure and mixed pixels in its description (Table 10.3). For example, a circular object (typically a tree) must have a diameter at least 8 times larger than the pixel size in order to be properly detected with at least 50% of pure pixels (i.e. with pure information over at least half of its surface). However, a good perception (e.g. 80% of pure pixels or more) is only possible with objects whose diameter is of 16 pixels or more. For example, Spot Panchromatic data (resolution 5 m) may provide good information on circular objects with diameters of at least 80 m (area ~0.5 ha). Similarly, a large tree with a canopy diameter of 20 m can only be well-detected with pixels <1 m. For analysing a shrub of two metres in diameter, pixels must be <10 cm. However, these recommendations are optimistic as tree canopies are almost never circular or homogeneous in reflectance, compared to agricultural plots. The presence of shadow in the canopy at the time of acquisition of the data also can confuse results. Gaps in the canopy introduce a signal from the understory vegetation or the soil into the spectral profile. This local heterogeneity is highly variable from one area to the other and cannot be estimated *a priori* in natural riparian woodlands where several vegetation strata typically exist (Figure 10.4). In a plantation plot, trees of the same age and the same species (often the same clone) are planted at regular intervals (typically every 4 or 5 metres) and therefore quantification might be more feasible.

10.3.3 Spatial/spectral equivalence for detecting changes

If one considers that a pixel of vegetation with a surface S has a reflectance R and is affected by an internal change (e.g. leafing, dieback, leaf fall, flowering etc.) characterised by a reflectance R' over a surface S', the resulting new reflectance R' of the pixel may be roughly approximated by:

$$R'' \times S = R' \times S' + R \times (S - S') \qquad (10.1)$$

The corresponding modification of the reflectance of the pixel by the disturbing factor is:

$$(R'' - R) = (R' - R) \times S'/S \qquad (10.2)$$

This equation simply reminds us that, in order to be detected by a pixel (S fixed), change relies both on the reflectance contrast (R'-R) of the disturbing factor and on the surface S' affected by this change. For example, doubling the reflectance change of the pixel requires doubling either the reflectance contrast of the disturbing factor, or its area, or the product of both. A change in reflectance can also be compensated by an inverse change in area and thus be undetectable. The equation also indicates that to detect the same change with a pixel size twice the size of another (e.g. using images of 20 m against 10 m

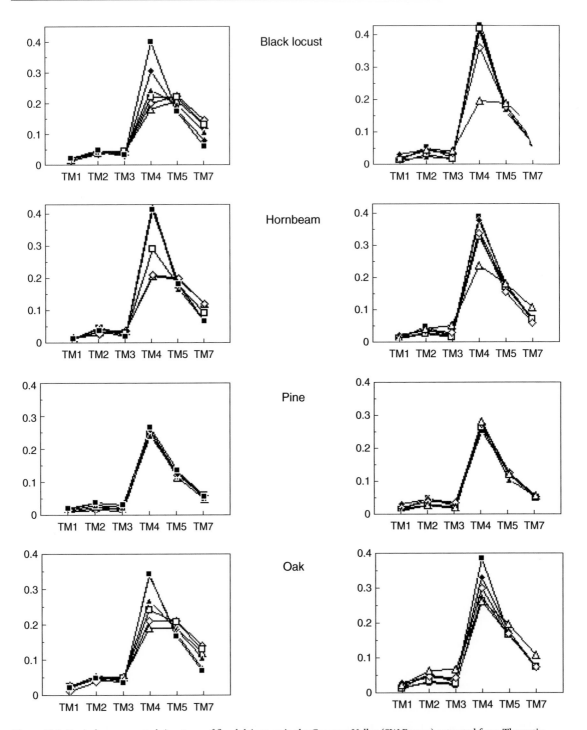

Figure 10.3 Typical mean spectral signatures of floodplain trees in the Garonne Valley (SW France) extracted from Thematic Mapper images at 12 different dates in a year. Reflectance values were corrected from atmospheric disturbances. TM4 seems efficient for discriminating most species, however high variations (not shown here) exist around the mean values and prevent an easy discrimination.

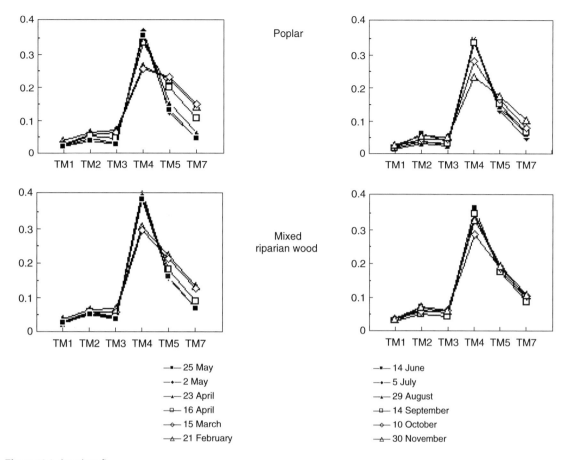

Figure 10.3 (*continued*).

resolutions) the corresponding disturbing factor must either cover an area $2^2 = 4$ times the size, or introduce an effective change in a reflectance that is 4 times the size, or the product of both must be increased by the same factor 4. Of course, this simplified approach does not take into account all of the physical factors that influence the actual radiative transfers. Rather, it simply reminds us that both the reflectance change and the surface affected by the change have to be considered in order to detect an effective modification in the image. Moreover, when the area affected by a slight change is close to the pixel area, the probability of detecting it is very low because it is only by chance that only one single pixel will be affected. The change will most probably affect several neighbouring pixels and the modifications in reflectance will then be too small to detect.

10.4 From scientists' tools to managers' choices: what do we want to know? And how do we get it?

Scientific developments in remote sensing and GIS science provide many useful tools for riparian vegetation studies: synoptic and repeated data collection, measurement of numerous biophysical parameters (notably vertical information such as tree height or floodplain roughness by combining optic and LiDAR data) and a broad-scale monitoring process. However, the utility and accessibility of these tools is still an open question; which ones are really useful and can be easily used by river managers and decision makers?

Table 10.3 Evaluation of the quality of the perception of a circular object as a function of its dimension and of the spatial resolution of the images.

	Quality of the Perception				
	Very Bad	Bad	Fair	Good	Very Good
diameter of the object (in m)	2	4	8	16	24
area of the object (in pixels)	4	16	60	216	484
% of pure pixels	0	25	53	76	86
% of mixed pixels	100	75	47	24	14

Pixel size of the image	Diameter of the object				
30 m	60	120	240	480	720
20 m	40	80	160	320	480
10 m	20	40	80	160	240
5 m	10	20	40	80	120
3 m	6	12	24	48	72
1 m	2	4	8	16	24

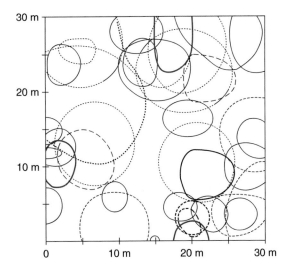

Figure 10.4 Example of the local diversity of riparian tree canopies following field measurements within a plot equivalent to one Spot or Landsat TM pixel.

10.4.1 Which managers? Which objectives? Which approach?

Remote imagery is used when there is a need for synoptic characterisation of an area, and hence an analysis on a broad spatial scale. The scientific and/or management objective is fundamental in choosing an approach

and imagery (Warbington et al., 2002) because available data vary in terms of nature, spatial resolution, spectral characteristics, temporal frequency of acquisition, age of database, homogeneity and cost (for example see Davis et al., 2002 and Johansen et al., 2010 for recommendations in image choice for riparian areas) (see Table 10.4).

We may distinguish, on the one hand, reach managers with one site or a collection of sites that work at local scale (i.e. at a reach length between a few hundred meters and several kilometres) and, on the other hand, managers that deal with the entire river network (thousands of kilometres in length). The reach managers are directly in charge of implementing and monitoring measurements on the ground. The river basin managers are principally in charge of political and strategical planning at network scale. This distinction, though arbitrary in many ways because the size of a given reach can also influence technical choices (Geerling et al., 2009), is nevertheless useful because the two different scales usually correspond to different objectives. The reach-scale manager needs more detailed information about species present on the land, the area occupied by a given vegetation type, the health condition of a given species, stand structure, exotic species extension and changes of the riparian community through time. Whereas at the network scale, managers traditionally need more general indicators of corridor structure (continuity, fragmentation, complexity) linked to the distribution of riparian vegetation along the network and human pressures. At such scale, all of these indicators need to be systematically evaluated across all contexts within the watershed (Cunningham et al., 2009; Claggett et al., 2010).

The distinction between scales is also useful as the best approach can depend on the scale used. For example Johansen et al. (2007a), compared field measurements to remotely-derived riparian zone characteristics such as percentage canopy cover, organic litter on the ground, canopy continuity, tree clearing, bank stability and flood damage. They showed that field measurements are more cost effective on a fine scale (1–200 km), whereas image analysis is superior on a larger scale (200–2000 km). At both scales, managers should always examine the advantages and the limitations of images with respect to their objectives and consider other approaches, notably field assessment, as an alternative or a complement.

10.4.2 Limitations of image-based approaches

With regards to riparian vegetation management, the reach scale is very dominant in scientific literature

Table 10.4 Examples of recommendation in image choice for riparian vegetation studies.

Parameter	Advice	Reference
Wavelength Bands	Three bands, centred within the wavelength regions of 0.53–0.54 μm, 0.66–0.67 μm, and 0.79–0.815 μm. Where four-band sensors are available, a preference for the fourth wavelength band centre should be either at 0.70 μm or near the 2.02–2.07 μm wavelength region, where different riparian species also display distinctly different amounts of reflectance.	Davis et al. (2002)
Spatial Resolution	Image data should be acquired at a resolution of about 20 cm per pixel to retain the textural information for typical riparian vegetation	Davis et al. (2002)
Scale of images	Adapt the scale to the target, for example for aerial photos 1/60 000 to 1/25 000 to map stand vs. 1/16 000 to 1/12 000 to map species.	Brohman and Bryant (2005)*
Time acquisition	Near summer solstice in order to minimise shading within vegetation	Davis et al. (2002)
	A winter mission can give a good access to water areas hidden by the canopy in summer, a distinction of the sempervirens vegetation (coniferous, Juniperus), and information about cultivated areas	Girel (1986)
	Differentiating between the targeted species and others can be easier in some specific conditions such as the season in the year when the species becomes senescent	Dipietro et al. (2002); Hamada et al. (2007)
	Multidate images to study phenological stages	Muller (1995)

*not specific to riparian areas

(Bendix and Stella, in press). Remotely-sensed imagery is commonly used at this scale by managers for some applications, including historical mapping, characterisation of inaccessible sites, or repetitive evaluation in some cover type areas. But is it always the best approach?

Firstly, the possibility to characterise something by image analysis does not necessarily mean that it is the best way to do it, especially for managers. Even if the data from other approaches (e.g. field survey) are not error-free, they can be less time-consuming to collect and process later, less expensive to acquire and/or easier to implement. Managers considering the use of remote imagery must balance the highest level of accuracy possible (Table 10.2), with the effort needed to acquire and process the information, particularly for small reaches or for specific information such as vegetation structure. In such cases, it could be more efficient to choose a field mapping approach. For example, Coroi et al. (2006) used a GPS field mapping technique for small streams and patches and considered it more efficient than image analysis since current image resolution was not high enough for them, and new image types were too expensive. Concerning vegetation structure, basic remote sensing approach usually provides coarse vegetation type categories, typically 5 to 10 categories (Muller, 1997). Obviously, it is possible to do a finer analysis on a local scale. For example Johnson et al. (1995) along the Snake River (USA) mapped 14 cover types from 1/7920 colour-infrared photographs and Dieck et al. (2004) provided a floodplain vegetation classification system for the Upper Midwest region (USA) that distinguished 31 vegetation types. Moreover, new tools such as object-oriented classification or combining images and LIDAR data now enhance the mapping process and provide very useful data (Boyd and Danson, 2005; Geerling et al., 2009). However, to reach such detail, specific problems emerge that can make it inappropriate for managers, including intensive field verification, the high level of technicality needed, the amount of time spent validating the results, or low transferability from one site to another.

Secondly, some data that are frequently needed by managers are difficult or impossible to collect through remote sensing (Rheinhardt et al., 2007), including the precise distribution of all tree species, tree age, composition of herbaceous layer (notably woody species regeneration) flood marks, overbank sediment thickness and grain size and evidence of grazing by herbivores. For example, the accuracy in distinguishing species is species dependent: Nagler et al. (2005b) proposed an efficient framework for cottonwood and willow characterisation but some other species such as saltcedar are more difficult to identify. Hopefully, the new images available (higher spatial, spectral and temporal resolution) provide opportunities to progress quickly in this domain and significant developments are expected in the forthcoming years.

For network-scale managers, until the 1990s, the practical choice was between broad-scene but low resolution or narrow-scene and good spatial resolution. In the first case, when riparian vegetation is a narrow strip along the river, an important part of the network can not be mapped. In the second, due to the form and spatial distribution of riparian vegetation, the number of images needed for broad-scale characterisation is problematic in terms of data treatment. Advances in classification procedures, increased spatial resolution, GIS integration and computing capacities all contribute towards increased accuracy in reach scale vegetation mapping. Furthermore, these are also used increasingly in integrated mapping studies that combine broad-scale and detailed vegetation type characterisation (Claggett et al., 2010; Wiederkher et al., 2010; Tormos et al., 2010; Johansen et al., 2010c). For example, with a combination of high spatial resolution images (QuickBird) and an object-oriented procedure, Johansen et al., (2007b) discriminated seven different vegetation types within a 10 km by 5 km area. There are some specific limitations at such scale, for example homogeneity, information extraction or cost of images. Some other limitations are the same as for the reach scale, but the objectives usually differ: it is a general but exhaustive description that is reached rather than a detailed one. For this purpose, images have good potential. In conclusion, image analysis can provide some complementary sets of data to combine with field information (Brooks et al., 2009) and objectives should be clearly identified before application.

10.5 Examples of imagery applications and potentials for riparian vegetation study

Documenting and analysing the structure and changes of riparian vegetation is an important step for river managers, forest services and conservation biologists involved in conservation and restoration projects (Van Looy et al., 2008; Brierley and Fryirs, 2008; Dufour and Piégay, 2009). It is also a fundamental point for hydraulic engineers that have to set up flood control plans and thus evaluate river and floodplain roughness (Kouwen, 2000; Straatsma and Middelkoop, 2007). Lastly, for network-scale managers, riparian buffers mapping is crucial to assess physical and chemical river quality (Claggett et al., 2010). Thus, we now present three examples of imagery use in riparian

vegetation studies, ranging from a simple and accessible application using aerial photography to more sophisticated ones using LiDAR and Radar data.

10.5.1 A low-cost strategy for monitoring changes in a floodplain forest: aerial photographs

A river restoration project, supported by the European community LIFE program, has been launched in order to manage and restore ecological, landscape and recreational elements of the remaining alluvial forest patches within the Arve River valley. The Arve River is an intra-mountain tributary of the Rhône River flowing from Mont Blanc (French Alps). Until the beginning of the twentieth century, the Arve was a braided river system, except locally where embankments had been previously constructed. In the aftermath of the Second World War, intensive gravel extraction resulted in active channel degradation, corridor fragmentation and subsequent changes of community composition in the floodplain forest. Remaining alluvial forest patches consisted of approximately 600 ha of woodland along a 50 km reach. The restoration project was led by the French Forestry commission (ONF), local districts, and river managers with scientific support of the CNRS, the French national organisation for scientific research. While carrying out the project, managers were confronted with common limitations including a lack of historical surveys, limited funding for field work and a complex mosaic of communities. As a result, historical imagery was employed as the primary tool to describe the temporal changes of the landscape and riparian vegetation.

10.5.1.1 Landscape changes: historical changes of the vegetation within the corridor

Between 1936 and 2004, 11 different aerial surveys were carried out over the Arve River. Of these, five dates were selected for comparison so as to limit the cost of the analysis. The choice was made (1) to select the earliest and latest dates (1936 and 2000), (2) to reject poor-quality and low-resolution images, and (3) to select dates with good temporal resolution (i.e., decadal sampling in 1961, 1973 and 1984) during the most dynamic periods (Greco and Plant, 2003). All of the photographs have been rectified and georeferenced using ArcGIS software, and spatial scales varied between 1/15 000 and 1/25 000. The photo series were black and white until 1973, infrared colour in 1984 and in colour in 2000. For each date, a map of the land cover was drawn by photo interpretation with classes defined by the active river zone (unvegetated bars and main channel),

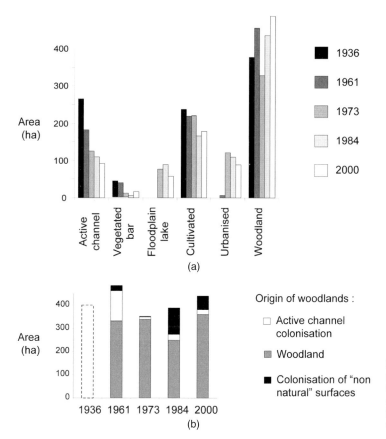

Figure 10.5 Changes of the Arve River (France) corridor from 1936 to 2000 (for the three most forested sites of the reach). a. Land cover trajectory obtained by photo interpretation, b. Changes of the origin of woodlands.

cultivated area, woodland, pioneer units, shrub units, floodplain lakes (i.e. former mining sites) and other 'non natural' units (i.e. gravel mining and urban areas).

Since 1936, the channel width of the Arve River has reduced by 200% (Figure 10.5). Two opposing riparian historical trends are apparent: (1) a decrease in woodland areas between 1961 and 1984 and (2) and increase in woodlands due to both active channel narrowing before 1961 and cultivated surface abandonment since 1984 (Figure 10.5).

10.5.1.2 Riparian vegetation description and monitoring: setting a sampling strategy

The second objective of the diagnosis phase was to describe the composition of riparian woodlands that colonised the Arve floodplain. To limit time and cost spent in the field research, a stratified sampling strategy was developed based on stand age. With the overlay of GIS data derived from photo interpretation, five types of vegetation units were identified (Figure 10.6): type 1,

vegetated since 1936 (i.e. woodlands older than 68 years old); type 2, vegetated in 1961 but not in 1936; type 3, vegetated in 1984 but before 1961; type 4, vegetated in 2000 but not before 1984; and type 5, vegetation that colonises surfaces that have been occupied by a human activity since 1936 such as gravel mining, urbanisation or cultivation (i.e. 'non-natural' vegetation). The extent of each type was mapped. Upon these surfaces, regularly spaced plots were implemented on a grid (cell size = 50 m). Each dot indicates the coordinates for a potential plot to survey (Figure 10.6). 10 plots were then randomly selected by type. For each (20 m by 20 m), a field survey allowed the description of the vegetation structure respecting a methodology following Dufour and Pont (2006): overbank sediment thickness, identification and measurement of trees with a DBH bigger then 7.5 cm, tree regeneration, dead wood density and fauna impacts.

This stratified sampling helped optimise the field survey for vegetation characterisation. A factorial analysis (DCA) on the complete set of vegetation samples (for trees with a DBH greater than 7.5 cm, i.e. 22 species) demonstrates

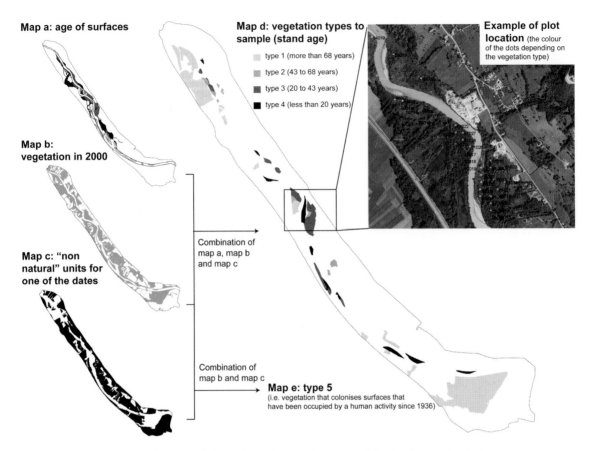

Figure 10.6 Example of a stratified approach for implementing sampling strategy. Map d and map e give the five types of vegetation to sample, and were obtained from the overlay of the map a (age of surface), map b (vegetation types in 2000) and map c ('non natural units'). For the signification of the five types, see the text.

a progressive shift in stand composition from younger stands characterised by *Salix spp.* and *Alnus glutinosa* to older stands dominated by *Fraxinus excelsior, Salix alba* and *Alnus incana* and individualised by *Acer spp.* and *Juglans regia* (Figure 10.7). They also provide information about which kind of vegetation should be expected at a given age. The stands that colonise formerly cultivated or urbanised surfaces are close to the stands that are between 20 and 43 years old.

10.5.1.3 Conclusions and limitations

To conclude, the approach developed along the Arve River is a very easy way to implement a historical analysis, a stratified strategy for the description of vegetation and a network of plots for a monitoring program (surveyed every 10 years). The feasibility at the reach scale for managers is good, because required data, software

and tools are now easily accessible outside of the scientific community. However, such an approach is not error-free. Indeed, it is based on the hypothesis that two units that look the same on two different photos at different dates are the same in the field. This is a fundamental limit of the physiognomic approach; behind the same physiognomy we can have very different communities and environmental conditions. Moreover, some aspects cannot be described or quantified solely using remotely sensed data. A field survey is complementary and necessary, for species composition analysis, for instance.

10.5.2 Flow resistance and vegetation roughness parametrisation: LiDAR and multispectral imagery

The objective of this case study is to produce a spatially explicit map of vegetation structure with relevant input

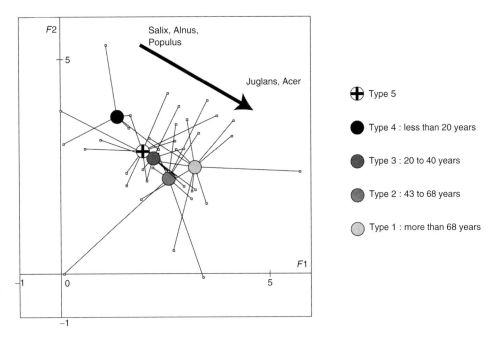

Figure 10.7 Factorial map that shows stand composition shift through time in the Arve River floodplain forests.

for hydrodynamic roughness. Hydrodynamic roughness, or flow resistance, determines the kinetic drag of the water flow. The higher the roughness, the slower the water will flow and, hence, the higher the water levels will reach, increasing the flood hazard. The roughness of the vegetation growing on the floodplains depends on vegetation structure and has been described by many different models that simplify different aspects of the interaction between water flow and vegetation. Kouwen (2000) lists the factors that are assumed to determine vegetation roughness, including vegetation height and density, rigidity of the stems, pattern of spacing of the stems, deformation of a plant under flow, density and orientation of foliage and the drag coefficient. In addition to vegetation roughness, the ground surface will also induce a friction force on the water. The problem is that no accurate, spatially distributed and quantitative method exists to parametrise the hydrodynamic roughness of the floodplains as the input for models. In practice, often a single roughness coefficient is used for the whole floodplain section, which is subsequently used as a calibration parameter to fit modelled water levels to measurements. This may seriously misrepresent local flow conditions relevant to sediment deposition and scour. More detailed methods consist of land cover classification, combined with a lookup table for a

roughness coefficient. Still, most of the variation in the 3D plant structure is not taken into consideration.

Various remote sensing data may provide information on vegetation type and structure including their dynamics. The key issue to overcome is the translation of remote sensing information, i.e. the translation of the intensity and pattern of reflected electromagnetic radiation into the relevant parameters in order to be able to register patterns of hydrodynamic roughness. Many studies have reported successful and accurate mapping of natural vegetation using multispectral or hyperspectral remote sensing data (e.g. Mertes, 2002). Recently, spectral information has been combined with height information in vegetation classification schemes (e.g. Dowling and Accad, 2003). Even though the spatial resolution and the level of detail of the classification varies with the type of remote sensing data, a lookup table is always required in order to convert the vegetation classes to vegetation structure values, which leads to the undesirable loss of within-class variation. In contrast, LiDAR enables the extraction of the structural characteristics of vegetation such as height, biomass, basal area, and leaf area index (see Straatsma and Middelkoop, 2007). However, LiDAR data shows noise with a standard deviation ranging from 1.5 to 4 cm making the application in meadows and pioneer vegetation unsuccessful.

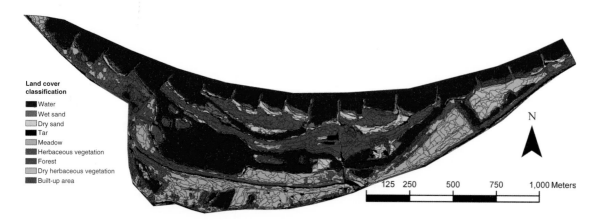

Figure 10.8 Classified land cover map using object-oriented segmentation and classification (river Waal, The Netherlands).

The combination of LiDAR and multispectral data showed promising results for the parameterisation of floodplain roughness. Straatsma and Baptist (2008) presented a method that fuses LiDAR data and airborne multispectral data of the river Waal in The Netherlands, the main distributary of the river Rhine. The fused data were subsequently used in an object-oriented segmentation using eCognition. Image objects were subsequently classified using a linear discriminant analysis, resulting in an overall classification accuracy of 81% (Figure 10.8). The classes 'meadow', 'unvegetated' and 'built-up' were assigned a roughness value based on a lookup table as no method for roughness determination exists for these classes. For the 'herbaceous vegetation' and 'forest' classes, vegetation height and density were determined based on LiDAR-derived vegetation structure predictors and regression equations. This included calibration of the relationships in order to predict vegetation structure based on LiDAR data using field reference data. Predictive quality for vegetation height ($0.74 < r^2 < 0.88$; $n = 42$) was higher than for vegetation density ($0.51 < r^2 < 0.66$; $n = 42$; Figure 10.9). Repeated field visits are a downside of the method's feasibility. Automatic calibration may be possible, but has not yet been worked out.

There were a number of difficulties in the implementation of the new method into the workflow of river managers with a mandate for flood control. Firstly, the current method, based on manual aerial image interpretation, already had the ability to provide highly detailed vegetation structure information, but no one knew exactly how accurate it was. Convincing the managers that the new method should be implemented proved to be difficult as there was no benchmark to compare the results to. Secondly, the required methods are not yet available in any commercial software, which means that quality control takes much time. In conclusion, we could say that data fusion of multispectral and LiDAR data has proven its ability to accurately predict vegetation structure for floodplain roughness parameterisation. Large opportunities exist since the total surface area of The Netherlands is mapped with LiDAR data with a 10 points/m² density. Before implementation, however, the workflow should be more automised in order to reduce production time of the maps, and the implications of prediction errors with respect to flood hazards should be determined.

10.5.3 Potential radar data uses for riparian vegetation characterisation

The first radar operational airborne system was developed in 1960 at the University of Michigan (USA) and in 1978, SEASAT, the first civil radar satellite, was launched by NASA. European Space Agency (ESA) followed in 1991 with ERS-1, ERS-2 and ENVISAT (2002), the Japanese Space Agency (JSA) developed J-ERS-1 in 1992 and the Canadian Space Agency (CSA) Radarsat-1 in 1995. At the same time, the first demonstration of fully polarimetric, multi frequency spaceborne was tested between April and October 1994 on a space shuttle. The latest satellites such as Alos-Palsar (Japan, 2005), TerraSAR-X (Germany, 2007) and Radarsat-2 (Canada, 2007) integrated this fully polarimetric acquisition and improved spatial resolution

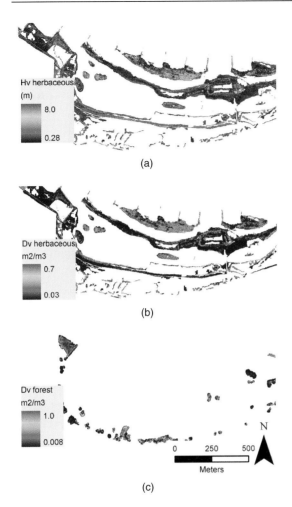

Figure 10.9 Vegetation structural characteristics based on LiDAR data. (a) Height of herbaceous vegetation (b) Density of herbaceous vegetation (c) Forest density.

for a specific product. Now, depending on the satellite, it is possible to order and buy different radar images (altimetry, interferometry, imagery, polarimetry) on the Spatial Agencies websites (ESA, NASA, CSA, and JSA) according to the objectives.

For riparian vegetation characterisation, data of active radar satellite, named SAR data (Synthetic Aperture Radar) as Radarsat-2, Alos-Palsar, or Envisat offer interesting potentialities because of their geometrical characterisations (high or very high spatial resolution, incidence angles . . .), their polarisation (fully polarimetric permits the decomposition of the scattering process) and their frequencies (bands used as the C-band or L-band are not sensitive to atmosphere conditions and can penetrate vegetation up to the ground). SAR specific characteristics for riparian vegetation identification are detailed below:

10.5.3.1 Geometrical characteristics

Three main geometrical characteristics for SAR data can be detailed here and appear useful for riparian vegetation characterisation: (1) incidence angles, (2) sensor orientation and (3) spatial resolution (for further information, see Spatial Agency websites). Generally, incidence angle vary between 20 to 60 degrees and offer interesting potential for vegetation characterisation, as with high incidence angles, a better extraction of the vegetation can be obtained. (Ulaby and Dobson, 1989). Besides, sensor direction (right or left) permits the identification of high and linear vegetation perpendicular to the beam. Lastly, resolution data of radar satellites which were until recently a significant limitation in the study of thin and complex geographical objects such as riparian vegetation (maximum: 20 meters for Radarsat-1) have considerably evolved. Sensors as TerraSar-X, Alos-Palsar, and Radarsat-2 have significantly improved their spatial resolution, with, for example, 1.6 m. in range and 2.8 m. in azimuth for Radarsat-2 (beam mode UltraFine, SLC) which allows the study and the monitoring of riparian vegetation not just in large floodplain environments (Townsend, 2001).

10.5.3.2 Polarisation characteristics

Several satellites offer just one polarisation, such as Radarsat-1 (Horizontal Emission–Horizontal Reception: HH), whereas others offer multi-polarisation, such as Envisat (HV, dual polarisation HH + HV) and recently, three sensors have acquired data in a polarimetric mode (TerraSar-X, Radarsat-2 and Alos-Palsar with the four polarisations HH, HV, VH and VV). Polarimetric data offer very interesting outlooks as polarimetric decomposition theorems (Cloude and Pottier, 1997, Touzi, 2007) let us differentiate backscattering characteristics of the studied area such as the contribution of the scattering volume, double-bounce component, the power scattered by the surface etc . . . For example, experiments show that multi-polarised and polarimetric data allow a fine characterisation and mapping of wetlands classes such as marsh, treed bog, shrub bog . . . (Brisco et al., 2008).

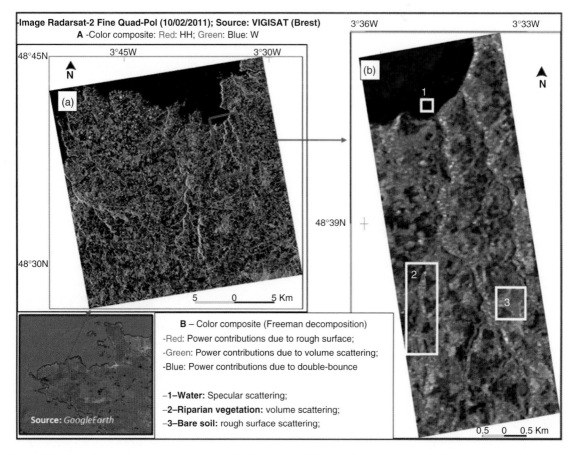

Figure 10.10 Riparian vegetation characterisation with a Radarsat-2 (©) image. Fine Quad-Pol on the Yar watershed in Brittany (France; 10/02/2010), at a. large and b. fine scales see text for detail. Radarsat-2 © image.

Figure 10.10 presents a composite colour of a polarimetric Radarsat-2 image in France (Brittany) at the end of February 2010. Bare soils are represented in magenta and blue because of high backscattering in HH amplitude band (sensitive to surface scattering). Vegetation, characterised by volume scattering, appears here in green, as the highest backscattering for high vegetation such as riparian vegetation is measured for the HV polarisation. With this type of data compounded with four polarisations, different polarimetric decomposition can be applied (Cloude and Pottier, 1997; Touzi et al., 2007, Freeman et al., 2007) and allow a good discrimination of geometrical parameters of the studied object as the type of scattering (surface, volume or double bounce). Inverse model can also be applied to retrieve soil or vegetation moisture, forest biomass etc (Beaudoin, 1994).

10.5.3.3 Frequencies

Three main wavelengths are generally used by SAR satellites: X band (2.4 to 3.75 cm) for cartography, C band (3.75 to 7.5 cm) for agricultural monitoring, land use and land cover changes and L band (15 to 30 cm) for forestry, geology, etc. The longer the frequency, the deeper the wavelength can penetrate the soil or the vegetation. Thus, each wavelength has its specificities, for example the L band present on the Alos-Palsar satellite deeply penetrates the canopy and provides useful information for the discrimination of different vegetation types (Hess et al., 1990; Izzawati et al., 2006). The signal of the radar back to the satellite is called the backscattering coefficient (measured in dB), it depends on the soil moisture, the roughness and the geometrical characteristics of the studied object. The analysis of backscattering coefficient permits the identification and characterisation of different types of land cover, such as wetlands or agricultural practices (Touzi et al., 2007).

Almost no specific studies have been realised on narrow strips of riparian vegetation because of the complexity of this geographical object and the available SAR data. Some studies of floodplain environments show that inundation and vegetation structure can be accurately characterised with radar data (Townsend, 2001; Townsend, 2002). Nevertheless, the latest radar satellite, which acquires polarimetric data in a high or very high spatial resolution, appears promising for small riparian corridor characterisation as it brings complementary information to classical data used for riparian studies (optical data, LiDAR, ground control campaign, soil maps etc). For example, with Radarsat-2 data, different types of scattering of the riparian vegetation can be extracted, bio physical parameters such as moisture and biomass can be retrieved and combined with other information sources, permitting a finer study and monitoring of riparian vegetation. Figure 10.10 presents two composite colours of a Radarsat-2 image (mode fine Quad-Pol; 10/02/2011) on the Yar watershed in Brittany (France). On the first colour composite (Figure 10.10a) which represents the entire scene, the colour red is associated with HH polarisation, the colour green with the HV polarisation and the colour blue with the VV polarisation. At this scale, we can distinguish quite precisely the vegetation (green) and bare soil (magenta) surfaces. In order to identify riparian vegetation we zoomed on the north of the Yar watershed (Figure 10.10b). We applied the Freeman decomposition (Freeman et al., 1998), which allows the extraction of the power contributions due to rough surface (Red), volume scattering (Green) and double-bounce (Blue). Bare soils are represented in red and magenta because of the high power contribution of the rough surface. Riparian vegetation, characterised by volume scattering, appears here in green and is clearly identified. Finally, urban surface appears on blue because of a main contribution of the double-bounce scattering. With this type of data compounded with four polarisations, different polarimetric decomposition can be applied (Cloude and Pottier, 1997; Touzi et al., 2007, Freeman et al., 2007 . . .) and permit a good discrimination of geometrical parameters of the studied object such as the type of scattering (surface, volume or double bounce).

Inverse model can also be applied to retrieve soil or vegetation moisture, forest biomass etc (Beaudoin, 1994).

10.6 Perspectives: from images to indicators, automatised and standardised processes

The efficiency of image use is linked to our capacity to provide a standard that can be specific to a given question, but also simple and relevant. There is a large panel of applied uses ranging from simple and general indicators (e.g. corridor width or vegetation continuity) to very complex or specific needs such as senescence characterisation (Lonard et al., 2000) or identification of potential restoration reaches (Mollot and Bilby, 2008). The efficiency is also linked to our ability to merge sensor data produced with field and ancillary data (Munne et al., 2003; González del Tánago and García de Jalón, 2006; Debruxelles et al., 2009). For river and forest managers the questions we need to address are : i) 'which indicators should we choose or define?' and, ii) 'how do we extract them?'. The first question needs an active dialogue between scientists (from several disciplines) and managers. To answer the second, we can use GIS and remote sensing tools that have been recently developed and can provide regularly spaced information for the assessment of riparian conditions (Tiner, 2004; Claggett et al., 2010; Wiederkehr et al., 2010; Tormos et al., 2010; Alber and Piégay, 2011). Indeed, besides all that the technical possibilities offer by remote sensing, image uses for riparian area understanding and management require some progress, notably with respect to automation and standardisation of processes and indicator calibration. They are very important both for river network managers in charge of national and regional planning (upscaling analysis from reach to network scale) and for technical transfer to reach-scale managers. For example, the combination of LiDAR data and satellite images has been shown as a very interesting approach at reach scale, but to be able to pass from one image to a set of images and then to the overall network, it needs specific tools in terms of data fusion, calibration and management (Hall et al., 2009; Arroyo et al., 2010). At the reach scale, there is a large gap between tools used or under development by the scientific community and standard approaches accessible for managers (e.g. historical analysis of aerial photos). Another question is thus how can managers have access to recent advances (structural parameters from LiDAR or radar data, very high resolution, ultra light vehicles)? For this, the (human and financial) means mobilised by managers are fundamental and concern funding for data acquisition and analysis, software, hardware, field work and data storage.

Acknowledgements

The authors are grateful to reviewers for their many useful suggestions and corrections. The authors are also grateful to the French Office National des Forêts for providing data (Arve River).

References

Adam, E., Mutanga, O., and Rugege, D. 2010. Multispectral and hyperspectral remote sensing for identification and mapping of wetland vegetation: a review. *Wetlands Ecol. Manage.* 18, 281–296.

Aguiar, F.C. and Ferreira, M.T. 2005. Human-disturbed landscapes: effects on composition and integrity of riparian woody vegetation in the Tagus River basin, Portugal. *Environmental Conservation* 32, 30–41.

Akasheh, O.Z., Neale, C.M.U., and Jayanthi, H. 2008. Detailed mapping of riparian vegetation in the middle Rio Grande River using high resolution multi-spectral airborne remote sensing. *Journal of Arid Environments* 72, 1734–1744.

Alber, A. and Piégay, H. 2011. Spatial disaggregation and aggregation procedures for characterizing fluvial features at the network-scale: Application to the Rhone basin (France). *Geomorphology* 125, 343–360.

Alencar-Silva, T. and Maillard, P. 2010. Assessment of biophysical structure of riparian zones based on segmentation method, spatial knowledge and texture analysis. In: Wagner W., Székely, B. (eds.): ISPRS TC VII Symposium – 100 Years ISPRS, Vienna, Austria, July 5–7, 2010, IAPRS, Vol. XXXVIII, Part 7B.

Alexander, R. and Millington, A.C. (2000). *Vegetation Mapping: From Patch to Planet*. Wiley, 339 p.

Antonarakis, A.S., Richards, K.S., and Brasington, J. 2008a. Object-based land cover classification using airborne LiDAR. *Remote Sensing of Environment* 112, 2988–2998.

Antonarakis, A.S., Richards, K.S., Brasington, J., Bithell, M., and Muller, E. 2008b. Retrieval of vegetative fluid resistance terms for rigid stems using airborne LiDAR. *Journal of Geophysical Research* 113, G02S07.

Antonarakis, A.S., Richards, K.S., Brasington, J., and Bithell, M. 2009. Leafless roughness of complex tree morphology using terrestrial LiDAR. *Water Resources Research* 45, W10401.

Antonarakis, A.S., Richards, K.S., Brasington, J., and Muller, E. 2010. Determining leaf area index and leafy tree roughness using terrestrial laser scanning. *Water Resources Research*, 46, W06510

Apan, A.A., Raine, S.R., and Paterson, M.S. 2002. Mapping and analysis of changes in the riparian landscape structure of the Lockyer Valley catchment, Queensland, Australia. *Landscape and Urban Planning* 59, 43–57.

Arroyo, L.A., Johansen, K., Armston, J., and Phinn, S. 2010. Integration of LiDAR and QuickBird imagery for mapping riparian biophysical parameters and land cover types in Australian tropical savannas. *Forest Ecology and Management* 259, 598–606.

Beaudoin, A., Le Toan, T., Goze, S., Nezri, E., Lopez, A., Mougin, E., Hsu, C.C., Han, H.C., Kong, J., and Shin R.T. 1994. Retrieval of forest biomass from SAR dat. *International Journal of Remote Sensing*, 15, 2777–2796.

Bawazir, A.S., Samani, Z., Bleiweiss, M., Skaggs, R., and Schmugge T. 2009. Using ASTER satellite data to calculate riparian evapotranspiration in the Middle Rio Grande, New Mexico. *International Journal of Remote Sensing* 30, 5593–5603.

Belward, A.S. and Valenzuela, C.R. (1991). *Remote sensing and geographical information systems for resource management in developing countries*. Springer, 520 p.

Bendix, J. and Stella, J.C. in press. Riparian vegetation and the fluvial environment: a biogeographic perspective. In *Treatise on Geomorphology 12: Ecogeomorphology (D. Butler and C. Hupp, Eds.)*. Elsevier, San Diego.

Boyd, D.S. and Danson, F.M. 2005. Satellite remote sensing of forest resources: three decades of research development. *Progress in Physical Geography*, 29: 1–26.

Brierley, G. and Fryirs, K. (eds) (2008). *River futures: an integrative scientific approach to river repair*. Island Press, Washington, DC, 328 p.

Brisco, B., Touzi, R., Van der Sanden, J.J., Charbonneau, F., Pultz, T.J., and D'Iorio, M. 2008. Water resource applications with RADARSAT-2 – a preview. *International Journal of Digital Earth* 1, 130–147.

Brohman, R., Bryant, L., (eds.) 2005. Existing vegetation classification and mapping technical guide. Gen. Tech. Rep. WO–67. Washington, DC: U.S. Department of Agriculture Forest Service, Ecosystem Management Coordination Staff. 305 p.

Brooks, R., McKenney-Easterling, M., Brinson, M., Rheinhardt, R., Havens, K., O'Brien, D., Bishop, J., Rubbo, J., Armstrong, B., and Hite, J. 2009. A Stream–Wetland–Riparian (SWR) index for assessing condition of aquatic ecosystems in small watersheds along the Atlantic slope of the eastern U.S. *Environ Monit Assess* 150: 101–117

Butera, M.K. 1983. Remote sensing wetlands. *IEEE Trans. Geosci. Remote Sensing* 21, 383–392.

Charlton, J. 2000. Kaw River Valley Scenes Revisited after 130 Years. *Transactions of the Kansas Academy of Science* 103, 1–21.

Claggett, P.R., Okay, J.A., and Stehman, S.V. 2010. Monitoring regional riparian forest cover change using stratified sampling and multiresolution imagery. *Journal of the American Water Resources Association* 46, 334–343.

Cloude, S.R. and Pottier, E. 1997. An entropy based classification scheme for land applications of Polarimetric SAR. IEEE Trans. *Geoscience Rem. Sens.* 35, 68–78.

Corenblit, D., Tabacchi, E., Steiger, J., and Gurnell, A.M. 2007. Reciprocal interactions and adjustments between fluvial landforms and vegetation dynamics in river corridors: A review of complementary approaches. *Earth-Science Review* 84, 56–86.

Coroi, M., Sheehy Skeffington, M., Giller, P., Gormally, M., and O'Donovan, G. 2006. Using GIS in the mapping and analysis of the landscape and vegetation patterns along streams in Southern Ireland. Biology and Environment: Proceedings of the Royal Irish Academy 106B, 287–300.

Congalton, R.G., Birch, K., Jones, R., and Schriever, J. 2002. Evaluating remotely sensed techniques for mapping riparian vegetation. *Computers and Electronics in Agriculture* 37, 113–126.

Corbane, C., Raclot, D., Jacob, F., Albergel, J., and Andrieux, P. 2006. Remote sensing of soil surface characteristics from a multiscale classification approach. *Catena* 75, 308–318.

Cunningham, S.C., Mac Nally, R., Read, J., Baker, P.J., White, M., Thomson, J.R., and Griffioen, P. 2009. A robust technique for mapping vegetation condition across a major river system. *Ecosystems* 12, 207–219.

Davis, P.A., Staid, M.I., Plescia, J.B., and Johnson, J.R. 2002. Evaluation of airborne image data for mapping riparian vegetation within the Grand Canyon. USGS Open-File Report 02–470, available on line (1/1/2011) at geopubs.wr.usgs.gov/openfile/of02-470.

De Rose, R.C., Barrett, D., Marks, A., Caitcheon, G., Chen, Y., Simon, D., Lymburner, L., Douglas, G., and Palmer, M. 2005. Regional patterns of riparian vegetation, erosion and sediment transport in the Ovens River basin. CSIRO Land and Water Client Report, Canberra, Australia, 26 p.

Debruxelles, N., Claessens, H., Lejeune, P., and Rondeux, J. 2009. Design of a watercourse and riparian strip monitoring system for environmental management. *Environ. Monit. Assess.* 156, 435–450.

Dieck, J.J. and Robinson, L.R. 2004. Techniques and methods book 2, Collection of environmental data, Section A, Biological Science, Chapter 1, General classification handbook for floodplain vegetation in large river systems: U.S. Geological Survey, Techniques and Methods 2 A–1, 52 p.

DiPietro, D., Ustin, S.L., Underwood, E.C., Olmstead, K., and Scheer, G.J. 2002. Mapping the invasive riparian weed *Arundo donax* (Giant Reed) using AVIRIS. in *Eleventh JPL Airborne Visible Infrared Imaging Spectrometer (AVIRIS) Workshop*. Jet Propulsion Laboratory, Pasadena, CA. Available on http://cstars.ucdavis.edu

Dowling, R. and Accad, A. 2003. Vegetation classification of the riparian zone along the Brisbane River, Queensland, Australia, using light detection and ranging (LiDAR) data and forward

looking digital video. *Canadian Journal of Remote Sensing* 29, 556–563.

Dufour, S. and Pont, B. 2006. Protocole de suivi des forêts alluviales: l'expérience du réseau des réserves naturelles de France. Revue Forestière Française LVIII, 45–60.

Dufour, S. and Piégay, H. 2009. From the myth of a lost paradise to targeted river restoration: forget natural references and focus on human benefits. *River Research and Applications* 25, 568–581.

Dunford, R., Michel, K., Gagnage, M., Piégay, H. and Trémelo, M.L. 2009. Potential and constraints of UAV technology for the characterisation of Mediterranean riparian forest. *International Journal of Remote Sensing* 30, 4915–4935.

Ehlers, M., Gähler, M., and Janowsky, R. 2003. Automated analysis of ultra high resolution remote sensing data for biotope type mapping: new possibilities and challenges. *ISPRS Journal of Photogrammetry & Remote Sensing* 57, 315–326.

Everitt, J.H., Fletcher, R.S., Elder, H.S., and Yang, C. 2007. Mapping giant salvinia with satellite imagery and image analysis. *Environ. Monit. Assess.* 139, 35–40.

Farid, A., Goodrich, D.C., and Sorooshian, S. 2006. Using airborne LiDAR to discern age classes of cottonwood trees in a riparian area. *West. J. Appl. For.* 21, 149–158.

Farid, A., Goodrich, D.C., Bryant, R., and Sorooshian, S. 2008. Using airborne lidar to predict Leaf Area Index in cottonwood trees and refine riparian water-use estimates. *Journal of Arid Environments* 72, 1–15.

Ferreira, M.T., Aguiar, F.C., and Nogueira, C. 2005. Changes in riparian woods over space and time: Influence of environment and land use. *Forest Ecology and Management* 212, 145–159

Franklin, S.E. and Wulder, M.A. 2002. Remote sensing methods in medium spatial resolution satellite data land cover classification of large areas. *Progress in Physical Geography* 26, 173–205.

Freeman, A. and Durden, SLA. (1998). Three-component scattering model for polarimetric SAR data. *IEEE Transactions on Geoscience and Remote Sensing*, 36(3): 963–973

Friedman, J.M. and Lee, V.J. 2002. Extreme floods, channel change, and riparian forests along ephemeral streams. *Ecological Monographs* 72, 409–425.

Ge, S., Carruthers, R., Gong, P., and Herrera, A. 2006. Texture analysis for mapping *Tamarix parviflora* using aerial photographs along the Cache Creek, California. *Environ. Monit. Assess.* 114, 65–83.

Geerling, G.W., Ragas, A.M.J., Leuven, R., van den Berg, J.H., Breedveld, M., Liefhebber, D., and Smits, A.J.M. 2006. Succession and rejuvenation in floodplains along the river Allier (France). *Hydrobiol.* 565, 71–86.

Geerling, G.W., Vreeken-Buijs, M.J., Jesse, P., Ragas A.M.J., and Smits A.J.M. 2009. Mapping river floodplain ecotopes by segmentation of spectral (CASI) and structural (LiDAR) remote sensing data. *River Research and Applications* 25, 795–813.

Genç, L., Dewitt, B., and Smith, S. 2004. Determination of wetland vegetation height with LiDAR. *Turk. J. Agric. For.* 28, 63–71.

Gilvear, D.J. and Bryant, R. 2003. Analysis of air photography and other remotely sensed data. In H. Piegay and M. Kondolf (Eds), *Tools in Fluvial Geomorphology*, Wiley, 133–168.

Girel, J. 1986. Télédétection et cartographie à grande échelle de la végétation alluviale: exemple de la basse vallée de l'Ain. In: Roux, A.L., (Ed.), Document de Cartographie Ecologique, recherches interdisciplinaires sur les écosystémes de la basse-plaine de l'Ain (France): potentialités évolutives et gestion, 29: 45–74.

Girel, J. and Manneville, O. 1998. Present species richness of plant communities in alpine stream corridors in relation to historical river management. *Biological conservation* 85, 21–33.

Goetz, S.J., Wright, R.K., Smith, A.J., Zinecker, E. and Schaub, E. 2003. IKONOS imagery for resource management: Tree cover, impervious surfaces, and riparian buffer analyses in the mid-Atlantic region. *Remote Sensing of Environment* 88, 195–208.

Goetz, S.J. 2006. Remote sensing of riparian buffers: past progress and future prospects. *Journal of the American Water Resources Association* 42, 133–143.

González del Tánago, M. and Garciá de Jalón, D. 2006. Attributes for assessing the environmental quality of riparian zones. *Limnetica* 25, 389–402.

Gonzalez, E., Gonzalez-Sanchis, M., Cabezas, A., Comin, F.A., and Muller, E. 2010. Recent changes in the riparian forest of a large regulated mediterranean river: implications for management. *Environmental Management* 45, 669–681.

Grams, P.E. and Schmidt, J.C. 2002. Streamflow regulation and multi-level f lood plain formation: channel narrowing on the aggrading Green River in the eastern Uinta Mountains, Colorado and Utah. *Geomorphology* 44, 337–360.

Greco, S.E. and Plant, R.E. 2003. Temporal mapping of riparian landscape change on the Sacramento River, miles 196–218, California, USA. Landscape Research 28, 405–426.

Greco, S.E., Fremier, A.K., Larsen, E.W., and Plant, R.E. 2007. A tool for tracking floodplain age land surface patterns on a large meandering river with applications for ecological planning and restoration design. *Landscape and Urban Planning* 81, 354–373.

Gurnell, A.M. and Gregory, K.J. 1995. Interactions between semi-natural vegetation and hydrogeomorphological processes. *Geomorphology* 13, 49–69.

Gurnell, A.M., Peiry, J.L., and Petts, G.E. 2003. Using historical data in fluvial geomorphology. In: Kondolf, M., Piégay, H., (Eds.), *Tools in fluvial geomorphology*. Wiley, pp. 77–103.

Hall, R.K., Watkins, R.L., Heggem, D.T., Jones, K.B., Kaufmann, P.R., Moore, S.B., and Gregory, S.J. 2009. Quantifying structural physical habitat attributes using LiDAR and hyperspectral imagery. *Environ. Monit. Assess.* 159, 63–83.

Hamada, Y., Stow, D.A., Coulter, L.L., Jafolla, J.C., and Hendricks, L.W. 2007. Detecting Tamarisk species (*Tamarix* spp.) in riparian habitats of Southern California using high spatial resolution hyperspectral imagery. *Remote Sensing of Environment*, 109: 237–248.

Hess, L.L., Melack, J.M., and Simonett, D.S. 1990. Radar detection of flooding beneath the forest canopy: a review. *International Journal of Remote Sensing*, 11, 1313–1325.

Hess, L.L. and Melack, J.M. 2003. Remote sensing of vegetation and flooding on Magela Creek floodplain (Northern Territory, Australia) with the SIR-C synthetic aperture radar. *Hydrobiologia* 500, 65–82.

Hohensinner S., Habersack, H., Jungwirth, M., and Zauner, G. 2004. Reconstruction of the characteristics of a natural alluvial river-floodplain system and hydromorphological changes following human modifications: the Danube River (1812–1991). *River Research and Applications* 20, 25–41.

Izzawati, Wallington, E.D. and Woodhouse, and I.H. 2006. Forest height retrieval from commercial X- Band SAR products. *GeoRS*, 44: 863–870.

Iverson, L.R., Graham, R.L., and Cook, E.A. 1989. Applications of satellite remote sensing to forested ecosystems. *Landscape Ecology* 3, 131–143.

Johansen, K., Tiede, D., Blaschke, T., Phinn, S., and Arroyo, L.A. 2010a. Automatic geographic object based mapping of streambed and riparian zone extent from LiDAR data in a temperate rural urban environment, Australia. Proceeding GEOBIA 2010-Geographic Object-Based Image Analysis. Ghent University, Ghent, Belgium, 29 June – 2 July. *ISPRS* Vol. No. XXXVIII-4/C7.

Johansen, K., Phinn. S., and Witte, C. 2010b. Mapping of riparian zone attributes using discrete return LiDAR, QuickBird and SPOT-5 imagery: Assessing accuracy and costs. *Remote Sensing of Environment* 114, 2679–2691.

Johansen, K., Coops, N.C., Gergel, S.E., and Stange, Y. 2007b. Application of high spatial resolution satellite imagery for riparian and forest ecosystem classification. *Remote Sensing of Environment* 110, 29–44.

Johansen, K., Phinn, S., Dixon, I., Douglas, M., and Lowry J. 2007a. Comparison of image and rapid field assessments of riparian zone condition in Australian tropical savannas. *Forest Ecology and Management* 240, 42–60.

Johansen, K., Phinn, S., and Witte, C. 2010c. Mapping of riparian zone attributes using discrete return LiDAR, QuickBird and SPOT-5 imagery: Assessing accuracy and costs. *Remote Sensing of Environment* 114, 2679–2691.

Johnson, W.C., Dixon, M.D., Simons, R., Jenson, S., and Larson, K. 1995. Mapping the response of riparian vegetation to possible flow reductions in the Snake River, Idaho. *Geomorphology* 13, 159–173.

Johnson, M.R. and Zelt, R.B. 2005. Protocols for mapping and characterizing land use/land cover in riparian zones: U.S. Geological Survey Open-File Report 2005–1302, 22 p.

Jones A.F., Brewer, P.A., Johnstone, E., and Macklin M.G. 2007. High-resolution interpretative geomorphological mapping of river valley environments using airborne LiDAR data. *Earth Surf. Proc. Land.* 32, 1574–1592.

Klimo E. 2008. Floodplain forest of the temperate zone of Europe. Lesnická práce, Prague, 623 p.

Kouwen, N. 2000. Friction factors for coniferous trees along rivers. *Journal of Hydraulic Engineering* 126, 732–740.

Latterell, J.J., Bechtold, J.S., O'Keefe, T.C., Van Pelt, R., and Naiman, R.J. 2006. Dynamic patch mosaics and channel movement in an unconfined river valley of the Olympic Mountains. *Freshwater Biology* 51, 523–544.

Lawson, T., Gillieson, D., and Goosem, M. 2007. Assessment of riparian rainforest vegetation change in tropical North Queensland for management and restoration purposes. *Geographical Research* 45, 387–397.

Lehman, A. and Lachavanne, J.B. 1997. Geographic information and remote sensing in aquatic botany. *Aquatic Botany* 58, 195–207.

Lejot, J., Delacourt, C., Piegay, H., Trémélo, M.L., and Fournier, T. 2007. Very high spatial resolution imagery for reconstructing channel bathymetry and topography from an unmanned controlled platform. *Earth Surf. Proc. Land.* 32, 1705–1725.

Loheide, S.P. and Gorelick, S.M. 2005. A local-scale, high-resolution evapotranspiration mapping algorithm (ETMA) with hydroecological applications at riparian meadow restoration sites. *Remote Sensing of Environment* 98, 182–200.

Lonard, R.I., Judd, F.W., and Desai, M.D. 2000. Evaluation of color-infrared photography for distinguishing annual changes in riparian forest vegetation of the lower Rio Grande in Texas. *Forest Ecology and Management* 128, 75–82.

Makkeasorn, A., Chang, N.B., and Li, J. 2009. Seasonal change detection of riparian zones with remote sensing images and genetic programming in a semi-arid watershed. *Journal of Environmental Management* 90, 1069–1080.

Malanson, G.P. (1993). *Riparian landscapes.* Cambridge University Press, Cambridge, UK, 296 p.

Marston, R.A., Girel, J., Pautou, G., Piégay, H., Bravard, J.P., and Arneson, C. 1995. Channel metamorphosis, floodplain disturbance, and vegetation development: Ain river, France. *Geomorphology* 13, 121–131.

Mason, D.C., Cobby, D.M., Horritt, M.S., and Bates, P.D. 2003. Floodplain friction parameterization in two-dimensional river flood models using vegetation heights derived from airborne scanning laser altimetry. *Hydrological Processes* 17, 1711–1732.

Mathieu, R., Aryal, J., and Chong, A.K. 2007. Object-based classification of Ikonos imagery for mapping large-scale vegetation communities in urban areas. *Sensors* 7, 2860–2880.

Mendonça-Santos, M.L. and Claramunt, C. 2000. An integrated landscape and local analysis of land cover evolution in an alluvial zone. *Computers, Environment, and Urban Systems* 25, 557–577.

Mertes, L.A.K., Daniel, D.L., Melack, J.M., Nelson, B., Martinelli, L.A., and Forsberg, B.R. 1995. Spatial patterns of hydrology, geomorphology, and vegetation on the floodplain of the Amazon River in Brazil from a remote sensing perspective. *Geomorphology* 13, 215–232.

Mertes, L.A.K. 2002. Remote sensing of riverine landscapes. *Freshwater Biology* 47, 799–816.

Miller, J.R., Schulz, T.T., Hobbs, N.T., Wilson, K.R., Schrupp, D.L., and Baker, W.L. 1995. Changes in the landscape structure of a southeastern Wyoming riparian zone following shifts in stream dynamics. *Biological Conservation* 72, 371–379.

Mollot, L.A. and Bilby, R.E. 2008. The use of geographic information systems, remote sensing, and suitability modeling to identify conifer restoration sites with high biological potential for Anadromous fish at the Cedar River municipal watershed in Western Washington, USA. *Restoration Ecology* 16, 336–347.

Muller, E. 1995. Phénologie forestière révélée par l'analyse d'images Thematic Mapper. C. R. Acad. *Sci. Sciences de la vie* 318, 993–1003.

Muller, E. 1997. Mapping riparian vegetation along rivers: old concepts and new methods. *Aquatic Botany* 58, 411–437.

Muller, E. and James, M. 1994. Seasonal variation and stability of soil spectral patterns in a fluvial landscape. *Int. J. Remote Sensing* 15, 1885–1900.

Muller, E. and Décamps, H. 2001. Modeling soil moisture reflectance. *Remote Sensing of Environment* 76, 173–180.

Muller E., Guilloy-Froget, H., Barsoum, N., and Brocheton, L. 2002. Populus nigra L. en vallée de Garonne: legs du passé et contraintes du présent. *C. R. Biologies* 325, 1–11.

Munné, A., Prat, N., Sola, C., Bonada, N., and Rieradevell, M. 2003. A simple field method for assessing the ecological quality of riparian habitat in rivers and streams: a QBR index. *Aquatic Conservation: Marine and Freshwater Ecosystems* 13, 147–163.

Nagler, P.L., Cleverly, J., Glenn, E., Lampkin, D., Huete, A., and Wan, Z. 2005a. Predicting riparian evapotranspiration from MODIS vegetation indices and meteorological data. *Remote Sensing of Environment* 94, 17–30.

Nagler, P.L., Glenn, E.P., Hursh, K., Curtis, C., and Huete, A. 2005b. Vegetation mapping for change detection on an arid zone river. *Environ. Monit. Assess.* 109, 255–274.

Naiman, R.J., Décamps, H., and McClain, M. (2005). *Riparia, ecology, conservation, and management of streamside communities.* Academic Press, Elsevier, San Diego 430 p.

Nakamura, F., Shin, N., and Inahara, S. 2007. Shifting mosaic in maintaining diversity of floodplain tree species in the northern temperate zone of Japan. *Forest Ecology and Management* 241, 28–38.

Narumalani, S., Zhou, Y., and Jensenb, J.R. 1997. Application of remote sensing and geographic information systems to the delineation and analysis of riparian buffer zones. *Aquatic Botany* 58, 393–409.

Neale, C.M.U. 1997. Classification and mapping of riparian systems using airborne multispectral videography. *Restoration Ecology* 5, 103–112.

Otahel, J., Feranec, J., and SuriI, M. 1994. Land-cover mapping of the Morava floodplain by application of color infrared aerial photographs and GIS spans. *Ekologia-Bratislava* 13, 21–28.

Ozesmi, S.L. and Bauer, M.E. 2002. Satellite remote sensing of wetlands. *Wetlands Ecol. Manage.* 10, 381–402.

Peinetti, H.R., Kalkhan, M.A., and Coughenour, M.B. 2002. Long-term changes in willow spatial distribution on the elk winter range of Rocky Mountain National Park (USA). *Landscape Ecology* 17, 341–354.

Pengra, B.W., Johnston, C.A., and Loveland, T.R. 2007. Mapping an invasive plant, Phragmites australis, in coastal wetlands using the EO-1 Hyperion hyperspectral sensor. *Remote Sensing of Environment* 10, 74–81.

Petit, S. 2006. Reconstitution de la dynamique du paysage alluvial de trois secteurs fonctionnels de la riviére Allier (1946-2000), Massif Central, France. *Géographie Physique et Quaternaire* 60, 277–294.

Piégay, H. and Salvador, P.G. 1997. Contemporary floodplain forest evolution, along the middle Ubaye river. *Global Ecology and Biogeography Letters*, 397–406.

Rheinhardt, R., Brinson, M., Brooks, R., McKenney-Easterling, M., Masina Rubbo, J., Hite, J., and Armstrong, B. 2007. Development of a reference-based method for identifying and scoring indicators of condition for coastal plain riparian reaches. *Ecological Indicators* 7, 339–361.

Ringrose, S. 2003. Characterisation of riparian woodlands and their potential water loss in the distal Okavango Delta, Botswana. *Applied Geography* 23, 281–302.

Rowlinson, L.C., Summerton, M., and Ahmed, F. 1999. Comparison of remote sensing data sources and techniques for identifying and classifying alien invasive vegetation in riparian zones. *Water SA* 25, 497–500.

Sakamoto, T., Van Nguyen, N., Kotera, A., Ohno, H., Ishitsuka, N., and Yokozawa M. 2007. Detecting temporal changes in the extent of annual flooding within the Cambodia and the Vietnamese Mekong Delta from MODIS time-series imagery. *Remote Sensing of Environment* 109, 295–313.

Sloan, J., Miller, J.R., and Lancaster, N. 2001. Response and recovery of the Eel River, California, and its tributaries to floods in 1955, 1964, and 1997. *Geomorphology* 36, 129–154.

Stella, J.C., Hayden, M.K., Battles, J.J., Piégay, H., Dufour, S., and Fremier, A.K. 2012. The critical role of abandoned channels as refugia for sustaining pioneer riparian forest ecosystems. *Ecosystems*, 14, 776–790.

Straatsma, M.W. and Baptist, M.J. 2008. Floodplain roughness parameterization using airborne laser scanning and spectral remote sensing. *Remote Sensing of Environment* 112, 1062–1080.

Straatsma, M.W. and Middelkoop, H. 2007. Extracting structural characteristics of herbaceous floodplain vegetation for hydrodynamic modeling using airborne laser scanner data. *International Journal of Remote Sensing* 28, 2447–2467.

Tabacchi, E., Correll, D., Hauer, R., Pinay, G., Planty-Tabacchi, A.M., and Wissmar, R. 1998. Role of riparian vegetation in the landscape. *Freshwater Biology* 40, 497–516.

Tal, M., Gran, K., Murray, A.B., Paola, C., and Hicks, D.M. 2004. Riparian vegetation as a primary control on channel characteristics in multi-thread rivers. In: Bennett, S.J.,

Simon, A., (Eds.), *Riparian vegetation and fluvial geomorphology: hydraulic, hydrologic, and geotechnical interaction*, American Geophysical Union Monograph, 16 p.

Thompson, S.D. and Gergel, S.E. 2008. Conservation implications of mapping rare ecosystems using high spatial resolution imagery: recommendations for heterogeneous and fragmented landscapes. *Landscape Ecology* 23, 1023–1037.

Tiner, R.W. 2004. Remotely-sensed indicators for monitoring the general condition of 'natural habitat' in watersheds: an application for Delaware's Nanticoke River watershed. *Ecological Indicators* 4, 227–243.

Tormos, T., Kosuth, P., Durrieu, S., Villeneuve, B., and Wasson J.G. in press. Improving the quantification of land cover pressure on stream ecological status at the riparian scale using High Spatial Resolution Imagery. *Physics and Chemistry of the Earth*.

Townsend, P.A. and Walsh, S.J. 1998. Modeling floodplain inundation using an integrated GIS with radar and optical remote sensing. *Geomorphology* 21, 295–312.

Townsend, P.A. 2001. Mapping seasonal flooding in forested wetlands using multi-temporal radarsat SAR. *Photogrammetric Engineering and Remote Sensing* 67, 857–864.

Townsend, P.A. 2001. Estimating forest structure in wetlands using multitemporal SAR. *Remote Sensing of Environment* 79, 288–304.

Townsend, P.A. and Foster, J.R. 2002. A synthetic aperture radar-based model to assess historical changes in lowland floodplain hydroperiod. *International Journal of Remote Sensing* 23, 443–460.

Touzi, R. 2007. Target scattering decomposition in terms of roll-invariant target parameters. *IEEE Trans. Geoscience Rem. Sens.*, 45.

Touzi, R., Deschamps, A., and Rother, G. 2007. Wetland characterization using polarimetric Radarsat-2 capability. *Canadian Journal of Remote Sensing* 33, 56–67.

Ulaby, F.T. and Dobson, M.C. (1989). *Handbook of radar scattering statistics for terrain*. Norwood, MA: Artech House.

Underwood, E.C., Mulitsch, M.J., Greenberg, J.A., Whiting, M.L., Ustin, S.L., and Kefauver, S.C. 2006. Mapping invasive aquatic vegetation in the Sacramento-San Joaquin delta using hyperspectral imagery. *Environ. Monit. Assess.* 121, 47–64.

Van der Nat, D., Tockner, K., Edwards, P.J., Ward, J.V., and Gurnell, A.M. 2003. Habitat change in braided floodplains (Tagliamento, NE Italy). *Freshwater Biology* 48, 1799–1812.

Van Looy, K., Meire, P., and Wasson, J.G. 2008. Including riparian vegetation in the definition of morphologic reference conditions for large rivers: A case study for Europe's Western Plains. *Environmental Management* 41, 625–639.

Walsh; S., Davis, F., and Peet, R.K. (editors). (1994). *Applications of geographic information systems and remote sensing in vegetation science*. Oppulus Press, Uppsala, 147 p.

Warbington, R., Schwind, B., Brohman, R., Brewer, K., and Clerke, W. 2002. Requirements of remote sensing and geographic information systems to meet the new forest service existing vegetation classification and mapping standards USDA Forest Service. available (01 march 2009) on http://www.fs.fed.us/r5/rsl/publications/pubs.

Wiederkehr, E., Dufour, S., and Piégay H. 2010. Location and semi-automatic characterisation of potential fluvial geomorphosites. Examples of applications from geomatic tools in the Dro me River basin (France). *Géomorphologie: relief, processus, environnement* 2, 175–188.

Yang, X. 2007. Integrated use of remote sensing and geographic information systems in riparian vegetation delineation and mapping. *International Journal of Remote Sensing* 28, 353–370.

11

Biophysical Characterisation of Fluvial Corridors at Reach to Network Scales

Hervé Piégay[1], Adrien Alber[1], J. Wesley Lauer[2], Anne-Julia Rollet[3] and Elise Wiederkehr[1]

[1]University of Lyon, CNRS, France
[2]Department of Civil and Environmental Engineering, Seattle University, Seattle, WA, USA
[3]University of Caen Basse-Normandie, Géophen, CNRS, France

11.1 Introduction

Previous chapters have discussed the extraction and classification of features representing in-stream habitat, vegetation patches, or other biophysical characteristics (bar grain-size, vegetation structure, water depth, temperature, relative elevation, etc.). The question we address here is how to expand local extractions to larger scales, and in particular how to evaluate the temporal trajectory of rivers at reach to network scale. Because of the ever increasing amount of spatial data and new technologies to analyse them, reach scale analysis offers a strategic perspective in terms of knowledge production (Ferguson, 2007). For example, classification of reach-scale patterns potentially allows rapid identification of ecologically interesting fluvial corridors or reaches adversely altered by human activities (Thorp et al., 2006). It can also provide basic physical data for better estimating the time necessary for a river reach to adjust to human pressure. This is a key issue of a geomorphic audit (Kondolf et al., 2003). Such analyses, often referred to as regionalisation studies (Bryce and Clark, 1996; Higgins et al., 2005), have gained increasing importance recently, particularly in Europe where the European Water Framework Directive implementation requires managers to plan restoration efforts at large hydrographic network scales even when only local scale processes are well characterised.

From a management perspective, there is a need for expanding the extent of spatial analysis from that of local sites of a few metres to a few hundred metres to that of river reaches of several km (e.g. the main stem of a large river) or of an entire stream network of a watershed. Network-scale analysis requires an extension of the reach-scale perspective, opening new methodological issues in terms of data availability, statistical analysis, and scale effects. It is now common to characterise fluvial corridor features, i.e. the channel and its natural margins, at the network scale. For example, the concept of a cascading sedimentary system is becoming a central tool in watershed management and explicitly considers sources, transfers and depositions of sediment in an integrated framework (Sear et al., 1995; Brierley et al., 2006). In addition to characterising natural processes, such approaches inform decisions that balance the interests of a range of stakeholders who may focus on natural resources management, ecosystem value, risk assessment and/or mitigation, and

Fluvial Remote Sensing for Science and Management, First Edition. Edited by Patrice E. Carbonneau and Hervé Piégay.
© 2012 John Wiley & Sons, Ltd. Published 2012 by John Wiley & Sons, Ltd.

the evaluation of impacts of human uses. Characterising biophysical features continuously along fluvial corridors is useful for identifying thresholds or discontinuities in longitudinal trends at a regional scale, and therefore targeting and planning management actions. Similarly, when continuous information is not available, a regional sample of geomorphic features, such as river reaches or meander bends, can also provide valuable information in terms of geographical variability. Physical processes are complex, and a better understanding of different existing geographical conditions provided by the observation of a set of contrasting cases is often useful for detecting process-domains and developing testable hypotheses (e.g. using flume experiments or numerical simulations) focused on understanding physics and predicting changes in specific geographical settings.

In this chapter, we aim to summarise how the use of images can be expanded to relatively large scales of interest to characterise fluvial corridors, highlight geographic patterns and develop an understanding of the spatial and temporal complexity of biophysical processes. We first detail the raw and sometimes quite heterogenous remotely-sensed data available for conducting regional-scale analysis. Then we focus on existing techniques for analysing longitudinal structures at these scales. Finally, while much information is available in images, there are some limitations when using them to cover long, continuous corridors. We address questions associated with locating and characterising images, dealing with the resolution in relation to the feature size, and addressing the uneven quality of image mosaics covering large spatial extents.

11.2 What are the raw data available for a biophysical characterisation of fluvial corridors?

The use of air photos for interpreting and characterising biophysical processes and features is based on a rich inheritance. Many of the methods introduced here are simply more automated forms of recognition of spatial objects that some researchers have long applied on shorter reaches. But capabilities are now transformed and tremendously increased by the technological changes and the availability of data.

For the past several years, online images have been available from websites such as Google Earth, NASA World Wind, Bhuvan (Indian earth observation visualisation) and Geoportail in France. Google Earth is presently the most powerful tool for visualising river systems because of its high resolution (from 15 to 1 meter) and extensive geographical coverage. Moreover, this platform also provides historical images. Ten to 15 dates are often available on river reaches in the USA, much less elsewhere, with urban areas having the best coverage. Unfortunately, even though such visualisation platforms integrate external information which can be superimposed on the images, the remotely sensed exploitation of the images is still at an early stage and it is surely here that future innovation can be expected in terms of both image management and geoprocessing tool development.

Researchers also have opportunities to access national imagery resources with better resolution, more homogeneous spatial coverage and better documentation of remote sensing procedures. Systematic acquisition at a national level began in earnest after the Second World War, although archived imagery for certain regions is available as far back as the 1920s. In the United States, historical imagery is archived at a national level by the Geological Survey (USGS) at the Earth Resources Observation and Science (EROS) Center (http://eros.usgs.gov). Coverage of the entire United States (excluding portions of Alaska) is available from the 1980s, with single aerial images and local mosaics available from about 1940. Notable data sources available for download at the EROS site include those produced at 1:40,000 scale as part of the National Aerial Photography Program between 1987 and 2004, the lower-resolution National High Altitude Photography program between 1980 and 1989, the USGS Digital Orthophoto Quadrangle mosaic which covers the entire United States at approximately 1-m resolution using images dating from the late 1980s through the 2000s, and sub-meter resolution Orthoimagery from the early 2000s for major US metropolitan areas. While many of these data are available from other sites, the EROS Earth Explorer webserver (http://edcsns17.cr.usgs.gov/EarthExplorer/) is helpful because it provides easy access to the photographic dates available from the EROS archive for a given locality. Similar data are available for Canada for a relatively nominal charge from the Natural Resources Canada National Air Photo Library website (http://airphotos.nrcan.gc.ca/). Another important source of more recent US imagery is the National Agricultural Imagery Program (NAIP), which provides 1-m or 2-m ground resolution orthorectified imagery for variable portions of the Lower United States each year starting in 2003. The imagery is generally flown during the summer growing season. NAIP photographs are available from the United States Department of Agriculture (USDA) Farm Service Agency (FSA) Aerial Photography Field Office (APFO) at www.apfo.usda.gov. Many historical images within the United States and Canada are archived at the state and provincial level, often at the map libraries of major public universities. These are often stored as hard-copy images and must be scanned and georeferenced before being useful for digital analysis.

Satellite imagery is available at a global level starting in 1972 with the launch of the LandSat mission. Individual LandSat scenes as well as mosaics from the mid 1970s, are available at the EROS Earth Explorer site. Ground resolution ranges from 57 m for earlier Multispectral Scanner (MSS) images to 14.25 m for Pan Sharpened ETM+ images. Most of the other satellite sources, notably the ones providing sub-metric resolution such as Ikonos, Quickbird or Worldview-1 are commercial and do not provide easy on line data access, and the archives, even if they exist, are often not extensive enough to cover entire regions (although a notable exception are SPOT satellite mosaics for Canada available at http://www.geobase.ca). In all cases, the horizontal precision of imagery should be carefully evaluated before making observations of relative geomorphic change.

Imagery can be combined with other data such as vectorial layers corresponding to land-use types (Corine Land-cover for example), hydrographic networks, road layers, and DEM from low (25 m for the French BD topo, 90 m for the SRTM available worldwide, 15–30 m for much of the U.S. National Elevation Dataset DEM, etc.) to high resolution (LiDAR). A national archive of publicly available LiDAR data in the United States is maintained by the USGS at lidar.cr.usgs.gov, but the archive does not necessarily provide access to datasets stored at the state and local level. Map libraries also often provide access to high resolution LiDAR DEM's, which are often distributed through state or local agencies. Historical maps, geological maps or discharge data may also be available from the Web (USGS NWIS; Other online data maintained by regional and local governments).

11.3 How can we treat the information?

11.3.1 What can we see?

From archived images, it is possible to perform a characterisation of the fluvial corridor based on the detection of

polygons having a biophysical meaning (water channel, islands, gravel/sand bars, active or abandoned channels) and the analysis of their geometric properties (area, length of edges, width, etc.) (see review of Gilvear and Bryant, 2003). Images are often not the only source of information and usually complement data provided from vectorial layers or from ground measures. Both are easily integrated using a GIS. Such external data (e.g. which are not provided by images) can also be introduced to validate or predict information from characteristics of observed features.

Planform geometry of meandering rivers has been covered by many authors since the 1970s, both at the bend and the reach scales (amplitude/wavelength of meander, radius of curvature, sinuosity, etc.) (Lewin, 1977, 1987, Hooke, 1984; Hooke and Redmond, 1989). Multivariate characterisations of meandering rivers have been explored by Howard and Hemberger (1991). Multi-threaded channels are also well studied (e.g., braided index) (Hong and Davies, 1979; Egozi and Ashmore, 2008). Common indices characterise the extent of braiding and/or anastamosing, the relative amount of edges, islands, gravel bars, etc. In specific contexts (e.g., very high resolution imagery, transparent flows free of surface distortion), pools and riffles, or, more generally, in-stream habitat can also be delineated (Figure 11.1).

When several image series are available, it is possible to observe changes in channel planimetry or channel pattern on annual to decadal timescales (Odgaard, 1987; Piégay et al., 1997; Gurnell, 1997; Gilvear et al., 2000; Winterbottom, 2000; Burge and Lapointe, 2005; Zanoni et al., 2008; etc.). It is also possible to predict discharge from water flow area, notably when water surface width is particularly sensitive to discharge change, which is the case of wandering or braided rivers.

Several methods have been developed to quantify lateral adjustments, based on linear rate of lateral shifting, rate of reworked areas, or elongation rate of channel centerlines (Aalto et al., 2008). Maps are used for getting a longer time period but aerial photo series are usually the main source of information when available. From the example of the Rillito Creek, Arizona, Graf (1984) provided an original contribution to the quantification of planform changes through the mapping of an erosion probability as a function of the lateral and upstream distance of a given location to the nearest active channel assuming continuation in the mobility trend using maps from 1871 to 1978. His method has been applied by Wasklewicz et al. (2004) on the lower Mississippi with a slight modification, assigning to each map a proportional weight according

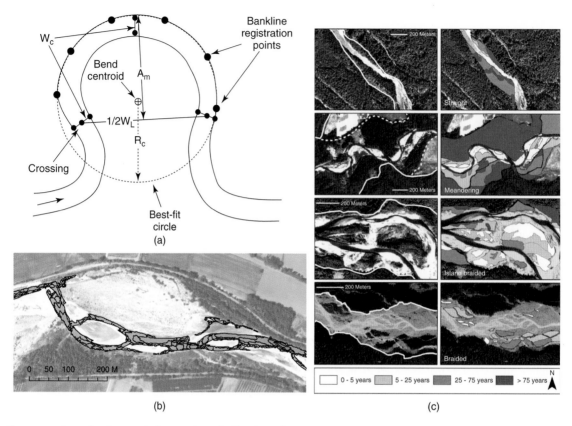

Figure 11.1 Example of geometrical properties and of biophysical polygons which can be extracted from aerial photos for a large scale geographical characterisation : a) Planimetric properties of a meander bend. Reproduced from Heo et al. (2009) Characterization & prediction of meandering channel migration in the GIS environment: A case study of the Sabine River in the USA. Environmental Monitoring and Assessment; 152(1):155, with permission from Springer-Verlag, b) Delineation of in-channel habitats (pools, riffles, lentic/lotic channels, bar bench). Reprinted from Geomorphology; 78(1–2), Beechie, T.J. et al. Channel pattern and river-floodplain dynamics in forested mountain river systems, p. 124. Copyright 2006, with permission from Elsevier, c) Geomorphic pattern types and age classes of floodplain patches (Beechie et al., 2006).

to the time of the map relative to the entire historic record (Figure 11.2a and b). The use of historical series of photos for overlaying channel positions and determining an erodible corridor and erosion sensitivity has then been discussed by Piégay et al. (2005).

Floodplain features can be detected from aerial photographs providing evidence of lateral shifting (Beechie et al., 2006), and opening new issues in terms of floodplain geomorphic classification (e.g. Nanson and Croke, 1992). Floodplain boundaries can also be detected when the analysis is augmented by a DEM (Belmont, 2011). Combined with topographic or bathymetric data, planimetric information can be extended to 3D mapping of in-channel features. Figure 11.2c shows a linear relationship between the measured areas of gravel

bars from maps and aerial photos mostly obtained at low flow and their estimated volume, illustrating that planimetric information can be a good proxy for estimating sediment accumulation (Church and Rice, 2009). Volume is calculated from a map of bed elevation changes developed from overlays of bathymetric surveys available for different dates. Sediment budgets are also increasingly established to quantify fluvial changes through estimation of volume exchanges of sediment per unit of time, as exemplified by McLean and Church (1999), Gaeuman et al. (2003), Rollet (2007), or Lauer and Parker (2008) at the reach-scale. Lane et al. (2010) also provided an analysis of changes in braided channel planform based on a DEM extracted from digital photogrammetry performed on archived images.

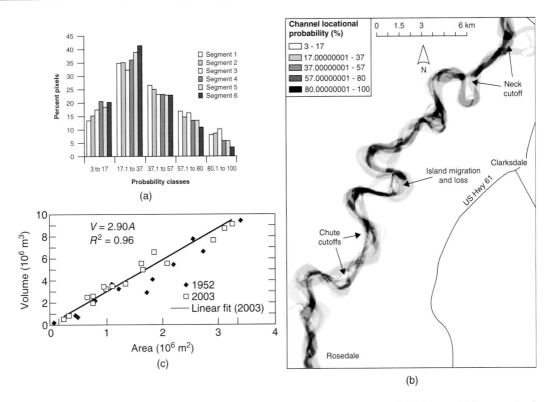

Figure 11.2 Example of analysis using different channel historical channel records to evaluate floodplain sensitivity to erosion (a and b) and estimating sediment volume (c). a) displays the variation in percent pixels representing a percentage of the total number of pixels along the Lower Mississippi River arranged into erosion probability classes and channel segments along the study reach. b) is an example of the maps showing some of the locations of higher channel stability. Reprinted from Geomorphology, 63(3-4), Wasklewicz, T.A. et al., Geomorphic context of channel locational probabilities along the Lower Mississippi River, USA, pp. 145–158. Copyright 2004, with permission from Elsevier, c) is an area-volume relation for bar platforms of gravel-bed reach of lower Fraser River, British Columbia, Canada calculated from historical records of bathymetric surveys and aerial photos. Reproduced from Church, M. & Rice, S.P. (2009) Earth Surface Processes and Landforms, pp. 1422–1432, with permission from John Wiley & Sons, Inc.

Remotely sensed imagery is also useful for estimating river discharge. Bjerklie et al. (2005) developed a relationship between channel discharge and average water surface width for in-bank flow conditions. The relationship is intended to apply to a range of river systems and requires an estimate of channel slope and average channel bankfull width, with width estimated by dividing total channel area by channel length for reaches at least one meander bend long. Site-specific calibration significantly improved the accuracy of the method, as did the incorporation of other variables such as meander wavelength (Bjerklie, 2007). Even better results are possible if local rating curves can be developed between discharge and reach-average water surface width. Once the rating curve is developed, it represents a functional relationship between water surface width and discharge. Since the water surface width at the

bankfull condition is often straightforward to identify on aerial imagery even when the water level is far below bankfull, such rating curves can be used to estimate the bankfull discharge in a reliable and repeatable way that does not depend on the identification of field indicators of bankfull geometry.

11.3.2 Strategy for exploring spatial information for understanding river form and processes

Typology and modelling are the two main issues justifying the spatial and temporal analysis of geographical features, notably for understanding and quantifying the scale of channel dynamics. The aim is to identify key spatial differences in feature types within the river corridor and, once

Table 11.1 A few examples of studies based on archives aerial and satellites photos.

Area Concerned	Time Series	Résolution	Data	Parameters Surveyed	References
Inter-site comparisons					
50 braided reaches	Present	50 cm	National Orthophotos Archives	Width, slope	Piégay et al., 2009
Pacific northwest region/42 reaches	Sequential photos 1939 to 2000	1:12,000 scale photos	National Orthophotos Archives	Channel shifting and pattern	Beechie et al., 2006
30 meanders/119 oxbows	Present	Variable	Google Imagery	Width, sinuosity, length	Constantine and Dunne, 2008
10 to 20 meanders belt of several 10th km	Present	Variable	Google Imagery	Meander geometry	Zhang et al., 2008
Study at reach scale					
Ain River (France)/40 km reach	1945 and 1991	50 cm	National Orthophotos Archives combined with LIDAR and field surveys	Erosion, floodplain vegetation	Marston et al. 1995; Rollet 2007
Rhône River (France)/500 km reach	83 dates : 1999 to 2009	60 m	Landsat	Surficial temperature	Wawrziniak et al. 2011
Tromie River (Scotland)/16 km reach	Fall 2008	3 cm	Airbone special survey aerial photographies (true color) combined with field surveys	Width, depth, grain size	Carbonneau et al., 2012
Drôme River (France)/106 km	Present	50 cm	National Orthophotos Archives	Width, sinuosity, length	Wiederkehr et al., 2010
Rhin River (France)/50 km reach	8 dates : 1949 to 2008	50 cm and 20 cm (recent orthophotos)	Airbone special survey : aerial photographies (color and black and white)	Morphological evolution	Arnaud et al., 2010
Brahmaputra River (Bangladesh)/ 200 km reach	14 dates : 1967 to 2002	10 m (1967); 80 m (1973 to 1987); 30 m (1989 to 2002)	Panchromatic photos images acquired by the American Corona satellite (1967); Landsat MSS (1973 to 1987); Landsat TM (1989 to 2002)	Width of the braided belt; number and width of channels, distribution of land cover attributes and land stability	Takagi et al., 2007

(continued overleaf)

Table 11.1 *(continued).*

Area Concerned	Time Series	Résolution	Data	Parameters Surveyed	References
Strickland River floodplains (Papua New Guinea)/330 km	1972, 1990/1993 and 2000/2001	14 to 57 m	Landsat	Channel shifting and width	Aalto et al., 2008
Sainte Marguerite River (Canada)/80 km reach	August 2002	3 cm and 10 cm	Airborne special survey	Grain size	Carbonneau et al., 2012
Brazos River (Texas, USA) and Lamar River (Wyoming, USA)/1 and 2 km reaches	August 1999 (Lamar River) and May 2002 (Brazos River)	1 m	Airborne special survey. Brazos River : (red, green, blue) digital aerial photographs and Lamar River : 128 band hyperspectral imagery	Depth (bathymetry models)	Fonstad and Marcus, 2005
Tay River and Tummel River (UK)/12.5 km reach	1971 to 1994	1:5,000; 1:7,500; 1:12,000; 1:24,000	Aerial photographies (color, false colour infrared and black and white)	Width, braided index and sinuosity	Winterbottom, 2000
Study at large scale					
Europe	Present	25 m	2 mosaics combined : Landsat ETM+ and Spot LISS	Riparian zone	Clerici et al., 2011
Hérault Bassin (France)/1,150 km	Present	50 cm et 10 m	National Orthophotos Archives and Spot 5 XS	Land use in riparian corridors	Tormos, 2010
Mekong River (China)/2,500 km	1998 to 2004	1 to 10 m	Ikonos and Spot 5	Channel forms	Gupta et Liew, 2007

identified, to model relationships between these types and the driving-processes. A range of strategies can be applied to extract information. As shown by Piégay and Schumm (2003), geographic patterns of hydrographic networks can be used to both highlight similarities between features at different scales (inter-object comparisons within a catchment or between catchments) and, by considering longitudinal trends, to provide information on sources and on local controls on sediment transfer (Table 11.1).

In the first case, it is possible to sample features and work on inter-object comparisons, sampling being driven by the working hypothesis. The comparison of geographic objects is not a new perspective, getting its roots in the hydraulic geometry approach and allogenic modelling (Church and Mark, 1980). In this area, research has mainly focused on channel planforms and notably on meandering geometry and cut-offs in a synchronic approach or in a diachronic way looking at channel shifting.

In the second case, the approach can be more inductive. One of the main benefits of images compared to field measures is the possibility to exhaustively analyse the longitudinal structure of corridors all along a channel or within a network (e.g., Piégay et al., 2000; Petts et al., 2000; Alber and Piégay, 2011), to locate specific features or thresholds between them, to define types of biophysical features, such as geomorphic reaches (Rollet, 2007), and to highlight regional organisation.

In terms of modelling, the aim is to provide regional models mainly linking width or other planimetric factors with catchment size and combined planimetric features to provide hydraulic meaning and process-based understanding. An example is provided by Lagasse et al. (2004) who explored prediction of meander migration rates based on GIS analysis of a vast database of single bends. Another is provided by Constantine and Dunne (2008) who worked with images of Google Earth to measure channel and oxbow-lake characteristics of 30 world rivers to identify the factors controlling oxbow lake development (Figure 11.3). They predicted the size-frequency distributions of lakes of five world river reaches using only channel sinuosity and the geometric mean oxbow lake length variance. The temporal rate of cut-off can be estimated using channel sinuosity, the fraction by which cut-off reduces channel length, and the rate at which the reach lengthens by meander growth.

When working in an inter-site comparative approach, scaling and size effects become a critical issue. It therefore appears that many parameters are scale-dependant, as recently stated by Dodov and Foufoula-Georgiou (2004). These authors notably showed law dependencies between channel morphology (sinuosity, meander wavelength, and radius of curvature) and consequent variations in channel cross-sectional shape with scale (e.g., mean velocity and discharge under different flow conditions) for at-station hydraulic geometry (HG) using between 70 and 100 stations in Nebraska, Kansas, Missouri and Oklahoma.

Following the increasing availability of aerial photo information, development of GIS tools to analyse geographical objects and their attributes is emerging as shown by Güneralp and Rhoads (2008) or Heo et al. (2009) to continuous characterisation of single-thread planform geometry or to continuous sampling of widths (Lauer and Parker, 2008). It is possible to draw or detect polygons and describe them by their radiometric values or by their geometry and size. The object can be basically characterised by a manual measure such as length (channel lengths mainly rated by different other lengths, braided index, sinuosity rate) or by advanced remote sensing

procedures. Generic geographical objects can be extracted from real ones having a biophysical meaning based on GIS rules (e.g., a centreline, an elementary polygon of constant length), mainly created as intermediate objects for geomatic treatments. Longitudinal polygons or linear features, such as the valley corridor or the active channel area can be also represented synthetically by a single line. These generic objects are used to provide spatial reference to create other metrics: centrelines (e.g., valley centreline, channel centreline, meander belt centreline, etc.), buffers of a given width on both sides of polygons or centrelines, or elementary segments with a systematic length that is proportional to the catchment size (see details in Alber and Piégay, 2011).

11.3.3 Example of longitudinal generic parameters treatment using unorthorectified photos

The use of historical records usually requires the rectification of images, a long and exhausting process that limits such effort both in time (to a limited number of dates) and space (to a limited reach). As a consequence, several approaches have focused on extracting information directly from the images without georeferencing them with great precision. It can be worthwhile when studying channel width, depth, sinuosity or planform characteristics). Lateral channel activity plays a prominent role in many of the classification systems described above. Most methods for quantifying lateral channel activity are sensitive to spatial error in the remotely sensed imagery because rectification error will lead to apparent shift even when the channel is laterally stable. These errors can be addressed to some extent by considering the centreline elongation that generally occurs as rivers migrate. On single-thread meandering channels, lateral migration causes the streamwise coordinate s (measured relative to a fixed starting coordinate) to elongate according to the equation (Seminara et al., 2001):

$$\frac{ds}{dt} = -\int_0^s CVds' \qquad (11.1)$$

where C is the local centreline curvature $d\theta/ds$, θ represents the angle relative to an arbitrary directional datum of a line tangent to the centreline, and V is the local lateral shifting rate orthogonal to the centreline (Figure 11.4A). Note that V is negative when the channel is shifting to the right with respect to the positive streamwise direction.

Aalto et al. (2008) showed that ds/dt was easily measured from two aerial photographs by comparing $s(t)$ with $s(t+\Delta t)$ at corresponding points along a river centreline,

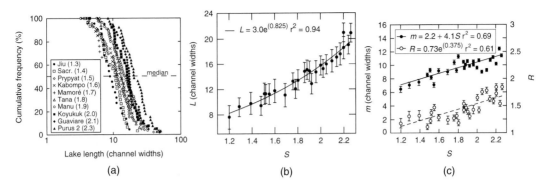

Figure 11.3 Example of illustrations provided by Constantine and Dunne (2008) within their study of 30 meandering reaches worldwide from Google Earth images : in (a) the cumulative size-frequency distributions of oxbow lake length, b) The geometric mean lake length (L) and c) The geometric mean meander length (m) and the ratio (R) of the geometric mean lake length to m both b) and c) versus the reach sinuosity (S). Reproduced from Constantine, J.A. & Dunne, T. (2008) Meander cutoff and the controls on the production of oxbow lakes. Geology; 36(1):23–26, with permission from the Geological Society of America.

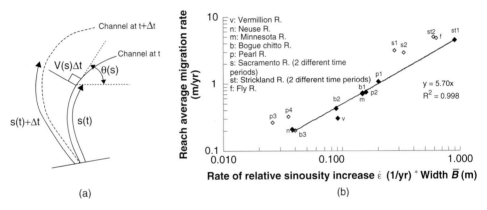

Figure 11.4 a) Measured variables, b) Relationship between reach-average lateral migration and relative sinuosity increase. The regression is for the best datasets only (represented here using filled symbols).

where corresponding points on the older and newer centerline were identified using short Bezier curves oriented normal to both axes (note that similar rates ds/dt could also be determined simply by finding the nearest point along the centerline arc at time $t+\Delta t$ to a given point at coordinate s and time t). The cumulative rate of change in channel coordinate, ds/dt, is then found by simple difference:

$$\frac{ds}{dt} \cong \frac{s(t+\Delta t) - s(t)}{\Delta t}. \qquad (11.2)$$

Note that ds/dt can also be approximated by accumulating a set of local rates if multiple aerial images with different dates are present for different portions of the reach.

However, the translation from an elongation rate or, nearly equivalently, a sinuosity change rate to a reach-average migration rate depends on the relative amount of bend elongation vs. lateral translation in a given reach. Dividing the reach elongation rate by reach length gives a rate of relative sinuosity increase $\dot{\varepsilon}$ for the reach. The variable $\dot{\varepsilon}$ is used rather than the actual rate of sinuosity increase (reach elongation rate/valley distance) because $\dot{\varepsilon}$ can be determined directly from channel centerlines, without developing a separate centerline for the valley. Developing a dimensionally homogenous relationship between $\overline{|V|}$ and $\dot{\varepsilon}$ requires a second length scale. We assume that reach-average channel width $\overline{B}$ is appropriate for this purpose and combine the variables into a single dimensionless relative sinuosity change rate as follows:

$$E^* = \frac{\overline{|V|}}{\overline{B}} \bigg/ \dot{\varepsilon}. \qquad (11.3)$$

In Figure 11.4b, cumulative up-channel elongation rates are clearly correlated with lateral migration and are thus a useful measure of lateral channel activity. Because elongation rate estimates are less sensitive to photograph alignment error than are simple lateral shifting measurements, they can be used to estimate reach-average lateral channel activity even when image rectification quality is poor. In Figure 11.4b, the reaches with data of lowest quality are plotted using open diamonds, while the other reaches used in the regression are plotted using filled diamonds. The slope of the regression line fit to the high-quality data only gives $E^* = 5.70 \pm 0.19$ at the 95% confidence level. Reaches that plot above the line are likely either migrating slowly enough that photograph alignment error causes an upward bias in lateral activity, or contain an anomalously large number of downstream translating bends.

In Figure 11.5 we present elongation rates and lateral migration rates computed by simple point-by-point lateral distance measurements for four separate sets of photograph pairs along two different river systems. All of the reaches illustrated in Figure 11.5 show some combination of elongation and translation, so their reach-average elongation rates are probably representative for a broad set of single-thread meandering river floodplains. For reaches not experiencing channel cutoff, elongation clearly depends both on the reach average absolute channel migration rate $\overline{|V|}$ and on total reach length (a long channel will elongate more than a short one, everything else being equal). As a method of characterising morphodynamic activity, they offer an added advantage over simple lateral shifting measurements in that cutoffs are plainly visible in the elongation data as abrupt shortening that occurs at a specific channel coordinate (e.g., at up channel coordinate 10 km on the Sacramento River dataset, in Figure 11.5b). Bends that migrate primarily in the downstream direction without changing form are also usually visible in the elongation-rate data. These bends are usually characterised by a rapid shortening followed by a relatively rapid elongation with little net change in channel length outside the bend (e.g., at up channel coordinate 137.7 on the older Sacramento River dataset, Figure 11.5a).

11.3.4 The aggregation/disaggregation procedure applied at a regional network scale

The process of disaggregation/aggregation described by Alber and Piégay (2011) is a way of considering a river continuum from a multi-scaled perspective, identifying meaningful geographical objects and characterising them based on photographs available at a regional scale. While the methods were originally developed for orthorectified images, many of the techniques described here, particularly those applicable to a single image set, could in principle be applied to poorly rectified images. Additional value is of course obtained when image rectification is sufficiently accurate to allow for temporal change to be detected.

In this approach, Unitary Geographical Objects (UGO) are defined as linear and polygonal spatial units that locate and delineate in space biophysical components of fluvial systems. Their degree of schematisation depends upon the spatial resolution of the raw data and of the method used for extraction. UGO are systematically disaggregated into elementary spatial units, so-called Disaggregated Geographical Objects (DGO), to provide linear referencing systems for attributes along stream networks that allow for characterisation of UGO continuously from upstream to downstream (Figure 11.6). DGO usually correspond to resolution-driven spatial units with a uniform space step. No clear rules dealing with the disaggregation processes have been established. It has been performed by many authors pragmatically considering segment lengths varying between 10 m and 1 km long depending on the channel size and the reach length. In any case, the spatial resolution should be higher than the characteristic size of the object of interest. Radiometric or planimetric information can be extracted at the scale of each DGO, and then statistical tests can be performed on the resulting longitudinal series in order to delineate biophysically meaningful spatial units, the so-called Aggregated Geographical Object (AGO) (Figure 11.6). AGOs result from spatial aggregation of DGOs that have been longitudinally differentiated into homogeneous groups. Spatial aggregation is based on statistical analysis of distributions generated by one or several attributes from the linear referenced DGOs. Several statistical techniques can be used depending on the number of attributes and the characteristics of their distribution (e.g., Heterogeneity tests such as Pettitt or Hubert methods, Contrast Enhancing, Spatial constraint clustering, or Hidden Markov Model ...) (see Leviandier et al., 2011). In this way, identification of AGO is objective and follows a logic of self-emergence (Parsons et al., 2004). AGO boundaries thus are likely to represent process boundaries as long as operators have selected classification criteria that are morphologically and/or functionally meaningful.

Such strategies offer new perspectives for characterising fluvial systems due to the generalisation of methods to large scale, such as planform measurement. The procedure for detecting change-points in the longitudinal

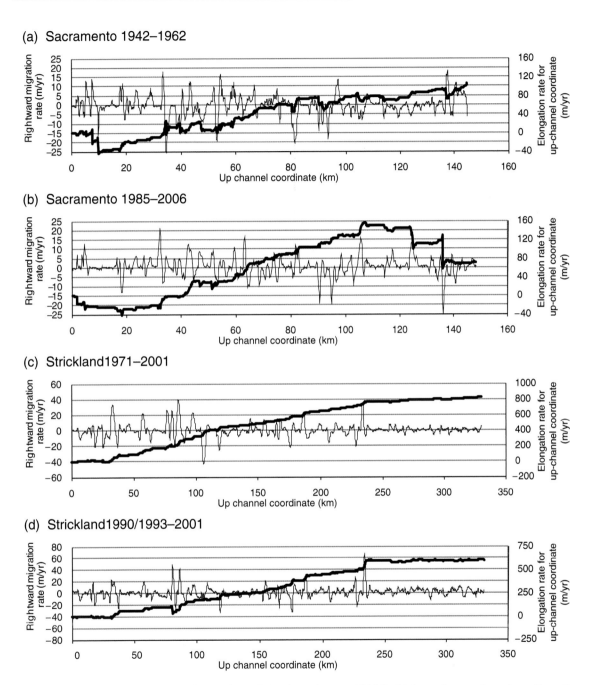

Figure 11.5 Lateral migration rates (thin line) and cumulative elongation rates (thick line) below a given up-channel coordinate. For the Sacramento (a and b), lateral migration rates were computed from air photo pairs according to the methods presented in Lauer and Parker (2008) and Aalto et al. (2008). For the two Strickland River datasets (c and d), data are identical to those presented in Aalto et al. (2008).

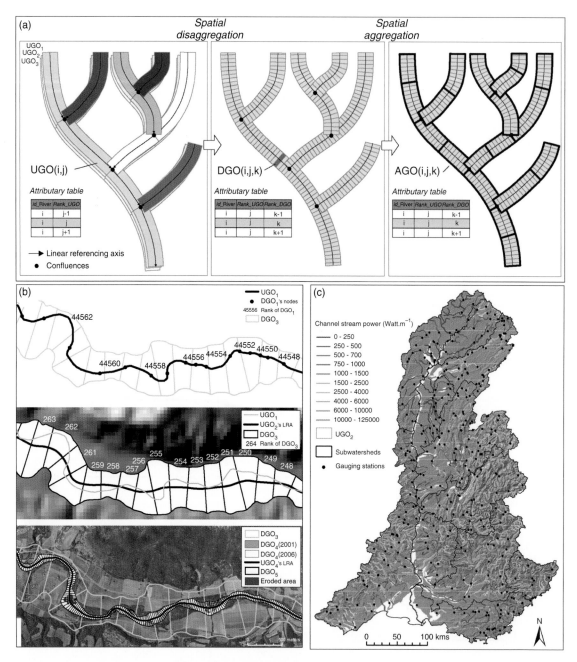

Figure 11.6 Examples illustrating the process of disaggregation and aggregation: a) Theoretical framework with associated tables, b) Geographical objects resulting from the disaggregation procedure established on the Drôme river, France: Elementary channel length shared at each inflexion point (DGO1), elementary floodplain segments (DGO3), elementary channel segments (DGO4) for 2001 and 2006 years, showing also eroded areas during the studied period, c) mapping of AGO channel stream power for biennal discharge at the Rhône network scale (40 000 km). Modified from Alber, A. and Piégay, H. Spatial disaggregation and aggregation procedures for characterizing fluvial features at the network-scale: Application to the Rhône basin (France). Geomorphology, 125(3), 343–360. Copyright 2011, with permission from Elsevier.

series leads to discussion of the hierarchy of geomorphic features. Segmented patterns are common when looking at channel planforms, which, for example, can change abruptly from a large braided pattern to an embanked channel. Similarly, in the case of a valley, the cross section can change from a V-shaped valley to an alluvial one. Such physical changes usually correspond with changes in AGOs, making it possible to use the aggregation/disaggregation process to detect forms at the appropriate spatial scale for measuring meaningful attributes by aggregating DGO-scale attributes (e.g. descriptive statistics,) or by combining other raw data or vectorial data.

The aggregation/disaggregation procedure is therefore a practical way to develop systematic characterisations of fluvial systems and, because other non-image based data can readily be disaggregated, to combine imagery with other sources of data. For example, data can be extracted in a systematic manner from longitudinal profiles, LiDAR or DEMs to providing attributes such as elevation, slope, drainage network and vertical channel changes (Reinfelds et al., 2004; Jain et al., 2006). In addition, data from oblique aerial photos, field campaigns (e.g. grain size distribution, hydraulic parameters) or other GIS resources (floodplain or land-use polygons, stream power/flow records using regionalized hydrological models) (Williams et al., 2000; Dawson et al., 2002; Nardi et al., 2004; Hall et al., 2007) can be incorporated into the analysis.

Nevertheless, the series are not always characterised by distinct segments, and other longitudinal structures can also emerge. Rather than showing distinct homogeneous reaches of different length separated by clear or transitional contacts, the longitudinal series can provide a longitudinal trend (constant change in channel width downstream) or a periodic organisation, well illustrated by the sequence of particular meso-habitats, such as pools and riffles. When looking at the longitudinal evolution of the flow channel width or of the channel depth, a periodic structure can be also observed as a proxy of the longitudinal changes in the geometry of meso-habitats. Other statistical procedures than the ones used to highlight homogeneous reaches, such as the Fourier transform, the spectral analysis, the spatial autocorrelation, or the wavelet analysis, can therefore be used to highlight these patterns. Moreover, the aggregation/disaggregation procedure applied at a network scale highlights a scaling issue. A scale dependency is then evident when looking at slope, discharge, active channel width and grain size as all are interdependent, with catchment area, discharge, width generally increasing when grain size and slope are decreasing. However, there are usually exceptions to the general scale dependencies, particularly for slope and grain size. Such exceptions represent interesting points in the network. Technical questions are then raised regarding the removal the size effects and exploration of the properties of reaches. Finally, the aggregation/disaggregation procedure opens new scientific questions on the nested and organised nature of biophysical features along a fluvial corridor, allowing re-exploration of the theoretical hierarchical framework of Frissell et al. (1986). We still do not know if the entities observed at a given scale level are nested in the ones observed at a higher level or if the entities observed at different levels are in fact independent.

11.4 Detailed examples to illustrate management issues

In this section, we present three contrasting case-studies in order to highlight how imagery, combined with additional information, can be exploited for providing insights for managers. First, we introduce a reach-scale retrospective approach that combines multiple sources of information on the Ain River collected in cooperation with local river managers and funded by a European Life project. The next two examples focus on network-scale approaches applied to a single image set at time t, one based on continuous data and the other on a discrete sample of reaches within a larger network. Both studies were supported by the Water Agency 'Rhône Méditerranée' responsible for the implementation of the Water Framework Directive in the Rhône district that requires regional data and planning tools for targeting conservation, restoration or mitigation measures. All these works utilised *Institut Géographique National (IGN)* aerial photograph resources. IGN covers the national French territory, with pioneer photos in the 1930s and more systematic surveys after 1945. Other national sources of aerial photos, such as those of the *Inventaire Forestier National* (National Forestry Inventory) are also available. All the territory is covered with orthorectified digital photos available as Orthophotography Database with dates starting in 1998. Older ones are archived on paper and are not yet available in a digital form. Scale is usually 1/20,000 with a spatial resolution of 50 cm.

11.4.1 Retrospective approach on the Ain River: understanding channel changes and providing a sediment budget

The lower valley of the Ain River (40 km long), which is a tributary of the Rhône River located 50 km north-east of Lyon, France, underwent important adjustments in terms of morphology and channel dynamics over the last century (see Marston et al., 1995). A diachronic study based on the analysis of historical aerial photographs (eight series for the study period ranging from 1945 to 2000) was performed to understand how dams continue to impact the current channel adjustment pattern and sediment dynamics, and then to assess a long term management plan for the river. The main hypothesis was that a dam built 10 km upstream from the alluvial reach in 1960 is interrupting bedload transport and is inducing a winnowing process downstream.

We identified by photo-interpretation on each of the aerial photograph series the wetted channel and the unvegetated bars and integrated the layers in vector format in a GIS environment. Then we disaggregated these layers into DGO of 10 m long to characterise the wetted channel and the bars from upstream to downstream of the study area. The longitudinal pattern of the cumulated bar surface revealed (Figure 11.7 a): i) a narrowing and bar surface decrease that propagated through time from upstream to downstream. A detailed DGO has been created to better distinguish the natural bar migration from the bar reduction due to progressive sediment winnowing. We did not observe bar migration but rather gravel bar disappearance and local retention of the sediment by channel-spanning infrastructure (weirs, bridges …). A propagation rate for the sediment deficit downstream from the dam was calculated by dividing the distance between the dam and the front of the sediment deficit identified by the progressive bar disappearance downstream, and the age of the dam. We obtained a minimum estimate of 500 m a year, which is coherent with other estimations of average annual bedload transport distance in the region (Liébault et al., 2005). This observation was then reinforced by conclusions provided from field data. We photographed the coarsest particle patch of each of the 109 bars and measured the median grain size from imagery based on an automated procedure (see technique explained in Chapter 15). Figure 11.7b shows a significant change in the median grain size at km 25 that we interpret as the front of a region of sediment deficit previously observed from photo analysis. These findings are critical in terms of river management. They provide

additional evidence of a sediment deficit progressing from upstream to downstream with social and ecological consequences, and allow for the development of a management strategy promoting, (i) the restoration of the reach still affected by the dam impact, and (ii) the preservation of the most interesting reaches downstream (e.g. Martinaz), not yet impacted, by acting on the critical process, the bedload transport. Bedload reintroduction was therefore promoted upstream from the section (Rollet, 2007). The option previously proposed by managers, an increase of morphological discharge frequency (e.g. frequent floods), was abandoned because of the risk to accelerate the propagation of sediment downstream without mitigating the gravel deficit.

Following these previous conclusions, and because the sediment deficit also impacts channel shifting with cascading ecological and social consequences downstream (see Piégay et al., 2005), new questions emerge about the sensitivity to change of the active shifting corridor due to the forthcoming sediment deficit. A major question relates to the capability of the river to recharge itself through bank erosion processes, so that it may slow the propagation of the bedload deficit. In such a context, the understanding of the sediment exchange between the active channel and the floodplain in the reach that is still not affected by the dam becomes an issue. We addressed this question by combining information derived from aerial photographs with an additional data source, the topography of the study area that results from a LiDAR survey achieved in 2008 complemented with bathymetry developed from a series of cross-section measurements. We quantified the volume of sediment stored in the point bar attached to each active meander bend due to the channel migration, and the one eroded from the floodplain between 1971 and 2008. The GIS overlays of the channel positions between 1971 and 2008, derived from the aerial photographs, provided the eroded and constructed floodplain surface areas. We calculated the volume of sediments for each of these polygons knowing its thickness from the DEM, the long profile of the channel defining a reference elevation, and subtracting the volume of fine sediment provided by overbank flows that we estimated based on field sampling (soil coring). For the eroded area, we determined an estimate of the volume by assuming that the bank height and the surface morphology eroded were similar to what we still presently observe along the eroding bank. This assumption was verified by the analysis of the 1971 aerial photos confirming the eroded surface is not different in textural characters from the uneroded one used for present field measures.

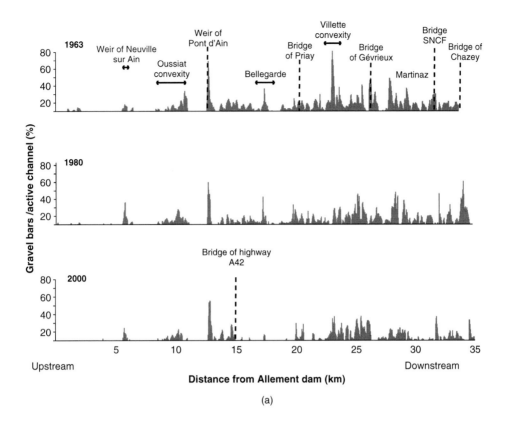

(a)

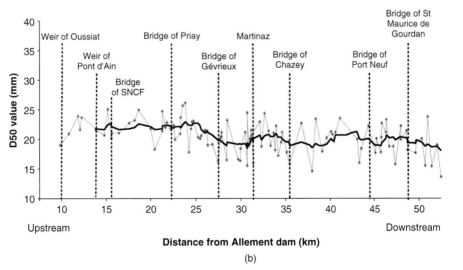

(b)

Figure 11.7 Analysis of the Ain river channel changes related to the Allement dam construction (1960) : aerial photos are combined with field data (grain size data provided from ground imagery). a) Longitudinal distribution of gravel bars in 1963, 1980 and 2000 showing the progressive disappearance of the gravel surface from 10 m long segments, b) Longitudinal distribution of gravel bar grain size showing the front of progressive erosion km 27 (dark line is a moving average, grey points are sampled bars).

We calculated the sediment budget at the bend-scale, rather than using standardised length DGO, to provide geomorphically meaningful information and to compare for each bend what it is reintroduced by bank erosion and transferred downstream with what is stored from upstream. This estimate is possible on the Ain River over three decades only because of the unidirectional channel shifting. If the channel had rejuvenated several times the surface over the period, the estimate would be low. Over the study reach from the bend 2 to the bend 32 (Figure 11.8), 2.9 million m^3 of sediment were eroded as opposed to 2.01 million m^3 stored, the difference being then very low 86,000 m^3 (5% of the storage). Annually 56,000 and 55,000 m^3 of sediment are locally introduced and stored, respectively, which is a very important contribution to transport processes within a fluvial corridor for which the mean annual bedload transport is estimated to 12 000 m^3/yr (Rollet, 2007). This analysis also shows that the budget is not uniform over the study area, with some of the bends trapping much sediment (bends 19, 21, or 24) whereas others are mainly affected by sediment introduction (bends 17, 22, 25). Very few bends have a balanced sediment budget. Interestingly, we do not observe a clear bend-to-bend bedload transfer over multiple decades as expected, the upstream bend erosion accumulating on the one immediately downstream following the theoretical schemes (e.g. Friedkin, 1945). Therefore, these results open interesting discussion on the integration of floodplain/channel sediment exchanges in the sediment transport modelling conducted on such shifting rivers.

11.4.2 The Drôme network: example of up- and downscaling approach using homogeneous geomorphic reaches

In this section, the geographical approach is illustrated in a multi-scaled perspective. All the development was based on geomorphic reaches identified using the DGO/AGO procedure. Practical questions are twofold: (i) Are the mesohabitat characteristics (e.g. fish habitats) different between natural (braided or wandering) and embanked reaches, or does step-pool frequency and spacing vary from one reach to another? (ii) Can we expect sediment reintroduced to prevent channel incision to reduce the availability of aquatic habitat at network scale?

The methodology is based on the identification of geomorphic reaches considering they are integrating features of biological communities following the nested-scale framework concept (Frissell et al. 1986). The geomorphic units were detected using the DGO/AGO procedure

(Wiederkehr et al., 2010a). The entire network was divided by 10 m (for water polygons) or 100 m (for active channel polygons) long DGO, and AGO units are then created based on valley width (provided from geomatic procedure applied to a DEM) and active channel width detected from orthophotography using an object-oriented remote sensing procedure which is more powerful than a pixel by pixel approach when using low spectral resolution images. The procedure resulted in the identification of 53 reaches along the main stem. In each of these reaches, geomorphic metrics were calculated (confinement index, mean active channel width divided by the catchment size, the sinuosity ratio of the active channel, the proportion of aquatic area within the active channel). A cluster analysis was then performed to identify AGO with similar geomorphic characteristics, ultimately classifying them into five unique geomorphic types.

This preliminary step was then used to characterise the aquatic meso-habitats within each homogeneous geomorphic reach and answer to the first question. Meso-habitat characterization is illustrated on the downstream part of the Drôme main stem where the aquatic zone is wide enough to be studied with a 50 cm pixel resolution. Aquatic in-channel features were detected by an object-oriented procedure based on radiometric parameters. 4100 polygons were then detected over 20 km. 410 polygons were then randomly selected and classified (e.g., pool, riffle, gravel bench, lentic and lotic channels). They were then used to perform a discriminant analysis and calculate a discriminant function allowing prediction of habitat type for each of the polygons. A map was then produced showing the position of in-channel habitats. Moreover, longitudinal patterns of meso-habitat occurrence were plotted, as were in-stream habitat statistics from each homogeneous reach type (e.g., braided, embanked or wandering). Results show that there are no obvious differences in habitat structure between the embanked, braided or wandering reaches (Figure 11.9).

The second approach illustrates an upscaling approach using the homogeneous reaches (Bertrand & Piégay, unpublished data). The previous procedure was applied to the entire network (597 km) and detected types corresponded well in characteristics and spatial distribution with the theoretical ones identified by Pont et al. (2009). The aim of the present analysis was to test sensitivity of geomorphic reaches to change following riparian forest harvesting intended to promote sediment reintroduction. Following Pont et al. (2009), we assume that the homogeneous geomorphic reaches represent good integrative units of biological potential, so called functional reach

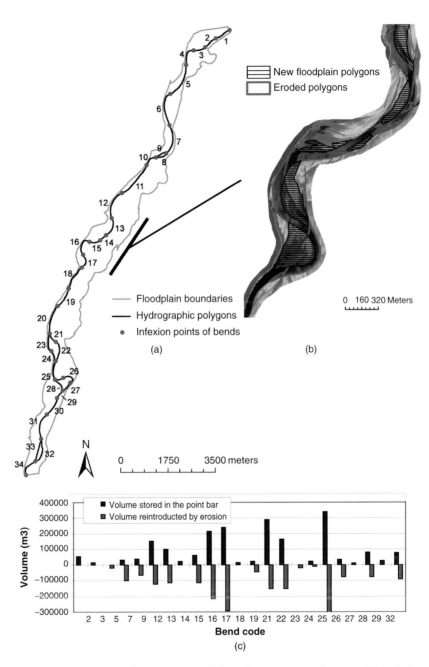

Figure 11.8 Sediment budget of the Ain river between Priay and Chazey between 1971 and 2008 combining information from LiDAR imagery and aerial photographies : a) Location of the studied bends, b) Example of the Villette reach showing detailed mapping of the relative DEM and eroded/constructed polygons built at the meander bend scale. c) Graphic values per bend of the sediment budget of the reach between 1971 and 2008 from bend 2 (upstream) to bend 32 (downstream).

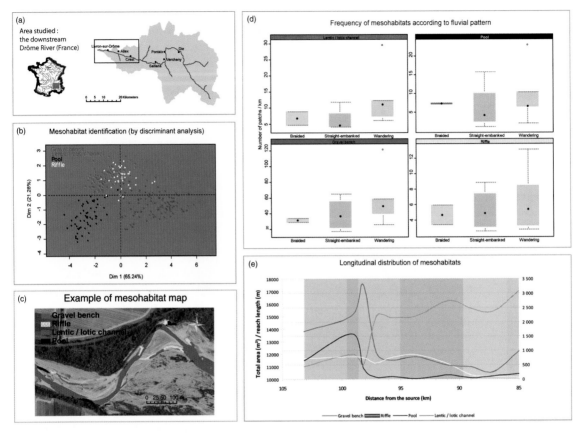

Figure 11.9 Characterisation of mesohabitats of the lower Drôme River (France) from orthophotos : a) Location of the studied reach, b) First factorial map of a Discriminant Analysis showing the mesohabitat types (Gravel bench, Lentic/lotic channel, Pool, Riffle), c) Mapping of mesohabitats, d) Number of mesohabitat types per km according to fluvial pattern, e) Total mesohabitat area (m²) rated by the reach length (m) for each of the mesohabitat types – color indicates different homogeneous fluvial patterns similarly to those in b). Left Y-axis units are accurate for Lentic/lotic channel and right Y-axis units for Pool, Riffle and Gravel Bench (Wiederkehr et al., 2010b).

types. Narrow, coarse grained and canopy covered reaches provide complementary habitats for fish and macroinvertebrate communities relative to that provided by the wider reaches characterized by active bedload transport and multiple flow channels. In this context, higher diversity in the set of function reach types present in a given region implies higher expected potential biodiversity.

In order to calculate functional reach type diversity at the network scale (e.g., 597 km), we used the Shannon index calculated using the relative contribution of each of the types (in % of the total network length). We used orthophotos from 2003 to detect valley bottom corridors from DEM information (elevation, slope) following the methodological procedure described by Alber and Piégay (2011) and for detecting biophysical

features such as gravel bars, water bodies, and riparian vegetation areas from which we determined the functional reach types following Wiederkehr et al. (2010b). Nine types of reaches were then detected using the procedure described above and applied to 333 initial reaches for which six geomorphic variables were measured and on which we applied a cluster analysis (Figure 11.10). The tested management strategies were based on the management recommendations of Liébault et al. (2008) that deal with the riparian forest harvesting for reactivating sediment reintroduction. Different areas of reactivation are likely to have varying downstream effects that depend on the geomorphic reach type. For the purposes of this study, potential changes in reach types following forest harvesting and sediment reactivation were classified as follows

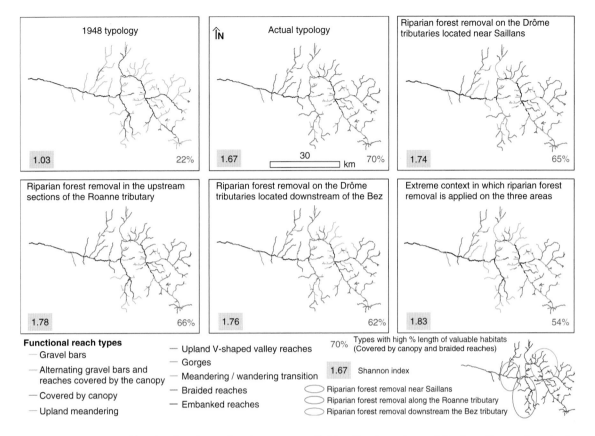

Figure 11.10 Sensitivity of functional reach type diversity as a proxy of habitat diversity to changes following sediment reintroduction (Bertrand and Piégay, unpublished data). A few situations are shown: the present state, the 1948 state calculated retrospectively, three management options (forest removal along tributaries near Saillans, along the Roanne tributary and downstream the Bez tributary) and an extreme option that combines the three basic options. Two indices are also calculated : the Shannon index based on the functional reach type frequency and the percentage of the network length occupied by braided and canopy covered reaches.

(in roughly down-channel order): "covered by canopy" and "alterning gravel bars and canopy covered" become "gravel bar reaches", and, further downstream, "meandering/wandering transition reaches" becomes "braided" reaches. Shannon index was then calculated based on the relative length of geomorphic types for the present state and after applying sediment reactivation in different sub-catchment areas. The procedure underlines that whatever the reactivation option, the changes in geomorphic types have positive consequences in term of functional reach type diversity at the network scale even in the case of the most ambitious scenario of reactivation (H = 1.67 in the present state against 1.74 to 1.83 for the different scenarios) (Figure 11.10). When using another habitat diversity proxy, such as the % length of valuable habitats considering only reaches covered by canopy and braided reaches,

the sensitivity analysis reveals that the reactivation has negative effects in terms of habitat diversity.

11.4.3 Inter-reach comparisons at a network scale

This third applied example focuses on the braided channel reaches of the Rhône basin (Figure 11.11). It is based on detailed case-studies from Piégay et al. (2009) and Belletti et al. (2010). The question of interest to water managers in charge of implementing the WFD is whether site-specific restoration or conservation programs can be developed for braided rivers based on the spatial variability of their planform properties and dynamics. There are two basic scales of analysis useful for answering the basic broad management question: (i) the scale of the active channel itself defined as the corridor of gravel

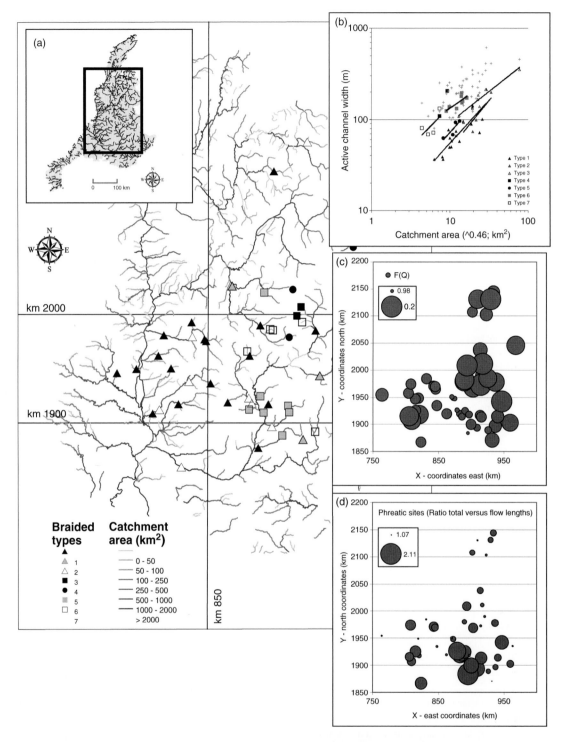

Figure 11.11 Example of inter-site comparison and associated allogenic model. a) Distribution of the seven braided river types within the Rhône basin (Each type corresponds to a particular elevation, average channel width, slope, and riparian forest width. Types are detailed in Piégay et al. 2009). b) Active channel width versus catchment size scatter on which are plotted the 50 observed reaches by type. c) Geographical distribution of studied braided reaches according to the discharge frequency observed at the date of the airborne survey (high F(Q), low discharge, low F(Q), high discharge) and d) Ratio between the total and the flowing accumulated aquatic channel lengths calculated for discharge observed in c) (Belletti et al., 2010).

transport, which is controlled by the vegetation resistance (the valley land-use) and bedload delivery which in turn are controlled by parameters changing at a decadal scale, and (ii) the scale of the aquatic network within the braided corridor whose architecture is fluctuating at a sub-annual timescale. Fifty braided reaches have been selected within the Rhône catchment, so that the entire area of interest is covered.

In the case of the first question, which focuses on a longer timescale, a typology of braided reaches was developed. It was based, as before, on an object-oriented remote sensing method applied to orthophotos to extract various landscape features: gravel, water and riparian vegetation patches. This information was combined with vectorial data provided from other GIS sources: the active channel width divided by catchment area, the channel slope divided by the catchment area, the mean elevation above sea level, and the percent of naturally recruited forest area in the floodplain corridor which is not used by human activities. Seven braided types were established using a cluster analysis (Figure 11.11).

A regional organisation was then observed (see detail in Piégay et al. 2009): western braided rivers have specific characteristics compared to the eastern ones, whereas both are located at similar elevations and positions in the catchment. Upstream to downstream position was also a discriminating factor. The upper reaches (type 4 for example) are characterised by a significantly narrower riparian corridor and wider active channel than the ones downstream. This research illustrates that there are several characteristic types of evolution for braided rivers within the Rhône basin. Types 4, 6 and 7 are active braided channels, much wider than the others. Types 4 and 6, located upstream, are in direct contact with available sediment sources whereas type 7 further downstream drains a region which is more sensitive to erosion due to lithology and climate conditions than the one located to the west. The western braided reaches may have also undergone narrowing caused by vegetation encroachment earlier than the eastern ones because change in human pressure occurred sooner. This regional context also implies that the western braided reaches presently have relatively unique riparian corridors which may need to be preserved. This approach also shows that braided rivers can be ordered from very active to inactive according to their active channel width. Their position can change through time because of flood series which impose inter-annual width fluctuations, but the primary factor is probably long term change in sediment delivery/availability which explains the more significant changes in width.

When considering the networkscape of aquatic water bodies in the braided corridor, it is expected to be linked to both discharge (Egozi and Ashmore, 2008) and groundwater upwelling and downwelling (Tockner and Malard, 2003). All the waterbodies of each of the 50 braided reaches were detected on the orthophotos using an object-oriented remote sensing method (see Belletti et al., 2010). From this raw information, two braided indexes were calculated. Indices included all the channels (total braided index) and just the portion containing flow (only branches connected upstream and downstream). We then compared these results with the discharge of each of the reaches during the survey. On Figure 11.11 are plotted the discharge frequency so that the different braided reaches can be compared to each others (11.11c). The figure also illustrates the difference between the total accumulated length for all channels and just those containing flow (11.11d). This shows that a set of reaches located 900 km east and 1900 km north (e.g., within a more focused area than that covered by the large map shown in Figure 11.11a) have a total aquatic channel length index much higher than the flowing index despite being at low flow. The results thus identify an interesting type of braided river with a large set of aquatic habitats observed at low flow. The type presumably depends on intense bedload transport and groundwater delivery, as detected at the regional scale, again providing river managers valuable information for designing a conservation strategy.

11.5 Limitations and constraints when enlarging scales of interest

The use of imagery for characterising biophysical characteristics of river corridors in space and time has limitations because of practical and methodological constraints associated with the need to work with multiple images. The main issues relate to: (i) image quality, (ii) treatment errors, (iii) heterogeneity of geographical areas, and (iv) treatment time.

The first limitation deals with the kind of imagery which is available to cover large networks/reaches. As shown by Wiederkehr et al. (2008), who did a detailed analysis of available sources, most such approaches are based on archived aerial photos (as opposed to archived satellite imagery) because it provides a resolution sufficient to include even narrow reaches within a regional context. As a consequence, the spectral resolution is often limited to one or in best cases three radiometric bands (panchromatic or red/green/blue). However, the infra-red colour is becoming more and more frequent over large areas so that it should provide more capacity to detect water and humid patches in the near future at this regional scale. In any case, when covering a large area, several sets of photos or sources of images have to be used, and it is not possible to avoid radiometric variations between tiles. Furthermore, there are artifacts such as shadows, overlighting, variations in spatial resolution and radiometry, particularly when using website images such as the one in Google Earth. These may vary from one mosaic tile to the next even within a larger mosaic. Furthermore, one of the problems with using sources such as Google Earth is associated with the difference in resolution from one regional area to another. Earth is mostly covered with a resolution of 15 m per pixel (Landsat images) with areas with better resolution reaching 2.5 m (Spot images), 0.5 m (Geo-Eye) or less when airborne information is provided by national suppliers. This may have practical consequences for automating classifications of polygons, and preliminary treatments may be needed to homogenise the light conditions (see discussions in Chapter 8). The problem is also present when comparing historic images with one another. Variability in lighting conditions, shadows, and discharge can lead to large changes in the patterns visible in the photographs (Figure 11.12).

When the classification is done automatically, preliminary tests must be done to identify the best parameters for the classification. An object-oriented procedure allowing incorporation of radiometric values as well as other parameters such as feature shapes, textures and context is becoming more and more popular for maximising the use of information available on low spectral resolution images such as the orthophotographs. In preliminary tests applied to the Rhône network, in particular on the Drôme River, a confusion matrix provided overall accuracy of 91.8% for gravel bars, flow channels and vegetation patches. The results varied according to the types of features as well as the active channel width. Gravel bars and vegetation patches were well identified, with respectively more than 90% and 95% of polygons detected whereas water areas were unevenly identified. Moreover the quality of the classification increased with the width of the active channel. Within the downstream part of the network, the rate of detection was around 85% of features. This rate reached only 50% of patches within the upstream part of the network. Parameters used to determine the water, gravel bars and vegetation polygons included a textural index, spectral indexes (mean radiometry of Red band and Green band and standard deviation of Blue band) and a shape index (area) (Wiederkehr et al., 2009). However, even though different classes could be predicted reasonably accurately within a set of photos at a regional level, the method cannot be applied directly at the 45,000 km network scale of the Rhône because of significant changes in light conditions. Consequently, several separate classifications were required at this scale. However, note that if an infra-red band is available, basic

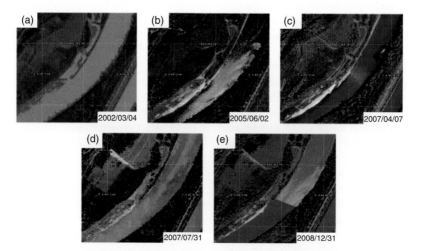

Figure 11.12 Images from Google Earth website of the Pierre-Bénite reach (Rhône River) downstream from Lyon on different dates. (a) corresponds to a flood period with turbid water, (b) and (e) have lightness diffusion with specular reflection, (e) also shows a tile boundary, (d) water surface is rough, due to very small waves and associated light diffusion, (c) is the only one date on which surface water is undisturbed, lightness conditions are good, and bathymetry can be predicted from image radiometry. Images of the Pierre-Bénite reach (Rhône River) sourced from Google Earth web site. © 2012 Google http://www.google.com/earth/index.html

classification as detailed above is much easier. Notably water areas and vegetation are well identified because of the absorption in infrared band of the water contained in the leaves (Jones et al., 2010). Better results have been obtained from combined near and mean IR channel from Spot 5 XS images and orthophotos (Tormos et al., 2009) and from IR colour orthophotos (Wiederkehr et al., 2010b). Tests performed on the Drôme River were based on an object-oriented method. Specific ratios were used for detecting water and vegetation patches. For water we used an average value of the infrared band below 80 and a maximum of the red band between 160 and 200 over a 0–256 scale. For vegetation we used the ratio of infrared band to the red band with a threshold between -0.7 and 0.1 to isolate the vegetation patches. Once these land-use classes were identified, the remaining polygons were classified as gravel bars. Detection rates for each class reached almost 100%. Note also that useful information can be developed from a non-automated analysis, and a comparison of the two approaches (visual versus automatic) must be done to evaluate which solution is the most effective.

The second problem is associated with errors in the imagery. Different ones have been identified all along the process dealing with georectification, photo-interpretation, and also water level measurement. Archived hard copies (photographic film or paper reproductions) need to be digitised and orthorectified. Old photos often have strong distortions that lead to significant rectification error. To prevent such problems, clear protocols must be defined to minimise the georectification error and the operator bias, notably because the errors accumulate when overlying different layers. This problem is critical when considering the channel shifting from channel overlays and associated sediment budgeting because image errors can be multiplied by errors associated with additional required field measurements (e.g., overbank fine sediment depth estimate, sometimes bank height estimate). Several authors have developed methods for estimating errors in quantifying channel migration rates that can be applied to GIS-based analysis of planform change (Mount and Louis, 2005; Hughes et al., 2006) and for estimating error in bankfull width comparisons from temporally sequenced raw and corrected aerial photographs (Mount et al., 2003). The effect of scanning resolution has been explored by Liebault (2003), and Toone (2009) considered the observer bias in the drawing of channel polygons over a range of photos whose scale varies from 1:30,000 to 1:17,000. She showed standard error in width of 0.5 to 0.8 m (3.2 to 4.7% of channel width) from repeat digitisations by four different operators over a set of 90 channel cross-sections.

Following Liebault (2003) who performed a precision test of the measured area, it is best to scan the archived photographs at a resolution of 700 to 1000 DPI, depending on their scale, to provide a final resolution of ca 0.5 m, similar to that of modern orthophotographs. In his case, the active channel surface area was slightly underestimated as resolution increased until a threshold was reached above 700 DPI. The threshold did not depend on the historical series used. After being georectified primarily using first order polynomial transformations, average RMS error was on the order of 2 to 3 m for each photograph.

Interpretation of features and, in the case of manual treatments, the associated drawing process, is also a major source of error. Scale variation between different photo series or light variation in a single photo series may affect the capacity of even a human observer for detecting features. For these reasons, it is sometimes worthwhile to advocate identifying feature boundaries by hand rather than using an image classification tool even if the manual approach can be particularly time-consuming. The operator has to choose a single set of rules regarding level of detail that is maintained throughout the drawing session. Particular care should be taken to work at a consistent photograph scale in order to minimise errors at this step (although it is appropriate to zoom in and out from time to time to help in interpretation). When considering historic change, it is important to realise that riverscape features are neither spatially nor temporally uniform and that this can influence feature detection. For instance, when a riparian forest is established along a channel, the canopy partly covers the channel. If the phenomenon is systematic in space and time, the channel width/area detection may be meaningful (at least in a relative sense), but if the canopy closes over time, it can provide both alignment error and classification error when overlaying different states. Comparing channel width calculation from photos and from field measures, Liebault (2005) found differences of 2.39 m (18% in average) between the two, the measures on photo usually underestimating the real width because of the canopy cover. This error also depends on the scale, and increases when resolution becomes coarser. Therefore, such analysis always needs to be checked for precision. There is not 'right' or 'wrong' way to digitise, but the digitisation is not particularly helpful unless it is checked by repeating the methods on other similar photos or having another person independently repeat the procedure on the single photograph in question.

Another limitation lies in the difficulty of getting the exact dates for images (day and hour are the best), and, when these are available, for associated discharge. This can have clear consequences for the detection of corridor features, especially if bankfull discharge is exceeded. This is particularly important on photo mosaics, which often have multiple dates within the mosaic. One needs to take care of the discharge/water level to compare features between dates (e.g., gravel bar area, oxbows) or when comparing features across a stream network (rainfall versus snow-melt or glacial hydrological regime, with high flows occurring in different seasons whereas aerial photographs are usually taken in summer). It is usually misleading to simply compare aquatic areas on two separate dates because discharge is almost never the same at the time the images were made. To address this problem, there is value in determining the discharge on the day and hour when the photo was taken. Furthermore, if discharge does vary from photograph to photograph, there is also a need to determine the error in width associated with the difference. This can be done by considering the stage-discharge relationships at the nearest gauging station or by calibration from field surveys. However, cross-section geometry is usually different in the area of interest compared to the nearest gauging station site. On the Rhine River, the error in channel width associated with a water level difference of up to 40 cm due to difference in discharge during the photographic acquisition was estimated from 12 selected cross-sections along the reach. The water level variation around a reference water level corresponds to an error between −5 and +5 m in flow channel width (e.g., +/− 5%) (Arnaud et al., 2010).

There is also a problem of contingency of the date at which aerial photographs are taken relative to flood history. It is sometimes difficult to distinguish if the observed patterns are associated with a long period of anomalously high or low flow or with a long term trend associated with other parameters, especially for lateral changes and interpretation of vegetation/pioneer unit coverage which is intrinsically linked to the temporal pattern of colonisation. It is thus difficult to study variation in channel shifting at the wider scale as photo sets used to highlight shifting may be associated with different time periods (effects of floods on the active channel width for example). Similarly, variation of water surface width at a range of flow conditions may be of interest for predicting discharge from water area. In this case, the observer will search for photos with contrasting discharges. However, this is not always possible because most photos are taken at low flow during the summer, thereby increasing the

difficulty of getting enough data for linking discharge and planimetric characteristics.

Even if some images have the advantage of covering large areas to provide regional biophysical representations of fluvial corridors, there are potential limitations associated with the inherent nature of fluvial features. Some of them are detectable by imagery only on a portion of a hydrographic network (Figure 11.13). Channels with a width smaller than the half diameter of riparian tree crowns are often difficult to delineate no matter what resolution imagery is being used. Due to the canopy coverage in the upstream sections or to the deep water column in the downstream sections, an exhaustive survey of in-channel features is usually not possible. On the Rhône catchments, 38 481 km of rivers are so narrow that it is impossible to classify in-stream features, 5 185 km combined exposed gravel patches and low flow channel, and 8 048 km are only water, showing that the analysis of in-channel features is possible on less than 10% of the channel network. There is also a size effect that complicates the detection of features that vary in either form or characteristic size from upstream to downstream. Floodplain patches along large rivers are different than very narrow floodplains observed along upper branches. Image resolution, however, does not usually increase in upland areas. As a consequence, image resolution provides some clear limitations when covering the entire channel. In many cases, a 20 to 50 cm pixel size, which is the most common size range of archived aerial photographs, limits the approach to rivers and large streams.

The treatment time for digitising, photo-interpreting, and computing is also a critical issue to consider when enlarging the spatial and temporal scale. Most historical images are presently archived as hard copies that require scanning and georectification. There is thus a need to develop automatic procedures for georectifying images and extracting features, and the associated computational requirements are not be trivial. In this domain new autorectification methods are in development which should minimise this effort, but error associated with the procedure needs to be evaluated across a range of contexts (Carbonneau et al., 2010). The classification procedure applied to multiple images also requires significant computer capacity to minimise the calculation time. In the case of the Rhône network (45,000 km long), the area of interest is covered by 6347 images (e.g, 1.9 Terabyte of information) and most of the imagery analyses are performed using several days of computational time on a computer cluster. In order to minimise the required computation time on the Rhône network, several steps were

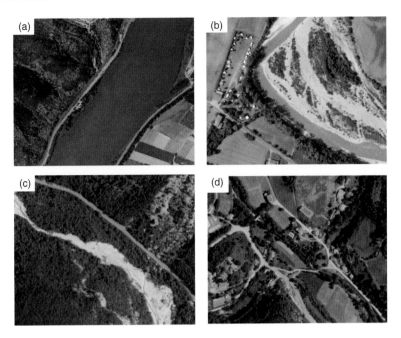

Figure 11.13 Examples of problems limiting the analysis of fluvial corridor features at a corridor to network scale (a) deep water preventing in-channel feature detection, (b) riparian canopy hiding in-channel features or providing shadow, (c) overlightening reducing the radiometric variation between water and gravel bars, (d) channel too narrow to be detected. Data from BD ORTHO®️ of the IGN, © IGN – France.

taken: (i) only the portion of the image covering the valley corridor as identified from a DEM analysis was extracted, (ii) within this corridor, the basic polygons (water, gravel, natural vegetation) were detected by an object-oriented procedure considering radiometry and channel size.

11.6 Conclusions

As in the 1960s, when major investments led to a new-found abundance of monitoring data related to water quality, river managers are faced with managing and evaluating new data describing the biophysical state of river corridors and incorporating the new data in the decision making process. This is particularly true under the new regulatory pressure of the WFD in Europe.

Moreover, it is now acknowledged that point-based typologies, such as RHS (River Habitat Survey) in the UK, are limited because they cannot easily detect longitudinal thresholds and discontinuities or downstream propagation of channel change. They should thus be complemented by continuous spatial information such as provided by images. While the development of a continuous description of the channel corridor is an important

investment, it can provide a significant source of data for multiple purposes, thereby allowing the development of risk assessment and prospective approaches based on various what-if scenarios. Biophysical information is needed to assess human pressures and impacts, develop predictive biological response models, identify riparian buffers that link physics and water chemistry, target conservation strategies, and plan restoration activities that link flood risk assessment and ecological improvement.

Following this technological revolution, scientists and managers are also facing questions related to data management and data sharing. It is clear that imagery web-platforms and online remote sensing data provide an opportunity to extract almost real-time information regarding biophysical change. Channel response to restoration measures will then be accessible to all from the web, allowing for improved linkage between public expectations and management actions.

Acknowledgements

The authors would like to thank the water agency RMC for its support to the ZABR research group within the braided river project and the imagery project. A part of this research is also supported by the ANR Gestrans and the research program 'eaux et territoires'/Project 'Créateur de Drôme' funded by CNRS, Cemagref and the Ministry of Environment. The authors also thank M. Bertrand and B. Belletti for kindly providing unpublished data for illustrating case-studies.

References

Aalto, R., Lauer, J.W., and Dietrich, W.E., 2008. Spatial and temporal dynamics of sediment accumulation and exchange along Strickland River floodplains (Papua New Guinea) over decadal-to-centennial timescales, *J. Geophys. Res.*, 113, F01S04, doi:10.1029/2006JF000627.

Alabayan, A.M. and Chalov, R.S. 1998. Types of river patterns and their natural controls. *Earth Surf. Process. Landforms*, 23: 467–474.

Alber, A. and Piégay, H. 2011. Spatial disaggregation and aggregation procedures for characterizing fluvial features at the network-scale: Application to the Rhône basin (France). *Geomorphology*, 125(3), 343–360.

Amos, Kathryn J., Jacky C. Croke, Andrew O. Hughes, Joanne Chapman, Ingrid Takken, and Leo Lymburner 2008. A catchment-scale assessment of anabranching in the 143 000 km2 Fitzroy River catchment, north-eastern Australia. *Earth Surf Process Landforms*, 33(8), 1222–1241.

Arnaud, F., Schmitt, L., Johnstone, K., Hoenen, D., Béal, D., Piégay, H., and Rollet, A.J. 2010. Etude historique des évolutions morphologiques du Vieux Rhin depuis 1800. Synthetic Report. UMR 5600 EVS-EDF CIH. 75 p.

Astrade, L. and Bravard, J. 1999. Energy gradient and geomorphological processes along a river influenced by neotectonics. *Geodinamica acta*, 12 (1): 1–10.

Brice, J. 1984. Planform properties of meandering river. In: CM. Elliott (Editor), *River Meandering: Proc. Conf. Rivers 1983*, New Orleans, pp. 1–15.

Beechie, T.J., Liermann, M., Pollock, M.M., Baker, S., and Davies, J. 2006. Channel pattern and river-floodplain dynamics in forested mountain river systems. *Geomorphology*, 78 (1-2): 124–141.

Belletti, B., Hervouet, A., Dufour, S., and Piégay, H. 2010. Caractérisation de la structure planimétrique du corridor fluvial. In: Piégay, H., B. Belletti, and F. Liébault [coord.], Typologie de rivières en tresses du bassin RMC. Unpublished report, Univ. Lyon, UMR 5600 CNRS, 105 p.

Belmont, P. 2011. Floodplain width adjustments in response to rapid base level fall and knickpoint migration. *Geomorphology*, 128 (1-2): 92–102.

Bjerklie, D.M., Moller, D., Smith, L.C., and Dingman, L. 2005. Estimating discharge in rivers using remotely hydraulic information. *Journal of Hydrology*, 309: 191–209.

Bjerklie, D.M. 2007. Estimating the bankfull velocity and discharge for rivers using remotely sensed river morphology information. *Journal of Hydrology*, 341: 144–155.

Brierley, G.J., Fryirs, K., and Jain, V. 2006. Landscape connectivity: The geographic basis of geomorphic applications. *Area*. 38.2, 165–174.

Bryce, S.A. and Clark, S.E. 1996. Landscape-level ecological regions: linking state-level ecoregion frameworks with stream habitat classifications. *Environmental Management*, 20: 297–311.

Burge, L.M. and et Lapointe, M.F. 2005. Understanding the temporal dynamics of the wandering Renous River, New Brunswick, Canada. Earth Surf. Process. *Landforms*, 30: 1227–1250.

Carbonneau, P.E., Bergeron, N., and Lane, S.N. 2005. Automated grain size measurements from airborne remote sensing for long profile measurements of fluvial grain sizes. *Water Resources Research*, 41, W11426.

Carbonneau, P.E., Dugdale, S.J., and Clough, S., 2010. An automated georeferencing tool for watershed scale fluvial remote sensing. *River Research and Application*, 26: 650–658.

Carbonneau, P.E., Fonstad, M.A., Marcus, W.A., and Dugdale, S.J. 2012. Making riverscapes real, *Geomorphology*, (2012), doi:10.1016/j.geomorph.2010.09.03

Church, M. Mark, D.M. 1980. On size and scale in geomorphology. *Progress in Physical Geography*, 4: 342–390.

Church, M. and Rice, S.P. 2009. Form and growth of bars in a wandering gravel-bed river. *Earth Surface Processes and Landforms*, 34: 1422–1432.

Clerici, N., Weissteiner, C.J., Paracchini, M.L., and Strobl, P. 2011. Riparian zones: where green and blue networks meet. Pan-European zonation modelling based on remote sensing and GIS. *JRC Scientific and Technical Reports*, 62 p.

Constantine, J.A. and Dunne, T. 2008. Meander cutoff and the controls on the production of oxbow lakes. *Geology*, 36(1): 23–26.

Dawson, F.H., Hornby, D.D., and Hilton, J. 2002. A method for the automated extraction of environmental variables to help the classification of rivers in Britain. *Aquatic Conservation: Marine and Freshwater Ecosystems*, 12: 391–403.

Dodov, B. and Foufoula-Georgiou, E. 2004. Generalized hydraulic geometry: Derivation based on a multi-scaling formalism, *Water Resour. Res.*, 40, W06302, doi:10.1029/2003WR002082.

Egozi, R., and Ashmore, P. 2008. Defining and measuring braiding intensity. *Earth surface processes and landforms*, 33: 2121–2138.

Ferguson, R. 2007. Gravel-bed rivers at the reach scale, In H. Habersack, H. Piégay, and M. Rinaldi *Gravel-bed River 6: From process understanding to river restoration*, Elsevier, Amsterdam. p. 33–60.

Fonstad, M.A. and Marcus, W.A. 2005. Remote sensing of stream depths with hydraulically assisted bathymetry (HAB) models. *Geomorphology*, 72: 320–339.

Friedkin, J.F. 1945. A Laboratory Study of the Meandering of Alluvial Rivers. U.S. Army Corps of Engineers Waterways Experimentation Station, Vicksburg, MS.

Frissell, C.A., Liss, W.J., Warren, C.E., and Hurley, M.D. 1986. A hierarchical framework for stream habitat classification: viewing streams in a watershed context. *Environmental Management*, 10: 199–214.

Gaeuman, D.A., Schmidt, J.C., and Wilcock, P.R. 2003. Evaluation of in-channel gravel storage with morphology-based gravel budgets developed from planimetric data. *Journal of Geophysical Research*, 108(F1), 6001, doi:10.1029/2002JF000002.

Gilvear, D., Winterbottom, S., and Sichingabula, H. 2000. Character of channel planform change and meander development: Luangwa River, Zambia. *Earth Surf. Process. Landforms*, 25: 421–436.

Gilvear, D. and Bryant, R. 2003. Analysis of aerial photography and other remotely sensed data. In G.M. Kondolf and H. Piégay (eds.): *Tools in Fluvial Geomorphology*, J. Wiley and Sons, Chichester, pp. 133–168.

Gupta, A. and Liew, S.C. 2007. The Mekong from satellite imagery: A quick look at a large river. *Geomorphology*, 85: 259–274.

Gurnell, A.M. 1997. Channel change on the River Dee meanders, 1946–1992, from the analysis of air photographs. *Regulated Rivers: Research and Management*, 13: 13–26.

Gurnell, A.M., Petts, G.E., Harris, N., Ward, J.V., Tockner, K., Edwards, P.J., and Kollmann, J. 2000. Large wood retention in river channels: the case of the Fiume Tagliamento, Italy. *Earth Surface Processes and Landforms*, 25(3): 255–275.

Graf, W.L. 1984. A probabilistic approach to the spatial assessment of river channel instability: *Water Resources Research*, 20: 953–962.

Güneralp, I. and Rhoads, B.L. 2008. Continuous characterization of the planform geometry and curvature of meandering rivers. *Geographical Analysis*, 40: 1–25.

Hall, J.E., Holzer, D.M., and Beechie, T.J. 2007. Predicting river floodplain and lateral channel migration for salmon habitat conservation. *Journal of the American Water Resources Association*, 43(3): 786–797.

Heo Joon, Trinh Anh Duc, Hyung-Sik Cho, and Sung-Uk Choi 2009. Characterization and prediction of meandering channel migration in the GIS environment: A case study of the Sabine River in the USA, *Environmental Monitoring and Assessment*, 152 (1-4): 155–165.

Higgins, J.V., Bryer, M.T., Khoury, M.L., and Fitzhugh, T.W. 2005. A freshwater classification approach for biodiversity conservation planning. *Conservation Biology*, 19: 1–14.

Howard, A.D. and Hemberger, A.T. 1991. Multivariate characterization of meandering. *Geomorphology*, 4: 1–186.

Hooke, J.M. 1984. Changes in river meanders: a review of techniques and results of analyses. *Progress in Physical Geography*, 8: 473–508.

Hooke, J.M. and Redmond, C.E. 1989. Use of Cartographic Sources for Analyzing River Channel Change with Examples from Britain. In: Petts, G.E. (ed.), *Historical Change of Large Alluvial Rivers: Western Europe*, John Wiley and Sons, Chichester, UK, p. 79–93.

Hong, L.B. and Davies, T.R.H. 1979. A study of stream braiding. *Geol. Society of America Bulletin*, 79: 391–394.

Hughes, M.L., McDowell, P.F., and Marcus, W.A. 2006. Accuracy assessment of georectified aerial photographs: Implications for measuring lateral channel movement in a GIS. *Geomorphology*, 74: 1–16.

Jain, V., Fryirs, K.A., Preston, N., and Brierley, G.J. 2006. Comparative assessment of three approaches for deriving streampower plots along long profiles in the upper Hunter River catchment, New South Wales, Australia. *Geomorphology*, 74: 297–317.

Jones, H.G. and Vaughan, R.A. (2010). *Remote sensing of vegetation: principles, techniques, and applications*. Oxford University Press, 353 p.

Kellerhals, R., Church, M. and Bray, D.I. 1976. Classification and analysis of river processes. *Journal of the Hydraulics Division*, American Society of Civil Engineers, 102: 813–829.

Kondolf, G.M., Piégay, H., and Sear, D. 2003. Chapter 21. Integrating geomorphological tools in ecological and management studies, in M.G. Kondolf, and Piégay H. (eds): *Tools in fluvial geomorphology*. J. Wiley and Sons, Chichester, U.K., pp. 633–660.

Lagasse, P.F., Zevenbergen, L.W., Spitz, W.J., and Thorne, C.R. (2004). *A methodology and Arcview Tools for predicting channel migration*. Ayres Associates, Inc. Fort Collins, Colorado.

Lane, S.N., Widdison, P.E., Thomas, R.E., Ashworth, P.J., Best, J.L., Lunt, I.A., Sambrook Smith, G.H., and Simpson, C.J. 2010. Quantification of braided river channel change using archival digital image analysis, *Earth Surface Processes and Landforms*, 35(8): 971–985.

Lauer, J.W. and Parker, G. 2008. Modeling framework for sediment deposition, storage, and evacuation in the floodplain of a meandering river: *Theory, Water Resour. Res.*, 44.

Leviandier, T., Alber, A., Le Ber, F., and Piégay, H. 2011. Comparison of statistical algorithms for detecting homogeneous river reaches within a longitudinal continuum. Geomorphology. 138(1): 130–144.

Lewin, J. 1977. Channel pattern changes. In Gregory, K.J. (ed.) *River channel changes*: Chichester, John Wiley and Sons, Inc., p. 167–184.

Lewin, J. 1987. Historical river channel changes. In Gregory, K.J., Lewin, J., Thornes, J.B. (eds.), *Palaeohydrology in practice: a river basin analysis*. Chichester, John Wiley and Sons, Inc., p. 161–176.

Leys, K.F. and Werritty, A. 1999. River channel planform change: software for historical analysis. *Geomorphology*, 29 (1-2): 107–120.

Liébault, F., Gomez, B., Page, M., Marden, M., Peacock, D., Richard, D., and Trotter, C.M. 2005. Landuse change, sediment production and channel response in upland regions. *River Research and Applications*, 21: 739–756.

Liébault, F., Piégay, H., Frey, P., and Landon, N. 2008. Chapter 12 Tributaries and the management of main-stem geomorphology. In S. Rice, A. Roy, and B.L. Rhoads (eds.): *River Confluences and the Fluvial Network*, John Wiley and Sons, Chichester, UK, pp. 243–270.

McLean, D.G. and Church, M. 1999. Sediment transport along lower Fraser River. 2. Estimates based on the long-term gravel budget. *Water Resources Research*, 35: 2549–2559.

Marcus,W.A., Legleiter, C.J., Aspinall, R.J., Boardman, J.W., and Crabtree, R.L. 2003. High spatial resolution, hyperspectral (HSRH) mapping of in-stream habitats, depths, and woody debris mountain streams. *Geomorphology*, 55(1-4): 363–380.

Marston, R.A., Girel, J., Pautou, G., Piégay, H., Bravard, J.P., and Arneson, C. 1995. Channel metamorphosis, floodplain disturbance, and vegetation development: Ain River, France. *Geomorphology*, 13: 121–131.

Michalkova, M., Piégay, H., Kondolf, G.M., and Greco, S. 2010. Longitudinal and temporal evolution of the Sacramento River between Red Bluff and Colusa, California (1942–1999). *Earth Surface Processes and Landforms*, 36(2): 257–272.

Mollard, J.D. 1973. *Air photo interpretation of fluvial features*. Fluvial Processes and Sedimentation. Proc. Hydrology Symp. Univ. Alberta. Natl. Research Council, Canada, Ottawa, pp. 341–380.

Mount, N.J., Louis, J., Teeuw, R.M., Zukowskyj, P.M., and Stott, T. 2003. Estimation of error in bankfull width comparisons from temporally sequenced raw and corrected aerial photographs. *Geomorphology*, 56: 65–77.

Mount, N.J. and Louis, J. 2005. Estimation and propagation of error in measurements of river channel movement from aerial imagery. *Earth Surface Processes and Landforms*, 30: 635–643.

Nanson, G.C. and Croke, J.C. 1992. A genetic classification of floodplains. *Geomorphology*, 4(6), 459–486.

Nardi, F., Vivoni, E.R., and Grimaldi, S. 2006. Investigating a floodplain scaling relation using a hydrogeomorphic delineation method. *Water Resources Research*, 42, W09409.

O'Connor, J.E., Jones, M.A., and Haluska, T.L. 2003. Flood plain and channel dynamics of the Quinault and Queets Rivers, Washington, USA. *Geomorphology*, 51: 31–59.

Odgaard, A.J. 1987. Streambank Erosion Along Two Rivers in Iowa. *Water Resources Research*, 23(7): 1225–1236.

Parsons, M., Thoms, M.C., and Norris, R.H. 2004. Using Hierarchy to Select Scales of Measurement in Multiscale Studies of Stream Macroinvertebrate Assemblages. *Journal of the North American Benthological Society*, 23(2): 157–170.

Petts, G.E., Gurnell, A.M., Gerrard, A.J., Hannah, D.M., Hansford, B., Morrissey, I., Edwards, P.J., Kollmann, J., Ward, J.V., Tockner, K., and Smith, B.P.G. 2000. Longitudinal variations in exposed riverine sediments: a context for the ecology of the Fiume Tagliamento, Italy. *Aquatic Conservation: Marine and Freshwater Ecosystems*, 10: 249–266.

Piégay, H., Cuaz, M., Javelle, E., and Mandier, P. 1997. A new approach to bank erosion management: the case of the Galaure river, France. *Regulated Rivers: Research and Management*, 13: 433–448.

Piégay, H., Salvador, P.G., and Astrade, L. 2000. Réflexions relatives à la variabilité spatiale de la mosaïque fluviale à l'échelle d'un tronçon Zeitschrift für Geomorphologie, 44(3): 317–342.

Piégay, H. and Schumm, S.A. 2003. System approach in fluvial geomorphology. in M.G. Kondolf and Piégay H. (eds): *Tools in fluvial geomorphology*. J. Wiley and Sons, Chichester, U.K., pp. 105–134.

Piégay, H., Darby, S.A., Mosselmann, E., and Surian, N. 2005. The erodible corridor concept: applicability and limitations for river management. *River Research and Applications*, 21: 773–789.

Piégay, H., Alber, A., Slater, L., and Bourdin, L. 2009. Census and typology of braided rivers in the French Alps. *Aquatic Sciences*, 71(3): 371–388.

Pont, D., Piégay, H., Farinetti, A., Allain, S., Landon, N., Liébault, F., Dumont, B., and Mazet, A. 2009. Conceptual framework and interdisciplinary approach for the sustainable management of gravel-bed rivers: the case of the Drôme River basin (SE France). Aquatic Sciences, 71(3): 356–370.

Reinfelds, I., Cohen, T.J., Batten, P., and Brierley, G. 2004. Assessment of downstream trends in channel gradient, total and specific stream power: a GIS approach. *Geomorphology*, 60: 403–416.

Richard, G.A., Julien, P.Y., and Baird, D.C. 2005. Statistical analysis of lateral migration of the Rio Grande, New Mexico. *Geomorphology*, 71(1-2): 139–155.

Rinaldi, M. 2003. Recent channel adjustments in alluvial rivers of Tuscany, Central Italy. *Earth Surface Processes and Landforms*, 28: 587–608.

Rollet, A.J. 2007. Etude et gestion de la dynamique sédimentaire d'un tronçon fluvial à l'aval d'un barrage: le cas de la basse vallée de l'Ain. Unpublished PHD, Université Lyon 3, 305 p.

Rust, B.R. 1978. A classification of alluvial channels. In: A.D. Miall (Editor), Fluvial Sedimentology. *Can. Sot. Petrol. Geol. Mem.* 5: 187–198.

Sear, D.A., Newson, M.D., and Brookes, A. 1995. Sediment related river maintenance: the role of fluvial geomorphology, *Earth Surface Processes & Landforms*, 20: 629–647.

Shields, Jr.F.D., Simon, A., and Steffen, L.J. 2000. Reservoir effects on downstream river channel migration. *Environmental Conservation*, 27: 54–66.

Seminara, G., Zolezzi, G., Tubino, M., and Zardi, D. 2001. Downstream and upstream influence in river meandering. Part 2. Planimetric development. *Journal of Fluid Mechanics*, 438: 213–320.

Surian, N., Ziliani, L., Comiti, F., Lenzi, M.A., and Mao, L. 2009. Channel adjustments and alteration of sediment fluxes in gravel-bed rivers of North-Eastern Italy: potentials and limitations for channel recovery. *River Research and Applications*, 25: 551–567. doi: 10.1002/rra.1231

Takagi, T., Oguchi, T., Matsumoto, J., Grossman, M.J., Sarker, M.H., and Martin, M.A. 2007. Channel braiding and stability of the Brahmaputra River, Bangladesh, since 1967: GIS and remote sensing analyses. *Geomorphology*, 85: 294–305.

Tockner, K. and Malard, F. 2003. Channel typology. In Ward, J.V. and Uehlinger U. (eds): *Ecology of a glacial floodplain*, 57–73.

Tormos, T. 2010. Analyse à l'échelle régionale de l'impact de l'occupation du sol dans les corridors rivulaires sur l'état écologique des cours d'eau, Agro Paris Tech. Doctorat, 426p.

Tormos, T., Kosuth, P., Durrieu, S., Villeneuve, B., and Wasson, J.G. 2010. Improving the quantification of land cover pressure on stream ecological status at the riparian scale using High Spatial Resolution Imagery. *Physics and Chemistry of the Earth*, Parts A/B/C.

Toone, J.A. 2009. Ecological implications of geomorphological discontinuities in a mixed bedrock-alluvial channel, River Drôme, France. Cotutelle PhD univ. Loughborough/Univ. Lyon 3.

Tooth, S., 2006. Virtual globes: a catalyst for the re-enchantment of geomorphology ? *Earth Surface Processes and Landforms*, 31: 1192–1194.

Thorp, J.H., Thoms, M.C., and Delong, M.D. 2006. The riverine ecosystem synthesis: biocomplexity in river networks across space and time. *River Res. Applic.*, 22: 123–147.

Verdu, J.M., Batalla Ramon, J., and Martinez-Casasnovas, J.A. (2005). High-resolution grain size characterisation of gravel bars using imagery analysis and geo-statistics. *Geomorphology*, 72: 73–93.

Walsh, S.J., Butler, D.R., and Malanson, G.P. 1998. An overview of scale, pattern, process relationships in geomorphology: a remote sensing and perspective. *Geomorphology*, 21: 183–205.

Wasklewicz, T.A., Anderson, S., and Liu, P.S. 2004. Geomorphic context of channel locational probabilities along the lower Mississippi River, USA. *Geomorphology*, 63: 145–158.

Wawrzyniak, V., Piégay, H., and Poirel, A. 2011. Longitudinal and Temporal Thermal Patterns of the French Rhône River using Landsat ETM+ Thermal Infrared Images. *Aquatic Sciences*, DOI: 10.1007/s00027-011-0235-2.

Wiederkehr, E., Dufour, S., and Piégay, H. 2008. Apport des techniques d'imagerie pour l'étude des réseaux hydrographiques. Synthèse des connaissances et évaluation d'indicateurs de caractérisation. Agence de l'eau RMC-ZABR-CEREGE, unpublished report, UMR 5600 CNRS.

Wiederkehr, E., Dufour, S., and Piégay, H. 2009. Caractérisation du corridor naturel alluvial du réseau hydrographique du basin du Rhône à partir des orthophotographies de l'IGN. Premiers retours d'expérience pour l'élaboration de modèles hydrogéomorphologiques prédictifs. Agence de l'eau RMC-ZABR-CEREGE, unpublished report, UMR 5600 CNRS.

Wiederkehr, E., Dufour, S., and Piégay, H. 2010. Intégration de données extraites des orthophotos de l'IGN pour la caractérisation et la modélisation de l'habitat aquatique. Agence de l'eau RMC-ZABR-CEREGE, unpublished report, UMR 5600 CNRS.

Wiederkehr, E., Dufour, S., and Piégay, H. 2010. Localisation et caractérisation semi-automatique des géomorphosites fluviaux potentiels. Exemples d'applications à partir d'outils géomatiques dans le bassin de la Drôme. *Géomorphologie*: relief, processus, environnement, 2: 175–188.

Winterbottom, S.A. 2000. Medium and short-term channel planform changes on the Rivers Tay and Tummel, Scotland. *Geomorphology*, 34: 195–208.

Winterbottom, S.A. and Gilvear, D. 2000. A GIS-Based approach to mapping probabilities of river bank erosion: regulated river Tummel, Scotland. *Regulated rivers. Research and Management*, 16: 127–140.

Williams, W.A., Jensen, M.E., Winne, J., and Redmond, R.L. 2000. An automated technique for delineating and characterizing valley-bottom settings. *Environmental Monitoring and Assessment*, 64(1): 105–114.

Zanoni, L., Gurnell, A.M., Drake, N., and Surian, N. 2008. Island dynamics in a braided river from analysis of historical maps and air photographs, *River Research and Applications*, 24(8): 1141–1159.

Zhang, G.B., Ai, N.S., Huang, Z.W., Yi, C.B., and Qin, F.C. 2008. Meanders of the Jialing River in China: Morphology and formation. *Chinese Science Bulletin*, 53(2): 267–281.

12 The Role of Remotely Sensed Data in Future Scenario Analyses at a Regional Scale

Stan Gregory[1], Dave Hulse[2], Mélanie Bertrand[3] and Doug Oetter[4]

[1]Department of Fisheries and Wildlife, Oregon State University, Corvallis, OR, USA
[2]Department of Landscape Architecture, University of Oregon, Eugene, OR, USA
[3]Irstea, Unité de Recherche ETNA, Saint-Martin-d'Hères, and University of Lyon, CNRS, France
[4]Department of History, Geography, and Philosophy, Georgia College and State University, Milledgeville, GA, USA

12.1 Introduction

Anticipating future effects of land use and climate change on dynamic, coupled natural-human systems has become an increasingly important challenge in science, planning and policy. Large rivers, their floodplains, and surrounding landscapes provide the foundation for the development of civilisation – integrating landscape patterns and processes, connecting terrestrial ecosystems to coastal margins, and meeting human needs throughout the world. Large river floodplains support some of the highest levels of biodiversity and habitat complexity of any ecosystem (Gregory et al., 1991, White et al., 1999) and are valued for their ecosystem services, such as water, food and fibre production, transportation, and recreation. Stream networks extend from the mountains to the lowlands, dissecting the landscape, creating corridors of riparian vegetation, providing migratory pathways, and transporting nutrients and sediments to lowlands and coastal margins. Mosaics of landscapes and riverscapes shape patterns of human land use, which in turn alter land

cover and landscape processes. Cities, towns, farms and ranches throughout the world develop along floodplains, benefiting from their productivity and resources, and then attempt to harden riverbanks and prevent the very channel dynamics responsible for the floodplain's productivity that led them to locate there in the first place. Assessment of river networks and their basins requires detailed information about land cover and land use that often is prohibitively costly to obtain by direct observation at large spatial extents. Remotely sensed information, such as aerial photography or data obtained from satellites, provides ecologically relevant data at basin extents while being more cost-effective than ground-based measurements.

One of the major challenges in regional decision making is creating spatially explicit and technically sound analyses of future patterns of landscapes, river networks, land uses, and resource availability. Future resources and landscape patterns of a river basin are shaped by thousands of decisions made each day by individuals, organisations and communities. One tool for

Fluvial Remote Sensing for Science and Management, First Edition. Edited by Patrice E. Carbonneau and Hervé Piégay.
© 2012 John Wiley & Sons, Ltd. Published 2012 by John Wiley & Sons, Ltd.

exploring the effects of future policies and land use practices on trajectories of ecological conditions is the assessment of alternative future scenarios (Steinitz et al., 1996, 2003, Hulse et al., 2004, Ahern 2001, Santelmann et al., 2001). These future scenarios can be created by a number of ways – citizen involvement, scientists, model projections – and each method of scenario development offers different strengths and weaknesses (Liu et al., 2007, Henrichs et al., 2010).

Recently, use of environmental change scenarios has increased, allowing scientists, managers, and the public to anticipate future environmental conditions and their effects on a wide variety of things people care about. Scenario-based approaches are often used in research, especially in multidisciplinary studies of landscapes and river networks. Scenarios also offer a conceptual framework for representing and modeling complex environmental systems, with which the socio-economic impacts of differing land and water uses can be assessed (Bouman et al., 1999, Hubacek and Sun 2001, Bolliger et al., 2007). As a result, these approaches are increasingly employed to explore decision-making options for managing land and water resources, as a means to integrate expert knowledge of environments into policy making, and to investigate the perceptions of local communities about their landscapes. Exploration and evaluation of future conditions also provide a basis for assessing the vulnerability of the landscape to future change and resource management policies (Hulse et al., 2009, Lautenbach et al., 2009).

Application of remotely sensed information in the analysis of rivers and their basins is guided by the time frame for the landscape characterisation, the method for projecting biophysical or human infrastructural characteristics of the landscape, and the purpose for conducting a scenario-based analysis of future landscapes. Landscape analyses can depict the current properties of the landscape and serve as a snapshot in time or they can depict past or future landscapes in retrospective analyses and explorations of future alternatives. All of these projections of landscapes past, present, and future are scenarios – abstract representations of mapped or remotely sensed information that inherently include assumptions about the built and biophysical features of the landscape. The purposes of such studies, the processes of depicting scenarios, and the methods of employing remotely sensed information all shape the resulting credibility and nature of uncertainty of the scenarios (Table 12.1).

12.1.1 The purposes of scenario-based alternative future analyses

Scenario-based analyses can serve different purposes: 1) to support scientific exploration and research, 2) to inform education and collaborative learning processes, or 3) to underpin decision processes and strategic planning (Henrichs et al., 2010). For each example presented in Table 12.1, we identified the most relevant purpose according to types defined by Henrichs et al. (2010).

Examples of scenario analysis conducted for multidisciplinary scientific research are Lawler et al. (2006), Aggett et al. (2009) or Baas et al. (2010). In this research context, sensitivity analyses of critical parameters are often employed (Lopez-Lopez et al., 2006). Risk and sensitivity analyses attempt to measure the probability and severity of adverse effects and assess how changes of one specific factor (input, parameter, state etc.) affect the response of a model.

A common type of scenario, which often is named 'business as usual' or 'plan trend', extrapolates future trends of environmental parameters from past observations under the assumption of continued application of current policies and management practices. Such scenarios are typically used for collaborative learning, such as Steinitz et al. (2003), Hunter et al. 2003), and Bolliger et al. (2007).

Management agencies and local and regional governments often use scenario analysis for planning and decision making. These analyses differ in several ways, in the methods used to develop and analyse scenarios, in the approaches for integrating management alternatives, and in the processes of stakeholder participation (USDA 1994, Baker et al., 2004, Kepner et al., 2004, or Brierley et al., 2009). The approaches may include public participation processes (Baker et al., 2004), decision support systems (Lautenbach et al., 2009), or multi-criteria assessments (Nelson et al., 2009).

12.1.2 Processes of depicting alternative future scenarios

Scenario analysis is a process for assessing the range of possibilities by considering differing alternatives of landscape evolution (Henrichs et al., 2010). Each scenario represents a plausible description of a state, which is not a forecast of the future state of a system, but an alternative representation of possible trajectories of change. The spatial scale of these scenario-based approaches can be

Table 12.1 Examples of the application of remotely sensed information in assessments of alternative landscape scenarios. Numbers listed under Purposes refer to 1) research, 2) education, and 3) decision making from Henrich et al., 2010. Names of models are provided in parentheses under Approaches.

Reference	Purpose	Approaches	Description	Response Parameters	Spatial Scale	Time Scale	Remote Sensing
Aggett et al. 2009	1	hydrological/hydraulic modeling (HECRAS)	flood scenarios; inundation potential of modeled events	flow regime, inundation zones	Naches River, Washington, USA (3.2 km)	retrospective (based on 1933)	LIDAR
Baas et al. 2010	1	modeling dune field formation and evolution (DECAL)	cellular automation modeling of dune field formation and evolution and subsequent vegetation development	dune formation, vegetation cover	coastal dune field near Rio São Francisco, Brazil	400 model years	air photo, DEM
Baker et al. 2004	1, 2, 3	land use/land cover classification, ecological response models	stakeholder defined future landscape and development patterns; mapping of land use and land cover, models based on relationships between land cover/land use and ecological phenomena	trajectories of change in vegetation, fish, invertebrates, wildlife, habitat, and water availability	Willamette River basin, Oregon, USA (30,000 km^2)	1850–2050	Landsat, air photo
Bolliger et al. 2007	2, 3	land cover classification and modeling habitat relationships	qualitative approach for socio-economic drivers impact on land use; scenarios of 1) business as-usual, 2) liberalization, 3) reduced agricultural production; species habitat distribution modeling	land cover; habitat quality for selected species	Switzerland (41,000 km^2)	1985–1997	air photo
Bouman et al. 1999	3	sustainable options for land use, biophysical and economic land use analysis at different scales (SOLUS)	analyze and evaluate land use scenarios; inputs and outputs of production systems; a linear programming model; GIS	land cover; economic trends; employment; soil N, P, K; biocide use; nitrogen loss	region of Costa Rica (4,470 km^2)	present	air photo

(continued overleaf)

Table 12.1 (continued).

Reference	Purpose	Approaches	Description	Response Parameters	Spatial Scale	Time Scale	Remote Sensing
Brierley et al. 2009	3	used geomorphic relationships and channel form evolution to inform river restoration planning at basin scales	trajectory of adjustment; relation reach/downstream patterns; extensive catchment scale analyses of geomorphic response to human disturbance	river forms and associated biodiversity	Bega & Upper Hunter catchments	retrospective (Holocene); future (50-100 yr) rehabilitation actions	air photo
Fullerton et al. 2009	1, 3	land cover classification and modeling habitat relationships and fish responses, modeling fish passage beyond barriers	modeled fish responses to scenarios of degradation controlled sprawl, and planned growth and consequences of restoration of habitat above dams	fish habitat, riparian habitat, fish survival, sediment, runoff	Lewis River basin, WA, USA (2,760 km²)	2000–2050	Landsat, air photo
Hemstrom et al. 2001	3	land cover classification and modeling succession of vegetation in response to land use	projected land cover changes; developed indices of landscape health based on vegetation cover	Forest condition, rangeland condition	Interior Columbia River basin, USA (580,000 km²)	2000–2100	Landsat, air photo
Hubacek et al. 2001	2, 3	input – output (I–O) model, biophysical linkages (land use and land cover change), Socioeconomic changes linked to types of land (land requirement coefficients associated with specific economic activities), AEZ	China's land use and land cover change; technical change, income growth, changing pattern in consumption and production, urbanization, and population growth. 7 regional models and 1 national model	land use suitability, land productivity	China (9,596,961 km²)	1992–2025	IIASA LUC for China
Hunter et al. 2003	1, 2, 3	land cover classification and modeling changes in vegetation and land use and subsequent habitat for vertebrates	projected changes in habitat conditions for 12 endangered species and 244 other species under scenarios of human population density	Biodiversity (7 amphibians, 44 reptiles, 62 mammals, 153 birds), habitat quality, human development	Mohave Desert, California, USA (74,000 km²)	1990–2020	Landsat, air photo

Reference	No.	Approach/methods	Application	Focus	Location	Time period	RS data
Kepner et al. 2004	3	hydrological modeling, GIS; AGWA; Soil Water Assessment Tool (SWAT); KINematic Runoff and EROSion Model (KINEROS2)	changes in land-use (population, urbanization, agriculture practices) and consequences for surface runoff	land-use 2020 (current land management and projected census growth), Surface runoff, channel discharge, percolation, sediment yield	San Pedro River basin (7600 km^2)	2000–2020	Landsat, air photo
Lautenbach et al. 2009	3	biophysical approach, biotic factor; decision support system	effectiveness of management actions (reforestation, treatment plant technology, buffer strips) under the influence of external constraints on climate, demographic and agro-economic changes to meet water management objectives (water quality standards and discharge control)	erosion, nutrient surplus, land use changes	Elbe River basin, Germany (148000 km^2)	2000–2055	Landsat, air photo
Lawler et al. 2006	1	Modeling current vegetation types (MAPSS) and future climate (HADCM2SUL); modeling of mammal response to change in land cover	predict climate-induced range shifts of 2954 species of birds, mammals, and amphibians	presence and absence of a species; future range shifts and extinctions	Western hemisphere	2000–2100	AVHRR satellite data
Magdaleno et al. 2011	3	Pressures and impacts (IMPRESS) analysis, hydrological planning process	evolution of riparian forests and channel morphology; future scenario of ecohydrological management	spatiotemporal modification of the channel; evolution of riparian vegetation	central reach of 106 km of the Ebro River, Spain	retrospective (1927–2003)	air photo
Märker et al. 2008	1	erosion modeling RUSLE	impact of climate change and agriculture practices on erosion, A2 future climate scenario	soil erosion susceptibility, land-use sensitivity	750 km^2 Albegna River basin, Italy	2070	air photo
Nelson et al. 2009	1, 2, 3	land use/land cover classification, ecological response models	based on alternative futures from Baker et al. 2004; modeled ecosystem services	soil erosion, carbon sequestration, biodiversity, commodity production	Willamette River basin, Oregon, USA (30,000 km^2)	1850–2050	Landsat, air photo

(continued overleaf)

Table 12.1 (*continued*).

Reference	Purpose	Approaches	Description	Response Parameters	Spatial Scale	Time Scale	Remote Sensing
USDA 1994	3	land use/land cover classification, ecological response models	developed 9 alternatives for management and modeled ecological outcomes for 100 yr	forest, riparian reserves, roads	Federal forest lands in Pacific Northwest, USA (100,000 km²)	100 yr	Landsat, air photo
Santlemann et al. 2006	1, 3	land use/land cover classification, ecological response models	used land use/land cover to project responses of wildlife to 3 scenarios of increased agricultural production, water quality, and biodiversity	biodiversity, habitat quality for 247 species of vertebrates (146 birds, 52 mammals, 29 reptiles, 12 amphibians) and 117 butterfly species	Walnut and Buck Creek basins, Iowa, USA (51 and 88 km²)	1994–2025	Landsat, air photo
Schluter et al. 2007	3	modeling habitat responses to scenarios of water allocation (TUGAI)	projected changes in habitat quality based on water allocation model and model of land cover dynamics	water availability, habitat suitability	Amudarya River Delta, Tajikistan	28 model years	Landsat
Steinitz et al. 2003	2, 3	public information, land use, land cover, and models of species relationships to habitat	used patterns of land cover and human development to model ecological responses to changes in habitat in alternative landscape scenarios	vegetation cover, groundwater and surface water impacts, species habitat availability, endangered species habitat, vertebrate richness, visual preference	San Pedro River basin and region (10,660 km²)	2000–2020	Landsat, air photo
Vaché et al. 2002	1, 3	land use/land cover classification, water quality response to land use (SWAT)	effect of management practices on surface water discharge and annual loads of sediment and nitrate; evaluated response to agriculture and water quality scenarios	sediment, nitrate	Walnut and Buck Creek basins, Iowa, USA (51 and 88 km²)	1994–2025	Landsat, air photo

Virkkala et al. 2008	1	Generalized additive models (GAM), probability of occurrence as a function of climate and land cover, climate scenarios	response of bird species to alternative climate scenarios; use of models of climate and land cover change and bird habitat relationships to predict changes in bird species distribution	bird species distribution	Finland (338000 km²)	2000–2100	Landsat, satellite images (CORINE)
Wolski et al. 2008	3	modeling hydrology and inundation; model vegetation responses to projected changes in water availability	changes in river hydrology and floodplain inundation based 3 climate models; model related hydroperiod to vegetation assemblages	vegetation cover classes	Okavango Delta, Botswana	50 model years	Landsat
Zezere et al. 2004	1, 3	relationships landslide inventory and landscape variables,	integration of spatial and temporal data for the definition of different landslide hazard scenarios	landslide hazard, amount-duration combinations of rainfall responsible for past landslide event, landslide return periods	Fanhões – Trancão basin, Portugal (20 km²)	different return periods	air photo

local (Aggett et al., 2009), regional (Bolliger et al., 2007), or hemispheric (Lawler et al., 2006). Various time scales can also be considered, with retrospective analysis (Bolliger et al., 2007, Brierley et al., 2009), middle term prospective (Kepner et al., 2004, Santelmann et al., 2006), long-term prospective (Fullerton et al., 2009, Lautenbach et al., 2009), or very long-term prospective analysis (Virkkala et al., 2008, Baas et al., 2010). In geomorphology, such approaches are often used to assess the impact of human activities (Zezere et al., 2004), land use change (Bouman et al., 1999, Kepner et al., 2004) or the effects of natural processes on the morphology of a river and its floodplain (Wolski et al., 2008, Magdaleno et al., 2011). Recently, these approaches have been increasingly used to assess the future quality of habitats in response to human activities (Hemstrom et al., 2001, Vaché et al., 2002, Fullerton et al., 2009), or climate change (Lawler et al., 2006).

12.1.3 Methods of employing remotely sensed information in alternative futures

Three major methods are commonly used to project biophysical or human infrastructural characteristics of the landscape from remotely sensed information. First, remotely sensed information can be classified or interpreted to explicitly represent landscape features, such as land cover (e.g., vegetation types, water, rock or soil) (Steel et al., 2004, Fullerton et al., 2009). Second, remotely sensed information can be enhanced with social information to increase the resolution of the classification, such as land use (e.g., crop type, residential and industrial zoning, transportation data, census data) (Baker et al., 2004). Third, algorithms or models can be used to estimate biophysical characteristics from remotely sensed information (e.g., fish and wildlife community structure, abundance, habitat quality) (VanSickle et al., 2004, Schumaker et al., 2004, Lawler et al., 2006).

The purposes of the analysis strongly influence the processes for applying remotely sensed information. Processes used to develop scenarios include definition of assumptions for scenarios and their interpretation by scientists, stakeholder groups, decision makers and managers.

The remainder of this chapter describes, as a case study, work conducted between 1995 and 2002 by the Pacific Northwest Ecosystem Research Consortium, a multi-university/federal agency partnership with support from the US Environmental Protection Agency. This effort consisted of a river-basin extent alternative future scenario analysis that employed remotely sensed information as a

cornerstone of its representation of the territory under study, western Oregon's Willamette River basin.

12.1.4 Alternative future scenarios for the Willamette River, Oregon as a case study

This chapter is based on the research of more than 30 scientists from Oregon State University, University of Oregon, and the US Environmental Protection Agency (EPA) who assembled ecological and social information about the Willamette River basin (WRB) to assess trajectories of environmental change from 1850 to 2050. Past, present, and future landscapes were depicted in maps to represent scenarios of landscape conditions and human actions that shape the trajectories of ecological conditions in this basin. Rather than having scientists define the assumptions and relationships for future scenarios, we used the recommendations of groups of stakeholders who met with us for more than two years to articulate and review the assumptions used to create spatially explicit characterisations of three future scenarios for 2050 (Hulse et al., 2004). This summary of environmental trajectories in the WRB is reported in detail in the Willamette River Basin Planning Atlas (Hulse et al., 2002) and a special issue of Ecological Applications (Baker et al., 2004 and associated papers).

This chapter aims to: 1) explore many of the aquatic and terrestrial resource responses that were informed by remotely sensed information and 2) illustrate an approach at nested spatial analyses within a large river basin. This basin scale assessment of trajectories of ecological change is based on spatially explicit representations of 1) historical land cover ca. 1850 prior to recent trends of land conversion and resource consumption, 2) current land cover and land use ca. 1990 based on remotely sensed data from Landsat images and aerial photography, 3) alternative future scenarios of land use and land cover ca. 2050 based on an interactive stakeholder process, and 4) modeled projections of major ecological characteristics of these landscape scenarios based on observed empirical relationships in the WRB.

The landscape assessment in this research required integration of remotely sensed data and ecological models to explore complex relationships between people and the changing landscapes that shape their futures. The 30,000-km^2 Willamette basin makes up 12% of the state of Oregon and supports more than two-thirds of Oregon's population, which is expected to double by 2050. The basin provides important agricultural and forest products for the region but also is one of the most ecologically diverse basins in the region. The Willamette River and its tributaries contain the richest native fish fauna in Oregon, as well as several species listed under the Endangered Species Act as threatened or endangered (spring Chinook salmon, steelhead trout, Oregon chub). Forests account for more than two-thirds of the WRB, including federal forests in higher elevation uplands and a mosaic of private forest lands distributed across the lower elevations of the Cascade Mountains and Coast Range. More than half of the Willamette Valley has been converted to agricultural use (43% of the valley area) and urban/rural residential development (11% of the valley area). Population growth, social policies, and individual choices have substantial potential to change the environment and livability for local communities (Baker et al., 2004). The research described in this chapter created a framework for quantifying historical resource loss and anticipating potential future trends while providing several mechanisms for citizen involvement in the assessment and prescription of their region.

Oregon is relatively unique in the United States because, in addition to resource management agencies, it has developed land use zoning regulations for all lands, urban growth boundaries for all cities, ecosystem restoration-oriented natural resource management agencies, and more recently citizen-based watershed councils to implement local conservation and restoration practices. The WRB contains 23 active watershed councils comprised of local citizens and land owners. In addition, the Willamette Livability Forum, a group of 150 regional community leaders, critiqued our representations of future scenarios and the assumptions on which they were based (Hulse et al., 2004).

The stakeholder advisory team defined plausible assumptions of future actions and policy decisions over the period from 1990 to 2050 for three different alternative future scenarios, each accommodating a doubling of the human population. The three scenarios encompassed a range of potential future trajectories (Figure 12.1), with the Plan Trend 2050 scenario representing the future landscape if we continue to implement our current policies and practices. The Development 2050 scenario represented a shift in current policies and practices that relaxed environmental regulations to provide greater flexibility for market forces. At the opposite end of the spectrum of plausible choices, the Conservation 2050 scenario depicted the future landscape with greater application of conservation and restoration actions to insure long-term ecological function. Future land use and land cover patterns under these three future

scenarios were determined by land allocation models for agriculture, forestry, urbanisation, rural residential development, natural habitat processes, and water use.

12.2 Methods

Maps of land use and land cover (LULC) change over time, created in part from remotely sensed information, provided an empirical source of evidence for the trajectories of landscape change. The spatial framework of coupled human and natural systems in the river basin required an extensive and detailed land cover representation (Hulse et al., 2004). Each LULC map was represented by 30 m by 30 m pixels and the 33 million pixels collectively represented the 30,000-km^2 WRB. Several sources of data were used to define the land cover or the land use depicted for each pixel. A pre-European settlement land cover map was created from surveys of the U.S. General Land Office (GLO) from plat maps and survey notes ca. 1851. LULC maps for ca. 1990 were created initially from image classification of multi-temporal Landsat Thematic Mapper (TM) satellite imagery.

Upland portions of the basin in the Coast Range or Cascade Mountain Ecoregions were mapped from a single-date 1988 TM data set. The map of the more heavily populated and agricultural valley region below 315 m elevation was created from an analysis of physiognomic changes in a multi-season data set consisting of five images from March, May, June, July, and August 1992 (Oetter et al., 2001). Each of the five 1992 images was georeferenced to the geocoded 1988 TM data set. We used an automated ground control point selection algorithm and a second-order polynomial, nearest neighbour resampling (Kennedy and Cohen, 2003). The root mean square error was less than one pixel for all dates.

Radiometric normalisation was an important consideration. We used a relative normalisation technique, whereby each date was adjusted to a common image. The June 1992 image served as the reference, and the other dates were adjusted to it. First a subset of image pixels containing water, forest, and urban areas were identified which were determined to be no-change pixels throughout all the image dates. The no-change pixels were selected by multiple isodata clustering of a seven-band image (six bands of TM differences plus Band 4 of the reference image). Then those pixels were subsampled using a random stratified method to represent equally the three types of land cover. Each band of the uncorrected images was then transformed using a single regression linear

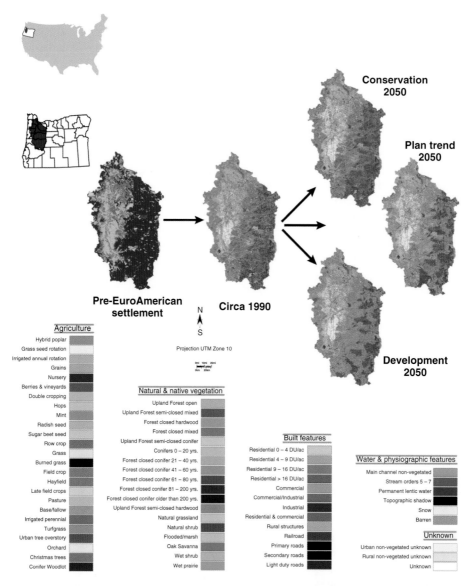

Figure 12.1 Maps of the Willamette River basin ca. 1850, 1990, and three alternative scenarios of the future – Plan Trend 2050 depicting landscape patterns that would emerge based on current policies and practices, Development 2050 depicting outcomes if environmental policies are relaxed, and Conservation 2050 depicting landscape patterns that would be likely if conservation policies and practices are strengthened and expanded. All future scenarios include a doubling of the human population.

equation relating the no-change pixels in the uncorrected image to the same pixels in the reference image (Oetter et al., 2001). Given the number of spectral bands across five different dates and the large geographic study area, each image was subset to the first three tassled cap transformation bands to save file space and allow for physical interpretation of band values.

The basin was sub-divided into four mapping units: 1) the forest area previously mapped in the 1988 Western Oregon forest cover project (Cohen et al., 2001), 2) urban area within the basin as defined by the state's urban growth boundary zoning distinctions, 3) other forest area mainly in the valley floor, and 4) non-forest agricultural and natural areas.

These analyses were used to determine 56 categories of land cover including crop type. Classification of agricultural land cover was augmented with information from aerial photographs, US Department of Agriculture (USDA) farm records, county farm statistics, and ground truthing (Enright et al., 2002, Berger and Bolte 2004), creating LULC maps used for model projections of trends in natural resource abundance and distribution.

12.2.1 Ground truthing

The ground reference data used to complete this study came primarily from aerial photographs and digital orthophotographs, especially:

1988 & 1990 Color 1:12000 U. S. government aerial photographs of U.S. Forest Service (USFS) and Bureau of Land Management (BLM) lands

1988 Color Infrared 1:60000 U. S. government 1988 National High-Altitude Program photographs (NHAP)

1993 Color 1:24000 Western Aerial Corporation (WAC) aerial photographs of Willamette basin

1992 Color 35 mm UUSDA Farm Service Agency (FSA) aerial slide photographs

1994-95 Black & White Spencer Gross digital orthophotographs of the Willamette river and its major tributaries

1996 Color digital orthophotograhs of metropolitan Portland, Oregon area

The USFS and BLM photographs were used to reference land cover conditions in the forested uplands of the basin. The NHAP images were used solely to identify a small number of testing polygons for the barren (lava) and permanent snow cover classifications. The colour digital orthophotograph set, taken from a six mi^2 area near Tigard, Oregon was used to reference the urban land cover classification. The Spencer Gross digital orthophotographs were used for cross-reference and spot-checking of forest, urban, and agricultural cover types. The 1993 WAC aerial photographs were used to reference forest cover conditions in and along the valley floor. Two hundred and eighty-four forest cover polygons were interpreted for percent canopy closure of conifer and hardwood tree species, particular to oak, orchard, or Christmas trees where discernable. This data set was divided into training and testing sets and used to reference the valley floor forest classification.

The majority of the non-forest portion of the basin was referenced using the Farm Service Agency 35 mm slides. These data are acquired annually by the FSA for their crop compliance program. They were not georeferenced nor subjected to strict photogrammetric requirements, yet proved very useful for the identification of crop types in 1992. Each of 369 slides was scanned into a tif format at 300 dpi, then imported to Imagine 8.3 and georeferenced to the Landsat TM imagery with a minimum of nine ground control points and an acceptable RMS error of 10 m. The slides were then mosaiced into 34 separate study areas, intentionally distributed throughout all nine counties in the basin, and across all land cover types.

To train the photo interpretation of the FSA slides, five study areas were interpreted for each property tract which participates in the FSA crop compliance program. Our interpretations were then compared to the crop reports filed by the farmers for the study year, which yielded a photographic interpretation error of less than 12%. Over 1100 polygons within the FSA study areas were interpreted for 56 different land use/land cover codes. This data set was then divided by class and geographic location into testing and training data sets.

The interpretation of individual plots was sometimes aided by inspection of ancillary GIS coverages, including the Environmental Protection Agency Pudding River riparian assessment, U. S. Fish & Wildlife Service National Wetlands Inventory, U. S. Geological Survey Willamette River Basin pesticide study, Oregon Department of Fish and Wildlife Ecological Application Laboratory natural vegetation inventory, county zoning and taxlot coverages, The Nature Conservancy Willamette River basin wetlands inventory and valley field visit database. In addition, inspection of the Landsat TM imagery was used to determine greenness curves throughout the 1992 growing season. Where possible, field visits were conducted during 1998 to examine land use types that may have remained consistent over the six-year interval since the photographs were acquired (Oetter et al., 2001).

An external review of the land cover classification concluded that aggregated land use/land cover classes derived from the Landsat TM data 'will likely yield correlation results comparable to more costly photo-interpreted data for certain aggregated LCLU classes and spatial domains (e.g., riparian zones). Forest, woody, woody+grass/forb, and agriculture classes appear strongly correlated across a broad range of lateral and longitudinal scales, suggesting that they may be good candidates for future model consideration. The built-up class variable was not as strongly correlated, but may have utility at the largest riparian band (150 m) or watershed levels' (Lattin and Peniston, 1999). Accuracies of spectral classification of land cover were reported in Lattin and Peniston, 1999 and Hulse et al., 2002.

In areas of extensive human development and land conversion, additional data sources were used to refine satellite data (e.g., tax lot data, 1990 U.S. Census data, road data from transportation agencies, aerial photographs, USGS topographic quadrangle maps). Rivers, streams, lakes, and wetlands were mapped using TM and U.S. Census data, USGS maps, digital elevation models, aerial photographs, and ground truthing. Maps of historical channels and riparian forests of the mainstem Willamette River were created from General Land Office survey maps from 1850–1860 and US Army Corps of Engineer river maps from 1895 and 1932 (Gregory et al., 2002d, 2002g, Oetter et al., 2004).

12.2.2 Use of remotely sensed data in the larger alternative futures project

Simply put, the map that resulted from the methods described above was used to depict existing conditions ca. 1990 for the Willamette River basin. This depiction became a baseline reference for comparing changes from the past to the present, and from the present to three alternative futures. These depictions were used to better understand and anticipate trajectories of change in human occupancy and natural resource condition from ca. 1850 to ca. 2050. In the Willamette River basin project, all three major purposes of scenario development identified by Henrichs et al. (2010) were served. During a 30-month period, we worked with lay and professional citizen groups to create, map, and refine a set of value-based assumptions about future policy in three scenarios concerning land and water use. The Plan Trend 2050 scenario represented the expected future landscape in 2050 if current policies are implemented as written and recent trends continue. Development 2050 reflected a loosening of current policies, to allow freer rein to market forces across all components of the landscape, but still within the range of what citizen stakeholders considered plausible. Conservation 2050 placed greater emphasis on ecosystem protection and restoration, still reflecting a plausible balance among ecological, social, and economic considerations as defined by the stakeholders. For the Conservation scenario, natural resource managers and scientists provided estimates for the area of key habitats required to sustain, in perpetuity, the array of dependent species. Spatially explicit analyses identified locations biophysically suited to meet the area targets. These locations, titled the Conservation and Restoration

Opportunity Areas, were mapped and then reviewed by a series of groups regarding the political plausibility of conserving or restoring them to the indicated vegetation types. The three alternative 2050 futures, as well as the 1850 past conditions, were then evaluated by an array of computerised evaluation models.

Maps of several natural resources – forest type, riparian forest, surface water – were used as ecological response measures in the assessment of alternative scenarios. In addition, the effects of each future on major terrestrial and aquatic communities were modeled based on observed relationships between habitat and community composition or species abundance in the WRB (Schumaker et al., 2004, Van Sickle et al., 2004). Models were used to project wildlife diversity, native fish richness, cutthroat trout abundance, fish habitat suitability indices, and EPT richness (numbers of taxa of Ephemeroptera, Plecoptera and Trichoptera; orders of aquatic macroinvertebrates that require high water quality). Models of future water availability were based on designated water availability basins and projected water use by different water use sectors (Dole and Niemi, 2004). Future environmental changes that are recognised but not analysed in this study of future scenarios include climate change, groundwater withdrawal, or new toxic substances in the environment. Past effects on these three phenomena are inherent in the observed relationships between land cover, land use, and biotic responses, but future changes in these environmental parameters may not be reflected in our analyses.

12.3 Land use/land cover changes since 1850

Land cover in the Willamette basin changed markedly from the time of settlement to the present, as illustrated by changes in mapped land cover from 1850 and remotely sensed land cover from 1990. During this 140-year period of EuroAmerican settlement, lowlands of the WRB were settled and converted to agricultural and urban land uses. Historically 50% of the lowlands were forested, and by 1990 forested land in the lowlands had been reduced to approximately 30% (Gregory et al., 2002a, 2002e). The uplands were almost completely covered with forests (98%), which largely remain the case (96%). Older forests (conifers >80 yr) once covered 60% of the WRB, but by 1990 these older forests occupied only 20% of the basin.

The land cover of the WRB changed markedly from 1850 to 1990, and riparian forests changed to an even greater extent. River floodplain and riparian corridors at the time of settlement were heavily forested by Oregon ash, black cottonwood, bigleaf maple, white alder, and several willow species in the lowlands. Montane rivers in the basin contained coniferous riparian zones mixed with red alder and willow. More than 80% of the riparian areas of the Willamette River were covered by mature forests, which extended 2–10 km wide along the mainstem Willamette River (Gregory et al., 1998, Gregory et al., 2002b). In 1850, prairies and oak woodlands that were major features of the Willamette Valley floor were minor components of the floodplain vegetation. Of the 445 km of mainstem river channels (both mainstem and side channels), forests and woodlands that contained hardwoods occupied 68% of the length, mixed forests 14% and conifer forests 7% (Gregory et al., 2002f). Prairie and wetland habitats have been reduced extensively throughout the Willamette Valley, but these open habitats historically accounted for only 6% of the riparian margin. By 1990, the riparian length of hardwood forests was 17%, mixed forests 18%, and conifer forest 2%, reducing wooded riparian forests from 89% of the riparian length to 37% of the riparian length.

Longitudinal patterns of the floodplain forest extent are illustrated in 1-km bands or slices of the floodplain from the upper mainstem river to its mouth (Figure 12.2). By 1990, the extent of floodplain forest was reduced to less than 10% of the historical floodplain area. Floodplain bands of 1-km in length generally contained more than 200 ha of forest in the more geomorphically complex middle and upper river. By 1990, not a single reach of the Willamette River floodplain contained 200 ha/km and most had been reduced to less than 50 ha/km (Gregory, 2002i).

In 1850, the Willamette River was physically more complex than in 1990 (Figure 12.3). Channelisation and bank hardening have straightened the mainstem, reducing total length of river channels by 22%, area of off-channel sloughs by 30%, and islands by over 60% (Gregory, 2007). Prior to settlement and river modification, the floodplains of the upper Willamette River contained as much as 11 km of channel length within a 1-km floodplain distance, and most of the reach included 4–8 km of channel per km of floodplain (Figure 12.3). By 1995, no reach exhibited more than 7 km of channel/km of floodplain and most were less than 2 km of channel/km of floodplain (Gregory et al., 2002h).

Along smaller streams, riparian areas and wetlands also changed greatly from 1850 to 1990. In the lowlands, riparian areas within 120 m of small streams (1st to 4th order) were 25% conifer, 25% hardwood, and 5% mixed forest (Gregory et al., 2002b). By 1990, these forest types were reduced to less than 10% for conifer and hardwood forests and increased to 13% for mixed stands, representing a loss of more than 40% of the riparian areas along small streams. Wetlands decreased from 14% to 1% of the riparian area.

Land conversion and fire suppression around homes, farms, and towns have altered terrestrial land cover, eliminating most of the prairie and oak savanna habitat, both of which are fire-dependent. Upland forests are in public and private ownership and are still dominated by coniferous forests. However, timber harvest has shifted the age of the upland forests, reducing the extent of conifers older than 80 yr by approximately 66%. Consumption of surface water for irrigation, municipal and industrial water uses removes more than 1,000 m^3 of water/day, dewatering an estimated 130 km of 2nd to 4th order streams in moderately dry summers. None of these streams would be expected to become dry in the absence of human withdrawals (Niemi et al., 2002).

Land conversion between 1850 and 1990 caused major changes in terrestrial and aquatic habitats in the WRB. All models of natural resources (fish, aquatic invertebrates, wildlife, riparian forests, forest cover, surface water availability) exhibited major declines from pre-EuroAmerican settlement to the present (Figure 12.4; Baker et al., 2004). Most of the natural resources we used as ecological indicators were 15 to 90% more abundant historically than today, depending on the specific resource.

12.4 Plan trend 2050 scenario

In spite of the projected doubling of the human population in the WRB from 2 million to roughly 4 million by 2050, land conversion from 1850 to 1990 was far greater than the additional land conversion projected to occur by 2050 regardless of the future scenario (Figure 12.4). Future landscape changes will, to a greater extent, shift land uses on already converted lands as opposed to converting intact native landforms and streams to additional urban and agricultural land use. Existing public lands and unimpacted, natural ecosystems are likely to be conserved. New urban or rural residential areas will be developed on lands that previously had been converted to agriculture. Though future land conversion is not projected to be as spatially extensive as historical land conversion, projected outcomes for alternative scenarios differed substantially

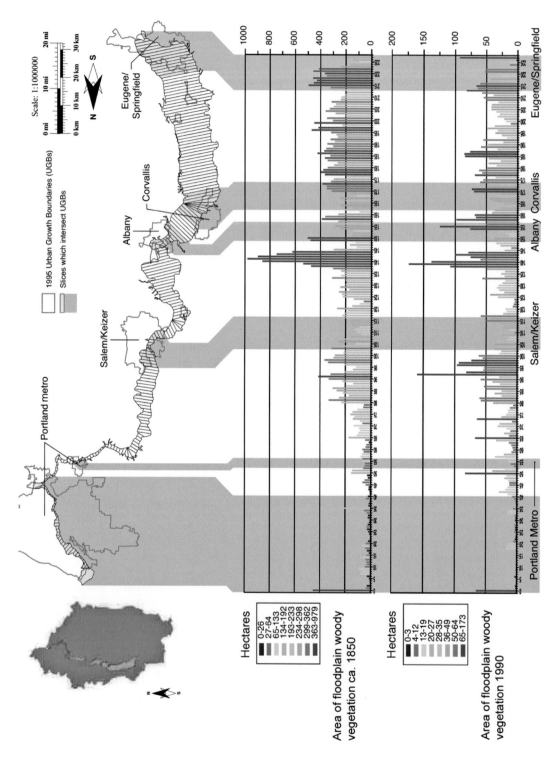

Figure 12.2 Longitudinal patterns of floodplain forests in 1-km bands perpendicular to the main axis of the floodplain along the mainstem of the Willamette River. Note change in scale of vertical axis. Reproduced from Willamette River Basin Planning Atlas: Trajectories of Environmental and Ecological Change, edited by David Hulse, Stan Gregory, and Joan Baker for the Pacific Northwest Ecosystem Research Consortium (Oregon State University Press, 2002), with permission from Oregon State University Press.

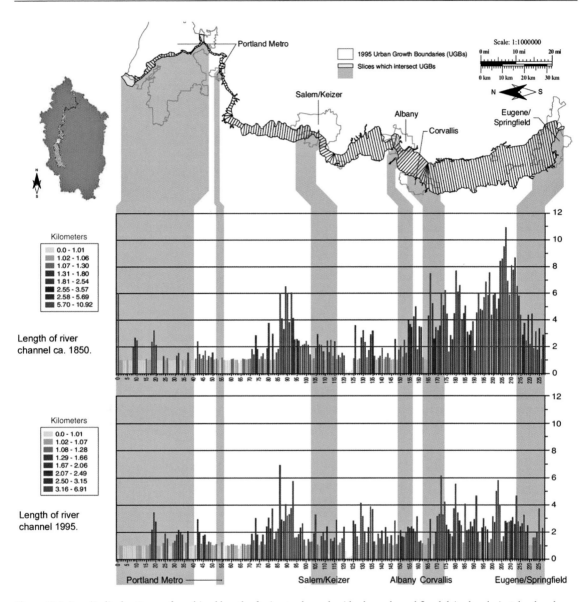

Figure 12.3 Longitudinal patterns of combined length of primary channels, side channels, and floodplain sloughs in 1-km bands perpendicular to the main axis of the floodplain along the mainstem of the Willamette River. Reproduced from Willamette River Basin Planning Atlas: Trajectories of Environmental and Ecological Change, edited by David Hulse, Stan Gregory, and Joan Baker for the Pacific Northwest Ecosystem Research Consortium (Oregon State University Press, 2002), with permission from Oregon State University Press.

because of the differences in intensity and distributions of land uses and human development.

The Plan Trend 2050 scenario fundamentally represented the continued application of existing land use policies and practices with an increasing human population. Several major land use policies in Oregon strongly shape land use patterns for the future. The Oregon Land Use Planning Program requires local government to develop comprehensive land use plans to constrain patterns of urban development, prevent loss of agricultural and forest lands, and maintain natural resources. These locally developed land use plans are then coordinated by state government. Private and state forests are regulated by the Oregon Forest Practices Act to maintain productive forest lands while protecting natural resources with riparian buffers and other conservation measures.

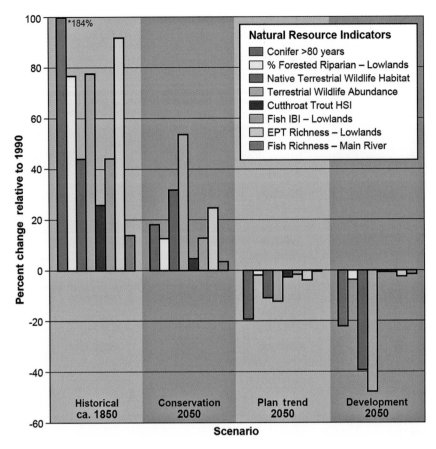

Figure 12.4 Percentage change in selected natural resources indicators in the Willamette River basin. Zero change represents resource conditions circa 1990. Reproduced from Baker et al. (2004), with permission from the Ecological Society of America.

Federal forest lands are managed by the Northwest Forest Plan, which established extensive networks of riparian reserves and late successional forest reserves to maintain regional biodiversity beginning in 1998. Riparian reserves account for 11% of the federal forest lands in the WRB, late successional forest lands account for another 30%, and wilderness areas and other congressionally withdrawn lands make up another 30%.

Projected human population growth and implementation of existing land use regulations resulted in a dramatic increase in population density within areas zoned for urban and residential development. Population densities increased from an average of 9.4 residents/ha in ca. 1990 to 18.0 in 2050 even though the land base for urban areas and residential areas increased by only 25%. Increased housing density accommodated the population increases while protecting more than 98% of the existing farm and forest land. Though the forest land base did not change

greatly, continued timber harvest reduced the amount of older forest land (age >80 years) by 19% on private and state forests.

Model projections indicate that land use changes under Plan Trend 2050 caused relatively small changes in aquatic communities and terrestrial wildlife (~10% relative to 1990) across the WRB. Local resources and certain species showed substantial changes, but the change in aquatic and terrestrial communities from 1990 to 2050 were far less than estimated losses from historical conditions to the present (Figure 12.4). Water consumption and water availability changed greatly because of the doubling of the human population. With almost a 60% increase in surface water withdrawals, 20% increase municipal and industrial uses and doubling of irrigated agriculture, more than 270 km of 2nd to 4th order streams went dry during dry seasons in contrast to 130 km in 1990. Oregon Water Resources Department tracks 178 subbasins for water availability (Figure 12.5). All Water Availability Basins had water at their outlets in 1990, but 8% of the WRB area would have near zero stream flow at their outfall in 2050.

12.5 Development 2050 scenario

Under the Development 2050 scenario, land use policies were relaxed and urban and rural residential land was permitted over larger areas. Population densities in urban growth boundaries (14.6 residents/ha) increased by 55% relative to 1990. Urban and rural developed land accounted for 10.4% of the total WRB area, in contrast to 6.7% ca. 1990 and 8.3% in Plan Trend 2050. Most new development occurred on agricultural lands in the halos around existing towns and cities, often impacting prime farmland. Almost a quarter of the prime farmland was lost in the Development 2050 scenario. Though forestry practices included more clear-cutting and less stream protection than in Plan Trend 2050, major relaxation of natural resource regulation of forest lands would be unlikely.

Terrestrial wildlife exhibited greater responses to the practices and policies of the Development 2050 scenario than aquatic communities (Figure 12.4). Of the 17 terrestrial wildlife species modeled for changes in population abundance, nine experienced a 10% or greater decline in abundance relative to 1990 and the coyote, a habitat generalist, was projected to increase by at least 10%. Aquatic resources modeled in the assessment declined more than the Plan Trend 2050 projections but only slightly more than resource conditions in 1990. The major cause for the lack of major differences in aquatic resource responses under the Plan Trend and Development scenarios is the similarity of adverse effects of agriculture and residential development in the resource models. Stream habitats on agricultural land were not projected to decline further with conversion to residential use. Water consumption increased under Development 2050, though not as much as Plan Trend 2050. In a dry summer in 2050, 230 km of 2nd to 4th order streams and 5% of the WRB area would be dry. This surprising difference in water use is a result of the conversion of irrigated agricultural land to residential land.

12.6 Conservation 2050 scenario

The Conservation 2050 scenario assumed more extensive conservation and restoration measures than are implemented currently within the basin. These practices and policies were based on methods and programs that existed in 1990 but would be implemented more extensively across the WRB. Future human growth was contained largely within urban growth boundaries and existing residential developments. Urban and residential developments retained more natural vegetation. By living in higher densities, the doubled human population by 2050 required only an 18% increase in urban and rural residential land. Almost no farmland and forest was lost to development (<2%), and conservation practices reduced prime farmland and forest present in 1990 by only 15%.

Plausible future conservation actions included 30-m riparian buffers along all streams, restoration of grasslands, wetlands, oak savannah, and bottomland forests, establishment of field borders, protection of wildlife habitat in environmentally sensitive areas, and a 10% increase in irrigation efficiency. Side channels and sloughs along the mainstem Willamette were reconnected and restored at the same rate they have been disconnected or lost since the 1930s. Floodplain forests along the Willamette River were restored to meet the extent of bottomland forest that stakeholders identified as plausible. Private forest practices were modified to include 30-m riparian buffers on all 2nd to 4th order streams, smaller clearcut sizes, and green tree retention.

These modifications to agricultural and forest land uses under the Conservation 2050 scenario increased the area of conifer forests aged 80 years and older by 17% relative to ca. 1990. In contrast, area of conifer forests aged 80 years and older decreased by 19% and 22% under Plan Trend 2050 and Development 2050, respectively. The conservation practices reversed the loss of older forests, but the extent of these forests under Conservation 2050 would still be less than half of what occurred at the time of EuroAmerican settlement.

Abundance and distributions of aquatic and terrestrial wildlife increased under the Conservation 2050 scenario (Figure 12.4). Aquatic resources responded positively to conservation measures, particularly in lowland streams where land conversion had altered the streams, rivers, riparian forests, and floodplains to a far greater extent than in the mountainous portions of the basin. Habitat suitability for cutthroat trout throughout the WRB improved as a result of riparian practices, and increased channel complexity and extent of floodplain forest caused fish community richness to increase in the mainstem of the Willamette River (Van Sickle et al., 2004).

In the lowlands, changes in land use practices along streams caused the index of biotic integrity for fish communities and macroinvertebrate EPT richness to recover

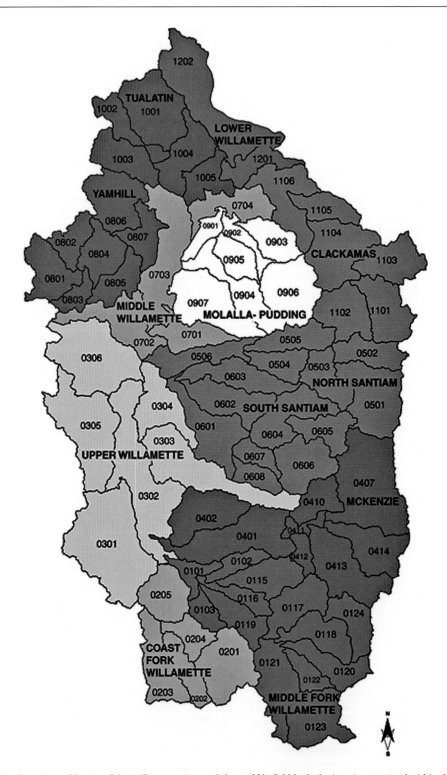

Figure 12.5 Twelve major subbasins of the Willamette River and the 77 fifth-field hydrologic units contained within the catchment. Reproduced from Willamette River Basin Planning Atlas: Trajectories of Environmental and Ecological Change, edited by David Hulse, Stan Gregory, and Joan Baker for the Pacific Northwest Ecosystem Research Consortium (Oregon State University Press, 2002), with permission from Oregon State University Press; and from US Geological Service (1997).

by 9–24% relative to ca. 1990. Instead of continued decline projected for the Plan Trend 2050 and Development 2050 scenarios, ecological endpoints for aquatic communities recovered 20–65% of the losses observed for these indicators since EuroAmerican settlement.

Terrestrial wildlife also responded positively under the Conservation 2050 scenario (Figure 12.4). More species (31% of the modeled species) gained habitat than lost habitat relative to ca. 1990 (Schumaker et al., 2004). The majority of wildlife species modeled for population abundance increased by at least 10% relative to ca. 1990, and only the mourning dove decreased by more than 10%. Like aquatic communities, most wildlife species not only slowed their rate of decline but also recovered a substantial amount of their historical abundance and distribution.

The Conservation 2050 scenario consumed less water than Plan Trend 2050 and Development 2050 with no water availability basins having zero flow. Though consumption was lower in this scenario, length of stream channels that dry in summer increased by 70% compared to ca. 1990 (225 km of 2nd to 4th order streams). The region is likely to face even greater water shortages in the future unless more stringent water conservation measures are adopted.

12.7 Informing decision makers at subbasin extents

We developed representations of ecological resource abundance for stream reaches with minimum mapping units of 30-m pixels for the five scenarios of past, present, and alternative future landscapes (Schumaker et al., 2004, Van Sickle et al., 2004, Hulse et al., 2002). One of the social challenges in illustrating data from remote sensing or ecological models at the scale of river basins is the 'My Backyard' response. When presented with spatial information, it is human nature to examine the map to see the characteristics of the portion of the landscape the observer knows best, often their own home, neighbourhood, or community. Even though overall error or uncertainty at the landscape scale is often relatively low, the potential for error in any single pixel or fine scale mapping unit is much greater. Regardless of the level of accuracy at the basin or landscape scale, most commonly the consistency between mapped representations and the individual's personal experience for a small subset of the basin determines their confidence in the maps, models, and landscape analyses. For that reason, we were reluctant to depict numerical

values for each stream reach or pixel. However, we knew that resource managers and decision makers faced key resource decisions in subbasins within the larger basin. Our overall projections of the consequences of alternative future scenarios would not be adequate for allocating conservation and restoration actions at smaller spatial extents. One alternative to reach-by-reach depictions of habitat conditions or resource abundance is subbasin analyses that highlight differences between major portions of a larger landscape.

We aggregated the analyses of resource responses for the landscape scenarios for the 12 major subbasins of the WRB (Figure 12.5; Branscomb et al., 2002). These 12 subbasins were comprised of 77 fifth-field hydrologic units (USGS 1997) and differed in land use, land cover, and population density. Two measures of aquatic habitat (habitat suitability index, large wood volume) and three biotic responses (fish richness, cutthroat trout abundance, abundance of EPT macroinvertebrate taxa) were projected for 4,045 reaches of the WRB (see Van Sickle et al., 2004 for model descriptions and empirical data on which models were based).

The habitat suitability index (HSI) model for cutthroat trout was an expert-based model derived from stream ecosystem function and habitat use by cutthroat trout. Separate HSI models were developed for lowland and upland streams with watersheds predominately in the Willamette Valley Ecoregion (lowland streams) versus streams with watersheds predominately in the Cascade or Coast Range Ecoregions (upland streams). HSI was calculated from a weighted model of stream gradient, annual mean flow, valley floor constraint, wood potential, closed forest in riparian network, % natural vegetation in riparian network, closed forest in the watershed, road density in the watershed, % development land in riparian network, and % agriculture in riparian network. Wood volume represented the potential contribution of large wood to the stream from the riparian area immediately adjacent to the reach, calculated as the weighted sum of 11 different forest vegetation classes with older conifers weighted most heavily (Van Sickle et al., 2004, Gregory et al., 2003). Native Fish Richness was a function of stream order, distance along the stream network from the sources, elevation, stream gradient, percent agricultural land in the 120-m riparian area, and percent developed land in the riparian area (Gregory et al., 2002c). Cutthroat trout abundance was a function of watershed area, percent agricultural land in the 120-m riparian area, and percent developed land in the riparian area. EPT richness was a function of lowland, stream power, percent agriculture and percent development.

All subbasins exhibited substantial declines on aquatic HSI (Figure 12.6), with the greatest decreases in the lower Willamette River subbasin near the Portland Metropolitan area. This portion of the WRB contains the highest population density of any subbasin and has experienced extensive land conversion. The model of average native fish richness in the subbasins projected fewer native species per stream reach under current landscape conditions than would have been expected in 1850 based on habitat conditions (Figure 12.6c). Greater native fish richness was projected in the lower Willamette River subbasin, but substantial declines also were noted in some of the upper subbasins, such as the Coast Fork of the Willamette and North Santiam River (Figure 12.6b). These same overall patterns were observed in estimates of cutthroat trout abundance per 1-km reach, with

these three subbasins exhibiting the greatest proportional decreases in trout abundance (Figure 12.6a). The lower Willamette River subbasin exhibited the greatest projected decline in cutthroat trout abundance, which is consistent with recent fish surveys. Though cutthroat trout were widespread in the lower Willamette River subbasin historically, recent surveys find almost no cutthroat trout in the mainstem Willamette River (Tom Friesen, Oregon Department of Fish & Wildlife, personal; communication; field sampling by Gregory) and low abundances in small tributaries of the subbasin. Aquatic macroinvertebrates show even greater relative declines in EPT richness (Figure 12.6d). The EPT orders – mayflies, stoneflies, caddisflies – require high water quality, substrate quality, and habitat complexity. Differences in habitats determined by remotely sensed land cover and land use cause the relative abundances of EPT species to be inherently greater in some subbasins, but all subbasins

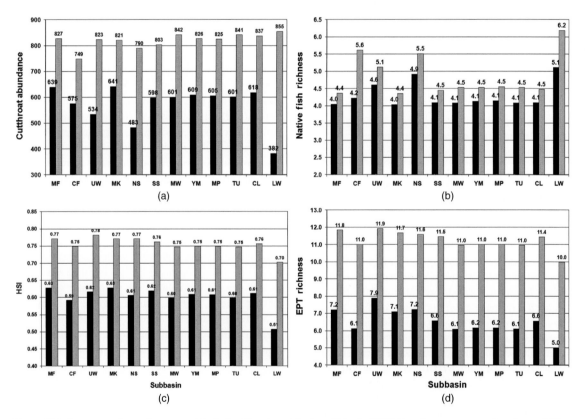

Figure 12.6 Responses of aquatic communities in the 12 major subbasins within the WRB. Grey bars represent values for ca. 1850 and black bars represent conditions in 1990. MF - Middle Fork Willamette, 3 533 km²; CF – Coast Fork, 1 725 km²; UW – Upper Willamette, 4 839 km²; NS – North Santiam, 2 012 km²; SS – South Santiam, 2 696 km²; MW – Middle Willamette, 1 807 km²; YM – Yamhill, 1 998 km²; MP – Mololla and Pudding, 2 272 km²; TU – Tualatin, 1 837 km²; CL – Clackamas, 2 436 km²; LW – Lower Willamette, 1107. See Figure 12.5 for map.

in the WRB have projected declines in EPT richness of approximately 40%. Again the lower Willamette River subbasin exhibited the greatest proportional decline and lowest projected EPT richness.

The spatial responses of aquatic communities and habitat characteristics can also be illustrated and contrasted between the lowland and upland regions of the WRB. HSI values for habitat conditions for cutthroat trout in 1850 were similar in uplands and lowlands (Figure 12.7d), which is consistent with cutthroat trout being distributed throughout lowland streams in the early to mid-twentieth century (Dimick and Merrifield, 1945). By 1990, habitat quality for cutthroat trout declined in both lowland and uplands, though the decrease in potential habitat quality was greater in the lowlands where land conversion was more extensive. Potential delivery of wood to streams and contribution to complex aquatic habitat was much greater in the more forested uplands where riparian areas were almost completely made up of older coniferous forests (Figure 12.7c). Loss of riparian and floodplain forests

described earlier account for the reduction of potential wood delivery by more than 30% in the lowlands and by 15% in the uplands. This loss of sources of large wood was responsible for a portion of the loss of complex habitat in the WRB. Reflecting these changes in habitat quality and sources of large wood for streams and rivers, the models projected reductions in cutthroat trout abundance in the lowlands of the WRB of more than 80%, but projected abundance reduced only slightly in the uplands (Figure 12.7a). Native fish richness also exhibited much higher diversity in the lowlands and much greater reductions in this portion of the basin where land conversion and habitat modification has been more substantial.

12.8 Discussion

This chapter explores the role of an empirically-based conceptual framework based on remotely sensed landscape information for tracking change over space and time in

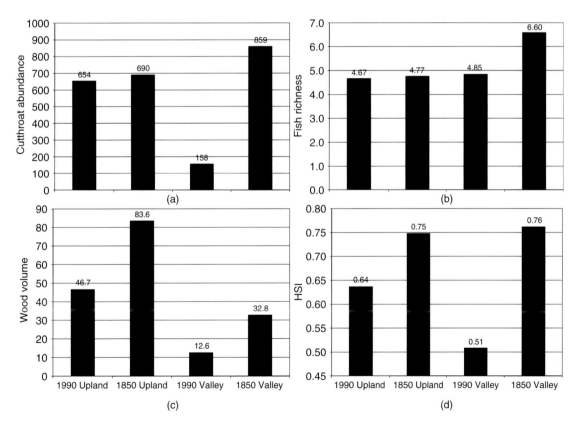

Figure 12.7 Responses of aquatic communities in the valley (Willamette Valley Ecoregion) and uplands (Cascade Mountain and Coast Range Ecoregion) of the WRB.

large river basins – one that integrates the geomorphic and hydrologic processes shaping landscapes and rivers, terrestrial and aquatic communities and processes across landscapes, and the human systems guiding their land and water use trajectories. Basin-wide analysis of aquatic and terrestrial ecosystems requires integration of remotely sensed and field-gathered data to track change and identify opportunities for conservation and restoration. We argue this is a necessary precursor to thoughtful basin management.

The notions of dynamism in complex adaptive systems (Levin, 1998, Holling and Gunderson, 2002) imply that dynamic features of landscapes, like rivers, should be allotted more territory than the historically wealth-motivated processes of human settlements have been willing to give (National Research Council, 1992, Cowx and Welcomme, 1998, Hulse and Ribe, 2000). Landscape planners and resource managers throughout the world are faced with the challenge of maintaining dynamic processes of channel meandering (both channel erosion and deposition) and floodplain inundation while allowing people to use these portions of the river floodplain and invest in infrastructure in this ever-changing portion of the landscape. The analyses at the scale of the Willamette River basin and its major subbasins demonstrate the importance of finding new strategies to gain more room for river networks in human-dominated landscapes, while acknowledging the real and largely irreversible investments that have already been made in these biologically and culturally important parts of human settlements.

The primary focus of tracking landscape and river network change for prioritising conservation and restoration efforts was to (1) spatially identify patterns of key resources as they vary over time and (2) enable regional decision makers to account for and anticipate the locations of increases and decreases in each resource and the potential consequences of environmental policies. Dynamic patterns of critical terrestrial and aquatic ecosystems juxtaposed with major human population centres and land use investments create a spatial context for identifying locations for conservation and restoration efforts and tracking their success, or failure, over time (Hulse and Gregory, 2001, 2004). Conservation opportunities are revealed in habitat matrices or corridors of ecologically functional land cover and native species richness. Restoration opportunities are highlighted in areas that demonstrate the greatest difference between current patterns and historical or reference conditions (Hulse and Gregory, 2004). Constraints and incentives for conservation or restoration created by human systems are

determined by (1) the patterns of human populations and structural development across the landscape and (2) the economic values and risks to productivity of the land base.

Five key lessons have emerged from this basin-scale analysis of trajectories of aquatic and terrestrial ecosystem change. First, the application of landscape scenarios based on remotely sensed land use/land cover and ecological models demonstrates that adopted policies and practices have great influence on the recovery or continued declines of natural resources. Plausible restoration practices and changes in land use policies can reverse long-term trends in natural resource declines even in the face of a doubling of the human population. Second, stream networks and large floodplain rivers strongly influence abundance and distribution of aquatic communities and terrestrial wildlife (Hulse and Gregory, 2003). The state of Oregon has used the findings and spatial information on aquatic communities to develop a Special Investments Partnership to conserve and restore habitats of the mainstem Willamette River and its floodplain, investing more than $2 million annually in conservation and restoration actions, and more than $30 million from other sources in 2010 alone. Third, there is a near-universal desire for production of both short-term wealth and long-term ecosystem services in landscapes where human settlement is a major force of landscape change (Hulse and Ribe, 2000, Chan et al., 2006). While the relative influence of these desires on realised landscape pattern and policy are in constant flux, our work indicates these desires must be reflected in any plausible future scenario. Scarcity of either of these desired goods, fear of it or attempts to avoid it motivates much intentional human action in the landscape (Schroter et al., 2005, Baumgartner et al., 2006). These actions occur at multiple societal levels of organisation, from individual people and their families to citizen groups, private corporations, and public agencies. Fourth, scientists and stakeholders are often self-limiting in defining future scenarios. Citizen groups may be reluctant or even unable to conceive dramatic shifts from existing policies and regulations because of the social upheaval accompanying such change. This is compounded by the fifth and final key lesson that inherent uncertainties are rarely explicitly represented in the products and decisions that derive from analyses of past, present and future conditions of large river basins.

We asserted earlier in this chapter that there are three purposes such alternative future efforts pursue: to support scientific exploration and research, to inform and educate, and to underpin decisions and strategic planning. For us, one of the most significant lessons of the

PNW-ERC was the need to assume, in pursuing all three of these purposes, that empirically-defined and model-based relationships from the present held true in both past and future landscapes. This projecting of present relationships onto the past and future brings with it a source of uncertainty that is different in kind and degree from the positional, classification and generalisation errors that are more familiar to those who work regularly with remotely sensed representations of contemporary land use and land cover (Branscomb, 2002). We are struck by the special significance this assumption has in light of current concerns regarding climate change, and the growing scientific consensus that, in ecologically important ways, the future may be unlike the past.

One of the most important outcomes of analyses of future scenarios is the realisation that the reversal of large scale environmental trends may require the development of radically different policies and practices to maintain a livable and more ecologically sound future. Most communities want to maintain the status quo to the extent possible to maintain social and economic stability. As an example in the Pacific Northwest region of the United States, the Northwest Forest Plan for federal forest lands became one of the largest integrated land use and conservation strategies in the world and was based on spatially explicit development of nine alternative scenarios based on remotely sensed information, ecological models, forest yield models, and socioeconomic models (USDA, 1994). The option selected by the United States government provided an intermediate level of protection of terrestrial and aquatic biodiversity and a reduction of the programmed timber harvest to less than 15% of the harvest level prior to the Plan. The Plan created enormous controversy and regional tension and continues to receive criticism from industry, environmentalists, and regional governments. The level of harvest in the first 10 years of the Northwest Forest Plan was even lower than planned due to lawsuits, challenges in implementation with reduced federal workers, and public controversy (Thomas et al., 2006). Environmental assessments a decade after adopting the plan found that old-growth forest extent was increasing at a rate slightly greater than forecasted, declines in Northern spotted owls were related to competition with an invading owl species rather than effects of timber harvest, and extent of forest-harvest roads and fish passage barriers was decreasing at expected rates (Bormann et al., 2006). In the midst of ongoing debate about environmentally and socially sound management of public lands, the use of remote sensing, ecological landscape models, and alternative future scenarios remain widely accepted

though contested tools for decision making in the region. Even with the uncertainty and concerns noted above, scientific projections of alternative future landscapes can explore and test highly probable trends in the livability of the landscape and the viability of important natural resources and ecosystem services.

The Alternative Futures Project for the Willamette River basin (Pacific Northwest Ecosystem Research Consortium) (Baker et al., 2004, Hulse et al., 2002) has been a less controversial application of remote sensing within the context of alternative future scenarios that has created new conservation and restoration opportunities. After the completion of the future alternative analysis, much of the land use and land cover information and related socioeconomic data was made available to the public and decision makers on a digital library website (http://willametteexplorer.info) to assist in local planning and conservation actions. The research provided the basis for developing an agent-based model of land use policies and ecological outcomes (Bolte et al., 2007), which now is used by several communities and agencies in North America for planning, decision making, and research. Regional planning projects for transportation, water availability, and local land management have used the databases of the PNW-ERC in their efforts. New studies of climate change and land management are building on the foundation provided by the landscape information and agent-based model (Steel et al., 2004, Nelson et al., 2009). The Oregon Watershed Enhancement Board adopted the spatial databases and ecological recommendations of the PNW-ERC as a framework for investing approximately $3 million annually to target conservation and restoration actions along the 245-km mainstem of the Willamette River (http://www.oregon.gov/OWEB/SIPWillamette.shtml). The value of alternative future scenario assessments emerges more from stimulating regional discussions about landscape management options and creating a spatially explicit framework for the evolving body of regional landscape information than the certainty of its future projections.

Plausible and scientifically defensible information can motivate communities to interactively develop innovative, transformational alternatives by anticipating the future rather than merely reacting to it. A recent review of river restoration projects across the United States concluded that an ecologically sound guiding vision is *the* essential characteristic of effective restoration planning (Palmer et al., 2005). For regional conservation and restoration, guiding visions are equally important at the scale of landscapes and river basins. Remotely sensed landscape information and ecological understanding of basin-scale resources and land uses make it possible for citizens and regional decision makers to anticipate trajectories of environmental change and, in so doing, develop more effective regional planning and community-based choices about the livability of their environment.

Acknowledgements

The research was funded originally by the U.S. Environmental Protection Agency (EPA). The scientists involved were part of the Pacific Northwest Ecosystem Research Consortium (URL: http://www.oregonstate.edu/dept/pnw-erc), a collaborative effort between EPA, Oregon State University, and the University of Oregon. Recent studies of the mainstem Willamette River have been funded by the Oregon Watershed Enhancement Board and the Meyer Memorial Trust.

References

Aggett, G.R. and Wilson, J.P. 2009. Creating and coupling a high-resolution DTM with a 1-D hydraulic model in a GIS for scenario-based assessment of avulsion hazard in a gravel-bed river. *Geomorphology* 113: 21–34.

Ahern, J. 2001. Spatial concepts, planning strategies, and future scenarios: a framework method for integrating landscape ecology and landscape planning. Pages pp 175–201 in Klopatek, J., Gardner, R. (eds) *Landscape ecological analysis: issues and applications.* Springer-Verlag, New York.

Baas, A.C.W. and Nield, J.M. 2010. Ecogeomorphic state variables and phase-space construction for quantifying the evolution of vegetated aeolian landscapes. *Earth Surface Processes and Landforms* 35: 717–731.

Baker, J.P., Hulse, D.H., Gregory, S.V., White, D., Van Sickle, J., Berger, P.A., Dole, D., and Schumaker, N.H. 2004. Alternative futures for the Willamette River Basin, Oregon. *Ecological Applications* 14: 313–324.

Baumgartner S, Becker, C., Faber, M., and Manstetten, R. 2006. Relative and absolute scarcity of nature: assessing the roles of economics and ecology for biodiversity conservation. *Ecological Economics* 59: 487–498.

Berger, P. and Bolte, J. 2004. Evaluating the impact of policy options on agricultural landscapes: an alternative futures approach. *Ecological Applications* 14: 342–354.

Bolliger, J., Kienast, F., Soliva, R., and Rutherford., G. 2007. Spatial sensitivity of species habitat patterns to scenarios of land use change (Switzerland). *Landscape Ecology* 22: 773–789.

Bolte, J.P., Hulse, D.W., Gregory, S.V., and Smith, C. 2007. Modeling biocomplexity – actors, landscapes and alternative futures. *Environmental Modeling and Software* 22: 570–579.

Bormann, B.T., Lee, D.C., Kiester, A.R., Busch, D.E., Martin, J.R., and Haynes, R.W. 2006. Synthesis – interpreting the Northwest Forest Plan as more than the sum of its parts, In R.W. Haynes, B.T. Bormann, D.C. Lee, and J.R. Martin, technical editors. *Northwest Forest Plan – the first ten years (1994–2003): synthesis of monitoring and research results.* General technical report PNW-GTR 651. U.S. Department of Agriculture Forest Service Pacific Northwest Research Station, Portland, Oregon.

Bouman, B.A.M., Jansen, H.G.P., Schipper R.A., Nieuwenhuyse, A., Hengsdijk, H., and Bouma, J. 1999. A framework for integrated biophysical and economic land use analysis at different scales. *Agriculture, Ecosystems and Environment* 75: 55–73.

Branscomb, A. 2002. Uncertainty and Error. Pages 156–157 in D.W. Hulse, S.V. Gregory, and J.P. Baker, editors. *Willamette River basin: trajectories of environmental and ecological change.* Oregon State University Press, Corvallis, Oregon, USA.

Branscomb, A., Goicochea, J., and Richmond, M. 2002. Stream network. Pages 16–17 in D.W. Hulse, S.V. Gregory, and J.P. Baker, editors. *Willamette River basin: trajectories of environmental and ecological change.* Oregon State University Press, Corvallis, Oregon, USA.

Brierley, G. and Fryirs, K. 2009. Don't Fight the Site: Three Geomorphic Considerations in Catchment-Scale River Rehabilitation Planning. *Environmental Management* 43: 1201–1218.

Chan, K.M.A., Shaw, M.R., Cameron, D.R., Underwood, E.C., and Daily, G.C. 2006. Conservation planning for ecosystem services. *PLoS Biology* 11: 2138–2152.

Cowx, I.G. and Welcomme, R.L. (eds). 1998. Rehabilitation of rivers for fish: a study undertaken by the European Inland Fisheries Advisory Commission of FAO. Blackwell Science Ltd. Malden, Mass. 02148.

Cohen, W.B., Maiersperger, T.K., Spies, T.A., and Oetter, D.R. 2001. Modeling forest cover attributes as continuous variables in a regional context with Thematic Mapper data. *International Journal of Remote Sensing* 22: 2279–2310.

Dimick, R.E. and Merryfield, F. 1945. The fishes of the Willamette River system in relation to pollution. *Engineering Experiment Station Bulletin Series* 20: 7–55. Oregon State College, Corvallis.

Dole, D. and Niemi, E. 2004. Future water allocation and instream values in the Willamette River basin: a basinwide analysis. *Ecological Applications* 14: 355–367.

Enright, C., Aoki, M., Oetter, D., Hulse, D., and Cohen, W. 2002. Land use/land cover ca. 1990. Pages 78–81 in D.W. Hulse, S.V. Gregory, and J.P. Baker, editors. *Willamette River basin: trajectories of environmental and ecological change.* Oregon State University Press, Corvallis, Oregon, USA.

Fullerton, A.H., Steel, E.A., Caras, Y., Sheer, M., Olson, P., and Kaje, J. 2009. Putting watershed restoration in context: alternative future scenarios influence management outcomes. *Ecological Applications* 19: 218–235.

Gregory, S., Oetter, D., Ashkenas, L., Minear, P., and Wildman, K. 2002f. Riparian areas. Pages 98–101 in D.W. Hulse, S.V.

Gregory, and J.P. Baker, editors. *Willamette River basin: trajectories of environmental and ecological change*. Oregon State University Press, Corvallis, Oregon, USA.

Gregory, S., Oetter, D., Ashkenas, L., Minear, P., Wildman, K., and Christy, J. 2002d. Pre-EuroAmerican scenario. Pages 92–93 in D.W. Hulse, S.V. Gregory, and J.P. Baker, editors. *Willamette River basin: trajectories of environmental and ecological change*. Oregon State University Press, Corvallis, Oregon, USA.

Gregory, S., Oetter, D., Ashkenas, L., Minear, P., Wildman, K., and Christy, J. 2002e. Natural vegetation. Pages 96–97 in D. Hulse, S. Gregory, and J. Baker, editors. Willamette River Basin planning atlas: trajectories of environmental and ecological change. Oregon State University Press, Corvallis, Oregon, USA.

Gregory, S., Oetter, D., Ashkenas, L., Minear, P., and Wildman, K. 2002h. Longitudinal patterns – channel. Pages 134–135 in D.W. Hulse, S.V. Gregory, and J.P. Baker, editors. Willamette River basin: trajectories of environmental and ecological change. Oregon State University Press, Corvallis, Oregon, USA.

Gregory, S., Oetter, D., Ashkenas, L., Minear, P., and Wildman, K. 2002i. Longitudinal patterns – vegetation. Pages 136–137 in D.W. Hulse, S.V. Gregory, and J.P. Baker, editors. *Willamette River basin: trajectories of environmental and ecological change*. Oregon State University Press, Corvallis, Oregon, USA.

Gregory, S., Oetter, D., Ashkenas, L., Minear, P., Wildman, K., and Christy, J. 2002a. Presettlement vegetation ca. 1851. Pages 38–39 in D.W. Hulse, S.V. Gregory, and J.P. Baker, editors. *Willamette River basin: trajectories of environmental and ecological change*. Oregon State University Press, Corvallis, Oregon, USA.

Gregory, S., Oetter, D., Ashkenas, L., Wildman, R., Minear, P., and Wildman, K. 2002g. Mainstem river. Pages 112–113 in D.W. Hulse, S.V. Gregory, and J.P. Baker, editors. *Willamette River basin: trajectories of environmental and ecological change*. Oregon State University Press, Corvallis, Oregon, USA.

Gregory, S., Ashkenas, L., Haggerty, P., Oetter, D., Wildman, K., Hulse, D., Branscomb, A., and Van Sickle, J. 2002b. Riparian vegetation. Pages 40–44 in D.W. Hulse, S.V. Gregory, and J.P. Baker, editors. *Willamette River basin: trajectories of environmental and ecological change*. Oregon State University Press, Corvallis, Oregon, USA.

Gregory, S., Wildman, R., Ashkenas, Wildman, K., and Haggerty, P. 2002c. Fish assemblages. Pages 44–45 in D.W. Hulse, S.V. Gregory, and J.P. Baker, editors. *Willamette River basin: trajectories of environmental and ecological change*. Oregon State University Press, Corvallis, Oregon, USA.

Gregory, S.V. 2007. Historical channel modification and floodplain forest decline: implications for conservation and restoration in a large floodplain river; Willamette River, Oregon. Pages 763–777 in *Gravel-bed Rivers*, 6th International Symposium, Lienz, Austria. John Wiley.

Gregory, S.V., Hulse, D., Landers, D., and Whitelaw, E. 1998. Integration of biophysical and socio-economic patterns in riparian restoration of large rivers. In *Hydrology in a Changing Environment*, H. Wheater and C. Kirby (eds.), vol. I, Theme 2 Ecological and hydrological interactions, D. Gilvear editor, pp. 231–247. John Wiley and Sons, Chicester.

Gregory, S.V., Swanson, F.J., McKee, W.A., and Cummins, K.W. 1991. An ecosystem perspective of riparian zones. *Bioscience* 41: 540–551.

Gregory, S.V., Meleason, M., and Sobota, D.J. 2003. Modeling the dynamics of wood in streams and rivers. Pages 315–336 in S.V. Gregory, K.L. Boyer, and A.M. Gurnell, editors. The ecology and management of wood in world rivers. American Fisheries Society, Symposium 37, Bethesda, Maryland.

Hemstrom, M.A., Korol, J.J., and Hann, W.J. 2001. Trends in terrestrial plant communities and landscape health indicate the effects of alternative management strategies in the interior Columbia River basin. *Forest Ecology and Management* 5504: 1–21.

Henrichs, T., Zurek, M., Eickout, B., Kok, K., Raudsepp-Hearne, C., Ribeiro, T., van Vuuren, D., Volkery, A. 2010. Scenario development and analysis for forward-looking ecosystem assessments, Ch. 5 in *Ecosystems and Human Well-Being: A manual for assessment practitioners*, Ash et al. (Ed's). Island Press. Wash. D.C. ISBN 978-1-59726-711-3.

Hubacek, K. and Sun, L. 2001. A scenario analysis of China's land use and land cover change: incorporating biophysical information into input-output modeling. *Structural Change and Economic Dynamics* 12: 367–397.

Hulse, D.W., S.V. Gregory, and J.P. Baker, editors. 2002. Willamette River basin planning atlas: trajectories of environmental and ecological change. Oregon State University Press, Corvallis, Oregon, USA. 178 p.

Hulse, D., Branscomb, A., Enright, C., and Bolte, J. 2009. Anticipating floodplain trajectories: a comparison of two alternative futures approaches. *Landscape Ecology* 24: 1067–1090.

Hulse, D., Branscomb, A., and Payne, S. 2004. Envisioning Alternatives: using citizen guidance to map future land and water use. *Ecological Applications* 14: 325–341.

Hulse, D. and Ribe, R. 2000. Land conversion and the production of wealth. *Ecological Applications* 10: 679–682.

Hulse, D. and Gregory, S. 2004. Integrating resilience into floodplain restoration. *Journal of Urban Ecology* 7: 295–314.

Hulse, D., Eilers, J., Freemark, K., White, D., and Hummon, C. 2000. Planning alternative future landscapes in Oregon: evaluating effects on water quality and biodiversity. *Landscape Journal* 19: 1–19.

Hulse, D.H. and Gregory, S.V. 2001. Alternative futures as an integrative framework for riparian restoration of large rivers. Pages 194–212 in V.H. Dale and R. Haeuber, editors, *Applying Ecological Principles To Land Management*. Springer-Verlag, New York, New York.

Hunter, L.M., Gonzalez, M., Stevenson, M., Karish, K.S., Toth, R., Edwards Jr., T.C., Lilieholm, R.J., and Cablk, M. 2003. Population and land use change in the California Mojave: Natural habitat implications of alternative futures. *Population Research and Policy Review* 22: 373–397.

Kennedy, R.E. and Cohen, W.B. 2003. Automated designation of tie-points for image-to-image coregistration. *International Journal of Remote Sensing* 24: 3467–3490.

Kepner, W., Semmens, D., Bassett, S., Mouat, D., and Goodrich, D. 2004. Scenario analysis for the San Pedro River, analyzing hydrological consequences of a future environment. *Environmental Monitoring and Assessment* 94: 115–127.

Lattin, P. and Peniston, B. 1999. Internal Report: Comparison of Land Cover/Land Use Characterizations of riparian areas and Watersheds Derived from Aerial Photography and LANDSAT Thematic Mapper imagery of Phase I Watersheds. Deliverable No. 3-09.010. U.S. EPA, Corvallis, OR. 97333.

Lattin, P.D., Wigington Jr., P.J., Moser, T.J., Peniston, B.E., Lindeman, D.R., and Oetter, D.R. 2004. Influence of remote sensing imagery source on quantification of riparian land cover/land use. *Journal of the American Water Resources Association* 40: 215–227.

Lautenbach, S., Berlekamp, J., Graf, N., Seppelt, R., and Matthies, M. 2009. Scenario analysis and management options for sustainable river basin management: Application of the Elbe DSS. *Environmental Modelling and Software* 24: 26–43.

Lawler, J.J., White, D., Neilson, R.P., and Blaustein, A.R. 2006. Predicting climate-induced range shifts: model differences and model reliability. *Global Change Biology* 12: 1568–1584.

Liu, Y., Mahmoud, M., Hartmann, H., Stewart, S., Wagener, T., Semmens, D., Stewart, R., Gupta, H., Dominguez, D., Hulse, D., Letcher, R., Rashleigh, B., Smith, C., Street, R., Ticehurst, J., Twery, M., van Delden, H., Waldick, R., White, D., and Winter, L. 2007. Formal scenario development for environmental impact assessment studies, in *State of the Art and Futures in Environmental Modelling and Software*, edited by Jakeman, A., Voinov, A., Rizzoli, A.E., and Chen, S., IDEA Book Series, Elsevier.

Lopez-Lopez, P., Garcia-Ripolles, C., Aguilar, J., Garcia-Lopez, F., and Verdejo, J. 2006. Modelling breeding habitat preferences of Bonelli's eagle (*Hieraaetus fasciatus*) in relation to topography, disturbance, climate and land use at different spatial scales., *Journal of Ornithology* 147: 97–106.

Luoto, M. Virkkala, R., and Heikkinen, R.K. 2007. The role of land cover in bioclimatic models depends on spatial resolution. *Global Ecology and Biogeography* 16: 34–42.

Magdaleno, F. and J. Anastasio Fernandez 2011. Hydromorphological alteration of a large Mediterranean river: relative role of high and low flows on the evolution of riparian forests and channel morphology. *River Research and Applications* 27: 374–387.

Märker, M., Angeli, L., Bottai, L., Costantini, R., Ferrari, R., Innocenti, L., and Siciliano, G. 2008. Assessment of land degradation susceptibility by scenario analysis: A case study in Southern Tuscany, Italy. *Geomorphology* 93: 120–129.

National Research Council. 1992. *Restoration of aquatic ecosystems*. National Academy Press. 552 pp.

Nelson, E.K., Mendoza, G., Regetz, J., Polasky, S., Tallis, H., Cameron, D.R., Chan, K.M.A., Daily, G.C., Goldstein, J., Kareiva, P.M., Lonsdorf, E., Naidoo, R., Ricketts, T.H., and

Shaw, M.R. 2009. Modeling multiple ecosystem services, biodiversity conservation, commodity production, and tradeoffs at landscape scales. *Frontiers in Ecology and Environment* 7: 4–11.

Niemi, E., Dole, D., and Whitelaw, E. 2002. Water availability. Pages 114–116 in Hulse, D.W., Gregory, S.V., and Baker, J.P., editors. *Willamette River basin: trajectories of environmental and ecological change.* Oregon State University Press, Corvallis, Oregon, USA.

Oetter, D.R., Ashkenas, L.R., Gregory, S.V., and Minear, P.J. 2004. GIS Methodology for Characterizing Historical Conditions of the Willamette River Flood Plain, Oregon *Transactions in GIS* 8: 367–383.

Oetter, D.R., Cohen, W.B., Berterretche, M., Maiersperger, T.K. and Kennedy, R.E. 2001. Land cover mapping in an agricultural setting using multi-seasonal Thematic Mapper data. *Remote Sensing of Environment* 76: 139–155.

Palmer, M.A., Bernhardt, E.S., Allan, J.D., Lake, P.S., Alexander, G., Brooks, S., Carr, J., Clayton, S., Dahm, C.N., Follstad-Shah, J., Galat, D.L., Loss, P., Goodwin, D.D., Hart, B., Hassett, R., Jenkinson, G.M., Kondolf, R., Lave, J.L., Meyer, T.K., O'Donnell, L., Pagano, and Sudduth, E. 2005. Standards for ecologically successful river restoration. *Journal of Applied Ecology* 42: 208–217.

Santelmann, M, Freemark, K., White, D., Nassauer, J., Clark, M., Danielson, B., Eilers, J., Cruse, R.M., Galatowitsch, S., Polasky, S., Vache, K., and Wu, J. 2001. Applying ecological principles to land-use decision making in agricultural watersheds. Pages 226–254 in: Dale, V.H., Haeuber, R. (eds) *Applying ecological principles to land management.* Springer-Verlag, New York.

Santelmann, M., Freemark, K., Sifneos, J., and White, D. 2006. Assessing effects of alternative agricultural practices on wildlife habitat in Iowa, USA. *Agriculture, Ecosystems and Environment* 113: 243–253.

Schröter, D., Cramer, W., Leemans, R., Prentice, I., Araujo, M.B., Arnell, N.W., Bondeau, A., Bugmann, H., Carter, T.R., Gracia, C.A., de la Vega-Leinert, A.C., Erhard, M., Ewert, F., Glendining, M., House, J.I., Kankaanpää, S., Klein, R.J.T., Lavorel, S., Lindner, M., Metzger, M.J., Meyer, J., Mitchell, T.D., Reginster, I., Rounsevell, M., Sabaté, S., Sitch, S., Smith, B., Smith, J., Smith, P., Sykes, M.T., Thonicke, K., Thuiller, W., Tuck, G., Zaehle, S., and Zierl, B. 2005. Ecosystem service supply and vulnerability to global change in Europe. *Science* 25: 1333–1337.

Schumaker, N., Ernst, T., White, D., Baker, J., and Haggerty, P. 2004. Projecting wildlife responses to alternative future landscapes in Oregon's Willamette Basin. *Ecological Applications* 14: 381–400.

Shluter, M. and Ruger, N. 2007. Application of a GIS-based simulation tool to illustrate implications of uncertainties for water management in the Amudarya river delta. *Environmental Modelling and Software* 22: 158–166.

Steel, E.A., Feist, B.E., Jensen, D.W., Pess, G.R., Sheer, M.B., Brauner, J.B., and Bilby, R.E. 2004. Landscape models to understand steelhead (*Oncorhynchus mykiss*) distribution and help prioritize barrier removals in the Willamette basin, Oregon, USA. *Canadian Journal of Fisheries and Aquatic Sciences* 61: 999–1011.

Steinitz, C., Arias, H., Bassett, S., Flaxman, M., Goode, T., Maddock, T., Mouat, D., Peiser, R., and Shearer, A. 2003. *Alternative futures for changing landscapes: The Upper San Pedro River Basin Arizona and Sonora.* Island Press, Covelo, California.

Steinitz, C., Binford, M., Cote, P., Edwards Jr., T., Ervin, S., Forman, R.T.T., Johnson, C., Kiester, A.R., Mouat, A.D., Olson, D., Shearer, F.A., Toth, R., and Wills, R. 1996. Biodiversity and landscape planning: alternative futures for the region of Camp Pendleton, California. Harvard University Graduate School of Design, Cambridge

Thomas, J.W., Franklin, J.F., Gordon, J., and Johnson, K.N. 2006. The Northwest Forest Plan: Origins, Components, Implementation Experience, and Suggestions for Change. *Conservation Biology* 20: 277–287.

U. S. Geological Survey (USGS). 1997. Hydrologic Unit Maps. USGS Water-Supply Paper 2294.

USDA (U.S. Department of Agriculture) Forest Service and BLM (Bureau of Land Management). 1994. Record of Decision for amendments to Forest Service and Bureau of Land Management planning documents within the range of the Northern Spotted Owl and standards and guidelines for management of habitat for late-successional and old-growth forest related

species within the range of the Northern Spotted Owl. USDA Forest Service, Portland, Oregon, and BLM, Moscow, Idaho

Vaché, K.B., Eilers, J.M., and Santelmann, M.V. 2002. Water quality modeling of alternative agricultural scenarios in the U.S. corn belt. *Journal of the American Water Resources Association* 38: 773–787.

Van Sickle, J., Baker, J., Herlihy, A., Bayley, P., Gregory, S., Haggerty, P., Ashkenas, L., and Li, J. 2004. Projecting the biological condition of streams under alternative scenarios of human land use. *Ecological Applications* 14: 368–380.

Virkkala, R., Heikkinena, K.R.K., Leikola, N., and Luotoc, M. 2008. Projected large-scale range reductions of northern-boreal land bird species due to climate change. *Biological Conservation* 141: 1343–1353.

White, D., Preston, E.M., Freemark, K.E., and Kiester, A.R. 1999. A hierarchical framework for conserving biodiversity. Pages 127–153 in Klopatek, J.M. and Gardner, R.H. (eds) *Landscape Ecological Analysis: issues and applications.* Springer-Verlag. New York.

Wolski, P. and Murray-Hudson, M. 2008. 'Alternative futures' of the Okavango Delta simulated by a suite of global climate and hydro-ecological models. *Water South Africa* 34: 605–610.

Zezere, J, Reis, E., Garcia, R., Oliveira, S., Rodrigues, M., Vieira, G., and Ferreira, A. 2004, Integration of spatial and temporal data for the definition of different landslide hazard scenarios in the area north of Lisbon (Portugal). *Natural Hazards and Earth System Sciences* 4: 133–146.

13 The Use of Imagery in Laboratory Experiments

Michal Tal[1], Philippe Frey[2], Wonsuck Kim[3], Eric Lajeunesse[4],
Angela Limare[4] and François Métivier[1]

[1]Aix-Marseille Université, France
[2]Irstea, Unité de Recherche ETNA, St-Martin-d'Hères, France
[3]University of Texas, Austin, Texas, USA
[4]Institut de Physique du Globe de Paris, Sorbonne Paris Cité, Paris, France

13.1 Introduction

Experimental-based research in fluvial geomorphology constitutes an important tool for studying processes occurring in natural rivers. Laboratory experiments are useful for inspiring questions, testing hypotheses, and identifying key processes and parameters. Insights from studies using laboratory stream-tables (relatively wide shallow containers in which the experimental river sets its own width and pattern) and flumes (relatively long narrow channels with fixed parallel side walls) underlie our understanding of processes such as bedload transport (e.g., Gomez and Church, 1989), controls on channel width (e.g., Ikeda et al., 1988), downstream fining (e.g., Seal et al., 1997), and the mechanics of flow through bends (e.g., Hooke, 1975), and have led to important insights about the underlying physics of large and chaotic systems such as braided rivers and deltas (e.g., Ashmore, 1991b; Sapozhnikov and Foufoula-Georgiou, 1997, Kim and Paola, 2007; Hoyal and Sheets, 2009).

Laboratory experiments lend themselves well to studies that are relevant to stream restoration and management because they provide a relatively inexpensive, fast, and simple way to collect large quantities of data under controlled conditions. Full control means that variables can be held fixed and the role of individual variables can be isolated. Experiments enable observation of long-term behaviour that can often only be predicted or inferred in the field. This is important because processes being studied in the field typically evolve well beyond the typical time frame of a grant or monitoring program and insight from long-term observation is often missed. Finally, the ability to collect data from an experiment using several different techniques permits researchers to draw conclusions based on an aggregate of measurements, while multiple runs allow for more rigorous statistics and greater confidence in results.

This chapter deals with the use of imagery to acquire data from laboratory experiments. Imagery in the laboratory provides the same advantages as in the field: data acquisition at very high temporal and spatial resolutions in an entirely non-invasive manner. Experiments have the intrinsic advantage offered by satellites and airplanes to study natural systems: a view of the system through a lens zoomed way out. Imaging techniques in the laboratory are enhanced by the high degree of control over lighting and materials which can both simplify image processing and greatly enhance the results. Despite these inherent

Fluvial Remote Sensing for Science and Management, First Edition. Edited by Patrice E. Carbonneau and Hervé Piégay.
© 2012 John Wiley & Sons, Ltd. Published 2012 by John Wiley & Sons, Ltd.

advantages, the technology to collect information to the full potential an experiment can offer has only recently become readily available, mainly in the form of inexpensive digital cameras. Today it is common to conduct experiments in which imaging is the only method of data collection.

We present several imaging techniques that offer a broad range of applications related to river management (Table 13.1). These techniques are based on the use of cameras ranging from simple digital cameras and video recorders to more sophisticated CDD (charge-coupled device) cameras and high-speed video recorders. The experiments presented were conducted in a wide range of flumes and stream tables and thus provide an idea of the diversity of setups and methods used in experimental geomorphology (Table 13.1). Our focus is on imagery techniques used in studies of river channels and deltas to measure (a) mass flux of sediment: sediment grain-size distribution, bedload transport rate, trajectories and velocities of individual particles, (b) local and regional properties of the flow field: depth, width, migration rates, sinuosity, braiding index, and (c) bed topography in river channels and deltas (Table 13.1). This chapter is not an exhaustive review of imagery techniques used in the laboratory. We concentrate on a few key examples which illustrate how the powerful combination of laboratory experiments and imagery can be used to advance our understanding of river processes and inform management. (Commercial software and brand names of equipment used in the studies presented here are included purely as examples based on the authors' experience. The methodologies described do not depend on any specific software packages or equipment and references to these do not constitute endorsement).

13.2 Bedload transport

Fundamental to the study of the dynamics of any river is characterising the rate (mass or volume per unit time) of sediment transported as bedload and the size distribution of the transported material. In addition, detailed data about bedload transport at the grainscale under different flow conditions are needed in order to improve our understanding of the physics governing bedload transport and derive physically based transport equations and models. However, direct measurements of bedload transport in rivers are difficult and can be dangerous during larger flows, which account for most of the sediment flux (Wilcock, 2001). In addition, direct measurements

yield imprecise estimates due to the inherent spatial and temporal variability in the processes of bedload transport (Diplas et al., 2008). Flume experiments have been used extensively to improve our understanding of sediment transport processes (e.g., Parker et al., 1982) and many of the most widely used bedload transport equations were derived from experimental data (e.g., Meyer-Peter and Müller, 1948; Wilcock and Crowe, 2003). Experimentally, measuring bedload transport rate and obtaining a grain-size distribution (GSD) of the material transported (in the case of non-uniform bed mixtures) typically consists of collecting the sediment at the flume outlet, manually sieving the sample, and weighing the individual size fractions. Needless to say, this is a cumbersome and time-consuming task.

Even more difficult to measure than sediment transport in the field is bed surface size distribution, which directly affects the observed transport (Wilcock, 2001). As in the case of sediment transport, the bed surface in natural rivers can often be sampled under low flow conditions, but is almost impossible to sample during high transport conditions. Several methods for automated grainsize estimation from images exist (Rubin, 2004; Graham et al., 2005; Buscombe et al., 2009, 2010). These techniques were developed for field applications but can be adapted to laboratory experiments. Because flow in a flume can be instantaneously shut off, experiments offer a unique opportunity to study the bed surface associated with active transport. This makes flume experiments particularly useful for addressing questions such as what happens to surface armour layers during high flow and improving our understanding of fractional transport. In a well-known set of experiments conducted in the 'bed of many colours' (BOMC), imagery was used in an innovative way to estimate bed surface grainsize distribution (and from this variations in grain mobility) during active transport in a way that was non-destructive to the bed (details in Wilcock and McArdell, 1993; Wilcock, 2001). In these experiments each size fraction in the sediment bed was painted a different colour. Photographs of the bed surface when the flow was turned off were used to measure the grainsize distribution using point counts.

Here we present two techniques for studying sediment transport dynamics in experimental channels. The first is an image-based technique which enables continuous in-line measurement of size and velocity of all transported particles at the downstream exit of a flume. This information is then used to determine the GSD and total sediment discharge. The second technique uses imagery to reconstruct 2D trajectories of individual grains

Table 13.1 Examples of experimental facilities used to study rivers and deltas. W, L, and D are the width, length, and depth respectively.

	Flume with adjustable width and slope typicaly used for experiments to study sediment transport. W = 0.096 m, L = 0.24 m, IPGP experimental geomorphology laboratory.
	Stream table with adjustable slope typically used for experiments to study channel morphodynamics, IPGP experimental geomorphology laboratory. W = 0.75 m L = 2 m
	Large stream table with fixed sidewalls typically used for experiments on channel morphology and interactions with riparian vegetation, St. Anthony Falls Laboratory. W = 2 m L = 16 m
	Experimental tank with subsiding floor designed to study the links between deltaic processes, subsidence, and base level, St. Anthony Falls Laboratory. W= 3 m L = 6 m D = 1.5 m

transported as bedload. This technique applies particle tracking velocimetry (PTV) algorithms commonly used in studies of fluid mechanics and, more recently, suspended sediment in flows.

13.2.1 Image-based technique to measure grainsize distribution and sediment discharge

The objective of the technique presented here is to characterise GSD by weight and total sediment discharge exiting the downstream end of a flume. The methodology is based on image analysis of sediment in the sand and gravel size range flowing across a light table (Frey et al., 2003; Figure 13.1). The equipment required for this technique are a light table that can be tilted independently of the flume (in the study described here, this consisted of an adjustable transparent ramp with a light source fixed underneath it) and a digital video recorder that can operate in backlighting mode. The aim is to have particles move across the table in a single layer with a low number of clusters and the thinnest possible flow. The tilt of the table should be adjusted to obtain the smoothest possible flow and dispersion of particles. For a given sediment concentration exiting a flume, the steeper the ramp the higher the velocity will be and the lower the number of clusters. In the experiments described here, the velocity of sediment moving across the ramp was typically 2–6 m/s. A camera that captured full-frame monochrome images with a resolution of 640×480 pixels and a frequency of 60 Hz was used (Bigillon et al., 1999). The range of medium diameters of the particles was initially restricted to about one order of magnitude, typically 2–20 mm, however the technique can be used to measure a larger grainsize

distribution by capturing images at higher frequency or using two cameras with different fields of view.

Assuming all the particles have the same density (ρ), the mass of each particle is calculated from its volume (V). Particles are described by their three principal dimensions: minimum diameter (the thickness), medium diameter (d), and the maximum diameter (D); the grainsize of each particle is characterised by the medium diameter (d). If (v) is the streamwise velocity of the particle (in pixels per second) and (h) the streamwise length of the image (in pixels), the total sediment discharge is the sum of the elementary contribution of each particle :

$$Q_s = \sum_{i \in particles} \frac{\rho \cdot V(i) \cdot v(i)}{h} \qquad (13.1)$$

where $V(i)$ is the volume of the ith particle. This sum is calculated on all particles for all images. The GSD is derived from the sum of elementary sediment discharges relative to one class of medium diameters divided by the total sediment discharge. If all particles have the same constant streamwise velocity on the backlit plate, the problem is simplified to the ratios of the volumes of particles of one class of medium diameters to the total volume, i.e.:

$$\frac{\sum_{i \in class} V(i)}{V_{tot}} \qquad (13.2)$$

In this case, streamwise velocity needs to be assessed in order to calculate total sediment discharge.

Calculating the volume of each particle is central to this methodology. To do this, image processing algorithms are first used to estimate the diameters and areas of the particles in the 2D image. Next, the volume is calculated

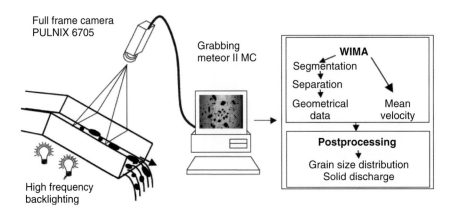

Figure 13.1 Experimental setup and processing chain to characterise the grainsize distribution and sediment discharge exiting the downstream end of a flume using image analysis of sediment flowing across a light table.

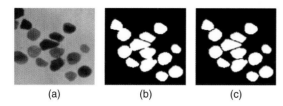

(a)   (b)   (c)

Figure 13.2 Sequence of image processing steps for segmenting particles (a) original image (b) result after segmentation (c) result after object separation.

by assessing the third dimension perpendicular to the image plane. In the study presented here, a user-friendly image processing software called WIMA, developed by the Laboratoire Hubert Curien (formerly TSI) of the University of St. Etienne, was used (Ducottet, 1994). Any image processing software with standard functions and which permits users to modify and add new functions can be used. The main goal of the image processing algorithms is to segment the particles in order to gain access to their dimensions. The three main image-processing steps required to achieve this goal are segmentation (i.e., detection), object separation, and object measurement (Figure 13.2). Edge detection techniques based on the use of Canny gradient operators and localisation of maximum gradient modulus were used (Jernot et al., 1982; Canny, 1986). Once these steps are achieved, the final step in the image analysis consists of extracting the boundaries of all detected particles in the image series. It is then possible to calculate areas as well as minimum and maximum principal diameters.

Assessing the volume of each particle is rather difficult for particles with variable shapes, and several assumptions have to be made in order to do this. First, it must be assumed that the particles flow in their stable state. Considering that particles are described by their three principal dimensions, it is possible to measure the medium (d) and maximum (D) dimensions (corresponding to the minimum and maximum diameters in the image respectively) as well as area (A) in the image plane. An ellipsoid shape is assumed for the particles, and thus grain thickness, which is perpendicular to the image plane, is a fraction of the medium diameter and is calculated using a shape coefficient (α). The shape coefficient (α) must be calibrated for the sediment being used. The volume of each particle then goes as:

$$V = \frac{\pi}{6} a \cdot d^2 D = \frac{2}{3} a \cdot d \cdot A \qquad (13.3)$$

Preliminary tests showed that calculating the volume using area rather than diameters yielded more accurate

results (Frey et al., 2003). Since GSD is derived from the ratios of volumes, if the dispersion around the mean value of the shape coefficient (α) for each class of diameters is small it is not necessary to know this value. On the other hand, the sediment discharge cannot be computed without assessing at least the mean shape coefficient. It was thought initially that velocities of particles could be highly dispersed across the light table. However, preliminary tests showed that dispersion around the mean velocity was low. Furthermore, velocity was not correlated with diameter of particles. Using sediment with grainsizes ranging from 2–12 mm, the velocity of one class of diameters was found to deviate no more than 10% from the mean velocity, which is then only required to compute sediment discharge. Because calculating sediment discharge requires both mean streamwise velocity and at least a mean shape coefficient, estimates of sediment discharge are less accurate than estimates of the GSD.

Numerous tests were carried out in order to compare medium diameters and masses obtained by image analysis with measured values (see Frey et al., 2003). Motionless particles with known shape coefficients were tested in a first series. Diameters and mass predictions were good, with uncertainties on the order of 3% and 6% respectively. In a second series, the light table was attached to the outlet of an experimental flume. The flume was 15 cm wide and had the same width as the transparent ramp. A sample of 400 particles were mixed into the flow slightly upstream of the outlet in order to simulate real experimental conditions. Results were improved when volume calculations considered differentiated shape coefficients, although results were remarkably good even when a uniform shape coefficient was used. Considering the diameters commonly used to describe grainsize: d_{30}, d_{50}, and d_{90}, where d_{xx} is the diameter for which xx% by weight of the sample is finer, the errors were 4%, 3% and 1% respectively. The minimum number of particles needed to achieve a significant statistical measurement – a proxy for the length of time over which measurements should be collected – for both GSD and sediment discharge was approximately 4,000 particles. With more than 2,500 particles the computed value was within 3% of the actual value and with more than 1,500 it was approximately 7%. A much lower number of particles (~1000) were required for low discharges in which coarser particles were dominant and the GSD was more uniform. Empirical testing is recommended in order to determine the optimal number of particles needed to achieve a good statistical fit. This method has been used successfully to investigate mechanisms responsible for bedload sheet production and migration (Recking et al., 2009). It has also been adapted and improved to study step-pools experimentally (see Zimmermann et al., 2008).

13.2.2 Particle trajectories and velocities using PTV

Particle tracking velocimetry (PTV) and a closely related technique – particle imagery velocimetry (PIV) were developed and are widely used in the field of fluid mechanics (for a review see Adrian, 2005 and Chapter 16, this volume). The use of these image-based techniques to study sediment transport has been steadily expanding; such techniques were first applied to studies of suspended sediment transport in turbulent flows and more recently to bedload transport. Tracking a single detected particle in a series of images is relatively easy and a number of authors have used image analysis to measure the trajectories of saltating and rolling particles (e.g., Niño et al., 1994; Hu and Hui, 1996; Lee et al., 2000; Ancey et al., 2002; Ancey et al., 2003). Tracking sediment flux is more complex and requires segmentation of all the particles in the image as a first step. Segmentation can be achieved by applying algorithms commonly used in PTV. Such algorithms were specifically developed to track tracers in a flow in regions where standard PIV algorithms based on cross-correlations were not well adapted because of large velocity gradients. These algorithms were then applied to tracking suspended sediments. Sechet and Le Guennec (1999) investigated the role of near wall turbulent structures. Nezu and Azuma (2004) building on a technique described by Okamoto et al. (1995) used PTV to characterise particle-laden free surface flows. Studies tracking coarse material are rare: Pilotti et al. (1997) analysed incipient motion of dark coloured grains on a light coloured smooth bed, Papanicolaou et al. (1999) tracked green glass beads over a layer of fixed transparent ones. In both cases the number of particles being tracked was low and the contrast between the grains and the bed was sufficiently high to allow segmentation using simple colour thresholding procedures. In these studies only the motion of segmented particles was tracked, rather than the entire bed. In a more sophisticated application, Capart et al. (1997) investigated water-sediment interaction following a dam-break using segmentation to track the motion of 6 mm plastic beads. More recently, Spinewine et al. (2003) extended the work of Capart et al. (2002) to the study of granular flows.

We present two flume studies that use PTV to reconstruct 2D trajectories of grains transported as bedload. In the first study, images of a 2D bed captured in plan-view were used to characterise the trajectories of bedload particles entrained by turbulent flow in a channel that was wide compared to the grainsize (Lajeunesse et al., 2010). In a second study, a high-speed camera was used to film the movement of particles flowing down a steep narrow channel only slightly wider than the grain diameter. In this study the 2D trajectories of the particles were captured in side-view of a transparent flume (Böhm et al., 2006).

13.2.2.1 2D trajectories in plan-view of individual grains transported as bedload

The goal of the study presented here was to investigate the motion of bedload particles over a flat bed of uniform grainsize under steady, spatially uniform, turbulent flow. The methodology consists of capturing images in plan-view of bedload particles moving across a bed using a high-speed camera mounted directly above the flume. From image analysis, particle velocities, step (or flight) lengths and durations, and the surface-density of moving particles can be measured (vertical velocity and saltation height cannot be measured in plan-view). Details of the study and the methodology can be found in Lajeunesse et al. (2010).

The experiments were carried out in a rectangular tilting flume with a width of 0.096 m and length 0.24 m. Water was injected by a pump at the upstream inlet of the flume and flowed over a sediment bed composed of quartz grains several centimeters thick. Three series of experiments were performed with a D_{50} of 1.92, 2.6 and 5.5 mm respectively. No additional sediment was fed at the upstream and as a result an erosion wave slowly propagated from the inlet towards the outlet of the flume. All the experiments were stopped well before the erosion wave reached the study reach situated in the middle of the flume and therefore it did not influence the results (this was also verified by measurements of the bed slope at the start and end of the experiment which showed that it remained constant). Approximately 10% of the sediment particles in the bed were dyed black. The rest were the natural colour of quartz (i.e., clear to white). A high-speed camera (250 frames per second, 1024×1024 pixels) mounted vertically directly above the bed was used to track the motion of the sediment particles. The position of the black particles could be tracked between successive frames with an interval of 0.04 s and the goal was to determine velocities and trajectories of each dyed bead from the temporal sequence of images (Figure 13.3). The spatial resolution of the camera was

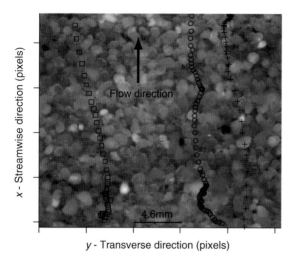

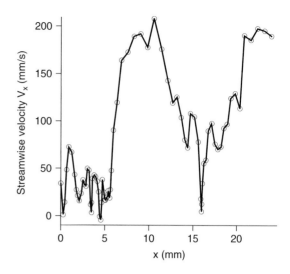

Figure 13.3 Example of three particles trajectories determined using particle tracking algorithms. The time interval between two successive positions is 0.004 s. The field of view is 23.5×23.5 mm^2 (particle diameter is approximately 50 pixels).

Figure 13.4 Streamwise velocity of a single particle through time as it moves downstream.

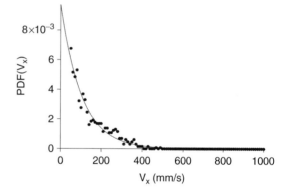

Figure 13.5 Distribution of particle streamwise velocities (Vx) measured using particle tracking algorithms. The PDFs decrease monotonically to zero and follow an exponential function.

such that the particle diameter was about 50 pixels and the field of view approximately 20×20 particle diameters. At these conditions, the position of a dyed particle could be determined with an accuracy of 0.05 mm. Particle tracking was performed using algorithms developed using the IgorPro data processing software (Lajeunesse et al., 2010). The algorithm consisted of two steps. In the first step each image was thresholded and binarised to create an image in which the dyed particle was black and the other particles were white. A particle detection algorithm was applied in order to detect and localise the center of each black particle. Lighting quality and the choice of threshold criteria are crucial for this first step. The second step consisted of reconstructing particle trajectories by tracking the black beads through the sequence of images (Figure 13.3). The data were then processed to calculate streamwise and transverse particle velocities and their distributions (Figures 13.4 and 13.5).

The main source of error in the experiments performed by Lajeunesse et al. (2010) was oscillation of the water surface which caused an apparent movement of the particles. This error can be corrected by determining the apparent velocity of a particle at rest for each experiment. Based on measurements of apparent velocity, a threshold value is defined below which particle velocities are discarded. Lajeunesse et al. (2010) had threshold values between 10–30 mm/s, depending on the flow rate. The same particle-tracking algorithm can be used to compute

lengths and durations of particle step-lengths. To do this, the size of the field of view had to be increased so that it is sufficiently larger than the characteristic particle step-lengths.

13.2.2.2 Tracking of beads entrained in a quasi two-dimensional channel

In this study PTV was performed from the side of a quasi-two dimensional tilting flume (2 m long and 0.2 m high) with transparent walls. The width of the flume was only slightly wider than the width of the particle diameters (6.5 mm). The goal of the experiments was to study the motion of coarse spherical glass beads entrained by a

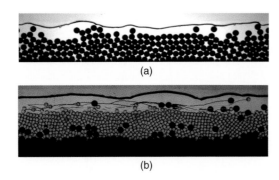

(a)

(b)

Figure 13.6 (a) Image of a unimodal experiment with 6 mm beads, (b) image from an experiment with a two-size mixture, bead-trajectories corresponding to the previous 30 images are superimposed.

turbulent supercritical flow down a steep channel (slopes were between 7.5–15%). Details can be found in Böhm et al. (2004, 2006). The experiments were performed with one-size and two-size sediment mixtures (Figures 13.6a and b respectively). The channel was supplied continuously with beads and measurements were made when transport was at equilibrium over the mobile bed (Böhm et al., 2004). Image processing was used to determine velocities and trajectories of each bead from a temporal sequence of approximately 8,000 images (Böhm et al., 2006). Numerous measurements permitted a thorough statistical description of unimodal sediment transport based on stochastic Markov-type processes (Ancey et al., 2006, 2008).

As in the previous studies, image processing consisted of first detecting and localising all beads and then reconstructing their trajectories by tracking them through the sequence of images. Here again, the image processing software WIMA in combination with several custom designed algorithms was used. In this study all the particles present in the measurement window were tracked (as opposed to only coloured particles in the previous study) making segmentation a crucial step. The quality of segmentation is largely dependent on flume setup, lighting, and material, and requires blending of know-how on image acquisition and knowledge of segmentation algorithms. As a general rule, particle tracking works best when applied to high quality well-contrasted images achieved through the right combination of lighting and material. Experiments should be setup with PTV needs in mind; bad quality images result in numerous false detections and can hamper the possibility to perform tracking at all. In the uniform size experiment (Böhm et al., 2006), using a two-dimensional channel slightly larger than the

dark beads was sufficient to make segmentation very easy. In the two-size experiment several trials were needed in order to choose smaller transparent beads (4 mm) to contrast with the 6 mm dark beads (Hergault et al., 2010).

Distinct segmentation algorithms were used to detect the two types of beads. The problem of tracking particles along a temporal sequence is not straightforward and is known as the point-tracking problem in the literature. Typically, points can be either small object centroids or interest points within larger rigid or unrigid objects. Depending on the application and on the assumptions made regarding point displacement, more or less complex approaches have been proposed (Hwang, 1989; Salari and Sethi, 1990). A particular application is to determine the velocity field of a fluid carrying small particles (Nishino et al., 1989; Fayolle et al., 1996; Udrea et al., 2000). In the studies reported here, Böhm et al. (2006) and Hergault et al. (2010) focused on individual particle motions. Particle velocities differed from those of the surrounding fluid since the particles were coarse and had a higher density than water. Although the use of spherical particles of uniform size made detection easier, the calculation of the trajectories was more difficult since it was not possible to distinguish particles based on their shape. The high frame rate of the camera resulted in a displacement of a particle between two images that was always smaller than the particle diameter. This was essential to achieving a good accuracy with the tracking algorithm as well as to reconstructing trajectories at high resolution (Figure 13.6b).

13.3 Channel morphology and flow dynamics

Timelapse imagery of dynamically evolving surfaces are an efficient means for measuring a broad-range of variables characterising channel dynamics and bed evolution at high temporal and spatial resolutions. Timelapse images can be readily compiled into movies showing the continuous evolution of a system. We present some general considerations in acquiring timelapse imagery and present examples on how they have been used in experimental studies of rivers and deltas.

When acquiring timelapse images it is recommended to use cameras that can be computer operated and allow automatic download. Computer operation is important because it helps prevent movement of the camera once it is fixed in place; this is especially important if the camera is mounted in a hard-to-reach location. Having images displayed in real time alerts the operator to any problems immediately and prevents a camera stopping mid-experiment due to a full memory card. Computer operation is particularly useful for operating multiple cameras; images captured simultaneously can be stitched together to capture a larger study area. The software development kit (SDK) that is included with most cameras only allows for one camera to be operated at a time (in other words, a separate computer is required to operate each camera). We are aware of only one commercially available software that permits simultaneous operation of multiple cameras from a single computer: Pine Tree Computing LLC. Many camera SDKs include an option for stitching images and batch processing. In order to successfully stitch images there must be sufficient overlap and the focal length of the images must be the same. It is also important to include stitch-points, i.e., clear points that appear in each pair of images to be stitched (assuming the setup does not change, stitch-points are required in only one set of images). PTGUI is a commercially available software for stitching images that is independent of a specific camera firmware. Finally, when capturing timelapse images it is recommended the camera be connected to an external power source to prevent it from entering sleep mode between shots or having a battery die mid-run.

Images typically need to be post-processed to rectify lens distortion (curvature) and camera angle (perspective) so that distances and areas are true everywhere in the image. The first step in correcting images is to acquire a calibration image of the study area under the exact conditions that will be used during an experiment. A calibration image consists of a grid composed of parallel and orthogonal horizontal and vertical lines of known dimensions placed parallel to the plane of the experimental surface. The grid can be constructed, drawn onto a board, delineated using stationary markers, etc. (it is recommended to do the calibration before starting an experiment). If the experimental setup (stream-table, camera position, focal length) remains constant the grid can be removed and the same calibration can be used to correct all the images. A new calibration is required whenever any changes are made to the experimental setup. For best results the grid should be as close as possible in size to the field of view of the camera. Once an image is properly corrected the grid-lines should be straight and parallel everywhere in the image and all distances should accurately scale with real-world distances. A computer program can be written to automate image correction for a series of images or widely available commercial image correction software

can be used – Andromeda LensDoc is an example of a commercial Photoshop plug-in that enables setting calibration parameters by simply clicking on different points in the calibration image. The parameters can then be used to batch process a series of images. It is recommended that all images be corrected before stitching.

Common techniques to better visualise naturally transparent flow include adding coloured dye (Winterbottom and Gilvear, 1997; Gran and Paola, 2001; Tal and Paola, 2007) and/or pigments in powder form such as titanium dioxide (TiO_2) to make it opaque (Martin et al., 2009). Adding colour/opacity to the flow makes it much easier to identify both qualitatively and quantitatively from images based on the colour value of each pixel (RGB and HSV; colour value can be read and manipulated using image processing software such as Photoshop and ImageJ or using a program that treats images as matrices such as Matlab). Jpeg image format compresses images and therefore requires less memory than Tiff or Raw, however the latter should be used whenever possible because the colour value of each pixel is preserved.

A constant dye concentration can be used to estimate the flow depth based on colour intensity (Winterbottom and Gilvear, 1997; Gran and Paola, 2001; Tal and Paola, 2007). This technique requires uniform light-coloured sediment; the optimal dye concentration and the variation of colour intensity with flow depth need to be calibrated for every experimental setup. Once a mathematical relationship between dye intensity and flow depth has been obtained, flow depth can be estimated anywhere in the image based on the pixel colour value. A simple way to calibrate flow depth based on colour intensity is to place tilted trays with sediment glued to the bottom and sides and filled with dye water in the camera's field of view. Images should be captured with a camera located directly overhead (normal to the flow). A polariser filter (lens) can be added to the camera in order to reduce glare from the flow. This can be further augmented by placing a polariser sheet in front of any lighting in order to achieve cross-polarisation. The dye-density technique is highly sensitive to variations in lighting which can pose a problem in the case of large experiments. Variations in lighting across the study reach should be identified and multiple calibrations (i.e., trays) should be used to minimise the error. Finally, many dyes photo-degrade (i.e., decay) over time and need to be replenished regularly to maintain a constant concentration. Once again, multiple and continuous calibrations help account for these changes.

13.3.1 Experimental deltas

The Experimental EarthScape (XES) facility at the St. Anthony Falls Laboratory, University of Minnesota is a research tank designed to study deltaic sedimentation associated with geological controls, e.g., changes in sediment supply, subsidence, and sea-level. Premixed sediment and water are introduced at the upstream end of the flume and flow for a short distance as a fluvial system. At the downstream end of the tank standing water forces sediment to deposit and form a delta. The tank is 3 m wide, 6 m long, and 1.5 m deep. In these experiments deltas self-organise in response to continuously evolving fluvial patterns. Details of these experiments can be found in Kim et al. (2006), Kim and Paola (2007), and Kim and Jerolmack (2008).

Throughout an experiment images of the evolving delta are captured every few seconds using an overhead camera. To better visualise the flow, water is dyed blue and made opaque by adding titanium dioxide. Flow opacity makes it possible to easily distinguish areas with active flow despite the use of two different colours of sediment (in the XES experiments quartz sand is white and crushed coal particles are black). Dye in these experiments was not used for measuring flow depths.

In order to create binary wet-dry images, the contrast between dry and wet areas was first enhanced by creating a greyscale image. For this set of images, differencing the red and blue colour bands produced a greyscale image that clearly highlighted the regions occupied by flow (Figure 13.7). This procedure can be done by, for example, using the 'calculations' option in Abode Photoshop. Once the greyscale map was obtained, a threshold value was used to convert the image into white (wet) and black (dry) – using the threshold option in Photoshop. The user should be able to identify approximately what flow depth a particular threshold value corresponds to and justify why it is an appropriate cutoff. A series of images can be batch processed once the algorithm has been established.

Analysis of the binary wet and dry images led to the observation that self-organised (autogenic) sediment storage and release are associated with changes in river-planform pattern (Figure 13.8; Kim and Jerolmack, 2008): the experimental deltas stored much of the sediment supplied from the upstream input and showed strong sediment deposition in the deltaic surface when the river pattern was tabular (i.e., sheet flow which corresponds to a large wetted fraction of the bed). In contrast, the delta released sediment stored from the surface through the shoreline to the ocean when the deltaic channels were

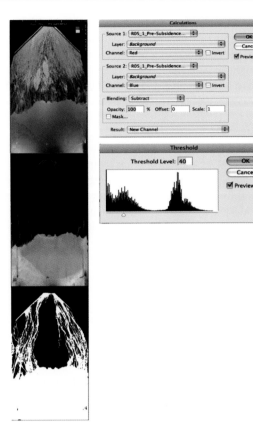

Figure 13.7 Example of an image analysis from the XES experiment (Kim et al., 2006; Kim and Paola, 2007; Kim and Jerolmack, 2008). The left image is the original colour image, the middle greyscale image is the result of differencing the red and blue colour bands, and the right binary image is based on a threshold value applied to the greyscale image to distinguish between wet (white) and dry (black) pixels.

confined to a narrow path and incised (i.e., channelised flow which corresponds to a small wetted fraction of the bed). The rate of seaward shoreline migration fluctuated as a result of changes in the sediment flux being released from the delta. Autogenic shoreline fluctuation measured from the images corresponded well with the fraction of the fluvial surface that was wet – measured from the binary images.

The analysis of the wetted fraction described above does not reflect the lateral mobility of a channel (i.e., if a channel migrates laterally while maintaining a constant width the wetted fraction remains the same). In order to characterise channel activity (planform pattern change + lateral mobility), Kim and Paola (2007) analysed the wet and dry binary images in a different way. The analysis consisted of accumulating the area of the deltaic surface

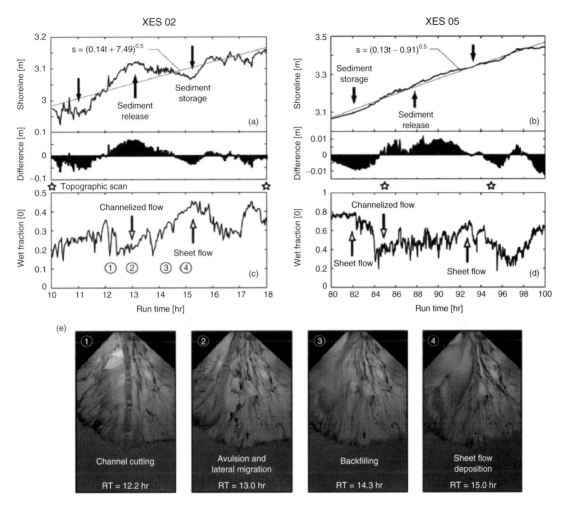

Figure 13.8 Shoreline and wet-fraction data from the XES experiment (Kim and Jerolmack, 2008). (a and b) Mean shoreline position averaged normal to the mean sediment transport direction and best fit curves. Graphs show shoreline position fluctuation after removal of long-term shoreline regression trend, (c and d) time series of wet fraction showing cyclic changes in fluvial pattern between sheet and channelised flow, (e) representative overhead images showing a cycle of changes in the wet-fraction. Reproduced from Kim & Jerolmack, Journal of Geology "The Pulse of Calm Fan Deltas" 116:4 (2008) Fig. 3, p. 319.

that became wet and the decay (decrease) in the area that remained dry over a set time interval (Figure 13.9). A faster decay represents stronger channel activity and vice-versa. Kim and Paola (2007) used this analysis to compare channel activity during various stages of the experiment that were subject to different external forcing.

13.3.2 Experimental river channels with riparian vegetation

In a separate set of experiments at the St. Anthony Falls Laboratory, Tal and Paola (2007, 2010) studied

the dynamic interactions between flow and vegetation. The experiments were designed to investigate whether riparian vegetation could cause a braided channel to evolve to a single-thread channel and were motivated by many field cases in which the encroachment of riparian vegetation was driving a change in planform. The experiments were conducted in a flume 16 m long by 2 m wide with steel walls. Alfalfa sprouts (*Medicago sativa*) were used as the experimental vegetation in the flume. The initial condition for the experiments was steady-state braiding in noncohesive sand under uniform discharge. From here, an experiment consisted of repeated cycles

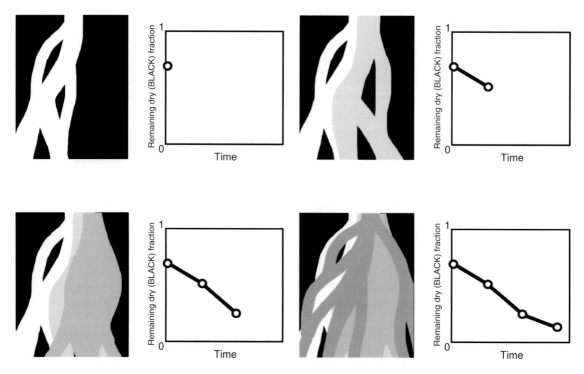

Figure 13.9 XES image analysis showing the evolution of the deltaic surface over a given time interval. White and grey are respectively the initial wet surface and surfaces that became progressively wet as a result of channel migration. Black is the surface that remained dry. A faster decay in the remaining dry surface corresponds to stronger channel activity (Kim and Paola, 2007).

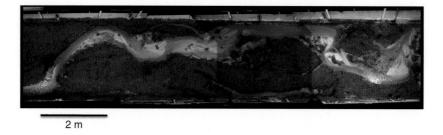

2 m

Figure 13.10 Experimental channel at the St. Anthony Falls Laboratory designed to study the interactions between flow and riparian vegetation (Tal and Paola, 2007; Tal and Paola, 2010). Image is stitched together from four simultaneous time-lapse images. Rhodamine dye is used to colour the flow.

alternating a long duration state of low discharge (3 – 6 days) during which the channel morphology remained relatively stable and plants grew on emergent river bars and banks, and a short-lived flood (typically 1 hour) with vigorous sediment transport in which physical processes predominated. Vegetation was added to the experiment by dispersing seeds manually over the entire bed at the end of each high flow with the water discharge set to its low value.

Four digital cameras (Olympus C-4000 Zoom) were mounted directly above the flume and equally spaced to capture images of a 10 m long by 2 m wide study reach (Figure 13.10). The images overlapped by ~0.5 m so they could be stitched together. The cameras captured images at a resolution of approximately 2 mm/pixel. Each camera had a polarising filter, and polariser sheets were hung underneath the halogen lighting to achieve cross-polarisation and minimise glare from the flow.

The cameras were controlled remotely using the camera controller software available from Pine Tree Computing LLC and captured simultaneous time-lapse images at intervals of 30-120 seconds during all of the high flows as well as once at the beginning and end of each low flow. Images were post-processed to correct for distortion associated with lens curvature and perspective using the Photoshop Andromeda Lensdoc plug-in. Each set of four simultaneous images was stitched using the commercial software PTGUI to create a single image.

Dye was added to the flow to enhance visualisation as well as to measure flow depth based on colour-intensity. Dyed water was made in batches by mixing 800 litres of water in a large tank with Rhodamine dye at a concentration of 2 ppm. The tank provided the only source of water during the floods in order to maintain a constant concentration of dye. Flow was recirculated with a pump. This procedure was used for all experiments that had a one-hour flood duration. In experiment with a four-hour high flow photodegradation of the dye did not permit flow recirculation. Instead, dye and water were fed in at a constant rate at the inlet. Tilted trays with sand glued to the bottom and sides were filled with dye water and used to calibrate depth variation with colour intensity (Figure 13.11). The calibration trays were placed at least once during each flood in the field of view of each of the four cameras and left permanently on the bed wherever possible without disturbing the flow dynamics. The minimum calibration error associated with the calibration trays was, on average, +/− 1 mm. In order to minimise calibration errors depth was estimated for each camera separately using the tray for that image only, and as close in time as possible to the time the image was captured.

Wetted widths and channel migration rates were measured from flow maps created from time-lapse images captured at 5 minute intervals during the floods. Colour images were first converted into binary images in which all the flow had a value of 1 (white) and all dry areas (sand + veg) had a value of 0 (black; Figure 13.12a and 13.12b). A threshold hue value of 0.8 and a saturation value of 0.3 were used to distinguish between wet and dry and the process was automated using Matlab. Data were extracted along transects normal to the images and spaced 0.05 m apart over the entire study reach (total of 226 transects). A cross-stream width of 0.06 m was set as a threshold for a group of wet pixels to be considered a channel. Wetted widths were calculated as the sum of wet pixels averaged over a downstream distance of 0.3 m (6 transects). Channel migration rates were calculated by summing the area (in pixels) that was converted from dry to wet between consecutive pairs of images and dividing this area by the total length of the image. Changes over 5-min intervals encompassed both gradual lateral migration as well as abrupt (<5 min) channel shifts from one location to another.

Analysis of flow maps for these experiments were used to measure how wetted width, number of active channels (braiding index), migration rates, and channel sinuosity evolved as the braided channels transitioned to single-thread channels (Tal and Paola, 2010). Similar to the analysis described above by Kim and Paola (2007) and using a technique developed by Wickert (2007), flow maps were used to measure the timescale for loss of pattern information, i.e. the time required for the entire bed to get reworked once by the flow. The number of pixels representing flow in the first image of a series that

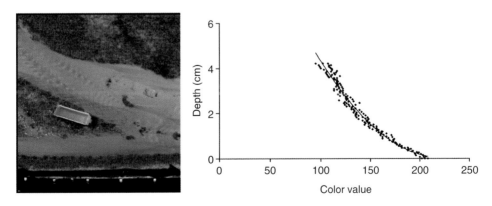

Figure 13.11 Flow with Rhodamine dye, calibration tray, and curve showing depth variation with colour intensity (Tal and Paola, 2007; Tal and Paola, 2010).

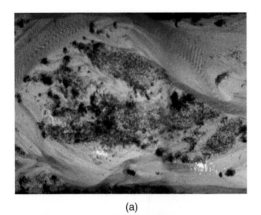

(a)

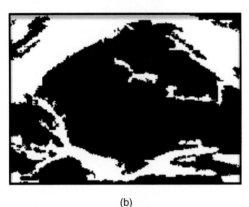

(b)

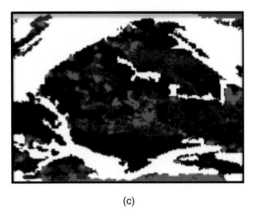

(c)

Figure 13.12 Example of part of a classified image (1 m × 1 m) (a) original image, (b) binary image, wet is white, black is dry, (c) area that was black in b classified into bare-sparse vegetation (black), dense vegetation (dark gray), very dense vegetation (light grey; Tal, 2008).

continued to contain flow at each subsequent time step was recorded (pixels that became dry, i.e., abandoned or vegetated and then wet again were not counted). The analysis demonstrated that vegetation slows the reworking time of the bed compared to unvegetated braiding. This reduction was attributed to a combination of slower erosion rates due to increased bank strength and the limited ability of opportunistic creation of new channels in areas not occupied by the flow due to the deterrence effect of plants (Tal and Paola, 2010).

Transition matrices were used to study feedbacks between flow and vegetation establishment. To do this, pixels representing dry riverbed in the binary flow maps were further classified into three additional classes representing different stages of vegetation cover (plant age and density; Figure 13.12c). The three main stages of vegetation cover were first visually identified in the images, then their distinct colour signature was measured (Figure 13.13). The colour value of the plants at different stages represented a combination of both plant age and density: dense clumps of young vegetation could have the same colour value as sparse patches of mature plants. Using the classified images transition matrices were computed for pairs of images at 5-minute intervals during all of the high flows as well as between high and low flows. The transition matrixes shed light on important feedbacks between flow and vegetation and the role of a fluctuating discharge. A key effect of the plants was to colonise abandoned (dry) areas during low flow and to deter the high flow from reoccupying these areas during the next flood, leading to progressive reinforcement of the low flow wetted width (Tal and Paola, 2010). The analysis also demonstrated that vegetation colonisation was a highly self-reinforcing process: areas where plants survived early provided a more stable surface for new plants to establish. As new vegetation established it further stabilised these areas and increased their chances of surviving subsequent floods (Figure 13.14). As a result, vegetated islands expanded through time and merged with other islands to form larger islands. This continuous process of accretion and amalgamation of islands eventually resulted in the formation of a permanently vegetated floodplain (Figure 13.10; Tal, 2008).

13.4 Bed topography and flow depth

Approaches to measuring bed topography range from simple manual point-gauges to sophisticated laser and ultrasonic scanners mounted on automated carriages

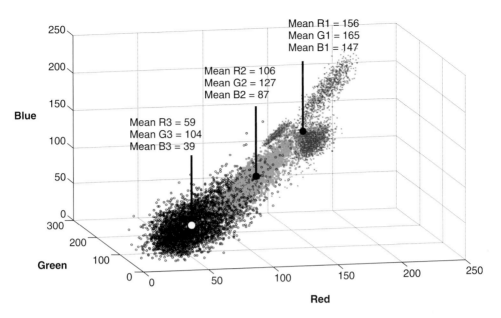

Figure 13.13 Pixels from vegetation samples plotted in colour space. Each colour/symbol represents the class that the sample corresponded to in the visual classification. The plot shows that the classes that were discerned visually do indeed fall into distinct classes based on their colour value (Tal, 2008; Tal and Paola, *in prep.*).

(e.g., Kim and Paola, 2007; Hoyal and Sheets, 2009). Another common technique consists of projecting single or multiple laser line(s) onto the bed and photographing it with a camera mounted at an oblique angle; vertical displacements of the line are calibrated to real-world changes in elevation (Figure 13.15; Leaf et al., 1993; Hasbargen and Paola, 2000; Lague et al., 2003). These point-by-point or line-by-line techniques are discontinuous in space and some can be very time-consuming, requiring long pauses in the experiment to perform each scan. Digital photogrammetry is one method that provides a way of measuring topography over a continuous surface (Chandler et al., 2001; Lane et al, 2001; Turowski et al., 2006). However, all these methods require a separate technique and setup to measure flow depths. For example, combinations of laser measurements and dye-density (discussed above) have been used to simultaneously measure both flow depth and bed topography (e.g. Huang et al., 2010).

Here we describe an optical method known as moiré for acquiring measurements of both bed topography and flow depth in laboratory experiments (Sansoni et al., 1999). The moiré projection method is part of a general family of techniques that use the projection of structured light to measure relief (Patorski, 1993) and enables image-based non-contact measurements over a continuous surface at very high spatial and temporal

resolutions. The moiré method has been successfully applied in metrology studies (Chiang, 1979), industrial inspection (Sansoni, 2000), human body mapping for medical diagnosis (Halioua, 1989; Kozlowski, 1997) and art inspection (Bremand, 2007).

A moiré method is based on projecting a fringe pattern (also known as a grating or grid) on the bed and analysing the deformation of the pattern caused by the topography with respect to a fringe pattern projected on a reference plane (i.e., a flat bed; Figure 13.16). The height of the object (i.e., topography) is encoded in the phase difference between the two patterns which is retrieved through a Fourier transform or phase shifting algorithms (Figure 13.17). While the mathematics behind the method is rather complex, the good news is the methodology is relatively easy to implement and user-friendly commercial software that perform the calculations automatically are available. For details about the theory and mathematical operations readers should refer to Sansoni et al. (1999), Pouliquen and Forterre (2002), and Limare et al. (2011). Here we present only the basics behind the method, considerations for implementing it, and an example from a study of braided channels.

The simplest way to implement a moiré method consists of capturing an image (or series of images) of a single

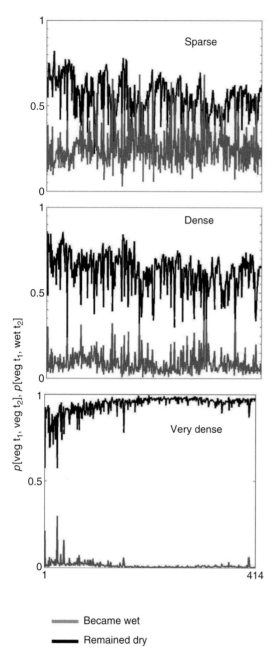

Figure 13.14 Probability by vegetation class of an area that was vegetated at t_1 to get eroded (become wet) at t_2 versus the probability that it will remain dry at t_2. Probabilities were calculated between images captured at 5-minute intervals during all of the high flows (Tal, 2008).

grid projected onto an evolving surface (using a video or more simply an overhead projector) and analysing the deformation of the grid with respect to the same grid projected on a flat plane (e.g., a smooth bed at the beginning of an experiment; Figure 13.16). The projection angle of the grid should be sufficiently small such that bed topography induces a significant deformation. Photos of the plane are recorded with a digital camera positioned vertically above the bed. The spectrum of grid patterns of the reference image and an image containing topography are analysed using the Fourier transform (Figure 13.17). The local shift of the grid lines (i.e., phase-difference) between the deformed pattern in the presence of topography and the reference pattern is contained in the width of the peak in the spectrum analysis and is proportional to the local thickness of the bed (i.e. topography). The phase difference is converted to bed elevation based on a calibration that relates the measured phase shift to a known change in elevation (Figure 13.17).

The simplified version of the moiré method described above has been used in laboratory physical experiments to study granular flows (Pouliquen and Forterre, 2002), incision dynamics of subaqueous channels (Lancien, 2005), and micro-scale braided rivers (Metivier and Meunier, 2003). The costs involved with this type of moiré setup include a digital camera, a computer, and a video-projector (or even less expensive – an overhead projector), the sum of which is substantially lower than of a laser scanner but without compromising precision. Data acquisition and processing are relatively fast so measurements can be made in quasi-real time and data processing can be automated.

A more elaborate but more precise implementation of the moiré method uses a procedure known as phase shifting to calculate the phase and robust phase unwrapping combined with grey coding to assess the geometric parameters (details in Limare et al., 2011). A more sophisticated phase-to-height conversion is based on a calibration using a plane tilted at a known incline rather than a stationary object (Figure 13.18). Rather than a single-grid projection, a series of different grid patterns are projected automatically in succession. Images are captured with a CDD camera connected to the same computer generating the grid patterns. The number of projected images can be varied making it possible to switch between high resolution and fast acquisition time according to the requirements of the experiment.

Regardless which moiré method is used, certain general requirements must be fulfilled in order to obtain the best raw images as input data. First, the grid projection must

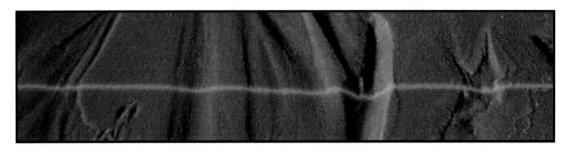

Figure 13.15 Example of a laser line (~0.4 m) projected onto a sediment bed and photographed with a camera mounted at an oblique angle (Tal, 2008).

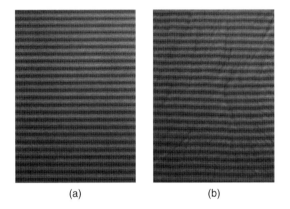

(a)   (b)

Figure 13.16 (a) Reference image (1 m long × 0.5 m wide) of a fringe pattern projected onto a flat bed and (b) a deformed fringe pattern due to bed topography formed by braided channels (Limare et al., 2011).

have a constant frequency across the experimental surface. If the projector has a significant incidence angle (either to project over a larger area or to reduce glare if water is present), it may be necessary to compensate for this angle when creating the grid to ensure a constant frequency across the entire plane. In general, the moiré method works well for light coloured, relatively uniform diffusive substrate. As in the other imaging methods discussed, glare should be reduced using polariser filters and sheets. Images should be corrected for lens and perspective distortion before any analysis.

A key advantage to the moiré method is the ability to estimate flow depths based on the refraction of light at the air/water interface. Bed topography reconstructed by any method based on grid deformation when water is present on the bed will be distorted relative to bed topography measured on a dry surface. This distortion can be used to estimate flow depth from images acquired with and without flow. Assuming the flow is locally

uniform (free surface parallel to the bed surface), light from the projector will bend normal to the surface and will produce a bright spot when hitting the surface. From the point of view of the camera, the bright spot is observed in the direction corresponding to the angle i_c with respect to the normal (Figure 13.19):

$$\frac{\sin i_p}{\sin r_p} = \frac{\sin i_c}{\sin r_c} = n_{water} \qquad (13.4)$$

where n_{water} is the refractive index of water. The difference in elevation between the apparent location of the bright spot (h) and the true depth (H) is obtained by trigonometry:

$$H = \frac{\tan i_p + \tan i_c}{\tan r_p + \tan r_p} h \qquad (13.5)$$

The difference between topographies obtained with and without flow will give a distorted bathymetry that is always shallower than the true bathymetry and a correction factor for the incidence angle needs to be applied. Incidence angles can be calculated for any point using simple trigonometry: $i_p, i_c = f(i, j)$. In the case of parallel projection and observation beams, the relation between true and distorted bathymetry has a simplified form: $H = n_{water} h$.

A moiré projection method was used to study the evolution of microscale braided channels (Figure 13.20; Limare et al., 2011). Experiments similar to the one here are described in detail in Metivier and Meunier, (2003). The experiments were conducted at the IPGP experimental geomorphology laboratory in a stream-table that was 1.5 m long and 0.75 m wide with a fully adjustable bed. The initial slope of the bed was approximately 0.05. The sediment in these experiments was composed of glass beads with a D_{50} of 250 micron and a density of 2500 kg/m³. The initial condition for each experiment was a flat bed with a straight channel (0.01 m deep and 0.02 m wide) carved down the middle. The flow was laminar and a

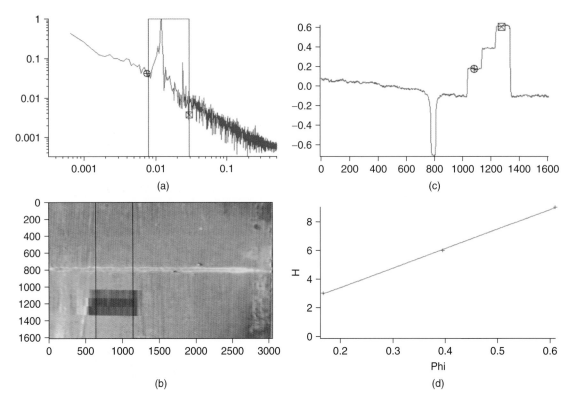

Figure 13.17 Steps involved in a moiré method analysis: (a) spectrum of the reference image obtained using the Fourier transform, (b) an image of the bed topography reconstructed from the phase shift between the reference image and an image with a straight channel carved into the bed and an object of known dimensions (consisting of three stairs), (c) a profile from top to bottom of one of the transects in b, (d) phase-to-height calibration based on the known heights of the three stairs.

fully braided morphology spontaneously developed and reached steady state (sediment input equaled sediment output) after a few hours. Sediment and water were supplied continuously at the upstream end of the flume at a constant rate. Flow discharge was around 2.5×10^{-5} m^3/s and sediment discharge was approximately 8×10^{-8} m^3/s. Glare caused by flow was reduced by using two cross axis polarisers placed in front of the projector and the camera respectively. A moiré method was implemented to measure bed topography and flow depth using a commercial software package called Light3D – developed at the Laboratory of Solid State Mechanics, University of Poitiers, France (Breque et al., 2004). The program uses phase shifting to calculate the phase and robust phase unwrapping combined with grey coding to assess the geometric parameters. The user can choose the number of phase-shifted patterns to be projected (3, 8, 16 or 32) and phase unwrapping can be performed with or without 8 grey code images. For the results shown here the

phase field was calculated with 8 phase-shifted images and robust unwrapping using a series of 8 grey code images. The computer generated patterns were projected from a Sanyo PLV-Z5 video-projector. A black and white μ eye Stemmer Imaging CCD camera (1280 × 1024 pixels) was used for image acquisition. Prior to each experiment a calibration procedure was carried out using a flat, white, highly diffusive board that could be tilted following the procedures outlined by the software. In addition to the phase-to-height conversion, the software uses the calibration to automatically calculate all the required geometrical parameters which can be difficult to measure accurately manually (Figure 13.18).

Over the course of an experiment bed topography and water depth were measured every 10 minutes in the following manner: 'wet' bed topography was measured with the flow on, the flow was turned off, 'dry' bed topography was measured after the flow was off for 1 minute (the time required for the bed to be uniformly drained). Water

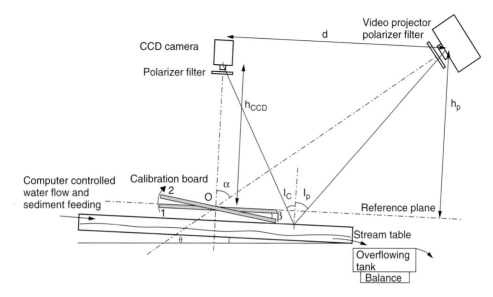

Figure 13.18 Setup of experiment with a moiré projection method to study microscale braided channels at the IPGP experimental geomorphology laboratory (Limare et al., 2011).

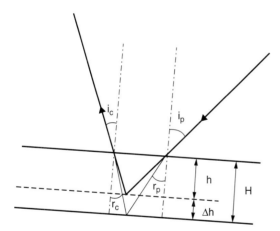

Figure 13.19 Light refraction into a layer of water (Limare et al., 2011).

depth was calculated by subtracting dry topography from wet topography taking into account water refraction and incidence angle as described above. The impact of stopping and restarting the flow on the bed topography was negligible. Figure 13.20 shows (a) bed topography and (b) flow depth along a cross-section approximately 0.5 m mid-way down the study reach. Several occupied and abandoned channels can be depicted along this transect. Dry bed topography in these microscale experiments ranged from −4 to + 4 mm relative to the initial flat bed,

maximum water depth was approximately 3 mm (flow depth increases in the negative direction).

Flow depths were systematically filtered to remove noise by analysing the PDF of flow depths for the entire study area. Average flow depth was −0.24 mm with a standard deviation of 0.43 mm respectively. We considered positive flow depth to be noise and assumed noise to be symmetrical around zero. Noise was subtracted from the original distribution to obtain a 'real' distribution of depths which had a mean and standard deviation of −0.49 mm and 0.41 mm respectively. The area of the flume occupied by flow after noise correction was 49%, which corresponds well with visual observations.

13.5 Conclusions

We have presented several imaging techniques used to study rivers in laboratory experiments and shown how these techniques provide efficient non-laborious means of acquiring high resolution data that enable researchers to describe the evolution and organisation of experimental rivers and deltas with increasingly robust statistics. Imaging techniques in the laboratory benefit from the high degree of control over lighting and colour of materials. The quality of the data collected using any imaging technique are optimised if the requirements of a given technique are considered in advance and implemented in

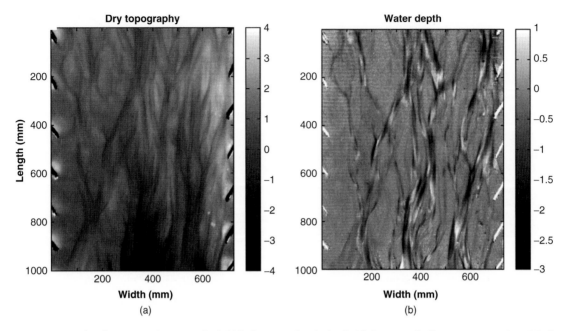

Figure 13.20 Results of a moiré projection method, (a) bed topography obtained with flow turned off, zero represents the original flat reference bed (b) flow depths obtained by subtracting images with and without flow (Limare et al., 2011). Flow increases in the negative direction. Zero represents dry bed. Positive flow depths represent noise that was later removed.

the experimental setup. The equipment required typically involves at least one digital camera or video recorder and computer to operate the camera remotely and download photos automatically. The type of camera that should be used depends on the method, e.g., basic digital camera for acquiring timelapse images (multiple cameras or a wide-angle can be used for large experiments), high-speed video recorder for PTV, and CDD camera for a moiré method. The latter also requires a video-projector. Cameras usually include software enabling control of only one camera at a time per computer, however commercial software is available for operating multiple cameras simultaneously from a single computer. Most of the difficult work involved in imaging techniques in conjunction with laboratory experiments is in the initial setup, assuming the setup remains constant thereafter. Once all equipment is placed in the optimum position (e.g., height and distance of the camera and projector), the optimal conditions have been established (e.g. lighting, dye concentration, colour of the sediment), and all necessary calibrations have been made (e.g., stitch points, image rectification), data acquisition is relatively fast and simple with minimal disruption to the experiment. While many image processing techniques are based on complicated equations and algorithms, user-friendly software are available to guide

users who do not wish to develop their own programs from scratch. Equipped with such software, excellent results can be achieved with only a basic understanding of the technique and adequate trial and error. Likewise, software is available to batch-process operations such as image rectification, cropping, stitching, colour thresholds, etc. for a large series of images.

Acknowledgements

We are grateful to Maarten Kleinhans and Richard Hardy for reviewing this chapter and for providing helpful comments.

References

Adrian, R.J. 2005. Twenty years of particle image velocimetry, *Exp Fluids* 39: 159–169.

Ancey, C., Bigillon, F., Frey, P., Lanier, J., and Ducret, R. 2002. Saltating motion of a bead in a rapid water stream. *Phys Rev E* 66 (3): 036306.

Ancey, C., Bigillon, F., Frey, P., and Ducret, R. 2003. Rolling motion of a bead in a rapid water stream, *Phys Rev E* 67 (1): 011303.

Ancey, C., Böhm, T., Jodeau, M., and Frey, P. 2006. Statistical description of sediment transport experiments. *Phys Rev E* 74 (1): 011302

Ancey, C., Davison, A.C., Böhm, T., Jodeau, M., and Frey, P. 2008. Entrainment and motion of coarse particles in a shallow water stream down a steep slope, *J Fluid Mech* 595: 83–114.

Ashmore, P.E. 1991b. How do gravel bed rivers braid?, *Canadian Journal of Earth Science* 28: 326–341.

Bigillon, F., Frey, P., Ducottet, C., Vierin, N., Ancey, C., Richard, D., and Lanier, J. 1999. Développement d'une technique d'analyse d'images pour la caractérisation du transport solide par charriage, *FLUVISU* 99, pp. 275–280, Toulouse, France.

Böhm, T., Ancey, C., Frey, P., Reboud, J.L., and Ducottet, C. 2004. Fluctuations of the solid discharge of gravity-driven particle flows in a turbulent stream, *Phys Rev E* 69(6): 061307.

Böhm, T., Frey, P., Ducottet, C., Ancey, C., Jodeau, M., and Reboud, J.L. 2006. Two-dimensional motion of a set of particles in a free surface flow with image processing. *Exp Fluids* 41: 1–11.

Brémand, F., Doumalin, P., Dupré, J., Hesser, F., and Valle, V. 2007. Optical techniques for relief study of analysis of Mona Lisa's wooden support, Proceedings of the 13th International Conference on Experimental Mechanics, Gdoutos, Greece.

Breque, C., Dupré, J.C., and Bremand, F. 2004. Calibration of a system of projection moiré for relief measuring: biomechanical applications, *Optics and Lasers in Engineering* 41: 241–260.

Buscombe, D. and Masselink, G. 2009. Grain-size information from the statistical properties of digital images of sediment, *Sedimentology* 56: 421–438.

Buscombe, D., Rubin, D.M., and Warrick, J.A. 2010. A universal approximation of grain size from images of noncohesive sediment, *Journal of Geophysical Research* 115: F02015.

Canny, J. 1986. A computational approach to edge detection, *IEEE Transactions on Pattern Analysis and Machine Intelligence* 8: 679–698.

Capart, H., Liu, H.H., Van Crombrugghe, X., and Young, D.L. 1997. Digital imaging characterization of the kinematics of water-sediment interaction, *Water Air and Soil Pollution* 99: 173–177.

Capart, H., Young, D.L., and Zech, Y. 2002. Voronoï imaging methods for the measurement of granular flows, *Exp Fluids* 32: 121–135.

Chandler, J.H., Shiono, K., Rameshwaren, P., and Lane, S.N. 2001. Measuring flume surfaces for hydraulics research using a Kodak DCS460, *The Photogrammetric Record* 17: 39–61.

Chiang, F. 1979. Moiré methods of strain analysis, *Experimental Mechanics* 19(8): 290–308.

Diplas, P., Dancey, C.L., Celik, A.O., Valyrakis, M., Greer, K., and Akar, T. 2008. The role of impulse on the initiation of particle movement under turbulent flow conditions, *Science* 322 (5902): 717–720.

Ducottet, C. 1994. Application of wavelet transforms to the processing of tomographic and holographic images of fluid flow, PhD Dissertation, University of Saint Etienne, France.

Fayolle, J., Ducottet, C., Fournel, T., and Schon, J.P. 1996. Motion characterization of unrigid objects by detecting and tracking feature points, *IEEE International Conference on Image Processing*, Lausanne, Switzerland, pp. 803–806.

Frey, P., Ducottet, C., and Jay, J. 2003. Fluctuations of bed load solid discharge and grainsize distribution on steep slopes with image analysis, *Exp Fluids* 35: 589–597.

Gomez, B. and Church, M. 1989. An assessment of bed load sediment transport formulae for gravel bed rivers, *Water Resources Research*, 25(6): 1161–1186.

Graham, D.J., Reid, I., and Rice, S.P. 2005. Automated sizing of coarse-grained sediments: Image- processing procedures, *Mathematical Geology* 37: 1–28.

Gran, K. and Paola, C. 2001. Riparian vegetation controls on braided stream dynamics, *Water Resources Research*, 37(12): 3275–3283.

Halioua, M. and Liu, H.-C. 1989. Optical three-dimensional sensing by phase measuring profilometry, *Optics and Lasers in Engineering* 11: 185–215.

Hasbargen, L.E. and Paola, C. 2000. Landscape instability in an experimental drainage basin, *Geology* 28 (12): 1067–1070.

Hergault, V., Frey, P., Métivier, F., Barat, C. Ducottet, C., Böhm, T., and Ancey, C. 2010. Image processing for the study of bedload transport of two-size spherical particles in a supercritical flow, *Experiments in Fluids* 49(5): 1095–1107.

Hooke, R.L. 1975. Distribution of sediment transport and shear stress in a meander bend, *Journal of Geology* 83: 543–565.

Hoyal, D.C. and Sheets, B.A. 2009. Morphodynamic evolution of experimental cohesive deltas, *Journal of Geophysical Research* 114: F02009.

Hu, C.H. and Hui, Y.J. 1996. Bed-load transport. 1. Mechanical characteristics. *J. Hydraul. Eng.* 122(5): 245–254.

Huang, M., Huang, A., and Capart, H. 2010. Joint mapping of bed elevation and flow depth in microscale morphodynamics experiments, *Experiments in Fluids* 49(5): 1121–1134.

Hwang, V.S.S. 1989. Tracking feature points in time-varying images using an opportunistic selection approach, *Pattern Recognition* 22: 247–256.

Ikeda, S., Parker, G., and Kimura, Y. 1988. Stable width and depth of straight gravel rivers with heterogeneous bed materials, *Water Resources Research* 24: 713–722.

Jernot, J.P., Coster, M., and Chermant, J.L. 1982. Model to describe the elastic modulus of sintered materials, *Phys. Stat. Sol..*(a). 72(1): 325–332.

Kim, W. and Jerolmack, D.J. 2008. The pulse of calm fan deltas: *Journal of Geology* 116 (4): 315–330.

Kim, W. and Paola, C. 2007. Long-period cyclic sedimentation with constant tectonic forcing in an experimental relay ramp: *Geology* 35 (4): 331–334.

Kim, W., Paola, C., Swenson, J.B., and Voller, V.R. 2006. Shoreline response to autogenic processes of sediment storage and release in the fluvial system, *J. Geophys. Res.* 111: F04013.

Kozłowski, J. and Giovanni, S. 1997. New modified phase locked loop method for fringe pattern demodulation, *Opt. Eng.* 36: 2025–2030.

Lague, D., Crave, A., and Davy, P. 2003. Laboratory experiments simulating the geomorphic response to tectonic uplift, *J. Geophys. Res.* 108 (B1): 2008.

Lajeunesse, E., Malverti, L., and Charru, F. 2010. Bed load transport in turbulent flow at the grain scale: Experiments and modeling, *J. Geophys. Res.* 115: F04001.

Lancien, P., Metivier, F., Lajeunesse, E., and Cacas, M. 2005. Incision dynamics and shear stress measurements in submarine channels experiments, *in River, Coastal and Estuarine Morphodynamics*, Parker and Garcia (eds), Taylor and Francis Group, London.

Lane, S.N., Chandler, J.H., and Porfiri, K. 2001. Monitoring river channel and flume surfaces with digital photogrammetry, *J. Hydraul. Eng.* 127: 871.

Leaf, R.B., Wilson, B.N., and Hansen, B.J. 1993. Field instrumentation for measuring soil topography, ASAE Summer Meeting, Spokane, Washington, Pap. 932111, 19 pp., Am. Soci. of Agric. Eng., St. Joseph, Michigan.

Lee, H.Y., Chen, Y.H., You, J.Y., and Lin, Y.T. 2000. Investigations of continuous bed load saltating process. *J. Hydraul Eng* 126: 691–700.

Limare, A., M.Tal, Reitz, M., Lajeunesse, E., and Metivier, F. 2011. Optical method for measuring bed topography and flow depth in an experimental flume, *Solid Earth* 2: 143–154.

Martin, J., Sheets, B., Paola, C., and Hoyal, D. 2009. Influence of steady base-level rise on channel mobility, shoreline migration, and scaling properties of a cohesive experimental delta, *Journal of Geophysical Research* 114: F03017.

Métivier, F. and Meunier, P. 2003. Input and output mass flux correlations in an experimental braided stream: Implications on the dynamics of bed load transport, *Journal of Hydrology* 271: 22–38.

Meyer-Peter, E. and Müller, R. 1948. Formulas for Bed-Load Transport, Proceedings 2nd Congress, International Association of Hydraulic Research, Stockholm, pp. 39–64.

Nezu, L. and Azuma, R. 2004. Turbulence characteristics and interaction between particles and fluid in particle-laden open channel flows, *J. Hydraul. Eng.* 130: 988–1001.

Niño, Y., Garcia, M., and Ayala, L. 1994. Gravel Saltation. 1. Experiments, *Water Resour. Res.* 30: 1907–1914.

Nishino, K., Kasagi, N., and Hirata, M. 1989. Three-dimensional particle tracking velocimetry based on automated digital image processing, *J. Fluids Eng.* 111(4): 384–392.

Okamoto, K., Hassan, Y.A., and Schmidl, W.D. 1995. New Tracking Algorithm for Particle Image Velocimetry. *Exp Fluids* 19: 342–347.

Papanicolaou, A.N., Diplas, P., Balakrishnan, M., and Dancey, C.L. 1999. Computer vision technique for tracking bed load movement, *Journal of Computing in Civil Engineering* 13: 71–79.

Parker, G., Dhamotharan, S., and Stefan, H. 1982. Model experiments on mobile, paved gravel bed streams, *Water Resources Research*, 18 (5): 1395–1408.

Patorski, K. 1993. *Handbook of the moiré fringe technique*, Elsevier, New York, 431 p.

Pilotti, M., Menduni, G., and Castelli, E. 1997. Monitoring the inception of sediment transport by image processing techniques, *Exp Fluids* 23: 202–208.

Pouliquen, O. and Forterre, Y. 2002. Friction law for dense granular flows: application to the motion of a mass down a rough inclined plane, *Journal of Fluid Mechanics* 453: 133–151.

Radice, A., Malavasi, S., and Ballio, F. 2006. Solid transport measurements through image processing, *Exp Fluids* 41: 721–734.

Recking, A., Frey, P., Paquier, A., and Belleudy, P. 2009. An experimental investigation of mechanisms responsible for bedload sheet production and migration, *J. Geophys. Res.* 114: F03010.

Rubin, D.M. 2004. A simple autocorrelation algorithm for determining grain size from digital images of sediment, *Journal of Sedimentary Research*, 74 (1).

Salari, V. and Sethi, I.K. 1990. Feature point correspondence in the presence of occlusion. IEEE *Transactions on Pattern Analysis and Machine Intelligence* 12: 87–91.

Sansoni, G., Carocci, M, and Rodella, R. 1999. 3D vision based on the combination of gray code and phase shift light projection: analysis and compensation of the systematic errors, *Appl. Opt.* 31: 6565–6573.

Sansoni, G., Carocci, M., and Rodella, R. 2000. Calibration and performance evaluation of a 3D imaging sensor based on the projection of structured light, *IEEE Trans. Instrum. Meas.*, 49: 628–635.

Sapozhnikov, V.B. and Foufoula-Georgiou, E. 1997. Experimental evidence of dynamic scaling and indications of self-organized criticality in braided rivers, 33 (8): 1983–1991.

Seal, R. and Paola, C. 1995. Observations of downstream fining on the North Fork Toutle river near Mount St. Helens, Washington, *Water Resources Research* 31: 1409–1419.

Sechet, P. and Le Guennec, B. 1999. The role of near wall turbulent structures on sediment transport. *Water Resources Research* 33: 3646–3656.

Spinewine, B., Capart, H., Larcher, M., and Zech, Y. 2003. Three-dimensional Voronoï imaging methods for the measurement of near-wall particulate flows, *Exp Fluids* 34: 227–241.

Tal, M. 2008. Interactions between vegetation and braiding leading to the formation of single-thread channels, PhD Thesis, University of Minnesota.

Tal, M. and Paola, C. 2007. Dynamic single-thread channels maintained by the interactions of flow and vegetation, *Geology* 35: 347–350.

Tal, M. and Paola, C. 2010. Effects of vegetation on channel morphodynamics: results and insights from laboratory experiments, *Earth Surface Processes and Landforms* 35(9): 1014–1028.

Turowski, J., Lague, D., Crave, A., and Hovius, N. 2006. Experimental channel response to tectonic uplift, *Journal of Geophysical Research* 111: F03008.

Udrea, D., Bryanston-Cross, P., Querzoli, G., and Moroni, M. 2000. *Particle tracking velocimetry techniques in Fluid mechanics and its application*, Kluwer Academic, pp. 279–304.

Wickert, A. 2007. Measuring channel mobility through the analysis of area-based change in analog experiments, with insights into alluvial environments, Bachelors thesis, Massachusetts Institute of Technology.

Wilcock, P.R. 2001. The flow, the bed, and the transport: Interaction in flume and field, *in Gravel-Bed Rivers*, V., edited by M. Mosley, N. Z. Hydrol. Soc., Wellington, pp. 183–219.

Wilcock, P.R. and Crowe, J.C. 2003. Surface-based transport model for mixed-size sediment, *Journal of Hydraulic Engineering* 129(2): 120–128.

Wilcock, P.R. and McArdell, B.W. 1993. Surface-based fractional transport rates: mobilization thresholds and partial transport of a sand-gravel sediment, *Water Resources Research* 29(4): 1297–1312.

Winterbottom, S.J and Gilvear, D.J. 1997. Quantification of channel bed morphology in gravel-bed rivers using airborne multispectral imagery and aerial photography, *Regulated Rivers: Research & Management* 13: 489–499.

Zimmermann, A.E., Church, M., and Hassan, M.A. 2008. Video-based gravel transport measurements with a flume mounted light table, *Earth Surface Processes and Landforms* 33: 2285–2296.

14 Ground based LiDAR and its Application to the Characterisation of Fluvial Forms

Andy Large[1] and George Heritage[2]
[1]School of Geography, Politics and Sociology,
Newcastle University, UK
[2]JBA Consulting, The Brew House, Wilderspool Park, Grenalls Avenue,
Warrington, UK

14.1 Introduction

While significant progress has been made in river research over the past two decades using digital elevation models (DEMs) to study stream processes, the accuracy of these traditional DEMs varies spatially as map contour interval and density dictate the resolution of information available to interpolate an elevation value for each pixel on the grid (Snyder, 2009). In particular, traditional DEMs miss many fine-scaled features, particularly those in low-relief terrain such as river valley floors. In contrast, both airborne and terrestrial laser scanning or LiDAR equipment and software permits the construction of very detailed DEMs that accurately represent such fine-scale landform surface variability and in turn offer an excellent opportunity to measure and monitor morphological change across a variety of spatial scales (e.g. Brasington et al., 2000; Lane and Chandler, 2003; Fuller et al., 2003; Hopkinson et al., 2005, 2009). As a result, new insights are emerging concerning river system function utilising digital elevation models of the riverine environment (e.g. Lane et al., 1994; Milne and Sear, 1997; Heritage et al., 1998; Brasington et al., 2000; Poole et al., 2002; Large and Heritage, 2007;

Hodge et al., 2009a, 2009b; Milan et al. 2010). The use of oblique laser scanners to generate detailed DEMs of landforms represents a major improvement on previous survey methods, both in aerial coverage and accuracy. This chapter demonstrates this primarily through a review of previously published material involving use of ground-based laser scanning in fluvial studies, covering spatial scales from centimetres to kilometres and daily to annual change timescales.

Areal coverage, data point density, data point accuracy and their relationship with field survey and post-processing time are particularly important when viewed alongside the range of spatial and temporal changes within fluvial systems where a general negative relationship exists between scale of change and rapidity of change (Knighton, 1998). Many studies are limited by area or resolution limitations due to a trade-off between spatial coverage and morphologic detail captured (Large and Heritage, 2009); techniques such as terrestrial photogrammetry produce high point density, accurate morphometric data, but aerial extent is restricted; aerial photogrammetry offers increased spatial coverage but reduced elevation

Fluvial Remote Sensing for Science and Management, First Edition. Edited by Patrice E. Carbonneau and Hervé Piégay.
© 2012 John Wiley & Sons, Ltd. Published 2012 by John Wiley & Sons, Ltd.

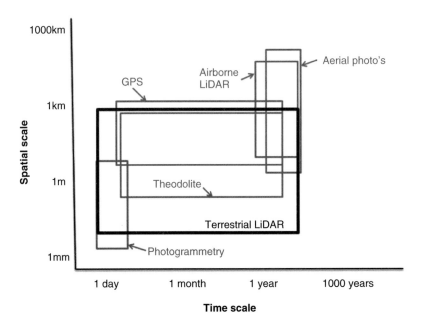

Figure 14.1 The range of surveying techniques available for monitoring fluvial systems across a range of spatial and temporal scales. Terrestrial laser scanning overlaps a number of other techniques and thus represents a versatile survey method of survey (After Heritage and Hetherington, 2007).

accuracy; EDM theodolite surveys suffer from long survey times resulting in reduced data density if large areas are measured (Figure 14.1).

14.1.1 Terrestrial laser scanning in practice

Since the development of the first terrestrial laser scanner in 1999 (Bryan, 2006), laser scanning technology has seen a continued phase of product development, growth and expansion into many areas of survey (Heritage and Large, 2009a). As a result, the development of sensors able to rapidly collect 3-D surface information has enabled high-density measurements to be made across landscapes that are unsuited to more conventional approaches due to their inaccessibility, hazardous nature or spatial extent (Lim et al., 2009). A major advantage is in the rate of data acquisition. First generation pulse-scanners consisted of long range, high precision machinery collecting on average 100 points per second, while later generations (from 2004) concentrated on speed of data acquisition and more on shorter ranges (Large and Heritage, 2009). The newest generation machines measure approximately 150 000 points per second and, as a consequence, collect a considerable amount of data during a survey. A common problem with LiDAR scanning techniques is that the resultant data file sizes can be difficult to manage.

The dataset, commonly referred to as a 'point cloud', can provide a 3-D impression or visualisation of the feature being measured. The maximum range achievable with a terrestrial LiDAR depends strongly on the meteorological visibility; at lower visibility, the maximum range is reduced due to atmospheric attenuation.

The conventional scanning methodology is to use measurements to a number of common targets (reflectors). This allows multiple scans to be related to each other, and to be georeferenced to an existing control network. In essence, the scanner is placed at one location and measurements taken to a number of targets as well as the actual area of interest (see case study below). The scanner is then moved to a second location and the process repeated, using at least three common targets from the first scanner location. This is necessary for linear features such as rivers where banks, vegetation and bends in the channel obscure scanning lines. The process is repeated until full coverage of the site is achieved filling as many shadow areas as possible. Some laser scanners do not require reflector targets to be located in the area being surveyed instead utilising internally referenced common features to allow meshing of scans to take place. This is especially advantageous in areas where there may be difficulty in gaining access e.g. opposite banks of rivers in spate.

At the smallest scale depicted in Figure 14.1, LiDAR data has been acquired and processed at densities sufficient to represent the surface at the grain scale (gravels range between 2 mm and 64 mm). The data point concentration at this scale has been estimated at between 4000 and 10 000 points m^{-2} by workers such as Lane et al. (1994). Studies at the sub-bar scale, identified as a significantly under-researched area by Charlton et al. (2003), have quantified morphological change over relatively short timescales showing the advantages such rapid data acquisition can provide in terms of monitoring short-term system alteration at this spatial scale. Monitoring of morphologic change at the reach scale and beyond requires the merging of separate scans and several studies employing differing methodologies are reported in this chapter to show the range of approaches that can be effectively used in river systems. In all cases we emphasise that, despite the obvious advantages of terrestrial laser scanning, great care needs to be exercised during data collection and processing in order to ensure that data accuracy is maximised. Table 14.1 provides a summary of the typical errors that are likely at various scanning scales and these should be considered carefully when designing a scan campaign. This chapter concludes with advice on survey protocols required in the light of a review of work by the authors and other workers that will help to minimise the potential errors listed in Table 14.1. Despite the seemingly endless potential of terrestrial LiDAR a word of caution is issued with regard to the indiscriminate use of laser-based scanning in fluvial systems.

Table 14.1 Potential error linked to terrestrial laser scanning.

Scale	Error
Sub-grain	Short distances – pulsed scanners unable to differentiate signal returns at very close range.
Grain/micro-topographic	Mixed pixel – issues with the size of the laser footprint and unrealistic areal averaging
Bar	Registration – error in tie-point recognition and relative positioning can affect point cloud registration. Scan angle – affects spread of laser footprint linked to unrealistic areal averaging
Reach	Reflectivity – more reflective surfaces can saturate the instrument sensor reducing the estimation of distance. Atmospheric conditions – aerosols will impact on scanner range. Registration – error in tie-point recognition and relative positioning can affect point cloud registration. Scan angle – affects spread of laser footprint linked to unrealistic areal averaging
Landscape	More suited at present to airborne systems but may suffer the same errors as for reach scale, particularly registration.

14.2 Scales of application in studies of river systems

14.2.1 The sub-grain scale

Madej et al. (2009) have used flume-based lasers with millimetric accuracy to model channel responses to varying sediment input. They concluded that such use of laser scanning in flume conditions can help confirm interpretations of observations that are limited by challenges of field conditions. It can also facilitate exploration of in-channel processes in more detail than would be the case in the real-life situation. Madej et al. (2009) observed channel response to varying sediment loads and flume patterns which included (i) overall channel form and degree of armouring; (ii) bed-surface fining and bedform smoothing during aggradation, and coarsening and bedform roughening during degradation and (iii) greater selective transport during low transport rates. Elsewhere, Michael and Gerhard (2006) have used a laser scanning system to examine benthic-relief to study the interaction of organisms with the surrounding flow regime, while Cui et al. (2008) have used lasers to investigate and model sediment transport in a flume with forced pool-riffle morphology.

14.2.2 The grain scale

At the grain scale, water flow level in river channels is moderated by the interaction with the roughness of the surface over which it flows. Milan et al. (2010) note that this interaction is highly complex and remains poorly understood due primarily to sample size and sample bias issues over extremely heterogeneous surfaces. They demonstrated the utility of oblique laser scan data in

Figure 14.2 Combined laser scan cloud image at the point bar scale demonstrating the utility of oblique laser scan data in defining gravel surface character. Site scanned is the River South Tyne at Lambley, Northumberland, UK. The bar extends approximately 100 m up and downstream and is 30 m wide close to the bridge.

defining gravel surface character across a 180 m² point bar located on the River South Tyne at Lambley, Northumberland, UK (Figure 14.2). The site offered excellent visibility across the surface from a road bridge and also valley side laser scan vantage points, ensuring that much of the bar surface was visible at acute scan angles. Scan point density was thereby maximised and laser pulse footprint elongation and associated point averaging minimised as interpolation between points becomes redundant. Multiple scanning of the surface from the same location also allowed an averaged scan surface to be computed from the mean of several scans of the same area. This minimises any instrument time of flight error.

The final dataset of the 180 m² bar surface contained 3.8 million points with a mean spacing of 0.012 m. Surface points were on average accurate to ±0.009 m when compared with 113 independent EDM validation points. The laser scan data were subsequently used to generate a 0.05 m resolution DEM of the point-bar surface. The local standard deviation (z) of the elevation data was calculated within a 0.15 m moving window (equivalent to the measured intermediate axis of the largest visible clast) The 0.15 m radius window was set to move at 2 cm intervals in plan (x and y) across the point cloud. A z value was then designated to each grid node, spaced every 2 cm. The small window size is assumed to be unaffected by regional surface trends. Each z value was then multiplied

by a factor of two to generate the effective roughness equivalent (Nikora et al., 1998; Gomez, 2003).

Point densities averaging 10 000 m⁻² allowed the surface to be gridded at 0.02 m intervals and data were extracted from the resultant grid for 0.5 m² sub-areas and the median and first standard deviation of each sub-unit were calculated to generate values for D_{50}, D_{16} and D_{84} respectively. These values were re-interpolated to generate facies maps of the bar surface (Figure 14.3).

Hodge et al. (2009b) have investigated the nature of gravel surfaces across exposed bar areas on the River Feshie and Bury Green Brook in the UK, the former being coarser in nature (Figure 14.4). A Leica HDS3000 laser scanner was used to collect data from 2 m² areas of the bar surfaces with an expected point position precision of 4 mm. Repeat scanning and scan overlap techniques were employed to further minimise error. Despite these precautions Hodge et al. (2009b) report erroneous points (averaging 1.3 mm standard deviation across the surface) across the surface due to instrument error, the size of the laser footprint and obscured surfaces at certain scan angles. Hodge et al. (2009b) argue that these errors can be further minimised using a variety of algorithms designed to eliminate outliers and smooth surfaces. Despite this the data were used to resolve individual grain inclination and orientation parameters enabling the authors to link this to the macro-morphology of the study sites.

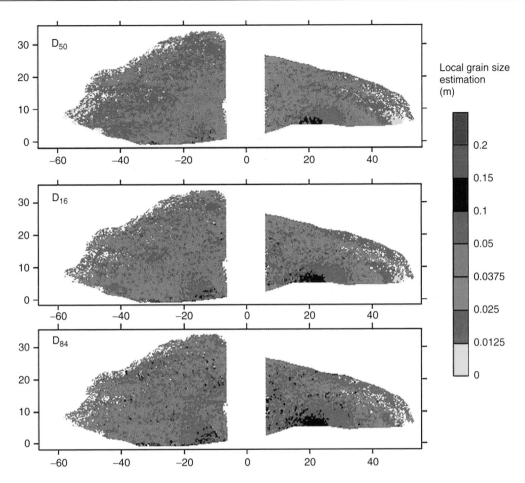

Figure 14.3 Facies maps of the bar surface on the River South Tyne at Lambley produced from data clouds with point densities averaging $10\,000\;\text{m}^{-2}$.

14.2.3 The sub-bar unit scale

Many of the early-generation terrestrial laser scanners were cumbersome, hence they were not usable in many field situations beyond localised sediment studies at the metre scale. More portable field scanners have been developed in the last decade thus offering (i) greater range (now over 2 km (ii) increased rapidity of data collection associated with an improvement in the angular resolution of instruments (down to +/−0.001 degrees) and (iii) improved ranging accuracy (of the order of a few millimetres). An early study using this more portable field scanning technology was conducted by Milan et al. (2007) on the proglacial outwash fan of Glacier du Mont Miné and Ferpècle situated in the Valais region of the Swiss Alps. Survey work concentrated on a 5881 m^2 reach of braided

gravel-bed channel, towards the tail of the outwash fan. The channel was fed primarily by melt-water originating from Glacier du Mont Miné, with the melt-water discharge regime displaying a strong diurnal signal during summer months. A LMS Z-210™ laser scanner manufactured by Riegl Instruments was used to collect a series of independent datasets recording range, height, surface colour and reflectivity. The instrument works on the principle of 'time of flight' measurement using a pulsed eye-safe infrared laser source (0 · 9 μm wavelength) emitted in precisely defined angular directions controlled by a spinning mirror arrangement. A sensor records the time taken for light to be reflected from the incident surface. Angular measurements are recorded to an accuracy of 0 · 036° in the vertical and 0 · 018° in the horizontal. Range error is 0 · 025 m to a radial distance of 350 m

Figure 14.4 Laser scans compared with photographs to show the nature of gravel surfaces across exposed gravel bar areas on the River Feshie, Scotland (a and c) and Bury Green Brook, UK (b and d). Reproduced from Hodge, R., Brasington, J. and Richards, K. (2009b) In situ characterization of grain-scale fluvial morphology using terrestrial laser scanning. *Earth Surface Processes and Landforms* 34: 954–968, with permission from John Wiley & Sons, Ltd.

(although this varies with surface reflectivity). The field laser scanner, connected to a Panasonic Toughbook™ computer, was used to conduct high-resolution scans of the study area on a daily basis for ten days (although no data could be gathered on day 8 because of fog). The study area was scanned in four component sections and subsequently merged using RiScan-Pro™ post-processing software. All four survey stations were slightly elevated from the river, one being on a roche moutonnée and the others on lateral and terminal moraines (Figure 14.5).

Survey control was facilitated by RiScan Pro™ survey software, capable of visualising point cloud data in the field. Scans were generally restricted to 240° in front of the scanner, and were collected with substantial overlap, ensuring that the surface of the study reach was recorded from several directions. The effect of this approach was to increase the point density across the surface and to reduce the occurrence of occluded areas due to the shadowing effect of roughness elements along the line of each scan.

Figure 14.5 demonstrates the detail captured during the scanning campaign. The survey data revealed a morphology where sub-bar units could easily be identified (Figure 14.5a), and allowed a useful comparison of sediment lobe sedimentology and dynamics with previously published studies. Slumped blocks were generated following small scale fluvial undercutting of the fine gravel bar edges can be seen in the enlarged box view (Figure 14.5b). Subtraction of repeat surface scans also documents the development of subtle chute channel features acting as a precursor to bar dissection through headward erosion of these surfaces (Figure 14.5c). A similar study using the same instrument covering a longer reach of the River Coquet, Northumberland, UK is reported by Entwistle and Fuller (2009). Despite use of fewer scans and reduced multiple scan overlap, the average point spacing across the 10 600 m² point bar surface was 0.04 m. Again the definition across the bar surface enables sub-bar features to be identified and grain scale maps were derived from local surface roughness measures (Figure 14.6).

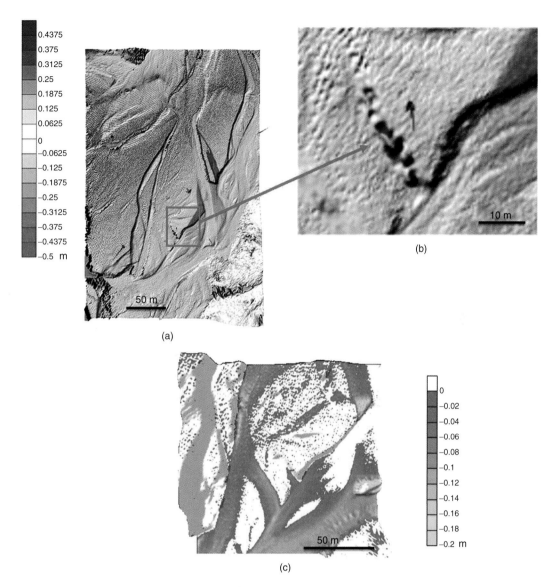

Figure 14.5 Laser scanned DEM (grey scale) of a 5881 m² reach of braided gravel-bed channel draining the proglacial outwash fan of Glacier du Mont Miné and Ferpècle situated in the Valais region of the Swiss Alps. (a) principal fine gravel bar features, (b) detail of eroding bar margin showing collapsed sediment blocks, (c) difference surface showing erosion and deposition at sub-bar unit scale (After Milan et al. 2007).

14.2.4 In-channel hydraulic unit scale

Large and Heritage (2007) have employed terrestrial laser scanning to characterise and map instream habitat in the form of hydraulic biotopes. Biotopes are stage-dependant, in-channel hydraulically-defined habitat units (e.g. Newson and Newson, 2000), and in UK rivers occur across scales similar to bar and sub-bar features. As mentioned above, the Riegl LMS Z-201 laser scanner is useful for investigating these hydraulic units as these scanners operate using a near-infrared laser (0.9 μm). While much of the energy at this wavelength is absorbed by still water, turbidity and surface roughness generate low energy returns generally below 20% of the initial

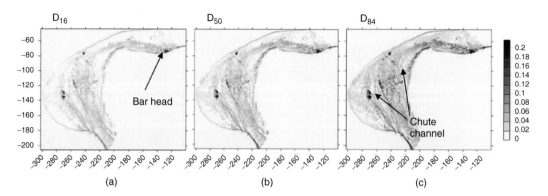

Figure 14.6 Point bar sedimentology (D_{16}, D_{50} and D_{84}) and sub-bar morphology mapped using terrestrial laser scanner on the River Coquet, Northumberland. Reproduced from Entwistle NS and Fuller IC (2009) Terrestrial laser scanning to derive the surface grain size facies character of gravel bars. In: GL Heritage and ARG Large (eds) Laser Scanning for the Environmental Sciences, with permission from Wiley-Blackwell.

laser energy. The extremely dense sampling of the water surface using terrestrial laser scanning coupled with the improved sensitivity of return pulse sensors allows data to be collected to accurately characterise water surface roughness (a primary determinant of biotope classification in the field).

The study site used by Large and Heritage (2007) on the South Tyne at Slaggyford is an upland cobble bed channel characterised by diverse bed morphology. The water surface data were initially transformed into a regular grid with 0.02 m spacing and this was subsequently used to determine the local standard deviation of sub areas across the surface. Figure 14.7 quantifies the biotope distribution at the site defined using the definition of Newson and Newson (2000), the most commonly used typology in the UK. It is clear that, at low flows, biotope unit variety is high and distribution complex, and it is argued by the authors that such complexity would be missed using observational techniques along the lines of Newson and Newson (2000) or that of the European Aquatic Monitoring Network (EAMN, 2004) as these techniques sub-divide reaches into larger components (at the scale of several square metres), whereas the laser scanning techniques gets down to the scale of water surface or bedform roughness – the scale of most relevance to the communities inhabiting these habitats.

Results obtained with TLS showed overlap between the riffle and run habitat types. This is to be expected, as stage rises and fall, these habitat types merge and change from one type to another. An implication from the point of view of instream hydraulics may be that these biotopes are providing very similar habitat to each other. The other feature apparent from Figure 14.7 is the inclusion of edge

areas in the biotope definition. This is intuitive, but the advance here is that using terrestrial laser scanning allows much better quantification of this critical in-channel component (edges *per se* have long been recognised as being of heightened importance from an ecological point-of-view).

14.2.5 Micro-topographic roughness units

A reach-based study of Kingsdale Beck in the Yorkshire Dales National Park (Entwistle and Fuller, 2009) used terrestrial laser scanner data to identify and map micro-topographic roughness element distribution across the bed of the river. The channel flows across limestone geology resulting in prolonged no-flow periods where the entire bed is exposed for scanning (an unusual situation nationally, but ideal for TLS studies of benthic environments!). The bed is composed of gravels and cobbles ($D_{50} = 0.09$ m, $D_{84} = 0.14$ m), and bedform clusters are typically of the order of 0.3 m wide and in excess of 0.5 m long. Due to the excellent exposure, the merged data file of the 90 m survey reach contained in excess of 20 million points with little occlusion. Topographic highs were isolated from the data set and plotted (Figure 14.8). The study was able to confirm the observations of Brayshaw (1984) that the location of micro-topographic roughness elements are strongly linked to the distribution of very large individual clasts (D_{99} and above).

14.2.6 The bar unit scale

A study by Milan et al. (2007) concentrated on a 5881 m^2 reach of braided gravel-bed channel, towards the tail of

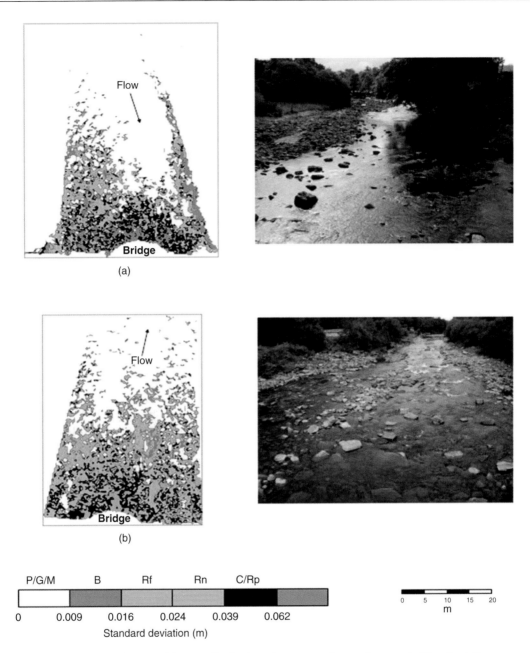

Figure 14.7 Use of TLS to spatially quantify biotope distribution using water surface roughness on the River South Tyne at Slaggyford, Northumberland, UK (a) upstream scan, (b) downstream scan.

the proglacial outwash fan of Glacier du Mont Miné and Ferpècle in Switzerland. The study was able to confirm that central bar deposition, transverse bar conversion, lobe deposition, chute cut-off, lobe dissection, channel constriction and channel choking were all important

processes influencing morphologic development in the reach (Figure 14.9). Perhaps of more importance were the results obtained concerning the impact of survey frequency on sediment budgeting. It is strikingly clear from Figure 14.9 that more infrequent survey fails to capture

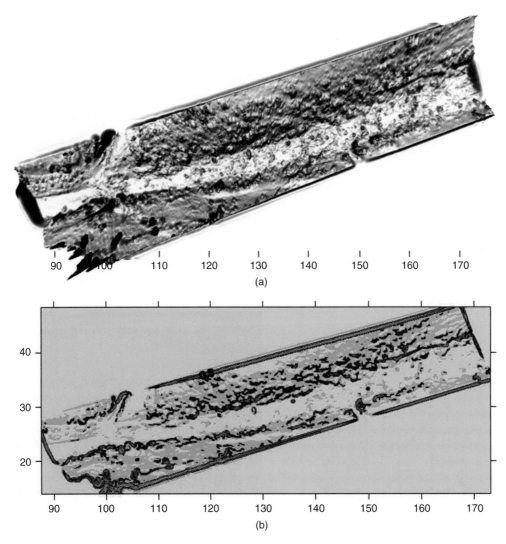

Figure 14.8 Reach-based study (90 m) of the Kingsdale Beck, Yorkshire containing in excess of 20 million points (a), allowing identification of roughness elements highlighted in (b) in the channel at the micro-topographic (individual clast) scale.

the sediment fluxes occurring over short timescales with deposition masking erosion to generate an inaccurate underestimate of material movement through the reach.

14.2.7 Reach-scale morphological analyses

The River Rede is a headwater tributary of the North Tyne (catchment area = 18 km^2) with a bankfull discharge of 8.5 m^3 s^{-1}. It is flashy in nature and experiences substantial bedload movement beginning at 23–35% bankfull (Milan et al., 2001). The mobile gravel and cobble material have created a well-defined morphology, consisting

of deeper pool areas associated with inner bank point bar deposits, sinuous reaches and shallow steep riffle sites (Figure 14.10).

Terrestrial laser scanning of the unregulated 250 m reach of the river upstream of Catcleugh Reservoir in Northumberland (Milan et al., 2010) utilised a rapid data acquisition protocol utilising surveyed reflectors; scanning the river from alternate sides ensured overlap with the previous scans. Six instrument setups generated a detailed composite point cloud image (Figure 14.11). It is immediately obvious from this point cloud dataset that

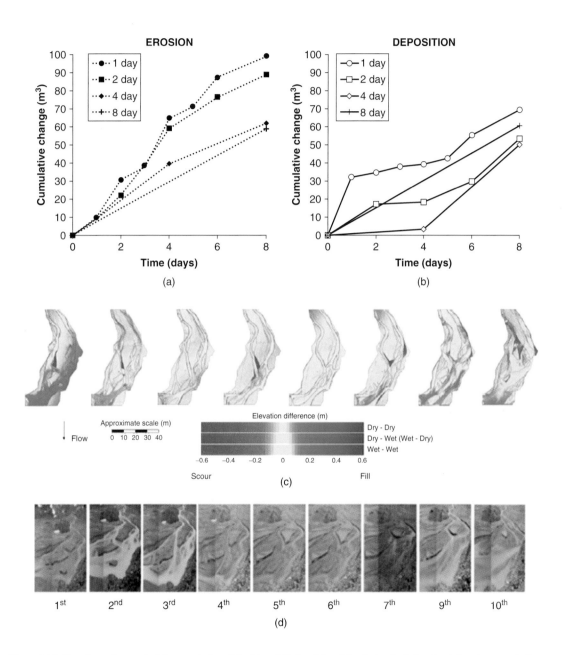

Figure 14.9 Data from the proglacial outwash fan of the Mont Miné and Ferpècle glaciers in Switzerland showing cumulative volumes of (a) erosion and (b) deposition calculated from DEM differencing for four survey frequencies, (c) composite DEMs of difference showing daily geomorphological change and (d) daily sequence of photographs over the 10 day survey period (After Milan et al. (2007)).

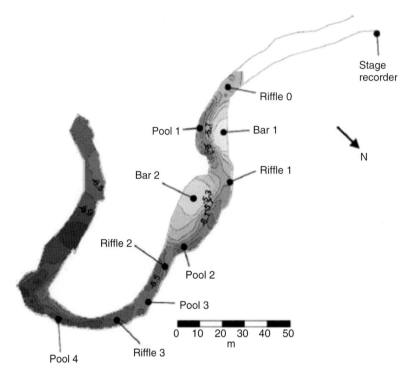

Figure 14.10 River Rede upstream of Catcleugh Reservoir, Northumberland, UK showing in-channel morphological unit distribution in the reach subjected to ground-based laser scanning.

Figure 14.11 Detailed composite point cloud image generated from meshing six different instrument scans of a 250 m reach of the River Rede, Northumberland, UK.

deeper pool areas, characterised by a slow flowing clear smooth water surface, are clearly differentiated from the rougher in-channel bar and riffle surfaces. As with the South Tyne case study from Slaggyford (see above), the Rede study demonstrated clear proof of concept for identification of within-channel, hydraulically-defined habitat units using terrestrial laser scanning at the bar unit scale.

14.2.8 Terrestrial laser scanning at the landscape scale

Moving beyond the reach scale, the scan range of most contemporary ground-based scanners (generally restricted to less than 500 m to ensure a dense and precise enough dataset) reduces their operational effectiveness with multiple overlapping scans and increased post-processing needed to cover larger areas. Nevertheless some studies have attempted mapping extensive areas using TLS in order to achieve greater detail and accuracy over airborne laser data.

Hodgetts (2009) used the Riegl LMSZ420i terrestrial laser scanner to collect 85 linked georeferenced scans of the Wadi Nukhul in the Sinai Peninsula. This generated a detailed structural surface model across a 12 km² area from which they derived geological strike and dip measurements and inferred palaeo-current direction data.

These types of study remain rare, and as such we conclude our discussion of laser scanning in river science with a number of studies at the landscape scale which utilise airborne LiDAR data, demonstrating its utility at this scale. Newer scanners offer longer range measurements over several kilometres and these will allow similar data sets to be generated to those reviewed below.

Aerial laser data of the floodplain of the River Ure at Jervaux Abbey, North Yorkshire (Figure 14.12a) has been analysed by the authors to identify floodplain palaeo-morphology (Figure 14.12b) as well as vegetation units (Figure 14.12c). The airborne data was supplied as a regular 2 m surface grid which contains information on the vegetation as well as the bare earth. A second data set provided by the EA was processed to remove the vegetation, effectively isolating the minimum elevation data recorded by the aerial LiDAR through a moving window technique. This is facilitated by the nature of the LiDAR data collection process as the footprint of the laser

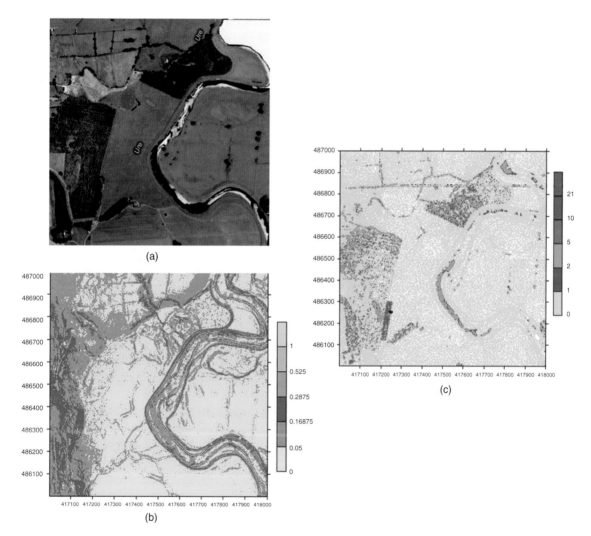

Figure 14.12 Aerial laser data of the floodplain of the River Ure at Jervaux Abbey (a), identifying floodplain palaeo-morphology (b) and vegetation units (c). Data supplied courtesy of the UK Environment Agency.

is often wide enough to penetrate to the ground through very dense vegetation (see Heritage and Large, 2009b; Crutchley, 2009; Danson et al., 2009 for more detailed discussion of the vegetation removal algorithm).

It is immediately clear from Figure 14.12c that a simple subtraction of the digital terrain model from the digital surface model is able to differentiate tree shrub and herbaceous vegetation types on basis of vegetation height, mapping plantation areas of differing ages, identifying mature riparian trees and differentiating managed floodplain pasture from bare earth and arable fields. The bare earth digital surface model data is used to identify locations on the floodplain where the local slope change is rapid using an edge detection algorithm. It is notable that a number of palaeochannels are revealed across the floodplain surface (Figure 14.12b), features which are more difficult to detect either from the aerial photograph or on the ground. Newer flood banks with sharper slope breaks are particularly well detected by the technology; however, where the surface is naturally variable (e.g. to the north in the image shown in Figure 14.12b) feature identification becomes more problematic.

14.2.8.1 Amblève Valley, Belgium

A similar set of techniques was applied to LiDAR data collected along the Amblève Valley in the Belgian Ardennes region. Figure 14.13a details the morphology of approximately 1 km² of the river valley showing a contemporary sinuous single thread channel set in a wandering channel situation as indicated by the numerous palaeo-channel and bar surfaces present across the floodplain. A shaded relief map generated from the LiDAR data of the valley bottom (Figure 14.13b) reveals the local morphological complexity. This simple technique may be rapidly applied to the digital terrain model and manipulation of the aspect and azimuth of the light ray permits detailed evaluation of subtle morphologic variation across the surface (see also Crutchley, 2009). While such a technique does not allow extraction of the location and metrics of any of the features apparent, calculating the local relative elevation change from the LiDAR data (Figures 14.13c, 13d) succeeds in differentiating channel and bar surfaces and allows extraction of data metrics based on the local elevation information.

14.2.9 Towards a protocol for TLS surveying of fluvial systems

The series of case study examples associated with ground-based laser scanning of river systems outlined above,

along with the brief comparison with studies involving airborne technologies, demonstrates the evolution of the discipline and its application in morphologically-similar environments. It also shows how the technology has potential value for feeding into monitoring and legislation programmes at a range of scales. A number of issues are raised with what are now well-established methods for the acquisition of precise and reliable 3-D geo-information, and overcoming these will be vital if laser scanning is to achieve its potential in the surveying and monitoring of river systems – systems which are spatially and temporally variable, and whose inherent variability poses particular challenges to the accurate collection and interpretation of geo-information.

TLS survey data remain subject to many issues that will generate inaccurate, misleading or inappropriate information if not considered. Error in TLS measurement is spatially variable, given the variation in survey range, laser footprint and incidence angle onto the target surface. Linear features such as river systems present particular challenges to surveying as they require consistent collection and integration of data from multiple viewpoints (Lim et al., 2009). The combination of separate scans, either spatially or through time, has the potential to introduce inconsistencies in the orientation, resolution and positioning of individual surveys. Probably because of it being in its relative infancy, of all survey techniques available for river corridors, TLS has the least standardised control; practices and error assessments (Lichti et al., 2005). Clear attention has to be given to planning, data collection and processing to minimise these disadvantages. In addition, most currently available laser scanners are not well specified regarding accuracy, resolution and performance (Hetherington, 2009) and only a minority are checked by independent institutes regarding their performance and whether they actually comply with manufacturer specifications (Boehler et al., 2003).

As Lim et al. (2009) emphasise, field procedures using TLS are concerned with two priorities absolute accuracy and the ability to locate the scanned data correctly in either local or global space. To ensure this, there is a need for clear and unambiguous surveying protocols for linear fluvial systems:

(a) consider the potential errors related to the scale of survey from Table 14.1;

(b) minimise the scan distance to ensure greater scan point density and ranging precision and to minimise footprint size;

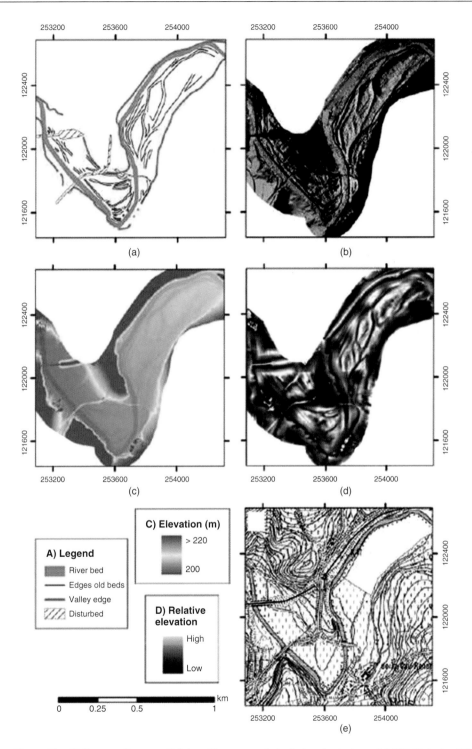

Figure 14.13 The *Amblève Valley* near Cheneaux, Belgium. The geomorphological setting is shown (a). LiDAR depicts hillslopes (b) and relative elevations (c). Fluvial features are also depicted (d), as compared to the paucity of information contained in a traditional contour map (e). After Notebaert et al. (2009).

(c) tilt the scanner towards the river channel to maximise the amount of data collected locally;

(d) select scan locations to minimise scan shadow effects caused by large obstructions;

(e) where possible, optimise the scan angle by setting the instrument well above the scanned surface;

(f) collect independent tiepoint/error check data to minimise systematic bias introduced during scan cloud merging;

(g) use manually selected tiepoints for more accurate scan merging due to the ability to select their location in the scan data with high precision;

(h) ensure that some reflectors/tiepoints are placed at the edges of the scanned area to minimise propagation of meshing errors;

(i) ensure a good variation in x, y and z dimensions when selecting tiepoints/reflector locations; this improves scan merging accuracy and reduces the possibility of 'chance' scan merging due to similar distances and elevations between tiepoints or reflectors;

(j) repeat scans from the same location to densify the data collected and potentially reduce extreme errors; and

(k) avoid low angle scans across water surfaces.

In conclusion, the ability to survey a river channel rapidly enough to capture changes between floods at a high spatial resolution is one of the biggest limitations when surveying morphology in the field, and this problem is particularly applicable to more active (e.g. braided) river systems. Using terrestrial laser scanning however, the issue of point distribution and potential operator bias (Lane et al., 2003) can be rendered obsolete as a dense cloud of meshed data points ensures that a surface is sampled many times. If proper collection and analytical protocols can be put in place, and data archiving and mining issues standardised, we may even get close to the situation described by Lane et al. 2003 where the authors state that 'just as theoretical debates ... are asking that we take a more holistic view of landform processes, so data collection systems are getting dangerously close to providing a resolution of data that can satisfy the reductionist tendencies of some without precluding this wider view'. Despite this assertion, data point quality may still potentially prove to be an issue for some studies aimed at the grain scale as the range error on current instruments may still lead to unacceptable inaccuracies in the DEM surface. New advances in ground-based LiDAR technology aim at addressing such outstanding issues.

References

Boehler, W., Bordas-Vincent, M., and Marbs, A. 2003. Investigating laser scanner accuracy, Proceedings of the 19th CIPA Symposium, 30th September – 4th October 2003, Antalya, Turkey.

Brasington, J., Rumsby, B.T., and McVey, R.A. 2000. Monitoring and modelling morphological change in a braided gravel-bed river using high-resolution GPS-based survey. *Earth Surface Processes and Landforms* **25**: 973–990.

Brayshaw, A.C. 1984. Characteristics and origin of cluster bedforms in coarse-grained alluvial channels. *Sedimentology of Gravels and Conglomerates* **10**: 77–85.

Charlton, M.E., Large, A.R.G., and Fuller, I.C. 2003. Application of airborne LiDAR in river environments: the River Coquet, Northumberland, UK. *Earth Surface Processes and Landforms* **28**: 299–306.

Crutchley, S. 2009. Using LiDAR in archaeological contexts: the English Heritage experience and lessons learned. 180–200 in Heritage, G.L. and Large, A.R.G. (eds). Laser Scanning for the Environmental Sciences. London: Wiley-Blackwell.

Cui, Y.T., Wooster, J.K., Venditti, J.G., Dusterhoff, S.R., Dietrich, W.E., and Sklar, L.S. 2008. Simulating sediment transport in a flume with forced pool-riffle morphology: Examinations of two one-dimensional numerical models. *Journal of Hydraulic Engineering-ASCE* **134**: 892–904.

Danson, F.M., Morsdorf, F., and Koetz, B. 2009. Airborne and terrestrial laser scanning for measuring vegetation canopy structure. 201–219 in Heritage, G.L. and Large, A.R.G. (eds). *Laser Scanning for the Environmental Sciences.* London: Wiley-Blackwell.

EAMN (European Aquatic Monitoring Network). 2004. *State-of-the-art in data sampling, modelling analysis and applications of river habitat modelling.* (Harby, A., Baptist, M., Dunbar, M.J., and Schmutz. S. (eds). COST Action 626 Report.

Entwistle, N.S. and Fuller, I.C. 2009. Terrestrial laser scanning to derive the surface grain size facies character of gravel bars. 102–114 in Heritage, G.L. and Large, A.R.G., (eds). *Laser Scanning for the Environmental Sciences.* London: Wiley-Blackwell.

Fuller, I.C., Large, A.R.G., and Milan, D.J. 2003. Quantifying channel development and sediment transfer following chute cutoff in a wandering ravel-bed river. *Geomorphology* **54**: 307–323.

Gomez, B. 1993. Roughness of stable, armoured gravel beds. *Water Resources Research*, **29**: 3631–3642.

Heritage, G.L. and Large, A.R.G. (eds) (2009a). *Laser Scanning for the Environmental Sciences.* London: Wiley-Blackwell. 278 p.

Heritage, G.L. and Large, A.R.G. (2009b). Principles of 3D laser scanning. 21–34 in Heritage, G.L. and Large, A.R.G. (eds). *Laser Scanning for the Environmental Sciences.* London: Wiley-Blackwell.

Hetherington, D. 2009. Laser scanning: Data quality, protocols and general issues. 82–101 in Heritage, G.L. and Large, A.R.G. (eds). *Laser Scanning for the Environmental Sciences*. London: Wiley-Blackwell.

Hodge, R., Brasington, J., and Richards, K. (2009a). Analysing laser-scanned digital terrain models of gravel-beds surfaces: linking morphology to sediment transport processes and hydraulics. *Sedimentology* **56**: 2024–2043.

Hodge, R., Brasington, J., and Richards, K. (2009b). *In situ* characterization of grain-scale fluvial morphology using terrestrial laser scanning. *Earth Surface Processes and Landforms* **34**: 954–968.

Hodgetts, D. 2009. LiDAR in the environmental sciences: Geological applications. 165–179 in Heritage, G.L. and Large, A.R.G. (eds). *Laser Scanning for the Environmental Sciences*. London: Wiley-Blackwell.

Hopkinson, C., Chasmer, L.E., Zsigovics, G., Creed, I., Sitar, M., Kalbfleisch, W., and Treitz, P. 2005. Vegetation class-dependant errors in LiDAR ground elevation and canopy height estimates in a Boreal wetland environment. *Canadian Journal of Remote Sensing* **32**: 191–206.

Hopkinson, C., Hayashi, M., and Peddle, D. 2009. Comparing alpine watershed attributes for LiDAR, photogrammetric and contour-based digital elevation models. *Hydrological Processes* **23**: 451–463.

Knighton, D. 1998. *Fluvial Forms and Processes: A New Perspective*. London: Edward Arnold. 383 p.

Lane, S.N. and Chandler, J.H. 2003. The generation of high quality topographic data for hydrology and geomorphology: new data sources, new applications and new problems. *Earth Surface Processes and Landforms* **28**: 229–230.

Lane, S.N., Chandler, J.H., and Richards, K.S. 1994. Developments in monitoring and modelling small-scale river bed topography. *Earth Surface Processes and Landforms* **19**: 349–368.

Lane, S.N., Westaway, R.M., and Hicks, D.M. 2003. Estimation of erosion and deposition volumes in a large gravel-bed, braided river using synoptic remote sensing. *Earth Surface Processes and Landforms* **28**: 249–271.

Large, A.R.G and Heritage, G.L. 2009. Laser scanning – the evolution of the discipline. 1–20 in Heritage, G.L., and Large, A.R.G. (eds). *Laser Scanning for the Environmental Sciences*. London: Wiley-Blackwell.

Large, A.R.G and Heritage, G.L. 2007. Terrestrial laser scanner based instream habitat quantification using a random field approach. Proceedings of the 2007 Annual Conference of the Remote Sensing & Photogrammetry Society, Mills, J, Williams, M. (eds), The Remote Sensing and Photogrammetry Society.

Lichti, D.D., Gordon S.J., and Tipdecho, T. 2005. Error models and propagation in directly georeferenced terrestrial laser scanner networks. *Journal of Survey Engineering* **131**: 135–142.

Lim, M., Mills, J., and Rosser, N. 2009. Laser scanning of linear features: Considerations and applications. 245–261 in Heritage, G.L. and Large, A.R.G. (eds). *Laser Scanning for the Environmental Sciences*. London: Wiley-Blackwell.

Madej, M.A., Sutherland, D.G., Lisle, T.E., and Pryor, B. 2009. Channel responses to varying sediment input: a flume experiment modelled after Redwood Creek, California. *Geomorphology* **103**: 507–519.

Michael, F. and Gerhard, G. 2006. Description of a flume channel profilometry using laser line scans. *Aquatic Ecology* **40**: 493–501.

Milan, D.J., Heritage, G.L., Large, A.R.G., and Charlton, M.E. 2001. Stage dependant variability in tractive force distribution through a riffle-pool sequence. *Catena* **44**: 85–109.

Milan, D.J., Heritage, G.L., and Hetherington, D. 2007. Application of a 3D laser scanner in the assessment of erosion and deposition volumes and channel change in a proglacial river. *Earth Surface Processes and Landforms* **32**: 1657–1674.

Milan, D.J., Heritage, G.L., Large, A.R.G., and Entwistle, N. 2010. Mapping hydraulic biotopes using terrestrial laser scan data of water surface properties. *Earth Surface Processes and Landforms* **35**: 918–931.

Milne, J.A. and Sear, D.A. 1997. Modelling river channel topography using GIS. *International Journal of Geographical Information Science* **11**: 499–519.

Newson, M.D. and Newson, C.L. 2000. Geomorphology, ecology and river channel habitat: mesoscale approaches to basin-scale challenges. *Progress in Physical Geography* **24**: 195–217.

Nikora, V.I., Goring, D.G., and Biggs, B.J.F. 1998. On gravel-bed roughness characterization. *Water Resources Research* **34**: 517–527.

Notebaert, B., Verstraeten, G., Govers, G., and Poesen, J. 2009 Qualitative and quantitative applications of LiDAR imagery in fluvial geomorphology. *Earth Surface Processes and Landforms* **34**: 217–231.

Overton, I.C., Siggins, A., Gallent, J.C., Penton, D., and Byrne, G. 2009. Flood modelling and vegetation mapping in large river systems. 220–244 in Heritage, G.L. and Large, A.R.G. (eds). Laser Scanning for the Environmental Sciences. London: Wiley-Blackwell.

Poole, G.C., Stanford, J.A., Frissell, C.A., and Running, S.W. 2002. Three-dimensional mapping of geomorphic controls on flood-plain hydrology and connectivity from aerial photos. *Geomorphology* **48**: 329–347.

Snyder, N.P. 2009. Studying stream morphology with airborne laser elevation data. *EOS, Transactions, American Geophysical Union* **90**(6): 45–46.

15 Applications of Close-range Imagery in River Research

Walter Bertoldi[1], Hervé Piégay[2], Thomas Buffin-Bélanger[3], David Graham[4] and Stephen Rice[4]

[1] Dipartimento di Ingegneria Civile e Ambientale, Universita degli studi di Trento, Trento, Italy
[2] University of Lyon, Platform ISIG/ENS, Lyon cedex 07, France
[3] Département de biologie, chimie et géographie, Université du Québec à Rimouski, Rimouski, Québec, Canada
[4] Department of Geography, Loughborough University, Leicestershire, UK

15.1 Introduction

The increasing spatial, temporal and spectral resolutions of imaging technologies is improving our ability to monitor river forms and processes, while more sophisticated analysis of the imagery obtained is providing new insights and understanding for river scientists. The speed with which the changes have occurred has dramatically increased in the last two decades. Introducing a recent Special Issue of Earth Surface Processes and Landforms dedicated to remote sensing of rivers, Marcus and Fonstad (2010) claimed that river remote sensing can be considered as an emerging new sub-discipline in the river sciences, promoted not only by the evolving methodologies, but also by the need for data describing a wide range of spatial and temporal scales, that help practitioners plan activities and target their actions.

Optical imagery is one of the most common remote sensing tools that can be used with different platforms and with different combinations of spatial resolution/covered area (e.g. Lane, 2000; Butler et al., 2001; Mertes, 2002; Gilvear and Bryant, 2003; Carbonneau et al., 2005; Feurer et al. 2008; Marcus and Fonstad, 2008, for recent reviews of the topic). Optical imagery has traditionally been collected from aerial platforms, mainly aeroplanes, but also more recently using unmanned aerial vehicles that allow surveys at sub-metre resolution (see Chapter 8) and from elevated ground positions.

Airborne image-based methods of data collection have a number of advantages over alternative field surveys (see for example Lane et al., 1993). Amongst them, we can cite: i) extended coverage with minimum effort in comparison to field surveys, especially in remote areas; ii) a well-developed methodology to obtain planimetric measurements and digital elevation models from vertical images (Lane et al., 1994; 1996); and iii) a resolution that has no real limitations (Carbonneau et al., 2003). Currently, conventional manned aerial platforms can efficiently collect vertical imagery at spatial scales up to that of the drainage basin and with a spatial resolution down to 10 to 50 cm whereas unmanned platforms now reach 1 to 10 cm.

However, aerial optical imagery has drawbacks that are related to the platforms used to acquire the images. While

Fluvial Remote Sensing for Science and Management, First Edition. Edited by Patrice E. Carbonneau and Hervé Piégay.
© 2012 John Wiley & Sons, Ltd. Published 2012 by John Wiley & Sons, Ltd.

vertical imagery is well suited for measuring planform characteristics, it is less appropriate for quantitative assessment of vertical features, such as river banks. Similarly, features of interest may not be visible from above, for example as a result of being obscured by a vegetation canopy. The cost of using an aerial platform can also be very high because of the cost of acquiring the data and of getting the platform to the field site. While manned aerial platforms may be very cost effective for infrequent surveys covering large areas, the cost may rapidly become prohibitive for high frequency surveys. Finally, the ability to secure an airborne platform at short notice may diminish the ability to document processes that are occurring with a pseudo random nature. In this context, airborne platforms become inappropriate for high frequency systematic monitoring and for spontaneous monitoring in response to unanticipated events. In river dynamics, short time scales (hours, days, weeks) are often relevant to document and describe significant events like floods that are rarely predictable, so these arguments are particularly apt for river geomorphology.

The use of optical imagery from the ground has rapidly increased over the last few decades as the availability of affordable instrumentation has improved and because it offers several key advantages that complement airborne remote sensing (Lane et al., 1994; 1996; Heritage et al., 1998; Butler et al., 2001, 2002; Chandler et al., 2002; Carbonneau et al., 2003; Graham et al., 2005). In particular, ground-acquired images have a significant advantage if a very high temporal or spatial resolution is required (see Lane et al., 1993). Indeed, ground imagery is the best that can be achieved in terms of spatial resolution because sub-centimetre resolutions depend on a limited distance between the camera and the surveyed object. As sensors and storage technologies develop, the spatial resolution obtained from airborne platforms may be expected to increase. However, where high spatial resolution is more important than wide areal coverage, surveys conducted from the ground or low-level unmanned platforms may be expected to retain an advantage.

In addition, river scientists can benefit from high frequency data capture to investigate rapid phenomena, like those occurring during individual floods, or to capture the exact location in space or time where an event occurs. Close-range image collection is rapid and can be automated, providing large amounts of information whilst minimising expensive field time (Lane et al., 1994). For events of short duration, or where change occurs rapidly, image-based data capture may be the only practical means of recording highly relevant information

including time of occurrence or the coincidence of events and controlling factors. Rapid changes in river morphology occur during floods, requiring a survey frequency that is not feasible with aerial platforms but that is readily feasible with automated ground cameras (Bertoldi et al., 2009). Because data are collected without physical contact, they do not result in any disruption to the objects of the study, making image-based methods ideal in monitoring studies. Moreover, they may record a wealth of information beyond that specifically sought for the project, making them a valuable archive and providing information that may help to explain behaviours and patterns that were not anticipated or can otherwise not be accounted for.

In recent years these factors, allied with rapid advances in digital imaging technology and parallel decreases in cost, have led to a widespread adoption of close-range imagery. This chapter provides an overview of recent trends in the use of close-range imagery in river research. First, we describe commonly used technologies and identify key technical and practical issues for close-range imagery. Second, we review numerous studies that have used either vertical or oblique close-range imagery to document fluvial objects or processes at a wide range of temporal and spatial scales. Several detailed field examples are described to illustrate the advantages of close-range techniques and, more importantly, to emphasise the type of knowledge that can emerge from their use. Most of the examples are based on ground acquisition, but one focuses on helicopter acquisition and combines the advantages of both airborne and close-range capabilities.

15.2 Technologies and practices

15.2.1 Technology

The photographic technology that is used is not particularly advanced, and commercial standard film or digital cameras are used, sometimes with a fisheye lens. The technical specifications of digital cameras are rapidly evolving so that there is little purpose in trying to describe possible configurations and technical options. Instead, we provide three examples of successful setups implemented in recent studies.

For example, a long term automatic monitoring system for the Tagliamento river (Italy) has been installed on two cliffs overlooking the braidplain in the piedmont area 2 km and 9 km upstream of Pinzano gorge, respectively (Figure 15.1) (see Bertoldi et al., 2010; Welber et al.,

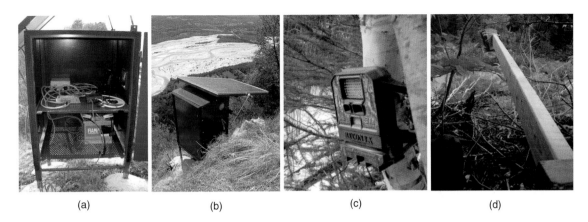

(a)   (b)   (c)   (d)

Figure 15.1 (a) and (b) The acquisition system on Monte Prat, overlooking the Tagliamento River, Italy; (c) a Reconyx camera installed on a tree on the Ouelle River and (d) a sliding arm holding a Reconyx Camera allowing it to overlook an eroding bank on the Ste-Pierre River, Québec.

2012, for more details). The aim is to survey the river configuration with a high temporal resolution, in order to investigate relationships between water, sediments, and vegetation processes at the scale of single flood events. Each monitoring system is composed of a digital reflex camera, with an 18 mm lens and a 6M pixel CCD sensor and pictures are acquired every hour during daylight. The choice of a wide angle lens has the advantage of increasing the extent of the area photographed, but at the cost of reduced spatial resolution. Given the distance of the camera to the river (1 km on average, ranging between 0.5 and 1.5 km), the ground pixel size is approximately 25–50 cm. Electric power is supplied by a photovoltaic solar panel, with two 12V batteries to guarantee a regular charge for up to four cloudy days. Specifically designed software allows full control of the camera, with the possibility to setup the desired acquisition time interval, and to store the pictures. The cameras were installed in 2008 and until the autumn of 2011, more than 10 flow pulses or floods were observed, with peak levels ranging from the threshold for sediment movement to bankfull. The temporal accuracy and the large range of discharge conditions allow the investigation of several geomorphological and hydrological processes at the reach scale (e.g. Bertoldi et al., 2010; Welber et al., 2012) and permit investigation of local processes such as bank erosion, tree transport and bifurcation evolution.

It would be relatively straightforward to equip each of these sites with two cameras so that stereo pairs could be acquired, which would enable extraction of a Digital Elevation Model using photogrammetric procedures. Building on the work carried out by Lane et al. (1994; 1996),

Chandler et al. (2002) used an automatic procedure of DEM extraction applied to digital stereo images of a 125 m long reach of the Sunwapta river, Canada. The monitoring setup proposed by Chandler et al. (2002) consisted of three digital reflex camera stations, with a resolution of 6 MB, positioned 125 m above the river channel. The survey produced daily orthophotographs and DEMs with each pixel providing ground coverage of approximately 5.5 × 2.5 cm, which made possible the analysis of morphological changes on the temporal scale of a single flood event with a high spatial resolution. This procedure needs a limited amount of ground data, but requires high image resolution to automatically recognise homologous points on the stereo images.

Recently, low cost digital cameras able to capture time lapse or triggered pictures have been developed by the hunting industry. These cameras are designed to capture the presence of animals when they wander in front of the camera. There are now several companies and models available on the market (Table 15.1). The cameras are generally mounted with a trigger sensor and an infrared flash to provide oblique or vertical scenes. The motion sensor triggers the camera when a movement occurs in the detection range and a series of pictures is taken at a predetermined frequency and number of repetitions. The infrared flash allows the acquisition of pictures at night for events occurring in the foreground. Some cameras allow capture of short videos. One such camera used by Buffin-Belanger in Quebec, is the Reconyx Hyperfire with a 2048 by 1536 pixels resolution. Ground resolution varies according to the distance of the camera to the area of interest. For example, at a distance of

Table 15.1 Specifications for three cameras developed for the hunting industry. Details were extracted from company websites, August 2011.

Specifics	Reconyx Hyperfire PC800	Trophy Cam Bushnell	Moultry Game Spy M-100
Image resolution	3.1 Megapixels	8.0 Megapixels	6.0 Megapixels
Camera	Custom focal distance, telephoto lens, custom motion detector lens, custom colour	LCD Display	LCD Display
Trigger mode configuration	From 2 pictures per second to 2 pictures per minute; up to 1 hour delay between triggers 1/5th second trigger speed	From 1 picture per second to 1 picture per hour; up to 1 hour delay between triggers	Maximum 3 pictures per 15 seconds; up to 1 hour delay between triggers
Time lapse configuration	Virtually any number of seconds, minutes, or hours	From 1 picture per minute to 1 picture per hour.	From 1 picture per 5 seconds to 1 picture per minute.
Video	Not available.	From 1 seconds to 1 minute	From 5 to 30 seconds
Battery	12 AA; battery life up to 40 000 pictures	12 AA; battery life up to 1 year	4 D-Cell

approximately 50 meters, the ground resolution of the Hyperfire camera is 5 mm/pixel. This resolution might be limiting for sophisticated grain size and DEM analysis. For an oblique photograph, resolution also varies from the foreground to the background, similarly to the ground video settings discussed in Chapter 16. The camera uses 12 AA batteries allowing collection of up to 40 000 pictures, depending on ambient temperature (as this affects battery charge), acquisition frequency and activation of the infrared flash. The Reconyx camera is self focusing and it automatically determines exposure time. This may lead to variable image quality, especially where contrast is high in the field of view. Nevertheless, the Reconyx camera can be successfully used to document the chronology of events occurring at the scale of a river reach for time periods varying from one day to one year. Amongst other applications, these cameras have been used by Buffin-Belanger and colleagues to document the activation of abandoned river channels, the formation and dislocation of river ice covers, the displacement of large woody debris, daily changes in proglacial stream discharge and morphology and erosion processes occurring on river banks.

15.2.2 Overview of possible applications

Close-range photographs can be acquired vertically or obliquely. Vertical acquisitions can be achieved at head height or from a tripod. To gain height and therefore capture larger ground areas, some researchers have

also obtained images from low-level cranes, extendable masts, or tethered balloons. A common application of ground-based vertical photography is the characterisation of surface grain size and bed roughness (Table 15.2, Figure 15.2), applications that are now being explored using ground-LIDAR (see Chapter 14). Considerable work has been published in this domain with the development of software such as the digital gravelometer that enables automatic grain-sizing from hand-held vertical photography (Graham et al., 2005a, b). Following this approach, Rollet (2007) collected 107 photographs of bed surface sediments from every gravel bar along a 40 km reach of the Ain River, France and identified two distinct reaches; an upstream reach with infrequent, coarse gravel bars affected by dam starvation and a downstream reach not yet affected, with much smaller particles. Additional applications of close-range vertical imagery of bed sediments include characterisation of morphometry and petrography (see details in section 4 of this chapter).

Oblique acquisition of close-range photos remains, however, the most frequent scenario and, again, several situations exist.

On the Waimakariri River, New Zealand, Hicks et al. (2002) explored the possibility of using video cameras, mounted 35 m above the river bed, to provide hourly imagery of the study reach. This technique produced an invaluable record of the river's morphodynamics and highlighted the coherence of morphologic features. This amount and quality of data, common in flume

Table 15.2 Examples of possible ground-based imagery applications.

	Types of photos	Area covered	References
Grain size	Vertical ground photo	Set of stations	Graham et al. 2005 a&b
Grain morphometry	Vertical ground photo	Set of stations	Roussillon et al. 2009
Grain petrography	Vertical ground photo	Experiment	Riquier 2009
Sand coverage over a gravel surface	Vertical ground photo		Carbonneau et al. 2005
Bank retreat using photogrammetry	Oblique ground photo	At a station	Lawler, 1993; Barker et al., 1997; Pyle et al., 1997
Wood raft area or isolated pieces detection	Oblique ground photo	At a station	Moulin and Piégay, 2004
Light under the canopy	Fish eye/ground photo	Set of stations	Digan and Bren, 2003; Ringold et al. 2003
Water level fluctuations; morphological changes	Oblique ground photo/timelaps video	At a station	Chandler et al., 2002; Hicks et al., 2002; Ashmore and Saucks, 2006; Luchi et al., 2007; Bertoldi et al., 2009; Bird et al., 2010
Wood censing, bank states	Oblique helicoptere photos	Continuous reach	Piégay and Landon, 1997
Riparian Landscape characterisation	Set of stations	Set of stations	Cossin et Piégay, 2001
Landscape changes between n dates	Oblique ground photo (present and archived)	Sampling landscape scenes	Start & Handasyde, 2002; Michel et al. 2010
River ice growth	Oblique ground	Set of stations	Dube, 2009
Subaerial processes on a river bank	Oblique ground photo	One river bank	Hamel, 2011
River sediment structures	Vertical from balloon or mast	River channel width, riffle	Church et al., 1998

experiments, is not easily found in field studies, particularly on large rivers. The Tagliamento River example discussed briefly above and more fully below, provides another case study where oblique photography from fixed locations has provided important reach-scale insights about a large braided river. A set-up similar to the one described by Chandler et al. (2002), with the acquisition of four daily images, allowed Ashmore and Saucks (2006) to assess the relationship between discharge and channel width in a braided network. The analysis showed that it is possible to use this method to obtain discharge data in a multi-channel system, where other direct measurements may be impossible.

Imagery can also be used to target smaller features and focus on a narrower scene. Some authors have used cameras 'at-a-station', focusing on specific river cross sections or on a morphological unit of interest. Examples of this include Dubé's (2009) quantification of river-ice growth rate on a riffle and pool sequence during the freezing season in Québec and Moulin and Piégay's (2004) examination of wood accumulations to evaluate wood trapping at reservoirs (Figure 15.3). In this last case, the reservoirs trap and accumulate migrating dead wood, which allowed the authors to determine the geographical origin and the transported volumes of dead wood in relation to river flow regime. In these examples, ground-based techniques are essential to guarantee a high temporal resolution, but even more importantly to accommodate the unpredictable nature of the documented processes.

Oblique close-range photography also has the advantage of providing information on sites or processes that are not seen by vertical images such as those beneath a vegetation canopy (e.g. bank profiles and channel beds in rivers with vegetated banks). For example, Bird et al. (2010) recently showed how close-range photogrammetry can overcome the limits of aerial surveys. In narrow

Figure 15.2 Contrasting examples of images acquired from ground technology: a) map of grain size across a river bar; b) river bank morphology; c) channel planform; d) canopy cover and light-conditions within a riparian corridor; e) ice jam formation; f) wood jam formation along an eroding bank.

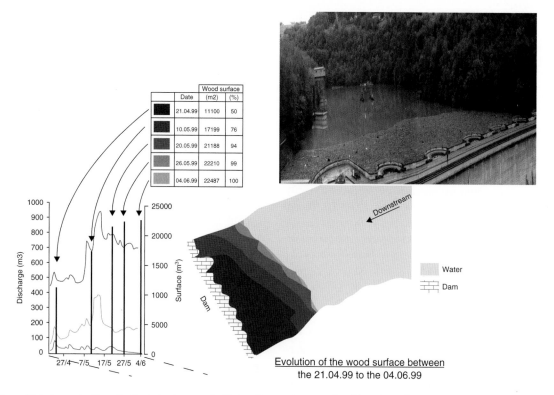

Figure 15.3 A wood raft on the Genissiat dam, on the Rhône, France: An example of deposits and weekly survey of the reservoir's wood coverage, between 21 April and 4 June 1999 (Modified from Moulin and Piégay, 2004).

channels (bankfull width <20 m), riparian tree canopy can hinder the survey of both the channel bed and banks. They acquired sequences of overlapping images from a vertically mounted non-metric camera suspended 10 m above the river by a unipod. This methodology allowed the generation of very accurate DEMs with a nominal ground resolution of 3 cm.

Other studies have focused on monitoring river bank erosion (Lawler, 1993; Barker et al., 1997; Pyle et al., 1997). These authors used terrestrial digital photogrammetry to reconstruct the bank profile through a high resolution DEM. For example, Pyle et al. (1997) obtained a spatial resolution of 2 cm, with an accuracy approximately equal to 1 cm, using oblique stereo pair images acquired about 15 m from the eroding bank. A great advantage of this technique is that it provides bank surveys at a very fine scale, avoiding the errors due to the interpolation process, as well as avoiding physical contact with the unstable slopes of the bank.

When acquired from a helicopter, oblique photos can provide qualitative information (e.g., bank characters/state) all along a channel course which is not possible by any other airborne/ground device (Piégay and Landon, 1997). Some researchers have used archived images and compared them with recent images to detect decadal landscape changes (Debusshe et al. 1999; Start and Handasyde, 2002). These authors compared a series of photographs, mostly taken between 1952 and 1990 to highlight changes in riparian vegetation characteristics due to the construction of dams. Cossin and Piégay (2001) used oblique photos to provide a typology of reaches according to riverscape character along a river corridor (see details in Chapter 18).

Close range imagery can also be applied to flow velocity measurements using particle image velocimetry (Hauet et al. 2006). This topic is addressed in Chapter 16.

15.3 Post-processing

The post-processing workflow of the acquired images may involve a wide range of tools and procedures, with different degrees of complexity depending on the aim of the investigation. Simple, qualitative observations are sometimes sufficient; for example when the process recognition is the main aim of the survey. In other applications, particularly when the objective is to extract quantitative information from the images, more sophisticated analyses are required. Here we consider approaches to obtaining grain size and particle morphometry from

vertical ground-based images and the rectification of oblique images.

15.3.1 Analysis of vertical images for particle size

The most common methods of measuring the grain-size distribution of surface sediments are variants of the Wolman (1954) procedure, where grains are sampled on a grid. Recent years have seen several attempts to develop and apply image-based analysis methods of grain-size analysis that substantially reduce field time (and thus cost) and do not disturb the sediment being studied. Two general approaches have been taken. The first utilises geostatistical techniques and empirical calibration to determine ensemble grain-size characteristics from a photograph of the sediment surface (empirical approach). The second is based on automated recognition and measurement of individual particles within an image (object-based approach).

Empirical approaches are based on statistical measurements of spatial variation in pixel intensity values, with high-frequency variability in intensity being indicative of small grains. The statistical measurements may be used to derive ensemble grain-size parameters (e.g. mean grain size) by empirical calibration against field-measured grain-size parameters. The most successful approaches employ measurements of the local semivariance (e.g. Carbonneau et al., 2004) and autocorrelation (e.g. Rubin, 2004; Buscombe and Masselink, 2009), both of which work in the spatial domain. Most recently, Buscombe et al. (2010) have utilised the spectral properties of images to estimate the median grain size, extending the spatial autocorrelation approaches into the frequency domain and requiring little calibration.

Empirical approaches have not been used widely for ground-based assessment of fluvial sediment surfaces, although all are potentially applicable. Local semivariance has primarily been applied to images collected from aerial platforms. Autocorrelation has most commonly been applied to beach sediments, although it is increasingly being developed for fluvial surfaces (e.g. Warrick et al., 2009; Buscombe et al., 2010). The principal limitation of empirical techniques is that they do not provide complete information about the grain-size distribution, being limited to estimates of particular grain-size percentiles, for which they have been shown to provide robust estimates (e.g. Warrick et al., 2009; Buscombe et al., 2010).

Object-based approaches use image processing methods to identify the boundaries of individual grains in

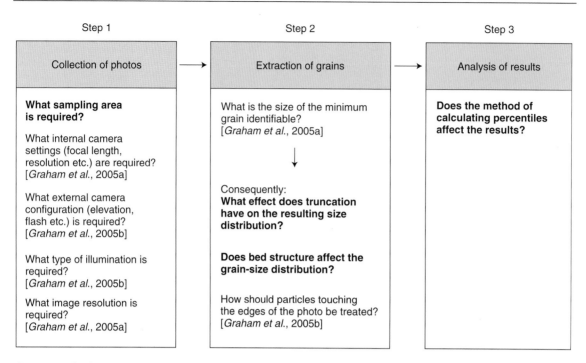

Figure 15.4 The three steps in deriving grain-size information from ground-based digital photographs, illustrating some key procedural questions associated with each. From Graham et al. (2010) Copyright 2010 American Geophysical Union.

the image and the image scale to determine their size (Figure 15.4). Such approaches are a development of earlier, manual, methods of analysis using analogue photographs (e.g. Adams, 1979; Ibbeken and Schleyer, 1986; Church et al., 1987; Diepenbroek et al., 1992; Diepenbroek and De Jong, 1994; Ibbeken et al., 1998). The development of automatic algorithms for extracting grain boundaries from digital images of in situ sediments is a comparatively complex problem because of the between- and within-site heterogeneity (e.g. of size, packing configuration, shape, surface texture and colour) of natural sediments and the difficulty of controlling environmental conditions (e.g. ambient lighting levels and light direction). Nevertheless, considerable success has been achieved, with errors comparable in magnitude to those associated with traditional field-based grid samples (Butler et al., 2001; Sime and Ferguson 2003; Graham et al., 2005a & b; Strom et al., 2010) whilst field time is reduced to a few minutes per sample. The key advantage of these methods over empirical techniques is that – because they measure each grain directly – they can provide complete grain-size distributions. They have also been shown to be transferable between sites with different petrography,

grain size and sorting, and packing configuration without the need for recalibration (Graham et al., 2005b).

Key limitations of object-based approaches are that lighting conditions need greater control than for empirical approaches (to maximise contrast; Graham et al., 2005a) and there is a minimum resolvable grain size that varies as a function of image resolution (resulting in truncation of the fine end of the grain-size distribution). Lighting is relatively easily controlled by shading the sediment from direct sunlight and using a camera-mounted flash. The truncation of small grains will become less significant as camera technology develops, but the effect on coarser percentiles (above D_{50}) is minimal except where the sediment contains more than 5% sand (Graham et al., 2010). It might be expected that sediment structure would have a significant effect on photographically-derived grain-size distributions (as a result of partial burial of grains, overlapping grains, and foreshortening). The magnitude of such effects is difficult to determine, but Graham et al. (2010) concluded that they are small relative to other sources of error.

One further issue is to ensure that photographs are sufficiently large to include a representative grain-size sample. Grain-size percentiles have been found to stabilise for grid samples of between 300 and 400 grains (Rice and Church, 1996), and for areal samples of between 100 and 400 times the area of the largest grain (Graham et al., 2010), indicating that these are the minimum sample sizes required to have confidence.

15.3.2 Analysis of vertical images for particle shape

Following Wentworth (1919) and Wadell (1932), well-known roundness indices have been proposed that utilise the radius of curvature of a clast's 'corners'. The ratio between the mean radius of curvature of the corners and the radius of the largest inscribed circle defines a roundness measure (Wadell, 1932). Making the relevant measurements for a large number of particles is time consuming, so visual charts were proposed by Krumbein (1941) for improving the efficiency of field measurements. More recently, ground photos have been used and different imagery procedures were tested to measure the roundness based on Fourier transforms (Diepenbroek et al., 1992), mathematical morphology (Drevin and Vincent, 2002) or discrete geometry (Roussillon et al., 2009) 3D laser scanning technologies have also been applied to this problem (Hayakawa and Oguchi, 2005). Roussillon et al. (2009) successfully calculated Wadell indices from images. They also provided a simple indicator of roundness which is fairly robust in comparison with the Krumbein chart: the ratio of the perimeter of the particle to the perimeter of the ellipse best fitting the particle (Figure 15.5).

When considering image analysis for particle morphometry, sampling strategy is critical because the relations between river position, particle size and particle shape remain somewhat obscure. In this case, we recommend that between sites or through time comparisons are based on examining similar size classes in space or time, respectively.

15.3.3 Analysis of oblique ground images

For oblique ground images, rectification is required in order to allow quantitative measurements and comparisons. The rectification process requires a set of ground control points and the procedure is generally more complex than for aerial imagery, where the vertical acquisition means the three orientation angles are close to zero. As a result, the ground control network has to be more accurate and include a larger number of points. It is advisable to exceed the minimum required number (four) by at least a factor of 2 or 3 (see for example Chandler et al., 2002). A further difference with aerial imagery is the pixel size that is quasi-unique in the vertical case, but can vary on a large range in oblique imagery. The spatial resolution of the rectified image depends on the largest pixel footprint (i.e. most distant from the camera). To limit this problem it is better to reduce the analysed area to the minimum required. Rectification procedures are detailed in Chapter 8.

Applications with ground-based oblique images may involve one single camera station (usually for temporal analysis) or stereo pair acquisition. In the latter case, photogrammetric techniques are possible that may lead to the construction of a digital elevation model. Further post-processing techniques may involve the automatic recognition of objects including channel edges, bar or bank characteristics or wood accumulation. The main problems with this type of analysis are related to variations

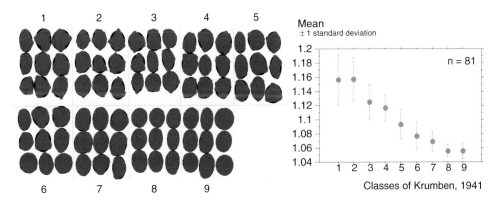

Figure 15.5 Conversion chart of Krumbein (1941) classes from 1 (very angular) to 9 (very rounded) (with in black the ellipse fitted and in grey the particle), and the average rP index (ratio of Particle/Ellipse perimeters) for each of the Krumbein classes.

in light conditions (as a consequence of the weather, time of day and time of year), reflections from the water surface and the presence of shadows, especially when the sun is close to the horizon. These may hinder an efficient, unsupervised procedure of object recognition, thus requiring time consuming work by a skilled operator.

Information obtained from close range photographs may also be used to complement other field measurements or to validate/calibrate numerical models. Time-lapse cameras, for example, can provide valuable information on flow levels that can then be used to inform boundary conditions for numerical modelling or to validate flow reconstruction at different discharge conditions.

15.4 Application of vertical and oblique close-range imagery to monitor bed features and fluvial processes at different spatial and temporal scales

15.4.1 Vertical ground imagery for characterising grain size, clast morphometry and petrography of particles

The physical properties of the channel substrate provide a fundamental control on bed stability/mobility, aquatic and riparian habitat availability and define the skin resistance to flow and the character of the boundary layer. A variety of substrate properties have been examined using imaging techniques, although these are inherently limited to examining the nature of the surface layer. The most widely studied substrate property is the grain-size distribution of the sediment, although this is often used as a proxy for roughness. More recently, attempts have been made to measure roughness directly.

Ground-based imagery has been used to examine spatial variations in bed material grain size at large and small scales. For example, Rice and Church (1998) used ground-based photography (alongside conventional sampling) to examine downstream fining along approximately 230 km of channel in the Pine-Sukunka River system of British Columbia. At such scales it is necessary to manage local-scale variability; for example, by consistently sampling the head of bars. At smaller scales, for example across individual bars, conventional sampling techniques, like Wolman sampling, do not detect within-site variations in grain size that may be important at that scale, unless great care is taken to first identify relevant structures and accommodate them using an

appropriate, stratified protocol (e.g. Wolcott and Church, 1991). To date, image-based sampling procedures have sought to replicate the approach of conventional sampling procedures, focusing on the generation of conventional grain-size percentile measures. However, one of the key benefits of image-based methods is that they facilitate the assessment of continuously varying trends in grain size at a variety of spatial scales. They therefore enable examination of scale-dependent variability in surface grain-size properties in a way that has not been achievable previously (Figure 15.6).

The morphometry of particles, which is controlled by petrography and abrasion, can also provide valuable information about sediment sources and particle maturity (a proxy for travel distance). Traditional methods of acquiring shape information are intensely time-consuming so it is encouraging that photographic methods are being developed to capture useful data. It can be used to evaluate the contribution of tributaries or adjacent hillslopes to the river bedload transport when an appropriate sampling scheme is adopted (Kuenen, 1956; Diepenbroek et al., 1992). This parameter is less sensitive to local hydraulic conditions than grain size (used for example with the same purpose by Rice, 1998 and Surian, 2002) and can be used in contexts where grain size is strongly disrupted by human activities.

The value of recent developments by Roussillon et al. (2009) was illustrated by applying them to bed sediments along the River Progo in Indonesia, over a reach of 140 km from the source to the ocean. The different roundness indicators did not show a simple longitudinal trend (Figure 15.7) but highlighted the location of the most significant changes in particle morphometry. In particular, a significant increase in angularity is observed between km 60 and 80, demonstrating local inputs of fresh and angular material from tributaries draining a recently active volcano.

This approach can be accurate for a given lithology where resistance to abrasion and the relation between distance travelled and roundness is consistent. In more complex lithological environments, it is necessary to distinguish the type of rocks before applying such morphometric measures. A combination of grain size, petrographic and particle morphometric information can then provide the most robust indication of particle sources along a river continuum. Some tests have been performed on pebbles collected on the Ouvèze River, a left side tributary of the middle Rhône River, France, characterised by a complex lithological basin with basalt, limestone, sandstone schist, and granite source areas

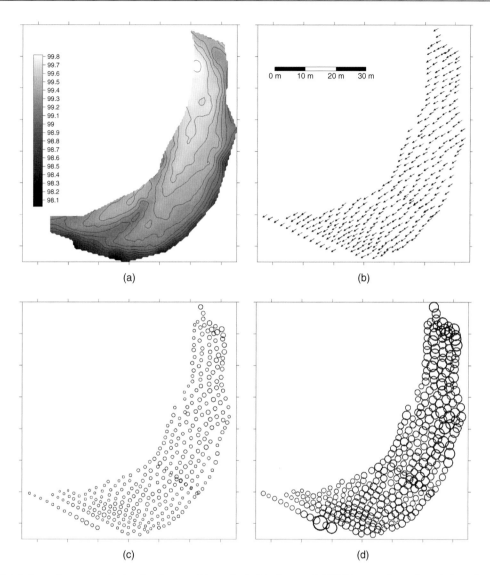

Figure 15.6 Information captured using the photographic method of Graham et al. (2005b) across the surface of a lateral point bar, Afon Elan, Wales, UK. The bar is approximately 80m long and 20m across (at its widest point). Analysis is based on 346 images, each with an area of 0.7 sq m. (a) Elevation obtained using a total station (in metres above an arbitrary datum). Contours are at 0.1 m intervals. (b) Flow direction inferred from mean grain orientation within each image. (c) Proportional symbols indicating the median grain size (D_{50}) for each photograph (min 14 mm; max 44 mm). (d) Proportional symbols indicating the D_{95} grain size for each photograph (min 31 mm; max 130 mm).

(Aloy, 2005; Riquier, 2009). Ground-photographs of particles were taken and their petrography was identified. A discriminant analysis was then performed on 24 selected particles for which the radiometric signature was measured using a spectro-radiometer (GER model 1500) to predict petrographic type according to radiometric characteristics (Figure 15.8). The analysis shows that the signature of basalts/schists, limestones, sandstones and granites are significantly different. In the visible wavelength (400 to 700 μm), the distinction between the classes is greatest, suggesting that petrographic classifications may be possible from conventional photos.

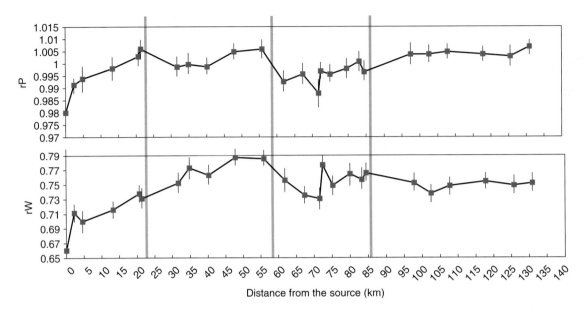

Figure 15.7 Pattern of the mean and confidence level (at $\alpha = 0.05$) of two pebble roundness indices measured "at-a-station" on samples of 100 particles along the main stem channel of the Progo from its source to the ocean : the Wadell index (rW) and the ratio of perimeters (pebble/ellipse (rP)). Modified from Roussillon et al. 2009. In grey, significant break-points in the longitudinal trends, identifying individual homogeneous river segments.

15.4.2 Monitoring fluvial processes

Historically, information about channel morphometry has been obtained by planform mapping (using field surveys, aerial photographs and published maps) combined with surveys of channel cross sections. Such data are highly valuable, but they capture only a snapshot, that is strongly dependent on the water stage, and, in general, a lot of information is not captured because of poor temporal resolution. In addition, bed elevation surveys are demanding of resources with the result that only a small number of cross sections can be measured. This type of data is therefore most appropriate for measuring changes that occur over large areas or that are due to a high magnitude process and will tend to overlook details between cross sections or small changes due to smaller magnitude processes. Moreover, 2D numerical models require continuous and high resolution input data and, similarly, produce continuous (in time and space) information on the flow field, bed topography and sediment transport, which needs suitable field data for validation.

The recent adoption of airborne and terrestrial laser scanning in river science may provide possible solutions to some of these problems, but their costs, and the impossibility to survey wetted areas (but see Smith et al., 2012

for recent developments) drastically limit the temporal resolution with which data can be collected. In order to maximise information in space and time, acquisition of oblique photography can be a valuable solution for monitoring temporal changes with an at-a-station setup.

Lane et al. (1994; 1996) and Chandler et al. (2002) demonstrated that digital oblique terrestrial imagery is an appropriate and affordable solution to the problem of frequent morphological survey of river reaches. This is particularly true when monitoring multi-channel rivers that are characterised by complex inundation dynamics (Lane et al., 1994; van der Nat et al., 2002; Ashmore and Saucks, 2006; Luchi et al., 2007; Bertoldi et al., 2009) and where morphological changes can happen on a daily/weekly timescale (Milan et al., 2007; Ashmore et al., 2011). A further valuable advantage of close range imagery is the possibility to pinpoint the exact spatial and temporal occurrence of infrequent but rapid processes, or those occurring in remote and dangerous environments.

Four examples are reported below that illustrate different possible applications for oblique close-range imagery; namely (15.4.3) subaerial bank processes, (15.4.4) braided rivers inundation dynamics, (15.4.5) river ice formation and dynamics, (15.4.6) characteristics of riparian structures and wood accumulation.

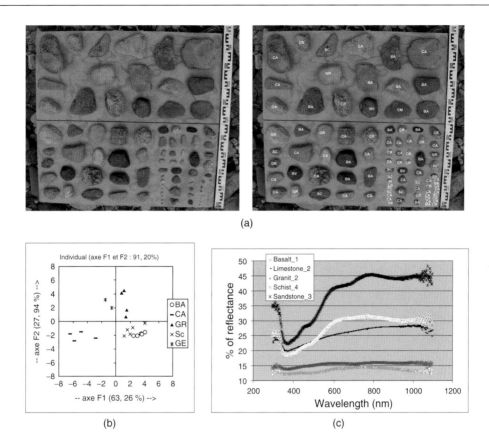

Figure 15.8 a) example of pebble samples obtained along the Ouvèze River, France for petrographic characterisation. On the left the initial sampling plate, on the right, the same set with a petrographic code on each of the particles (Unpublished data, Aloy, 2005) (CA = limestone, GR = Granits, BA = Basalt, SC = Schist, CR = Sandstone). b) Result of the discriminant analysis performed on 24 particles showing the first factorial map with a clear distinction of the petrography according to the % of reflectance calculated on 23 wavelength classes from 300 to 1100 nm. c) Example of wavelength signature of five pebbles, each having a different petrography (Unpublished results, J. Riquier, 2009).

15.4.3 Survey of subaerial bank processes

The lower part of the Ouelle River in Québec, Canada is a highly sinuous sandy river that is prone to intense gravel aggradation downstream from a slope change delimiting the meandering sandy section from a wandering gravel river reach. Recently, an intensification of the frequency at which the river bed has to be dredged to protect a highway bridge has increased. There is no indication of increasing discharge in recent years, so attention has been focused on the frequency and magnitude of subaerial processes (Couper and Maddock, 2001) that affect several eroding banks in the wandering reach of the Ouelle River (Hamel, 2011). To document the nature and the chronology of subaerial processes affecting one of these banks (Figure 15.9a), a Reconyx camera was deployed to capture

a picture every 60 minutes, from September to April to include the autumn, the winter and the spring periods. The photographic time series provided important information on several subaerial river bank erosion processes: their date and hour of occurrence (during daylight), their location on the bank and their relative magnitude. Erosion processes (rock falls, gully erosion, small scale debris flows) were documented by directly observing the process on a picture or were inferred from observed differences between successive pictures. Process locations were identified by overlaying a grid onto the photographs that defined the upper, the middle and the lower bank. Magnitude was evaluated using a visual appraisal of the spatial extent of each event. Figure 15.9b–d illustrates the time series of the key parameters for three subaerial processes that were extracted from the photographs. The

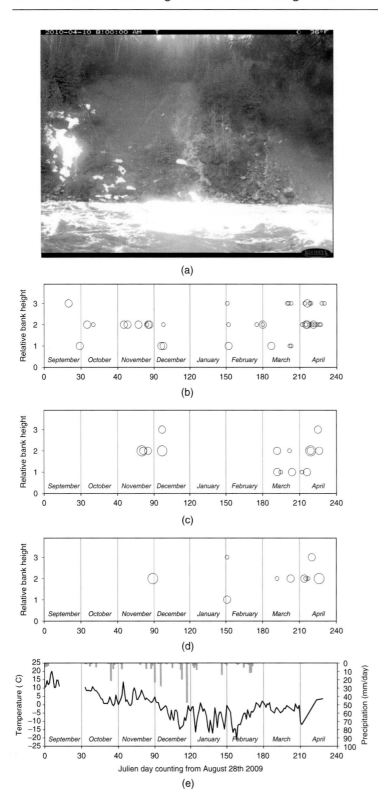

(a)

(b)

(c)

(d)

(e)

Figure 15.9 a) Picture of the eroding bank on the Ouelle River from the Reconyx Camera. Numbers indicate the upper (3), middle (2) and lower (1) sections of the bank. b, c, d) Occurrences, location and magnitude of subaerial processes : b) rock falls; c) gully erosion; d) debris flows. e) Precipitation and temperature for the selected time period.

figure reveals features that only the deployment of such a camera could provide. Among other things, it can be seen that falling rocks and boulders are the most frequently observed events, accounting for more than 50% of the total observed movements. Also, the vast majority of events (80%) occur in the upper and middle bank section causing some impact on the lower bank section. Finally, when compared with local meteorological data (Figure 15.9e), the results indicate the observed subaerial processes occur mainly when the air temperature oscillated around zero. The movements are also more frequent during fall and spring time.

The deployment of a ground based camera in this case, provided information that would be difficult to obtain in another way. Erosion pins, for example, allow the erosion or accumulation of sediment on a river bank to be measured, but it remains difficult to establish which process causes the morphological change (Lawler, 1993). In this context, the use of time lapse cameras become highly valuable when combined with other means of measurements, such as erosion pins. In her study, Hamel (2011) used monthly LIDAR surveys to extract volumetric changes on the bank. The camera complemented the study by giving key information about the physical processes that accounted for the volumetric changes and their timing.

15.4.4 Inundation dynamics of braided rivers

Monitoring braided river systems is a challenging task, which requires high spatial resolution and rapid execution to describe changes occurring on scales ranging from a single branch and node up to the whole braidplain (Westaway et al., 2003). Frequent temporal surveys are resource intensive, particularly if topographical and geomorphic information are required during (or after each) flood. However, the ground-based automated image acquisition system developed for the Tagliamento River (Italy) and introduced above, shows that it is possible to document the morphological evolution of a large braided channel at an hourly temporal scale and to use that information to understand, for example, inundation dynamics.

The Tagliamento River is a large, gravel-bed braided river located in the Friuli Venezia Giulia region, northeast Italy, flowing from the southern limit of the Alps to the Adriatic Sea. It still maintains a high degree of naturalness, where morphological and ecological processes can be observed (Tockner et al., 2003; Bertoldi et al., 2009). The main stem of the Tagliamento River flows through an immense active zone of about 150 km² and the floodplain width is up to 1.5 km. Most of the active floodplain is

characterised by a braided pattern with a highly variable cover of vegetated islands, and is fringed by continuous riparian woodland.

We selected two different study reaches: i) a bar-braided reach named the 'Cornino reach', located near Cornino just upstream of the bridge of the same name; and ii) an island-braided reach, 2 km upstream of the Pinzano bridge, near Flagogna. These two reaches have been subject to an intense multidisciplinary investigation carried out by various European research groups (Bertoldi et al., 2009). The first reach is about 1000 m long and 850 m wide and lies at an elevation of about 210 m above sea level. The mean bed slope is 0.4%, and the sediment size calibre ranges from fine sand to cobbles, with a D_{50} of about 40 mm. and D_{90} ranging from 50 to 150 mm. The study area is dominated by Monte Prat on the right side of the valley. The morphology is bar braided, with a large area of exposed gravel and very few vegetated stable areas. The reach is laterally unconstrained and does not show evident human impacts apart from the small embankment on the left side and a groin on the right side to protect the bridge.

The 'Flagogna' reach is narrower, with an average braidplain width of 700 m. The longitudinal bed slope is about 0.3%. Sediment size is quite similar to the 'Cornino' reach, with a larger fraction of sand that reduces D_{50} to about 30 mm. The reach is dominated by Monte Ragongna, on the left side of the valley. This reach is characterised by the presence of large, well-established vegetated islands, formed mainly by Poplar and Willow species. Occurrence of difference size vegetated patches leads to complex interactions between vegetation growth, sediment movement, and river morphology (Bertoldi et al., 2011).

Inundation dynamics were studied through the ground-based camera system described in section 2.1 (see also Figure 15.1). Approximately 70 images have been orthorectified and classified, to highlight the planform differences between exposed gravel and inundated areas for various hydraulic conditions. To establish the interior and exterior orientation parameters of the images the digital photogrammetric system ERDAS IMAGINE OrthoBASE Pro™ was used. The geometric distortion inherent in imagery was removed using the Ortho Resampling process of OrthoBASE Pro™. In order to fix the camera orientation 20 ground control points were identified and surveyed with a standard DGPS technique (Chandler et al., 2002). The precision of the orthorectified images have ranges between 2 m and 6 m, depending on distance from the camera. Figure 15.10

Figure 15.10 Two examples of orthorectified images of the Flagogna reach on the Tagliamento River (flow is from right to left). Flow levels at the Villuzza hydrometric station are 1.2 m and 2.25 m respectively.

shows two examples of the Flagogna reach at medium and high flow.

Georectified images have been manually digitised to produce maps of the water bodies, differentiating between fully connected channels, upstream disconnected channels and ponds. A wide range of flows, between minimum and bankfull levels has been examined, and this is entirely due to the use of the automated photograph acquisition. Figure 15.11 reports the results of the analysis, as a function of the water level measured at the Villuzza gauge station, 3 km downstream of the Flagogna reach. A clear relationship exists between the occupation of the braid plain by water and the river stage. Figure 15.11a shows the two reaches are characterised by a similar trend, with the wetted proportion linearly increasing from 10% of the braid plain width up to 90% at a water level corresponding to approximately a return period of two to three years. The island braided Flagogna reach is more likely to have a lower wetted percentage than the Cornino reach, particularly at higher flow levels and in spite of the narrower braid plain. This is due to the larger presence of vegetated patches that force the water into a smaller number of deeper branches in the Flagogna reach. A large difference is highlighted by the braiding index, i.e. the average number of branches in a cross section (Figure 15.11b). The braiding index peaks in both cases at a water level ranging between 1 and 1.5 m at the Villuzza gauge, but reaching an average maximum of 7 at Cornino, whereas in most cases no more than four branches are observed at the Flagogna reach. Braiding index then decreases toward 1 for a water level approaching 2.5 m.

Finally, we measured the proportion of upstream disconnected water bodies (ponds, groundwater fed branches, backwaters), as an indicator of slow flow habitats. The proportion of disconnected channels is up to 20% at low flow, but rapidly decreases when the flow level increases up to 1.5 m (see Figure 15.11c). The two sites have a slightly different behaviour, with the Flagogna reach showing a larger proportion of disconnected water bodies. The presence of vegetation induces more complex bed topography, with larger scours and deposits, and therefore a greater possibility to find disconnected branches up to a water stage of 2 m, which represents a return period of roughly one year. This fraction of wet habitat is particularly significant, because it represents possible refugia for most aquatic species, particularly during moderate flow pulses. These water bodies are characterised by completely different temperature conditions, ensuring warmer (ponds), or colder (groundwater fed branches) spots that greatly enhance habitat diversity.

15.4.5 River ice dynamics

The study of river ice dynamics remains a difficult task because of the dangerous nature of the processes involved and their strong dependence on hydro-meteorological conditions. It is, for example, difficult to obtain direct measurement of river ice freeze-up or break-up because of the fragility of the ice cover at those periods but also because of the destructive nature of the moving ice blocks. As a result and in spite of several advances in fluvial ice studies (Buffin-Belanger and Bergeron, 2011), some key questions still remain. The use of close-range photographic time series has helped the investigation of some of these questions.

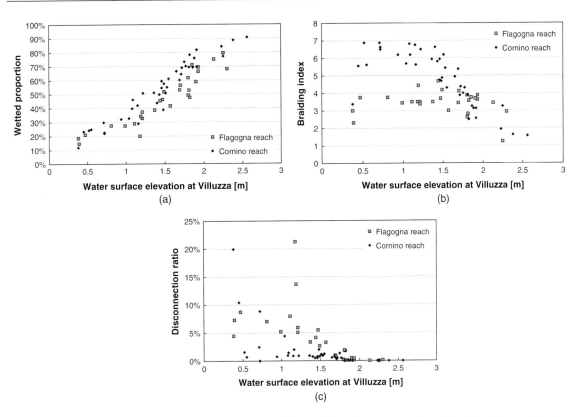

Figure 15.11 Inundation dynamics in two study sites on the Tagliamento River: a) water proportion, b) braiding index and c) ratio between upstream disconnected and connected water bodies, in the two study sites.

The effect of river morphology on river ice growth was recently documented using a long series of pictures taken during the freeze-up period on the Rimouski River, Canada (Dubé, 2009). In rivers where riffles and pools are present, the spatio-temporal ice growth varies according to complex interactions between river morphology, hydro-meteorological conditions and flow discharge. Dubé (2009) aimed to quantify ice cover growth rates at a riffle and pool scale to examine the effect of river morphology and of hydroclimatic conditions on ice cover growth. To achieve these objectives, 24 digital ground based photos were taken manually on a daily basis at 8 locations along a riffle and pool section of the Rimouski River from the 30/11/2005 to the 9/03/2006 (100 days). Figure 15.12a–f shows two series of pictures (for the pool and riffle) taken on the same days. The pictures clearly reveal differences in the ice cover growth between the pool and the riffle but also identify differences in ice types and the processes that lead to complete ice cover.

A challenge with ground based photographs is to convert the visual observation into quantitative measurements that can be used to compare sites or processes. Here, we used the daily pictures to compute daily ice cover growth rates. Because of the obliquity of the picture, the pixel ground size varied within the images. Instead of orthorectifying all of the pictures, we produced a rating curve to determine the daily width of the ice cover. Ten ground control points were used to compute a rating curve between the pixel rows and the ground distance for each morphology. The rating curves followed exponential models with R^2 above 0.99 in both cases. These models were then used to estimate the daily width of the ice cover. The change in ice cover width between two consecutive days was then used to compute the proportion of river closed by ice on a day by day basis.

Figure 15.12g presents the ice growth curves for the pool and the riffle that were extracted from the photographic time series. The curves reveal that the early growth of the ice cover is synchronous in pools and riffles and that

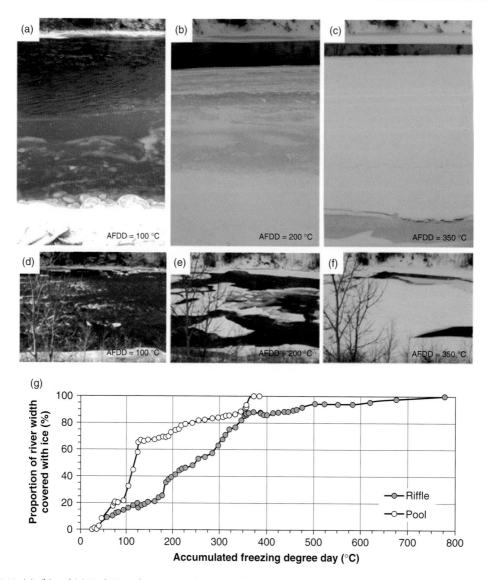

Figure 15.12 (a), (b) and (c) Evolution of an ice cover in a pool. (d), (e) and (f) Evolution of an ice cover in a riffle. (g) Ice cover growth curves for a pool and a riffle. Curves are obtained from daily pictures taken perpendicular to a riffle and to a pool in the Rimouski River, Québec, Canada. AFDD = Accumulated freezing degree day in °C.

both morphologies present growth curves with distinct growing periods. For both morphologies, there is an initially rapid growth period followed by a slower growth period leading to the complete closure of the ice cover (Dubé, 2009). There are, however, differences in the growth rates between the two morphologies, the pool having higher growth rates and leading more rapidly to full closure of the ice cover. Combined with qualitative information from the daily photographs, these curves

are associated with distinct freeze up processes. In the riffle, frazil ice accumulation and anchor ice dominate the formation of the ice cover during the early stages but border ice processes dominate to reach full closure. In the pool, frazil ice and border ice develop during the whole growing period.

In this example, the use of daily close-range pictures helped significantly to document a process that is rather difficult to measure in any other way. Measurements of river ice widths can be hazardous and time consuming to realise, especially in remote locations at temperatures far below zero Celsius. In this example, the 24 pictures were taken everyday in less than 30 minutes, minimising time in the field and maximising data analysis time back in a warmer laboratory.

15.4.6 Riparian structure and dead wood distributions along river corridors

Close range imagery allows the analysis of riparian canopy structure and in particular gives the possibility to assess vegetation density and therefore channel shading, which is an important aspect of in-channel habitat. The canopy affects water temperature by intercepting and emitting radiation and also influences local humidity, air flow and air temperature (Kelley and Krueger, 2005). In alluvial forest, light controls species composition and regeneration and interacts with flow and morphological disturbances to control vegetation dynamics (Hall and Harcombe, 1998). There is then a clear need to evaluate light transfer under the forest canopy to understand undergrowth vegetation patterns and distinguish the respective contributions of available light and erosion or sedimentation disturbances in controlling species distributions. This question is also meaningful when working on aquatic communities and evaluating the effect of canopy shading on invertebrate communities (Behmer and Hawkins, 2006). Davies-Colley and Payne (1998) highlighted the lack of quantitative assessments of channel shading and reviewed a variety of alternative approaches.

A variety of methods have been developed for measuring canopy cover, and thus inferring the level of shading in the channel. Kelley and Krueger (2005) assessed two traditional methods for measuring canopy cover (a clinometer, used to measure the angle between the horizon and the edge of the canopy, and a densitometer, a convex mirror used to estimate the proportion of the sky that is obscured) against a digital imaging method. The digital imaging method utilised a camera fitted with a $180°$ fish-eye lens pointed vertically to capture a complete hemispheric image of the canopy from below. Specialist image analysis software (HemiView; delta-t.co.uk) segments the image to separate areas of canopy from sky and, with the addition of camera orientation, location, elevation and date, models the proportion of total available solar radiation that reaches the water surface during the course of a year. Kelley and Krueger (2005) concluded that hemispherical image analysis was both more consistent and required smaller sample sizes than the alternative

methods. This method reduces the amount of field work necessary to survey a large area along a river corridor and allows a rapid assessment of the spatial variability of riparian vegetation.

A second example of using close-range imagery to understand riparian forest processes exploits the advantages of oblique photographs and the rapid coverage of long river reaches from a helicopter platform. For management purposes, accurate geomorphic description is often required for long lengths of river channel in order to target specific actions. Traditional aerial photographic surveys and/or lidar surveys cannot provide high quality information about surface topography with a high rate of elevation change, such as natural river banks, riprap emplacements or dikes. Moreover, some detailed features such as boulders, rock outcrops or wood pieces are not easily detected from aerial photos for which the resolution is 20 to 50 cm. In such a context, oblique close-range photos taken from helicopter can be a valuable source of data. Such monitoring campaigns were conducted along the Ain River (40 km long, in 1989 and in 1999), the Drome River (95 km long, in 1995), and the Eygues River (100 km long, in 2000). Photographs of one bank moving from upstream to downstream and of the other one when returning were acquired continuously providing 400 to ca 1500 photographs during a single flight (4–15 photos per km). Visual observation of these images plus linear measurements, as well as object identification and counting can then provide information about longitudinal patterns of geomorphic features (Piégay and Landon, 1997; Landon et al., 1998; Piégay et al., 2000; Lassettre et al., 2007).

On the Drôme, the campaign was taken at a mean daily discharge of 18.4 m^3 s^{-1} (120 days/year), providing information on bars and banks. The survey quantified the distribution of boulders (>256 mm), pools, in-channel and in-bar wood pieces (greater than 20 cm in diameter and 10 m in length), in-channel bedrock surfaces and also provided continuous information on bank characteristics (e.g. floodplain or valley wall, eroded or embanked). This data set, divided in homogeneous segments 500 m long, was used to compute a Habitat Richness Index per section (see details in Piégay et al. 2000). Figure 15.13 (a and b) shows examples of oblique photos on which bank states were observed and mapped. This information was tabulated for 500 m length sections to provide overall statistics and a synoptic view of longitudinal changes. Eroded bank cumulated lengths (right versus left banks) were calculated from km 0 (the Rhône confluence) to km 95 upstream. Figure 15.13c indicates that the reach

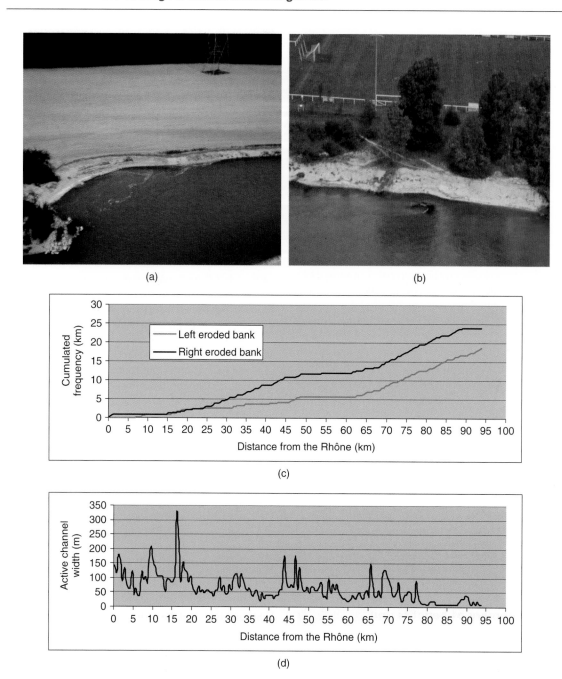

Figure 15.13 Two examples of photographs showing bank erosion on the Drome River, France (a and b). (c) Cumulated frequency of erosion along the banks of the Drôme in April 1995 (surveyed per segment of 500 m – over 188 segments) and (d) its average active channel width per 500 m long segment.

is characterised by more than 43 km of eroded banks over a total of 190 km of banks, the right bank being significantly more eroded than the left, mainly between km 30 and 47. This raises interesting questions about the cause of bank erosion and a number of possible explanations were investigated including the role of neotectonics, engineering infrastructure and sediment delivery from tributaries or coupled hillslopes. Some reaches are very stable on both banks because of embankment and other protection (e.g., km 50–60). This longitudinal trend can be compared with active channel width (Figure 15.13d) showing that unstable banks upstream of km 65 occurred all along the channel course whatever the width which demonstrates the strong disruption of the channel associated with degradation following a period of intense gravel-mining.

The segment procedure is useful for combining information from different sources, each of them being coded or measured for each channel segment. Information provided from the oblique air photos can then be linked with other information provided from long profiles, field measurements (e.g. grain size) but also aerial photos. Lassettre et al. (2007) showed that the frequency of in-channel wood pieces deposited in a river section increased linearly with the local amount of wood introduced. The information was calculated for 250 m long segments, the wood pieces were identified from oblique helicopter photographs whereas the wood delivery was based on an estimate of floodplain eroded surface area provided from an overlay of two sets of aerial photos and an estimate of wood volume from field work (Figure 15.14). This approach suggested that the accumulation of wood is not linked to upstream sources but to local ones, the sites of in-channel wood sources and deposits being usually the same.

15.5 Summary of benefits and limitations

In this chapter we showed how close-range images are complementary to other measurement techniques for studying river corridor changes and processes. We described and illustrated different techniques that use close range photographs to maximise information at high spatial or temporal resolutions and to monitor the occurrence of unpredictable events and to characterise their biophysical effects.

From the examples presented herein, the benefits of using close-range photographs are numerous. This is a low-cost technology that:
- can provide both qualitative and quantitative information;
- generates original spatial information that may be undetectable using standard, vertical aerial photography because of the small, sub-centimetre scales of interest or rapid changes in elevation;
- can be used to obtain time lapse or triggered sequences of photographs or videos over long time periods, thereby increasing the temporal resolution of field observations and avoiding the cost of lengthy fieldwork in remote or dangerous areas;
- can reveal the timing, location and amplitude of significant events complementing other measurements and facilitating expert interpretation of processes and/or forms;
- allows reliable physical measurements when proper ground control points are surveyed;
- and provides valuable data for developing predictive models when combined with other field measures.

Close-range image acquisition also has limitations. Acquisition and analysis of oblique photographs requires specific methodologies and a larger number of ground control points than standard vertical aerial imagery. Moreover, the spatial resolution and the angle of declination with respect to the surface are variable within the image, depending on the distance between the camera and the surveyed object. This implies that process detection and measurements cannot be performed with the same level of precision and resolution across the entire scene.

Measurements and observations are often made challenging by changes in image lighting and contrast (sunny, cloudy, shadows) and this must be considered when setting up the cameras, for example to avoid focusing towards the sun or its reflection on snow or water. When using time lapse pictures, continually changing light conditions and the occurrence of sun reflections are significant impediments to the possibility of automatically extracting planimetric parameters. Furthermore, picture acquisition is often limited to daylight, so that temporal sequences do not include events that happen during the hours of darkness. In this regard, there is a need to make use of technologies for obtaining night-time images, for example infrared sensors.

Most of the limitations listed above are not absolute and new technological developments could help to improve further the versatility of close-range camera systems.

Figure 15.14 Results of the wood survey on the Ain River, France. (a) Relationship between the mean number of wood pieces counted in oblique helicopter photos in 1989 and 1999 and the mean volume of wood recruited during the previous decade (1991–2000 to be compared with 1999 observations, 1980–1991 for 1989 observations) per homogeneous depositional reach (Modified from Lassettre et al., 2007). (b) and (c) examples of some of the photos used for this sensing.

15.6 Forthcoming issues for river management

Research based on large sets of ground photographs is just emerging and advances in this domain are evidently expected in the next 10 years. Technological improvements in sensor quality, data storage, sensor programming and power supply are likely to increase the spatial and temporal resolution of imagery and facilitate longer, unsupervised deployment periods.

New developments are necessary for automating procedures in order to permit the extraction of quantitative information from pictures of dissimilar quality. Developing interfaces between physical sensors and close-range

cameras would also provide more efficient triggering options while increasing the duration of battery life. Sub-minute acquisitions are needed for characterising changes but also understanding their controls with a higher accuracy. New frontiers will also be crossed including underwater acquisition. Waterproof video cameras have recently been used to monitor suspended sediment on the Colorado River by the USGS's Flying Eyeball (see http://walrus.wr.usgs.gov/posters/flyingeyeball.html).

Recently, Fonstad et al. (2011) and Westoby et al. (2011) showed it is possible to extract point clouds from a set of overlapping close range photographs, though a detailed ground control network is necessary to register the cloud. The proposed 'structure from motion' photogrammetric approach uses ordinary photographs (that can be acquired with different distance to the object, orientation, even date) and takes advantage of free and open-source software. This procedure is capable of producing a low-cost 3D mapping that can be used for a wide range of applications in geophysics.

Methodological and technological progresses are evidently expected since ground imagery represents one of the main emerging fields in remote sensing. In terms of river management, these approaches are particularly relevant as they improve our understanding of the morphological processes, helping us to manage the river environment using an understanding of real-time causes not only the observed (post flood) consequences. Moreover, the development of web-platforms could provide easy access to online, remote information, with relevant advantages in public communication (a picture may be worth 1000 words from a scientist) and security (e.g., development of flood warning systems).

Acknowledgements

The authors kindly thank Norbert Landon, Anne Citterio, Marylise Cottet, Christophe Aloy, Jeremie Riquier who participate to the field effort and the data production on the Drôme, the Ouvèze and the Ain Rivers, Jérôme Dubé and Valéry Hamel on the rivers of Québec, and Luca Zanoni, Martino Salvaro and Matilde Welber for their help in the Tagliamento river field measurements and image processing. Stuart Lane and an anonymous referee provided useful comments and suggestions and helped us to clarify and improve the chapter.

References

Adams J. 1979. Gravel size analysis from photographs. *Journal of the Hydraulic Division of the American Society of Civil Engineering*, 105, 1247–1255.

Aloy C. 2005. Origine et conditions de transit de la charge de fond d'un cours d'eau sub-méditerranéen: le cas de l'Ouvèze (Ardèche) Implications en termes de gestion du bassin versant. Unpub. Master Thesis, Master Interface Nature – Société/Travail d'Etude et de Recherche, Univ. Lyon 2.

Ashmore P. and Sauks E. 2006. Prediction of discharge from water surface width in a braided river with implications for at-a-station hydraulic geometry. *Water Resources Research*, 42, W03406 1-11.

Ashmore P., Bertoldi W., and Gardner J.T. 2011. Active width of gravel-bed braided rivers. *Earth Surface Processes and Landforms*, 36, 1510–1521.

Barker R., Dixon L., and Hooke J. 1997. Use of terrestrial photogrammetry for monitoring and measuring bank erosion. *Earth Surface Processes and Landforms*, 22, 1217–1227.

Behmer D.J. and Hawkins C.P. 2006. Effects of overhead canopy on macroinvertebrate production in a Utah stream. *Freshwater Biology*, 16(3), 287–300

Bertoldi W., Gurnell A.M., Surian N., Tockner K., Zanoni L., Ziliani L., and Zolezzi G. 2009. Understanding reference processes: linkages between river flows, sediment dynamics and vegetated landforms along the Tagliamento River, Italy. *River Research and Applications*, 25, 501–516.

Bertoldi W., Zanoni L., and Tubino M. 2010. Assessment of morphological changes induced by flow and flood pulses in a gravel bed braided river: the Tagliamento River (Italy). *Geomorphology*, 114, 348–360.

Bertoldi W., Gurnell A.M., and Drake N. 2011. The topographic signature of vegetation development along a braided river: results of a combined analysis of airborne LiDAR, colour air photographs and ground measurements. *Water Resources Research*, 47, W06525.

Bird S., Hogan D., and Schwab J. 2010. Photogrammetric monitoring of small streams under a riparian forest canopy. *Earth Surface Processes and Landforms*, 35, 952–970.

Buffin-Belanger T. and Bergeron 2011. Advances in river ice sciences. *River Research and Applications*, 27.

Buscombe D. and Masselink G. 2009. Grain size information from the statistical properties of digital images of sediment. *Sedimentology*, 56, 421–438.

Buscombe D., Rubin D.M., and Warrick J.A. 2010. A universal approximation of grain size from images of noncohesive sediment. *Journal of Geophysical Research*, 115, F02015.

Butler J.B., Lane S.N., and Chandler J.H. 2001. Automated extraction of grain size data from gravel surfaces using digital image processing. *Journal of Hydraulic Research*, 39, 1–11.

Butler J.B., Lane S.N., Chandler J.H., and Porfiri E. 2002. Through-water close range digital photogrammetry in flume and field environments. *Photogrammetric Record*, 17, 419–439.

Carbonneau P.E., Lane S.N., and Bergeron N.E. 2003. Cost-effective nonmetric close-range digital photogrammetry and its application to a study of coarse gravel river beds. *International Journal of Remote Sensing*, 24, 2837–2854.

Carbonneau P.E., Lane S.N., and Bergeron N.E. 2004. Catchment-scale mapping of surface grain size in gravel bed rivers using airborne digital imagery. *Water Resources Research*, 40, W07202.

Carbonneau P.E., Bergeron, N.E., and Lane S.N. 2005. Texture-based image segmentation applied to the quantification of superficial sand in salmonid river gravels. *Earth Surface Processes and Landforms*, 30(1), 121–127.

Couper P.R. and Maddock, I.P. 2001. Subaerial river bank erosion processes and their interaction with other bank erosion mechanisms on the River Arrow, Warwickshire, UK. *Earth Surface Processes and Landforms*, 26(6), 631–646.

Chandler J., Ashmore P., Paola C., Gooch M., and Varkaris F. 2002. Monitoring river channel change using terrestrial oblique digital imagery and automated digital photogrammetry. *Annals of the Association of American Geographers*, 92(4), 631–644.

Church M.A., McLean D.G., and Wolcott J.F. 1987. River bed gravels: Sampling and analysis. In: *Sediment Transport in Gravel-Bed Rivers*, edited by C.R. Thorne et al., pp. 43–79, John Wiley, Hoboken, N. J.

Church M., Hassan, M.A., and Wolcott J.F. 1998. Stabilizing self-organized structures in gravel-bed stream channels: Field and experimental observations. *Water Resources Research*, 34(11), 3169–3179.

Cossin M. and Piégay H. 2001. Les photographies prises au sol: une source d'information pour la gestion des paysages riverains des cours d'eau. *Cahiers de géographie du Québec*, 45(124), 37–62.

Davies-Colley R.J. and Payne G.W. 1998. Measuring stream shade. *Journal of North American Benthological Society*, 17, 250–260.

Debusshe M., Lepart J., and Dervieux A. 1999. Mediterranean Landscape Changes: The Old Postcards Testimony. *Global Ecology and Biogeography*, 8(1), 3–15.

Diepenbroek M., Bartholoma A., and Ibbeken H. 1992. How round is round? A new approach to the topic 'roundness' by Fourier grain shape analysis. *Sedimentology*, 39, 411–422.

Diepenbroek M. and de Jong C. 1994. Quantification of textural particle characteristics by image analysis of sediment surfaces – examples from active and paleo-surfaces in steep, coarse-grained mountain environments. In: Ergenzinger, P. and Schmidt, K.-H. *Dynamics and Geomorphology of Mountain Rivers*. Springer: Berlin/Heidelberg. 314 p.

Digan P. and Bren L. 2003. A study of the effect of logging on the understory light environment in riparian buffer strips in a south-east Australian forest. *Forest Ecology and Management*, 172, 161–172.

Drevin G.R. and Vincent L. 2002. Granulometric determination of sedimentary rock particle roundness. In: *International Symposium on Mathematical Morphology*, 315–325.

Dubé 2009. Effet de la morphologie d'une succession seuil-mouille sur la croissance et l'évolution d'un couvert de glace dans un tronçon de la rivière Rimouski, Est-du-Québec. Mémoire de maîtrise en sciences de la terre, Institut national de recherche scientifique, 101 pages.

Feurer D., Bailly J.S., Puech C., Le Coarer Y., and Viau A. 2008. Very high resolution mapping of river immersed topography by remote sensing. *Progress in Physical Geography*, 32(4), 1–17.

Fonstad M.A., Dietrich J.T., Courville B.C., Jensen J., and Carbonneau P. 2011. Topographic Structure from Motion. Oral presentation at AGU Fall Meeting 2011, 5-9 December 2011, San Francisco, US.

Gilvear D.J. and Bryant R. 2003. Analysis of aerial photography and other remotely sensed data. In *Tools in Fluvial Geomorphology*, Kondolf G.M., Piegay H. (eds). Wiley: London; 133–168.

Graham D.J., Reid I., and Rice S.P. 2005a. Automated sizing of coarse-grained sediments: Image-processing procedures. *Mathematical Geology*, 37(1), 1–28.

Graham D.J., Rice S.P., and Reid, I. 2005b. A transferable method for the automated grain sizing of river gravels. *Water Resources Research*, 41, W07020.

Graham D.J., Rollet A.J., Piégay H., and Rice S.P. 2010. Maximizing the accuracy of image-based surface sediment sampling techniques. *Water Resources Research*, 46, W02508.

Hall R.B.W. and Harcombe P.A. 1998. Flooding alters apparent position of floodplain saplings on a light gradient. *Ecology*, 79, 847–855.

Hamel (2011) Les mouvements subaériens sur une berge de la rivière Ouelle, Québec, Canada. Mémoire de maîtrise, Université du Québec à Rimouski, 81.

Hayakawa Y. and Oguchi T. 2005. Evaluation of gravel sphericity and roundness based on surface-area measurement with a laser scanner. *Computers & Geosciences*, 31, 735–745.

Hauet A., Creutin J.D., Belleudy P., Muste M., and Krajewski W.F. 2006. Discharge measurement using large scale PIV under various flow conditions - Recent results, accuracy and perspectives. In: *Conference Proceedings of River Flow*. Ferreira R., Alves E., Leal J., and Cardoso A.H. (Eds), Lisboa, Portugal.

Heritage G.L., Fuller I.C., Charlton M.E., Brewer P.A., and Passmore D.P. 1998. CDW photogrammetry of low relief fluvial features: accuracy and implications for reach-scale sediment budgeting. *Earth Surface Processes and Landforms*, 23, 1219–1233.

Hicks D.M., Duncan M.J., Walsh J.M., Westaway R.M., and Lane S.N. 2002. New views of the morphodynamics of large braided rivers from high-resolution topographic surveys and time-lapse video. In: *Structure, function and management implications of fluvial sedimentary systems*. Dyer F.J., Thoms M.C., and Olley J.M. (Eds). September 02-06 2002, ALICE SPRINGS, AUSTRALIA. Pp. 373–380.

Ibbeken H. and Schleyer R. 1986. Photo-sieving: A method for grainsize analysis of coarse-grained, unconsolidated bedding surfaces. *Earth Surface Processes and Landforms*, 11, 59–77.

Ibbeken H., Warnke D.A., and Diepenbroek M. 1998. Granulometric study of the Hanaupah Fan, Death Valley, California. *Earth Surface Processes and Landforms*, 23, 481–492.

Kelley C.E. and Krueger W.C. 2005. Canopy cover and shade determinations in riparian zones. *Journal of the American Water Resources Association*, 41(1), 37–46.

Krumbein W.C. 1941. Measurement and geological significance of shape and roundness of sedimentary particles. *Journal of Sedimentary Petrology*, 11(2), 64–72.

Kuenen P.H. 1956. Experimental abrasion of pebbles: 2. Rolling by currents. *Journal of Geology*, 64, 336–368.

Landon N., Piégay H., and Bravard J.P. 1998. The Drôme River incision (France): from assessment to management. *Landscape and Urban Planning*, 43, 119–131.

Lane S., Richards K., and Chandler J. 1993. Developments in photogrammetry - The geomorphological potential. *Progress in Physical Geography*, 17(3), 306–328.

Lane S., Chandler J., and Richards, K. 1994. Developments in monitoring and modeling small-scale river bed topography. *Earth Surface Processes and Landforms*, 19(4), 349–368.

Lane S., Richards K., and Chandler, J. 1996. Discharge and sediment supply controls on erosion and deposition in a dynamic alluvial channel. *Geomorphology*, 15(1), 1–15.

Lane SN. 2000. The measurement of river channel morphology using digital photogrammetry. *Photogrammetric Record*, 16(96), 937–957.

Lassettre N., Piégay H., Dufour S., and Rollet A.J. 2007. Temporal changes in wood distribution and frequency in a free meandering river, the Ain River, France. *Earth Surface Processes and Landforms*, 33(7), 1098–1112.

Lawler D.M. 1993. The measurement of river bank erosion and lateral channel change: a review. *Earth Surface Processes and Landforms*, 18, 777–821.

Luchi R., Bertoldi W., Zolezzi G., and Tubino, M. 2007. Monitoring and predicting channel change in a free-evolving, small alpine river: Ridanna Creek (North East Italy). *Earth Surface Processes and Landforms*, 32(14), 2104–2119.

Marcus W.A. and Fonstad M.A. 2008. Optical remote mapping of rivers at sub-meter resolutions and watershed extents. *Earth Surface Processes and Landforms*, 33, 4–24.

Marcus W.A. and Fonstad M.A. 2010. Remote sensing of rivers: the emergence of a subdiscipline in the river sciences. *Earth Surface Processes and Landforms*, 35(15), 1867–1872.

Mertes L.A.K. 2002. Remote sensing of riverine landscapes. *Freshwater Biology*, 47, 799–816.

Michel P., Mathieu R., and Mark A.F. 2010. Spatial analysis of oblique photo-point images for quantifying spatio-temporal changes in plant communities. *Applied Vegetation Science*, 13, 173–182.

Milan D., Heritage G..L, and Hetherington D. 2007. Application of a 3D laser scanner in the assessment of erosion and deposition volumes and channel change in a proglacial river. *Earth Surface Processes and Landforms*, 32, 1657–1674.

Moulin B. and Piégay H. 2004. Characteristics and temporal variability of large woody debris trapped in a reservoir on the River Rhone (Rhone): implications for river basin management. *River Research and Applications*, 20, 79–97.

Piégay H. and Landon N. 1997 Promoting an ecological management of riparian forests on the Drôme River, France., *Aquatic conservation: Marine and Freshwater Ecosystems*, 7, 287–304.

Piégay H., Thévenet A., Kondolf M.G., and Landon N. 2000. Physical and human factors influencing fish habitat distribution along a mountain river continuum, Drôme river, France. *Geographiska Annaler*, 82 A, 121–136.

Pyle C.J., Richards K.S., and Chandler J.H. 1997. Digital Photogrammetric Monitoring of River Bank Erosion. *The Photogrammetric Record*, 15, 753–764.

Rice S.P. 1998. Which tributaries disrupt downstream fining along gravel-bed rivers? *Geomorphology*, 22, 39–56.

Rice S.P. and Church M. 1996. Sampling surficial fluvial gravels: The precision of size distribution percentile estimates. *Journal of Sedimentary Research*, 66, 654–665.

Rice S.P. and Church M. 1998. Grain size along two gravel-bed rivers: statistical variation, spatial pattern and sedimentary links, *Earth Surface Processes and Landforms*, 23, 345–363.

Riquier J. 2009. Méthodologies de reconnaissance pétrographique de galets fluviaux à partir d'images numériques. Unpub. Master Thesis, Master Interface Nature – Société/ Travail d'Etude et de Recherche, Univ. Lyon 3.

Ringold P.L., Van Sickle J., Rasar K., and Schacher J. 2003. Use of hemispheric imagery for estimating stream solar exposure. *JAWRA Journal of the American Water Resources Association*, 39(6), 1373–1384.

Rollet A.J. 2007. Etude et gestion de la dynamique sédimentaire d'un tronçon fluvial à l'aval d'un barrage: le cas de la basse vallée de l'Ain. Unpublished PhD, Université Lyon 3, 305 p.

Roussillon T., Piégay H., Sivignon I., Tougne L. and Lavigne F. 2009. Automatic computation of pebble roundness using digital imagery and discrete geometry. *Computers and Geosciences*, 35(10), 1992–2000.

Rubin D.M. 2004. A simple autocorrelation algorithm for determining grain size from digital images of sediment. *Journal of Sedimentary Research*, 74, 160–165.

Sime L.C. and Ferguson R.I. 2003. Information on grains sizes in gravel-bed rivers by automated image analysis. *Journal of Sedimentary Research*, 73, 630–636.

Smith M., Vericat D., and Gibbins C. 2012. Through-water terrestrial laser scanning of gravel beds at the patch scale. *Earth Surface Processes and Landforms*, 37, 411–421.

Start A.N. and Handasyde T. 2002. Using photographs to document environmental change: the effects of dams on the riparian environment of the lower Ord River. *Australian Journal of Botany*, 50 (4), 465–480.

Strom K.B., Kuhns R.D., and Lucas H.J. 2010. Comparison of Automated Image-Based Grain Sizing to Standard Pebble-Count Methods. *Journal of Hydraulic Engineering*, 136, 8, 461–473.

Surian N. 2002. Downstream variation in grain size along an Alpine river: analysis of controls and processes. *Geomorphology*, 43, 137–149.

Tockner K., Ward J.V., Arscott D.B., Edwards P.J., Kollmann J., Gurnell A.M., Petts G.E., and Maiolini B. 2003. The Tagliamento river: a model ecosystem of European importance. *Aquatic Sciences*, 65, 239–253.

Van der Nat D., Schmidt A., Tockner K., Edwards P., and Ward J. 2002. Inundation dynamics in braided floodplains: Tagliamento River, Northeast Italy. *Ecosystems*, 5, 636–647.

Wadell H. 1932. Volume, shape, and roundness of rock particles. *Journal of Geology*, 40, 443–451.

Warrick J.A., Rubin D.M., Ruggiero P., Harney J., Draut A.E., and Buscombe D. 2009. Cobble Cam: Grain-size measurements of sand to boulder from digital photographs and autocorrelation analyses. *Earth Surface Processes and Landforms*, 34, 1811–1821.

Welber M., Bertoldi W., and Tubino M. 2012. The response of braided planform configuration to flow variations, bed reworking and vegetation: the case of the Tagliamento River, Italy. *Earth Surface Processes and Landforms*, in press.

Wentworth C.K. 1919. A laboratory and field study of cobble abrasion, *Am. J. Geol.*, 27, 507–521.

Westaway R.M., Lane S.N., and Hicks D.M. 2003. Remote survey of large-scale braided, gravel-bed rivers using digital photogrammetry and image analysis. *International Journal of Remote Sensing*, 24 (4), 795–815.

Westoby M.J., Glasser N.F., Brasington J., Hambrey M., and Reynolds J.M. 2011. 'Structure-from-Motion': a high resolution, low-cost photogrammetric tool for geoscience applications. Oral presentation at AGU Fall Meeting 2011, 5-9 December 2011, San Francisco, US.

Wolcott J. and Church M. 1991. Strategies for sampling spatially heterogeneous phenomena: the example of river gravels', *Journal of Sedimentary Petrology*, 61(4): 534–543.

Wolman M.G. 1954. A method of sampling coarse river-bed material. *Eos Trans. AGU*, 35, 951–956.

16 River Monitoring with Ground-based Videography

Bruce J. MacVicar[1], Alexandre Hauet[2], Normand Bergeron[3], Laure Tougne[4] and Imtiaz Ali[4,5]

[1]Department of Civil and Environmental Engineering, University of Waterloo, Ontario, Canada
[2]EDF - DTG - CHPMC, Toulouse, France
[3]Institut National de Recherche Scientifique, Centre Eau Terre et Environnement, Québec, Canada
[4]Universitéy de Lyon, Bron cedex, France
[5]Optics Labs, Islamabad, Pakistan

16.1 Introduction

Videography refers to the capture of a series of images. It is distinguished from the capture of still images by the fact that the images are taken at a sufficient frequency to allow the movement of objects within the image frame to be distinguished. The key challenge for scientists and river managers is to develop computer algorithms that emulate the mental processes by which we identify objects and visualise movement so that displacement vectors can be quantified and other useful information about the objects in the frame extracted. Particle Image Velocimetry (PIV) and other image analysis techniques have been used for almost three decades in laboratory experiments to measure instantaneous velocity vectors in laboratory flows (Adrian, 1991, Raffel et al., 1998). Videography is increasingly implemented to study processes in natural environments (Jodeau et al., 2008, Muste et al., 2008, MacVicar et al., 2009). However, there are a number of difficulties related to the weather, image angle, illumination, and contrast that must be overcome to take

advantage of the inherent advantages of videography for the monitoring of natural processes in the field.

In rivers, the use of videography for measuring surface water velocities has been intensively investigated. In mid 1990s, Japanese teams from the University of Kobe (Fujita and Komura, 1988, 1994, Aya et al., 1995, Fujita and Aya, 2000, Fujita et al., 2007a, Fujita et al., 2007b) first used image velocimetry for velocity measurements in a riverine environment. Major developments have been realised by teams at the University of Iowa, USA (Bradley et al., 2002, Muste et al., 2004, Muste et al., 2005, Muste et al., 2008) the national engineering research center (Institut National Polytechnique) in Grenoble, France (Creutin et al., 2003, Fourquet, 2005, Hauet, 2006, Hauet et al., 2006, Hauet et al., 2007, Hauet et al., 2008a), Cemagref, France (Jodeau et al., 2008), and other groups (e.g. Muller et al., 2002, Harpold and Mostaghimi, 2004). As most of the measurements were taken over surfaces much larger than those in traditional PIV, the technique was dubbed Large-Scale PIV (LSPIV) (Fujita and Hino, 2003). Applications of this approach include

Fluvial Remote Sensing for Science and Management, First Edition. Edited by Patrice E. Carbonneau and Hervé Piégay.
© 2012 John Wiley & Sons, Ltd. Published 2012 by John Wiley & Sons, Ltd.

the characterisation of velocity distributions, transport processes related to mixing, and discharge measurements.

Videography has also been successfully implemented for the quantification of other processes in a variety of natural environments. Examples include the use of satellite imagery to track atmospheric cloud movements (Leese et al., 1971) or sea ice (Ninnis et al., 1986), and the use of land-based cameras to quantify near shore wave patterns (Holland et al., 1997) or the movement of subaqueous nearshore ripples (Becker et al., 2007). Within rivers, efforts have also been made to track the velocities and concentrations of ice floes (Ferrick et al., 1992, Ettema et al., 1997, Jasek et al., 2001, Bourgault, 2008) and the volume of wood in transport and accumulations (Lyn et al., 2003, Moulin and Piégay, 2004, MacVicar et al., 2009).

Numerous advantages are driving the development of videography for river monitoring. First, the images are obtained remotely. Measurements are possible during floods when it is too laborious or dangerous to attempt them by other means, particularly when large floating objects such as ice and wood are a concern (Fujita and Hino, 2003, Hauet et al., 2008a). Velocities are also measured simultaneously over a large area, which is not possible with other instruments. Second, the recording of images is continuous, which means that relatively rare events will be recorded. This advantage is especially important in small basins where the time of concentration is short. A third advantage is that videographic systems can be connected to local and global networks. Digitally recorded videos can be transmitted from remote field sites to a central location for storage and analysis, significantly reducing the cost of monitoring. Finally, the proliferation of video monitoring for public security has driven down both the cost of video monitoring systems and the technical expertise required for their installation and operation.

This chapter presents three case studies in which videography was implemented to quantify stream discharge, water velocity vectors, and the flux of floating wood. These techniques rely on the detection of material floating on the surface of the water, called the 'seeding', to determine movement vectors. The estimation of the flux of floating wood further requires that the types of seeding material be distinguished using an object's secondary properties such as color, linearity, and direction of movement. This chapter is organized in four sections. General considerations for a videographic system are presented in Section 16.2 followed by a case study on stream gauging using a mobile LSPIV system in Section 16.3, a case study in which algorithms are developed to filter poor quality

LSPIV measurements in Section 16.4, and a case study in which videography was used to measure wood flux case study in Section 16.5. Further research questions and applications are also discussed.

16.2 General considerations

16.2.1 Flow visualisation and illumination

Image-based velocimetry requires high quality measurements of object displacements. Laboratory investigations of PIV have shown that it is necessary to consider the concentration of particles, their size with respect with the image processing parameters, and the anticipated particle displacement to ensure accurate measurements (Adrian, 1991). Unfortunately, ideal conditions are rarely obtained for field applications of videography. Illumination is not constant due to solar effects, weather conditions, and the time of day. As the angle of the sun changes, light reflections, lens flare, shadows and glare can reduce the quality of the recorded image (Kim, 2006, Hauet et al., 2008b). Fog, rain and snow can reduce the visibility of the surface. Luminosity is obviously greatly reduced at night. In general, adverse lighting effects can be minimised with a propitious selection of camera position. In the case studies presented in this chapter, mobile cameras were placed on masts or tripods and fixed cameras were installed at a high point on a bridge or bank to maximise the view of the river and angle of the camera to the horizontal. Lens shades were used and cameras were oriented away from the sun, preferably towards the north or where trees were likely to prevent direct sunlight from hitting the lens. For nighttime illumination infrared assist options were tested, including the installation of a high-powered infrared light at one site. As part of this chapter, algorithms are introduced that were used to reduce the errors from illumination problems related to surface reflections, shadows and other effects.

16.2.2 Recording

The capability to distinguish movement between two image pairs is commensurate with the size of the images, their resolution, seeding, and the time interval between images. The studies presented in this chapter used commercially available video cameras with image sizes less than 720×576 pixels and measurement frequencies between 5 and 30 Hz. The size of pixels in real coordinates was highly variable because river widths varied from 10 to 70 m and camera heights and angles were adjusted to capture different fields of view. The average pixel width for the presented case studies varied from 2 to 15 cm. Such image resolution and frequency parameters are inferior to the sophisticated high definition video systems that are typically employed in conventional PIV. In the field applications, tradeoffs were necessary between image quality, speed of data transmission, required storage, and cost. An additional variable to consider is surface tracers. Accurate measurements of movement vectors depend on the presence, visibility, and convection of tracers such as foam, wood, or water waves (Creutin et al., 2002, Fujita and Hino, 2003) (Figure 16.1a, b). These tracers need to be as big or larger than the pixel resolution. When natural tracers are insufficient, artificial tracers such as soap, which produces a floating foam, can be added. As a general note, it is necessary to confirm the accuracy of the installed videographic system for a given application.

16.2.3 Image ortho-rectification

Ground-based cameras have an oblique view angle (Figure 16.1a), which means that pixel size is variable and distortion can be an important effect (Hauet et al., 2008a). As a general rule, images are ortho-rectified prior to analysis. Ortho-rectification refers to the process by which image distortion is removed and the image scale is adjusted to match the actual scale of the water surface (Figure 16.1b – 1c). Ortho-rectification is accomplished by applying an appropriate image photogrammetric transformation (Mikhail and Ackermann, 1976) using known coordinates of ground control points (GCPs) in the real (X, Y, and Z) and the image (x and y) coordinate systems. The mapping relationships between the two systems can be expressed as (Fujita et al., 1998):

$$X = \frac{L_1 x + L_2 y + L_3 z + L_4}{L_9 x + L_{10} y + L_{11} z + 1} \tag{16.1a}$$

$$Y = \frac{L_5 x + L_6 y + L_7 z + L_8}{L_9 x + L_{10} y + L_{11} z + 1} \tag{16.1b}$$

where the eleven mapping coefficients $L_1 - L_{11}$ can be determined by the least square method using the known GCPs coordinates. A minimum of 6 GCPs are needed for conducting the transformation. While more GCPs are useful for checking errors in survey data, the uncertainty caused by pixel size is usually greater than that caused by survey error, and additional points are not necessary for the analysis (Bradley et al., 2002). Fewer points can be used if the image coordinates are assumed to be on the same vertical plane. The size of the non-distorted image should be nearly the same as that of the original image. In addition, a reconstruction of the pixel intensity distribution is made to obtain the ortho-rectified image. Following Muste et al. (1999), intensity at a pixel in the transformed image is obtained using a cubic convolution interpolation of the intensity in 16 neighbouring pixels of the original image.

16.3 Case 1 – Stream gauging

16.3.1 Introduction

The first case study will apply LSPIV to estimate discharges in a system prone to flash-flooding where gauging has been historically difficult. Flash-floods occur in the southern Mediterranean region of France and can result

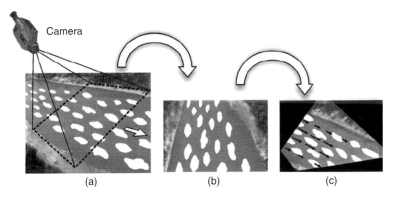

(a)　　　(b)　　　(c)

Figure 16.1 LSPIV measurement sequence: a) imaging the area to be measured (white patterns indicate the natural or added tracers used for visualization of the free surface); b) the distorted raw image; c) the undistorted image with estimated velocity vectors overlaid on the image.

in significant casualties and economic impacts. The flood on September 8-9, 2002 in the Gard River, for example, killed 24 people and caused damages estimated at 1.2 billion € (Delrieu et al., 2005). The understanding of flood generation and propagation processes requires reliable discharge estimates throughout the river network, in real time. The standard procedure is to use current meter and acoustic velocimeters to obtain point measurements of velocity across the cross-section and calculate discharge using the velocity-area method. Assuming that the channel bathymetry is known, point estimates of discharge are calculated by multiplying the point velocities by the area of the channel cross-section for which they are representative. The sum of the point estimates of discharge provides an estimate of the total discharge for the channel. The water level or stage is recorded and discharge is estimated at various stages to construct a stage-discharge rating curve.

A number of problems make this standard procedure difficult or impossible to apply to large or flashy floods: i) high flow velocities and floating debris endanger the operators and the equipment, ii) the river level and discharge can vary during the measurement period, compromising the quality of the discharge estimate, iii) the standard procedure is time consuming, labour intensive, and contributes to the high cost of river monitoring, iv) the accuracy of results at high discharges is reduced due to poor quality or missing data and may necessitate the extrapolation of rating curves beyond their verified ranges, and v) a developed rating curve may change in time due to geomorphologic changes or to modification of the hydraulic control downstream. More continuous and real-time monitoring of river discharge is desirable.

16.3.2 Field site and apparatus

This section reports on tests that were conducted with a mobile Large-Scale Particle Image Velocimetry (LSPIV) system. The mobile LSPIV system has been developed for preliminary tests and flood-triggered measurements at gauging stations that are not equipped with a fixed LSPIV system (Kim et al., 2008). The objectives of these experiments were to improve the monitoring of Mediterranean flash-floods throughout the Ardèche River catchment, to assess the accuracy and reliability of LSPIV discharge estimation, and to improve the extrapolation of rating curves to high ungauged discharge values.

The field tests were conducted at the outlet of the Ardèche River at the Sauze-St Martin Compagnie

Nationale du Rhone (CNR) gauging station. The reach at Sauze-St Martin has a catchment area of 2240 km^2 and a mean annual discharge of 63 m^3/s. The instantaneous discharges (Q_T) for return period T years are: $Q_2 = 1830$ m^3/s, $Q_5 = 2770$ m^3/s, and $Q_{10} = 3390$ m^3/s. The stage-discharge rating curve used by CNR at Sauze-St Martin is well documented, with 39 direct measurements of discharge between December 2003 and February 2009 at flow rates up to 2700 m^3/s with a stable stream morphology. Natural tracers such as bubbles, surface ripples and vegetal debris were sufficient for the LSPIV measurements. Tests were conducted on November 22–23 2007 at discharges that reached a maximum of 760 m^3/s. To verify the method, discharge was calculated from simultaneous Acoustic Doppler Current Profiler (ADCP) measurements for two discharges at approximately 330 m^3/s. ADCP measurements were acquired with a Teledyne RDI RioGrande 600 kHz instrument.

In this study, a video camera was set on a mobile lightweight telescopic mast for which the height could be varied from 2 to 10 m. A relatively cheap commercially-available digital video camera (Canon MV750i) with sufficient resolution and frame rates was selected for this application. Video was recorded in mini-DV format at a rate of 25 frames per second with an image resolution of 720x576 pixels. Estimates of mean velocity were made using the LSPIV technique from a 2 minute series of images with a time interval (Δt) of 0.2 s for a total of 600 images. Mapping coefficients were calculated using image ortho-rectification (Equation 16.1), with 10 Ground Control Points (GCP) measured along both sides of the river using a total station and white square targets that were visible in the image.

16.3.3 Image processing

The LSPIV algorithms for estimating velocities are the same as those used in conventional PIV (Adrian, 1991, Fujita et al., 1998, Bradley et al., 2002). They are based on the same concept as human vision in which pattern-recognition is used to follow the displacements of water surface tracers in consecutive images. An index of similarity is calculated between a small Interrogation Area (IA) in the first image and multiple IAs of the same size within a larger Search Area (SA) in the second image. A vector from the IA in the first image to the IA in the second image with the highest similarity index is assumed to be the displacement vector of the water surface over the time interval (Δt) between the capture of the two images. Velocity is calculated by dividing the displacement distance by Δt.

This searching process is applied successively to a grid of IAs on the first image to calculate an instantaneous vector field. The vector field can then be used to estimate spatial and temporal features of the flow, such as the mean velocity and streamlines, as well as other velocity-derived quantities such as vorticity, strain rates, fluxes, and dispersion coefficients due to shear for a single image pair. Displacement vectors from multiple image-pairs can be averaged to obtain mean velocity vectors.

The cross-correlation coefficient (R_{ab}) was used as the similarity index (Fujita and Komura, 1992, Fujita et al., 1998), defined as:

$$R_{ab} = \frac{\sum\limits_{i=1}^{MX}\sum\limits_{j=1}^{MY}\left\{\left(a_{ij} - \bar{a}_{ij}\right)\left(b_{ij} - \bar{b}_{ij}\right)\right\}}{\left\{\sum\limits_{i=1}^{MX}\sum\limits_{j=1}^{MY}\left(a_{ij} - \bar{a}_{ij}\right)^2 \sum\limits_{i=1}^{MX}\sum\limits_{j=1}^{MY}\left(b_{ij} - \bar{b}_{ij}\right)^2\right\}^{1/2}} \quad (16.2)$$

where MX and MY are the sizes of the interrogation areas, and a_{ij} and b_{ij} are the distributions of the grey-level intensities in the two interrogation areas separated by the time interval Δt. The overbar indicates the mean value of the intensity for the interrogation area. To improve measurement accuracy, a sub-pixel fitting method such as Gaussian or parabolic fitting is applied to the cross-correlation distribution (Fujita et al., 1998). The pair of interrogation areas showing the maximum cross-correlation coefficient is selected as the most likely vector.

The selected image processing algorithm is similar to the Correlation Imaging Velocimetry algorithm (Fincham and Spedding, 1997). Both algorithms use a variance normalised correlation, in which each pixel in the IA is equally weighted, such that the background is just as important as the particle images. Consequently, the algorithm can estimate velocities from low resolution images, such as those captured by standard mini-DV video cameras. Another important feature of the cross-correlation algorithm is that the interrogation area is decoupled from its fixed location in the first image to any arbitrary location in the second image. This process eliminates the velocity bias error (Adrian, 1991). It also greatly improves the signal-to-noise ratio in the presence of large displacements, significantly extending the dynamic range of the velocity measurement. More importantly, the algorithm allows the use of relatively small sampling areas, which significantly increases the available spatial resolution and reduces the errors due to spatial averaging when measuring high-vorticity flows.

16.3.4 Stream gauging

To compute discharge across a cross-section from the surface velocities measured by LSPIV, information on the bathymetry of the channel in the measurement section and the relation between depth-averaged and surface velocities is required. The channel bathymetry was measured prior to the LSPIV measurements using a total station. It was assumed that the bathymetry was stable. Surface velocities at several points along the surveyed cross section (V_i) were computed by linear interpolation from neighbouring grid points of the LSPIV-estimated surface velocity vector field. Missing velocities along the wetted width of the cross-sectional profile were interpolated or extrapolated by assuming a constant Froude number across the width of the channel (Fulford and Sauer, 1986). The depth-averaged velocity at each vertical was related to the free-surface velocity by a coefficient (α) that depends on the shape of the velocity profile. The coefficient is thus a function of the aspect ratio of the channel, Froude and Reynolds numbers, micro and macro bed roughness, and the relative submergence of large-scale roughness elements. Polatel (2005) showed that α varied between 0.789 and 0.928 in a series of laboratory experiments with varying velocities over smooth beds and beds roughened with dunes and ribs. Values of α were higher for smooth beds and larger flow depths. A value of $\alpha = 0.85$ is generally accepted by the hydraulic community for flows in rivers (Costa et al., 2000). The total discharge across the section was computed following the standard velocity-area method (Figure 16.2).

16.3.5 Results

A mean value of $\alpha = 0.90$ was calculated from the vertical velocity profiles measured by the ADCP at a discharge

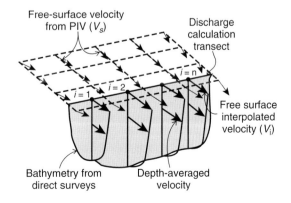

Figure 16.2 LSPIV-based discharge measurement procedure.

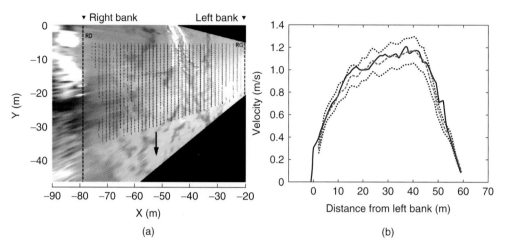

Figure 16.3 a) Ortho-rectified image from Sauze St. Martin showing instantaneous velocity vectors, b) comparison between corrected LSPIV velocities (solid blue line, $\alpha = 0.90$) and ADCP depth-averaged velocities (dashed lines, with $\pm10\%$ interval).

of approximately $300\,\mathrm{m^3/s}$. This value was consistent with a larger sample size of 29 current-meter gaugings conducted at Sauze-Saint Martin. This value is thought to be reasonable given that is close to the 'standard' value of $\alpha = 0.85$ (Costa et al., 2000) and, considering the $>50\,\mathrm{m}$ width of the channel and its relatively smooth bed profile, towards the upper end but within the range identified by Polatel (2005).

Mean velocities estimated with the LSPIV technique show good agreement with ADCP measurement (Figure 16.3). When the LSPIV surface velocities were corrected by the velocity profile coefficient ($\alpha = 0.90$), the overall difference between the two estimates was less that 10% across the entire cross-section. Larger deviations are visible at distances 0 and 50 m from left side and may be the result of errors in spatial sampling. A comparison of the discharges estimated from the ADCP and LSPIV with the discharge estimate from the established rating curve show that both methods are close to the rating curve (less than 5% difference) and are consistent between themselves (maximum difference of 2.6%) (Table 16.1).

On November 23, LSPIV measurements were made during the peak flow. Due to the very high surface velocities (up to 3 m/s), it was not possible to safely measure velocities using the ADCP at the same time. From the LSPIV measurements ($\alpha = 0.9$), the estimated discharge was $825\,\mathrm{m^3/s}$ – an overestimation of 8% when compared to the established rating curve ($760\,\mathrm{m^3/s}$).

The sources of errors in the LSPIV stream gauging technique include: image sequences with poor quality

Table 16.1 Comparison between ADCP/LPIV discharge measurements and discharges estimated using the rating curve.

Gauging Method	Gauging 1 ($\mathrm{m^3/s}$)	Difference (%)	Gauging 2 ($\mathrm{m^3/s}$)	Difference (%)
Rating curve	320		343	
ADCP	327	+2.1	345	+0.6
LSPIV	331	+3.4	336	−2.1

due to external conditions or instrumental limitations; an insufficient number of averaged image pairs taken during a limited time span to avoid significant discharge variations; possible variability in the velocity index (α) as flow conditions change; possible variability of the cross-section with stage; and an unquantified effect of using the Froude-constant technique to interpolate missing or truncated velocities. Improvements should aim at quantifying and reducing these errors, and defining criteria for discarding poor quality image pairs.

16.4 Case 2 – Filtering bed and flare effects from LSPIV measurements

16.4.1 Introduction

The utilisation of LSPIV in the field is complicated by uncontrolled environmental conditions such as wind and water surface reflections. This case study was concerned with two sources of error: the 'bed effect', related to the visibility of the channel bed and the 'flare effect', caused by light reflecting or glinting off the water surface. The bed effect occurs when LSPIV similarity index is biased by immobile bed particles. This situation arises when the water is clear and shallow and when the concentration of tracer particles at the water surface is low. Both of these conditions frequently occur in the field during moderate or low flows. The result is an error characterised by a zero or near-zero flow velocity. The flare effect occurs when the PIV algorithm momentarily tracks water surface reflections and interprets them as flow velocity vectors having a magnitude and orientation that corresponds with the characteristics of the flare. The result of this flare effect is an error characterised as random noise. In this section a filtering method is introduced that allows the removal of erroneous instantaneous velocity vectors from the record. The effectiveness of the filtering technique is illustrated with a field study conducted in two contrasting river environments in the United Kingdom and Canada.

16.4.2 Field site and apparatus

The first field site was on the River Wharfe, a small meandering gravel bed river located near Leeds, UK. The active channel width of the studied section was approximately 10 meters and the maximum water depth was 1 m. The weather on the sampling day was cloudy and the water was dark and turbid with abundant naturally occurring tracing elements floating at the water surface (foam and leaves)(Bérubé et al., 2004). No additional tracer material was added to the water surface. The second field site was on the St-Charles River, a larger gravel bed river located near Québec City, Canada. The active channel width of the studied section was approximately 30 m and the maximum flow depth was 0.6 m. The weather was sunny and the water was clear. Due to an absence of natural tracer particles, biodegradable shampoo foam was used to increase seeding.

Oblique digital video records of the water surface were obtained using a 3-CCD mini-DV digital video camera (Sony VX-1000, 480×720 pixels) mounted on a 1.5 m-high tripod positioned on the river bank. The height of the riverbank on the Wharfe and St-Charles rivers was respectively 1.5 and 3 m. The video records were obtained for a period of 1 min at 30 Hz on the River Wharfe and 2 min at 15 Hz on the St-Charles River. Videos were saved in mpeg4 compression format. At each site, a minimum of five ground control points were measured at the water

surface within the camera field of view using a DGPS (Leica RTK DGPS) on the River Wharfe and a total station (Leica TC 307) on the St-Charles River.

16.4.3 Data filtering

Georeferenced mean flow velocities were measured approximately one centimetre below the water surface over a period of 60 seconds using a Valeport current meter (model 002) on the River Wharfe (n = 14) and a Marsh-McBirney current meter (model Flow Mate) on the St-Charles River (n = 19). These measurements were used as reference values for flow velocities at the water surface.

Instantaneous measurements of surface flow velocity were obtained using LSPIV within interrogation areas of 40×40 pixels centered on each location in the image where velocities were measured with a current meter. As in the previous example, the cross-correlation coefficient was used as the similarity index. The resulting instantaneous velocity vectors in pixels/s were orthorectified using the control points of the water surface in order to yield estimates of instantaneous velocities in cm/s (Fujita et al., 1998).

Instantaneous velocities were filtered based on the construction and analysis of the spatial probability density function (PDF) (Figure 16.4a). For each interrogation area, the spatial PDF was constructed by discretising the velocity vector data space into 1×1 cm/s bins and computing the density of data points in each tile.

$$PDF(v_x, v_y) = \frac{N_{xy}}{N_t} \qquad (16.3)$$

where v_x is the streamwise velocity, v_y is the lateral velocity, N_{xy} is a count of the instantaneous velocity estimates in the $v_x v_y$ bin and N_t is the total number of instantaneous velocity estimates. Velocity estimates were filtered by removing all values where the PDF was less than a given percentile. In these tests a threshold value of the 90th percentile was found to be sufficient to preserve instantaneous velocity vectors due to natural turbulent fluctuations of the flow while removing erroneous velocity vectors created by bed and/or water surface flare effects (Figure 16.4b). The result of this filtering procedure is to isolate the primary peak of the spatial PDF from which the mean velocity vector is then calculated.

16.4.4 Results

Significant correlations were obtained at both sites between current meter and unfiltered PIV measurements of surface flow velocity (River Wharfe, $R^2 = 0.76$, p < 0.001; St-Charles River, $R^2 = 0.80$, p < 0.001). However, the slopes of the ordinary least square regressions were less than one (River Wharfe, slope = 0.43; St-Charles River, slope = 0.58), indicating a consistent underestimation in the LSPIV measurements (Figure 16.5a). Using the current meter measurements as the true velocities, the overall accuracy and precision of the unfiltered PIV measurements were respectively −4.29 cm/s and ±8.44 cm/s on the River Wharfe and −30.92 cm/s and ±9.96 cm/s on the St-Charles River. On the St-Charles River, the large underestimation of surface flow velocities by the PIV method is thought to be related to the bed effect due to the clear and shallow water. This bed effect was less important on the River

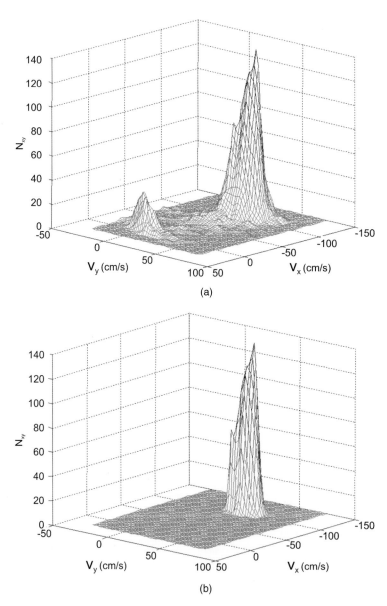

Figure 16.4 The spatial probability density function (PDF) of instantaneous velocity vectors in the x (streamwise) and y (lateral) directions: a) prior to filtering; and b) after filtering.

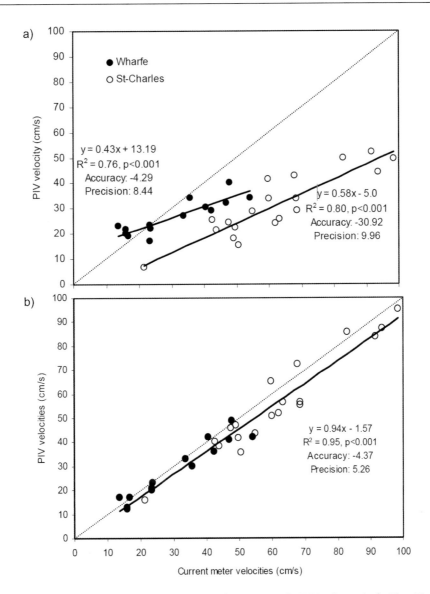

Figure 16.5 Comparison between current meter measurements and time-averaged LSPIV estimates in the River Warfe and the St. Charles River: a) unfiltered LSPIV data; and b) filtered PIV data.

Wharfe due to a high-suspended load. Flare effects were observed at both sites but were more pronounced on the St-Charles River due to the clear sky conditions that favoured the reflection of sunlight on the water surface.

Post filtering, the PIV measurements of both sites showed a strong and significant relationship with current meter measurements, with slope values approaching one (River Wharfe, $R^2 = 0.92$, p < 0.001, slope = 0.86;

St-Charles River, $R^2 = 0.92$, p < 0.001, slope = 1.02). When combining the two data sets, a strong significant relationship ($R^2 = 0.95$, p < 0.001, slope = 0.94) was obtained between current meter and PIV measured velocities, with PIV measures showing an accuracy of −4.37 cm/s and precision of ±5.26 cm/s (Figure 16.5b). These results are comparable to those obtained by Bradley et al. (2002) who found differences between PIV and current meter velocity measurements of less than 6 cm/s.

16.5 Case 3 – At-a-point survey of wood transport

16.5.1 Introduction

Wood is an integral component of river systems that has a strong influence on stream ecology, sediment transport, and geomorphology (Bisson et al., 1987, Montgomery et al., 1996, Gurnell and Petts, 2006). It also represents a risk for flooding and human infrastructure (Bradley et al., 2005, Comiti et al., 2006). There is a need for data to calibrate wood budgets and examine processes related to the transport of wood during floods (Benda and Sias, 2003, Hassan et al., 2005). Existing methods that use tagging or repeat surveys of stored wood volumes are ineffective and labour intensive in large rivers and do not monitor the transport or accumulation of wood during floods when access to the river is most hazardous (Lyn et al., 2003, Moulin and Piégay, 2004, MacVicar et al., 2009). This section describes field tests of the use of videography to detect and measure floating wood transported on the surface of a river.

16.5.2 Field site and apparatus

The camera was installed with a view of the Ain River, a piedmont river with a drainage area of $3500 \, \text{km}^2$ near the town of Chazey sur Ain (France). Bank erosion and wood mobilisation has been extensively studied in this river and a high frequency of wood transport events were expected (Piégay, 1993, Piégay et al., 1997, Lassettre et al., 2008). The particular location was chosen because a gauging station is already present and equipped with internet and telephone access. In addition, the site is located on a high embankment on the outside of a bend where the majority of wood was expected to pass with an unobstructed view of the water surface. The river is approximately 70 m wide at this location.

An Axis 221 Day/Night™ camera was installed at the gauging station on the Ain River in early 2007. A Profiline™ infrared light projector was also installed to increase luminosity at night. Videos are transferred via the web to remote servers and saved in 15 minute segments in mjpeg format. The image resolution is 640x480 pixels and the image frequency is 5 Hz. Video was recorded during 12 floods between May 2007 and February 2009. Of the recorded floods, five events were at or greater than the bankfull discharge ($Q_{bf} = 530 \, \text{m}^3/\text{s}$), the largest of which occurred in April 2008 when the discharge was twice the bankfull rate ($Q = 1080 \, \text{m}^3/\text{s}$). The sample

videos for the development of an automatic wood detection algorithm were taken from different positions in the hydrograph of a flood on November 22 – 24, 2007 that reached a maximum flow rate just over the bankfull ($Q = 550 \, \text{m}^3/\text{s}$).

16.5.3 Manual detection and measurement

To test and develop an automatic wood detection algorithm, it was necessary to manually scan videos and visually identify floating wood. A semi-manual algorithm written in Matlab™ was developed to assist with this procedure. Using the semi-manual algorithm, video playback was stopped by the user when wood was visually detected. The end and side points of wood pieces were recorded using the screen cursor and the pixel locations transformed into real coordinates using Equation (16.1). The length, diameter, and mean position were computed from the real coordinates. Wood volume was calculated by assuming a cylindrical shape for the wood pieces. The presence/absence of roots and branches were also noted. The video was then advanced several frames and the endpoints of the wood were relocated using the screen cursor to allow the calculation of velocity and rotation (Figure 16.6). The number of manually detected wood objects was assumed to represent the actual number of wood objects in the river. This ignores any submerged wood, but the objective at this stage was to test an automatic detection algorithm and the manual detection procedure was felt to be the best data to assess missed or false detections.

Preliminary results from the semi-manual algorithm demonstrate the non-linear response of wood transport to flow rate (Figure 16.7). A transport threshold exists at approximately $Q = 200 \, \text{m}^3/\text{s}$, below which negligible wood transport rates occur on the rising limb of the flood. The peak in both the frequency and volume of wood transport occurs before the peak of the flood ($Q \sim 425 \, \text{m}^3/\text{s}$). Transport rates on the falling limb are much lower than those on the rising limb and transport appears to be negligible below a flow rate of approximately $420 \, \text{m}^3/\text{s}$. This result matches with an expectation that wood debris is picked up from the bars, banks and overbank areas as flood stage rises. These results indicate that this effect of hysteresis is very strong and the volume of transported material on the falling limb of the hydrograph is almost negligible in most videos. A larger sample size is required to accurately characterise the variability of this complex process. However, the time required to analyse videos using the semi-manual program is prohibitive and a fully

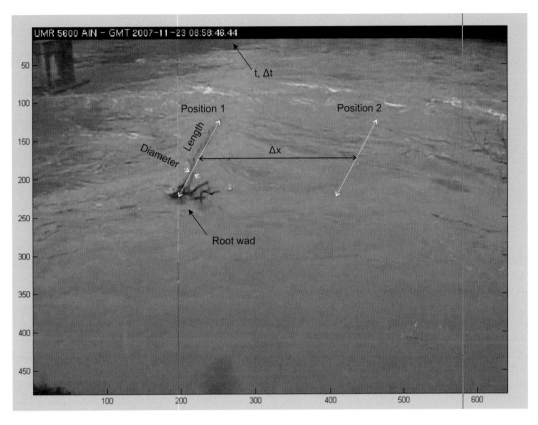

Figure 16.6 A greyscale video frame from the camera on the Ain River. A piece of floating wood is shown at Position 1. The length, diameter, and position 2 of the wood piece are shown to demonstrate the calculation of wood volume and velocity from the semi-manual image analysis procedure.

automatic computer algorithm was developed to process the videos and calculate wood transport frequency. The automatic detection algorithm and results are described in the following sections.

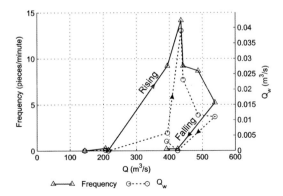

Figure 16.7 Wood frequency and volume as a function of discharge for a flood on the Ain River.

16.5.4 Image segmentation and analysis

Seven video segments (total duration of 36 minutes) were used to develop an algorithm to detect and count wood objects on the surface of the river. This algorithm was developed by breaking the larger problem into three tasks: 1) detection and recognition of objects on water surface (image segmentation); 2) agglomeration of objects in close proximity into a single object; and 3) distinction between wood and other types of objects such as water waves (Ali and Tougne, 2009) (Figure 16.8).

16.5.4.1 Detection and recognition of objects

Histogram thresholding is among the most popular techniques for identifying objects (segmentation) in grey-level images (Fu and Mui, 1981, Pal and Pal, 1993). Using this technique, histograms of the grey-level image intensity are calculated and regions with similar values are identified as objects or regions within the image. The Fisher

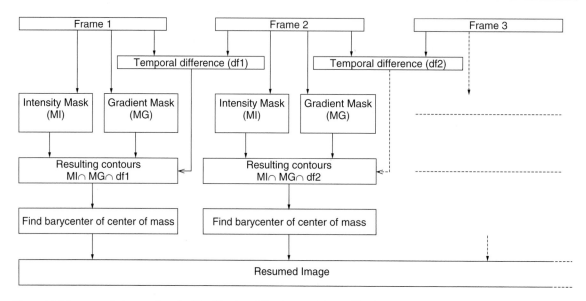

Figure 16.8 Image analysis procedure for identification of floating wood from video imagery.

linear discriminate technique, a standard method used in statistics for data clustering and pattern recognition, was applied to distinguish intensity clusters and obtain a intensity mask (MI) of objects or regions in each image frame (Haralick and Shapiro, 1985, Jain et al., 1995). This technique produced very good segmentation of images and identification of regions in the absence of direct sunshine (Example 1, Figure 16.9b) but was compromised by shadows on the water surface (Example 2, Figure 16.9b). In addition, water waves frequently had similar intensity values to wood, resulting in classification errors.

Due to possible errors associated with histogram thresholding, it was necessary to integrate spatial features of the image with spectral features. For this reason, images were also analysed using an edge-detection algorithm in which the local gradients in image-intensity values were used to define boundaries between regions within the image (Chapron, 1997, Zugaj and Lattuati, 1998, Zhao, 2008). The resulting image is called a gradient mask (MG) and is obtained for each image frame (Figure 16.9c). The advantage of a gradient mask is that it can be used to detect objects even when illumination is not constant over the entire image due to shadows from trees and the bridge. However, due to the roughness of the water surface, both water waves and wood have strong gradients in intensity values, and a large number of false detections occur if this method is used exclusively.

To reduce the number of false detections, an additional mask was calculated from the temporal difference between two consecutive frames (df). This mask was applied based on the principle that wood will be present in consecutive video frames while the majority of water waves will be stationary or dispersed between images. A final segmented image was calculated from the intersection of the intensity mask (MI), the gradient mask (MG), and temporal inter-frame difference (df). This final image is a binary matrix that identifies all detected objects within the frame (Figure 16.9d).

16.5.4.2 Calculating a 'meta-centroid' for objects in close proximity

An additional problem from the image segmentation procedure is that not all of the detected objects are distinct. For example, a single wood object can be made of a number of parts such as roots and branches. Part of the trunk may be submerged, resulting in the appearance of distinct objects. In addition, the size and shape of wood objects can change from one frame to the next due to water waves and the motion of the wood. Multiple objects identified in the image segmentation must be grouped together to match real objects and avoid false detections. This was accomplished by calculating the distances between all object centroids in an image and grouping close objects and calculating their 'meta-centroid' based on a criterion for the minimum distance between centroids.

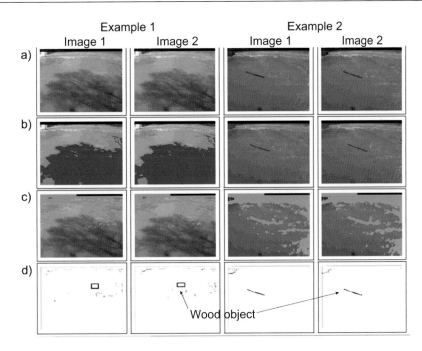

Figure 16.9 Image segmentation steps involved in floating wood identification procedure for a small object in shadows (Example 1) and a larger object in sunlight: a) original images, b) intensity masks, c) gradient masks, and d) resulting combinations of all segmentations.

16.5.4.3 Distinction of wood and water waves through object tracking

Despite the use of the temporal difference mask to reduce false detections, a number of water waves were present and in motion for consecutive frames. To distinguish between wood and water waves it was necessary to utilise some additional property of the wood. Given that water travels from left to right in the image frame, it was reasoned that wood must also travel in the same direction while the direction of wave propagation will be more random and would tend to disappear over time. To track the movements of objects, their meta-centroids were represented at each time step in a summary image. A multi-segment vector was formed in the summary image due to the movement of objects in a series of frames (Figure 16.10). The algorithm distinguished wood objects as those for which the meta-centroids were present in a number of consecutive images and moved continuously from left to right. Parameters were tested for different type lighting situations and different length of wood objects. It was determined that the optimum number of consecutive images to distinguish wood objects from waves was five images, or a total time of 1 second for the video recording at 5 Hz.

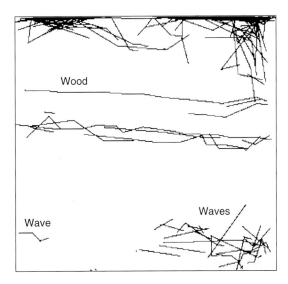

Figure 16.10 Example of summary image showing typical object displacement vectors.

16.5.5 Results

A comparison of the number of detected wood (N_d) to the number of wood pieces identified from the manual analysis of the video (N) show that there is an approximately 90% agreement between the two methods (Table 16.1). An additional 8% of wood pieces were detected with the algorithm but were not counted because they were not present for five consecutive frames in the video. Some floating wood pieces sometimes disappear for one or two frames as they bob slightly at the surface or momentarily reflect the light. Such types of wood pieces are difficult to count accurately using the current method. A comparison of the number of missed detections of wood (N_{md}) and the number of false detections of wood (N_{fd}) shows that these errors are in the range of 10 to 15% (Table 16.2). False detections occur due to occasional water waves that last more than five consecutive frames, while missed detections occur because the intensity of wood pieces can be very close to that of the water.

An infrared sensitive camera used in combination with an infrared light projector on the Ain River was not sufficient to allow wood detection at night. While wood could be visually identified in some cases, it was difficult to reliably distinguish wood from waves in the majority of cases. The main problem is that the projector does not illuminate the entire viewing area of the camera. The reduced area reduces the number of image frames with which to identify wood. A secondary problem is that live organisms are emitters of infrared radiation. As a result, the spiders that frequently construct their webs in the sun shade of the camera appear as very bright objects that move around the image frame throughout the night and obstruct the view of the river.

16.6 Discussion and conclusion

The presented case studies demonstrate that videography is a viable method to estimate water surface velocities, river discharge and wood transport. Water velocity estimates were within 5% of values measured using more conventional techniques. The method provides new information on the instantaneous flow field from which a number of parameters such as shear and vorticity can be calculated. Discharge estimates were within 8% of the established rating curve and the methods can be used to estimate discharge during floods where conventional methods are impractical or unsafe. For wood, there are no conventional techniques with which to measure transport during floods. The presented method thus provides a first look at the timing and nature of wood that is being moved through a watershed. This information will greatly assist the calibration of wood and carbon budgets and improve understanding of wood-related processes in rivers.

Continued technological advances will continue to improve the accuracy of videographic methods for river monitoring. Relatively low resolution cameras were implemented in the case studies here, but higher resolution cameras are now being manufactured. Image frequency is currently limited by internet connection speeds or computer memory. High frequency systems ($\sim$30 Hz) are more suitable for motion analysis but have only been implemented for systems in which the videos are stored locally and then downloaded

Table 16.2 Quantitative evaluation of proposed algorithm comparing the number visually detected wood pieces (N) to the number that were detected using the computer algorithm (N_d), the number of wood pieces that were missed by the algorithm (N_{md}), and the number of false detections made by the algorithm (N_{fd}).

Video	Total Frames	Manual Detected N	Algorithm Detected N_d	%	Missed Detections N_{md}	%	False Detections N_{fd}	%
1	650	82	78	95	4	5	5	6
2	900	73	67	92	6	8	22	30
3	860	21	17	81	4	19	4	19
4	750	40	36	90	4	10	3	8
5	550	38	29	76	9	24	1	3
6	800	28	26	93	2	7	8	29
7	880	36	33	92	3	8	7	19
Total	5390	318	286	90	32	10	50	16

manually. Camera model and internet transmission speeds restricted the system on the Ain River to image frequencies of 5 Hz, which was not optimal because it reduced the correlation between images and made the identification of floating wood more difficult. Higher frequencies should become the norm as internet transmission speeds improve.

The analysis of video for monitoring of discharge and floating wood is a developing scientific field and sources of error are still significant. Errors were reduced using filtering algorithms based on statistical techniques and secondary properties of the objects. However, not all errors were removed and it was not possible to get accurate measurements under all conditions. More techniques are necessary to accurately filter data and to establish criteria for discarding poor quality estimates. A significant data hole remains at night, where illumination is not sufficient to use conventional cameras. Despite the use of infrared sensitivity, nighttime readings were of poor quality, while fog, direct sun and some reflections caused an unfilterable hole in the data. Continued research on the improvement of analysis techniques is warranted. Standardisation of analysis procedures will increase the ease of application for river management.

Continued development of videography for river management could lead to some exciting new applications. One area for study is the measurement of velocity and estimation of discharge from a helicopter (Fujita and Hino, 2003, Fujita et al., 2007a). Such a development would allow the rapid characterisation of hydraulic habitats for fish and improve morphological models of shear stress and erosion during floods. Another possible application is to implement wood detection algorithms in real-time. Such an application would allow videography to be used as an early warning system where infrastructure or other areas of interest are at risk of flooding or failure due to wood accumulations. The ability to measure river discharge, velocity and wood transport in a continuous and remote fashion at a moderate cost should ensure that videographic systems are widely implemented for river management in the future.

References

Adrian, R.J. 1991. Particle-Imaging Techniques for Experimental Fluid-Mechanics, *Annual Review of Fluid Mechanics, 23*, 261–304.

Ali, I., and Tougne, L. 2009. Unsupervised video analysis for counting of wood in river during floods, in Conference Proceedings of *5th International Symposium on Visual Computing*, 578–586, Springer, Las Vegas, USA.

Aya, S., Fujita, I., and Yagyu, M. 1995. Field observation of flood in a river by video image analysis, in Conference Proceedings of *Hydraulic Engineering* 447–452, Japan Society of Civil Engineers.

Becker, J.M., Firing, Y.L., Aucan, J., Holman, R., Merrifield, M., and Pawlak, G. 2007. Video-based observations of nearshore sand ripples and ripple migration, *Journal of Geophysical Research-Oceans, 112*, C01007.

Benda, L.E., and Sias, J.C. 2003. A quantitative framework for evaluating the mass balance of in-stream organic debris, *Forest Ecology and Management, 172*, 1–16.

Bérubé, F., Smith, J.C., and Bergeron, N.E. 2004. Development and use of a particle image velocimetry (PIV) application for aquatic habitat mapping, in Conference Proceedings of *Fifth International Symposium on Ecohydraulics - Aquatic Habitats: Analysis and Restoration*, 1223–1226, Madrid.

Bisson, P.A., Bilby, R.E., Bryant, M.D., Dolloff, C.A., Grette, G.B., House, R.A., Murphy, M.L., Koski, K.V., and Sedell, J.R. 1987. Large woody debris in forested streams in the Pacific Northwest: past, present and future., in *Streamside Management; Forestry and Fishery Interactions*, edited by E.O. Salo and T.W. Tundy, 143–190, University of Washington Institute of Forest Resources, Seattle, WA.

Bourgault, D. 2008. Shore-based photogrammetry of river ice, *Canadian Journal of Civil Engineering, 35*, 80–86.

Bradley, A.A., Kruger, A., Meselhe, E.A., and Muste, M.V.I. 2002. Flow measurement in streams using video imagery, *Water Resources Research, 38*, 1315.

Bradley, J.B., Richards, D.L., and Bahner, C.D. 2005. *Debris Control Structures – Evaluation and Countermeasures*. FHWA-IF-04-016, Hydraulic Engineering Circular No. 9, 179 pp., U.S. Department of Transportation.

Chapron, M. 1997. A chromatic contour detector based on abrupt change techniques, in Conference Proceedings of *International Conference on Image Processing*, 18–21, Santa Barbara, CA.

Comiti, F., Andreoli, A., Lenzi, M.A., and Mao, L. 2006. Spatial density and characteristics of woody debris in five mountain rivers of the Dolomites (Italian Alps), *Geomorphology, 78*, 44–63.

Costa, J.E., Spicer, K.R., Cheng, R.T., Haeni, P.F., Melcher, N.B., Thurman, E.M., Plant, W.J., and Keller, W.C. 2000. Measuring stream discharge by non-contact methods: A proof-of-concept experiment, *Geophysical Research Letters, 27*, 553–556.

Creutin, J.D., Muste, M., Bradley, A.A., Kim, S.C., and Kruger, A. 2003. River gauging using PIV techniques: a proof of concept experiment on the Iowa River, *Journal of Hydrology, 277*, 182–194.

Creutin, J.D., Muste, M., and Li, Z. 2002. Traceless quantitative alternatives for measuremetns in natural streams, in Conference Proceedings of *Hydraulic Measurements & Experimental Methods*, ASCE-IAHR, Estes Park, CO.

Delrieu, G., Ducrocq, V., Gaume, E., Nicol, J., Payrastre, O., Yates, E., Kirstetter, P.E., Andrieu, H., Ayral, P.A., Bouvier, C., Creutin, J.D., Livet, M., Anquetin, S., Lang, M., Neppel, L., Obled, C., Parent-du-Chatelet, J., Saulnier, G.M., Walpersdorf, A., and Wobrock, W. 2005. The catastrophic flash-flood event of 8-9 September 2002 in the Gard region, France: A first case study for the Cevennes-Vivarais Mediterranean Hydrometeorological Observatory, *Journal of Hydrometeorology*, 6, 34–52.

Ettema, R., Fujita, I., Muste, M., and Kruger, A. 1997. Particle-image velocimetry for whole-field measurement of ice velocities, *Cold Regions Science and Technology*, 26, 97–112.

Ferrick, M.G., Weyrick, P.B., and Hunnewell, S.T. 1992. Analysis of river ice motion near a breaking point, *Canadian Journal of Civil Engineering*, 19, 105–116.

Fincham, A.M., and Spedding, G.R. 1997. Low cost, high resolution DPIV for measurement of turbulent fluid flow, *Experiments in Fluids*, 23, 449–462.

Fourquet, G. 2005. Développement d'un système hydrométrique par analyse d'images numériques. Evaluation d'une année de fonctionnement continu sur l'Isère à Saint Martin d'Hères., Ph.D. thesis, Institut National Polytechnique, Grenoble, France.

Fu, K.S., and Mui, J.K. 1981. A survey on image segmentation, *Pattern Recognition*, 13, 3–16.

Fujita, I., and Aya, S. 2000. Refinement of LSPIV technique for monitoring river surface flows, in Conference Proceedings of *Water Resources Joint Conference on Water Resources Engineering and Water Resources Planning and Management*, Minneapolis, MN.

Fujita, I., and Hino, T. 2003. Unseeded and seeded PIV measurements of river flows videotaped from a helicopter, *Journal of Visualization*, 6, 245–252.

Fujita, I., and Komura, S. 1988. Visualization of the flow at a confluence, in Conference Proceedings of *3rd International Symposium on Refined Flow Modelling and Turbulence*, 611–618.

Fujita, I., and Komura, S. 1992. On the accuracy of the correlation method, in Conference Proceedings of *6th International Symposium on Flow Visualisation*, 858–862.

Fujita, I., and Komura, S. 1994. Application of video image analysis for measurements of river-surface flows, *Annual Journal of Hydraulic Engineering, JSCE*, 38, 733–738.

Fujita, I., Muste, M., and Kruger, A. 1998. Large-scale particle image velocimetry for flow analysis in hydraulic engineering applications, *Journal of Hydraulic Research*, 36, 397–414.

Fujita, I., Tsubaki, R., and Deguchi, T. (2007a), PIV measurement of large-scale river surface flow during a flood by using a high resolution video camera from a helicopter, in Conference Proceedings of *Hydraulic Measurements and Experimental Methods - ASCE & IAHR*, 344-349, Lake Placid, NY.

Fujita, I., Watanabe, H., and Tsubaki, R. (2007b), Development of a non-intrusive and efficient flow monitoring technique: The Space-Time Image Velocimetry (STIV), *International Journal of River Basin Management*, 5, 105–114.

Fulford, J., and Sauer, V. 1986. Comparison of velocity interpolation methods for computing open-channel discharge, *U.S. Geological Survey Water Supply Paper*, 2290, 139–144.

Gurnell, A., and Petts, G. 2006. Trees as riparian engineers: The Tagliamento River, Italy, *Earth Surface Processes and Landforms*, 31, 1558–1574.

Haralick, R.M., and Shapiro, L.G. 1985. Image segmentation techniques, *Computer Vision Graphics and Image Processing*, 29, 100–132.

Harpold, A.A., and Mostaghimi, S. 2004. Stream discharge measurement using a large-scale particle image velocimetry prototype, *EOS Transactions AGU*, 85,

Hassan, M.A., Hogan, D.L., Bird, S.A., May, C.L., Gomi, T., and Campbell, D. 2005. Spatial and temporal dynamics of wood in headwater streams of the Pacific Northwest, *Journal of the American Water Resources Association*, 41, 899–919.

Hauet, A. 2006. *Discharge Estimation and Velocity Measurement in River Using LSPIV*. Institut National Polytechnique, Grenoble, France.

Hauet, A., Belleudy, P., and Muste, M. 2007. Estimation of the bathymetry of a channel using LSPIV, in Conference Proceedings of *32nd IAHR Congress*, Venice, Italy.

Hauet, A., Creutin, J.D., and Belleudy, P. (2008a), Sensitivity study of large-scale particle image velocimetry measurement of river discharge using numerical simulation, *Journal of Hydrology*, 349, 178–190.

Hauet, A., Creutin, J.D., Belleudy, P., Muste, M., and Krajewski, W.F. 2006. Discharge measurement using large scale PIV under various flow conditions - Recent results, accuracy and perspectives, in Conference Proceedings of *River Flow*, edited by R. Ferreira, E. Alves, J. Leal, and A.H. Cardoso, Lisboa, Portugal.

Hauet, A., Kruger, A., Krajewski, W.F., Bradley, A., Muste, M., Creutin, J.D., and Wilson, M. (2008b), Experimental system for real-time discharge estimation using an image-based method, *Journal of Hydrologic Engineering*, 13, 105–110.

Holland, K.T., Holman, R.A., Lippmann, T.C., Stanley, J., and Plant, N. 1997. Practical use of video imagery in nearshore oceanographic field studies, *IEEE Journal of Oceanic Engineering*, 22, 81–92.

Jain, R., Kasturi, R., and Schunck, B.G. 1995. *Machine Vision*, 549 pp., McGraw-Hill, New York, NY.

Jasek, M., Muste, M., and Ettema, R. 2001. Estimation of Yukon River discharge during an ice jam near Dawson City, *Canadian Journal of Civil Engineering*, 28, 856–864.

Jodeau, M., Hauet, A., Paquier, A., Le Coz, J., and Dramais, G. 2008. Application and evaluation of LS-PIV technique for the monitoring of river surface velocities in high flow conditions, *Flow Measurement and Instrumentation*, 19, 117–127.

Kim, Y., Muste, M., Hauet, A., Krajewski, W.F., Kruger, A., and Bradley, A. 2008. Stream discharge using mobile large-scale particle image velocimetry: A proof of concept, *Water Resources Research*, 44, W09502.

Kim, Y.S. 2006. Uncertaintly Analysis for Non-Intrusive Measurement of River Discharge Using Image Velocimetry, Ph.D. thesis, 209 pp., University of Iowa, Iowa City.

Lassettre, N.S., Piégay, H., Dufour, S., and Rollet, A.J. 2008. Decadal changes in distribution and frequency of wood in a free meandering river, the Ain River, France, *Earth Surface Processes and Landforms*, 33, 1098–1112.

Leese, J.A., Novak, C.S., and Clark, B.B. 1971. An automated technique for obtaining cloud motion from geosynchronous satellite data using cross correlation, *Journal of Applied Meteorology*, 10, 118–132.

Lyn, D.A., Cooper, T., Yi, Y.K., Sinha, R., and Rao, A.R. 2003. *Debris Accumulation at Bridge Crossings: Laboratory and Field Studies*. FHWA/IN/JTRP-2003/10, Indiana Department of Transportation and Federal Highway Administration, 59 pp. Purdue University, West Lafayette, Indiana.

MacVicar, B.J., Piégay, H., Henderson, A., Comiti, F., Oberlin, C., and Pecorari, E. 2009. Quantifying the temporal dynamics of wood in large river: field trials of wood surveying, dating, tracking, and monitoring techniques, *Earth Surface Processes and Landforms*, 34, 2031–2046.

Mikhail, E.M., and Ackermann, F. 1976. *Observation and Least Squares*, 497 pp., Dun-Donnelley Publishing, New York.

Montgomery, D.R., Abbe, T.B., Buffington, J.M., Peterson, N.P., Schmidt, K.M., and Stock, J.D. 1996. Distribution of bedrock and alluvial channels in forested mountain drainage basins, *Nature*, 381, 587–589.

Moulin, B., and Piégay, H. 2004. Characteristics and temporal variability of large woody debris trapped in a reservoir on the River Rhone (Rhone): Implications for river basin management, *River Research and Applications*, 20, 79–97.

Muller, G., Bruce, T., and Kauppert, K. 2002. *Particle Image Velocimetry: a simple technique for complex surface flows*. D. Bousmar and Y. Zech, editors. International Conference on River Hydraulics, 1227–1234, Lisse, The Netherlands.

Muste, M., Fujita, I., and Hauet, A. 2008. Large-scale particle image velocimetry for measurements in riverine environments, *Water Resources Research*, 44, W00D19.

Muste, M., Schone, J., and Creutin, J.D. 2005. Measurement of free-surface flow velocity using controlled surface waves, *Flow Measurement and Instrumentation*, 16, 47–55.

Muste, M., Xiong, Z.G., and Kruger, A. 1999. Error estimation in PIV applied to large-scale flows, in Conference Proceedings of *International Workshop on Particle Image Velocimetry*, Santa Barbara, California.

Muste, M., Xiong, Z.J., Schone, J., and Li, Z.W. 2004. Validation and extension of image velocimetry capabilities for flow diagnostics in hydraulic modeling, *Journal of Hydraulic Engineering*, 130, 175–185.

Ninnis, R.M., Emery, W.J., and Collins, M.J. 1986. Automated extraction of pack ice motion from advanced very high resolution radiometer imagery, *Journal of Geophysical Research-Oceans*, 91, 10725–10734.

Pal, N.R., and Pal, S.K. 1993. A review on image segmentation techniques, *Pattern Recognition*, 26, 1277–1294.

Piégay, H. 1993. Nature, mass and preferential sites of coarse woody debris deposits in the lower Ain Valley (Mollon Reach), France, *Regulated Rivers-Research & Management*, 8, 359–372.

Piégay, H., Citterio, A., and Astrade, L. 1997. Interactions between large woody debris and meander cut-off (example of the Mellon site on the Ain River, France), *Zeitschrift Fur Geomorphologie*, 42, 187–208.

Polatel, C. 2005. Indexing free-surface velocity: A prospect for remote discharge estimation, in Conference Proceedings of *International Association of Hydraulic Research*, Seoul, Korea.

Raffel, M., Willert, C., and Kompenhans, J. 1998. *Particle Image Velocimetry: a Practical Guide*, 448 pp., Springer, New York, NY.

Zhao, A. 2008. Robust histogram-based object tracking in image sequences, in Conference Proceedings of *Digital Image Computing Techniques and Applications*, 45–52, Canberra, Australia.

Zugaj, D., and Lattuati, V. 1998. A new approach of color images segmentation based on fusing region and edge segmentations outputs, *Pattern Recognition*, 31, 105–113.

17 Imagery at the Organismic Level: From Body Shape Descriptions to Micro-scale Analyses

Pierre Sagnes

Université Claude Bernard, Lyon 1, Villeurbanne Cedex, France

17.1 Introduction

As a complement to remote sensing methods operating at various spatial scales, photographic imagery is widely used at the organismic level. In freshwater studies, image analyses (hereafter abbreviated as IA) range from the observation of microscopic structures through to morphological investigations regarding the potential body shape of Nessie.[1] Quantitative IA (i.e. the extraction of information from pictures) is often time-consuming and based on subjective concepts. However, the use of image processing methods has many advantages: (1) they allow a repeatable approach to formally and mathematically describe morphological traits; (2) they sometimes avoid the manipulation of injured or fragile organisms; (3) images or videos can be stored and consulted whenever it is required (e.g. for complementing or correcting the data); (4) IA methods release humans from 'routine identification' processes so that more emphasis can be placed upon determining rarer patterns (Weller et al., 2006).

Over time, more and more complex (and accurate) imagery techniques have been used centered on (1) the way to acquire data (from direct measurements on organisms to photos or video recordings), (2) the type of data collected (from linear distances to 3D coordinates or outline descriptors), and (3) the way to analyse the data collected (from linear regressions through to more and more complex algorithms that sort and synthesise information). For instance, the morphology of aquatic organisms has mainly been examined using conventional metric approaches that consist of linear distances – 'classical' body lengths, or 'truss network' (see Strauss and Bookstein, 1982). However, such methods do not account for the overall form and when shape is complex some crucial information may be lost. To improve on these techniques, morphometric methods relying on the analysis of coordinates of homologous landmarks (in 2D or 3D) have been developed (Bookstein, 1986). 2D or 3D coordinates of body landmarks can be used to characterise shape after using superimposition methods, which remove translation, rotation and size parameters (see Zelditch et al., 2004). These geometric morphometrics methods derived from ideas expressed early in the twentieth century by Thompson (1917), stating that morphometric variation can be explained by simple transformations of homologous features in coordinate space. Currently, body deformations from one shape to another (e.g. during growth, between taxa)

---

[1] Pet name of the famous Loch Ness Monster.

Fluvial Remote Sensing for Science and Management, First Edition. Edited by Patrice E. Carbonneau and Hervé Piégay.
© 2012 John Wiley & Sons, Ltd. Published 2012 by John Wiley & Sons, Ltd.

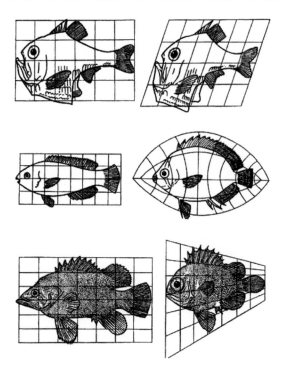

Figure 17.1 Morphometric variation of fishes, expressed as regular deformations of rectangular coordinate systems. Reproduced from Thompson, D.W. (1917) *On Growth and Form*, with permission from Cambridge University Press.

can be visualised through Thin Plate Spline (TPS) analysis (see Zelditch et al., 2004 and Figure 17.1).

Shape characterisation based on landmarks has produced valuable results; however, the main difficulty of this approach is that landmark points cannot be located very accurately all of the time. In this case, other methods, such as outline analyses, can be used. Most of these methods consist in expressing outlines in periodic signals. Using Fourier transform, such signals are fitted by a sum of trigonometric functions (or harmonics) that have different amplitudes and phases (see a review of shape descriptors in Zhang and Lu, 2004).

Imagery methods are currently applied, for various purposes, on almost all biological model organisms (from viruses to vertebrates). Observations can be automated and are sometimes made *in situ*. At the species (or multi-species) level, imagery was used to detect the presence of specimens and/or estimate their body lengths, abundance and biomass; methods were also developed for recognising and separating various species from each other. At the infra-species (i.e. within species or sub-species) level, IA has helped to differentiate sexes, ontogenetic stages

or stocks, to characterise life-history traits, to describe behaviours, to detect diseases, to make chronic stress diagnostics or to relate morphology and ecology.

This chapter presents selected examples of how imagery techniques are currently (or could be) used in freshwater studies at the level of the organism.

17.2 Morphological and anatomical description

17.2.1 Identification

Automated (or semi-automated) systems of shape identification with IA almost always follow the same sequence of procedures: (1) background elimination, (2) image segmentation (i.e. location of objects of interest and often shift to a binary image, Figure 17.2), (3) focus check, (4) object feature extraction, (5) feature selection and measurement, (6) feature analysis, (7) classification and (8) estimation of system performance. Such protocols were widely used to recognise and classify various taxa, but other techniques (e.g. recognition of colouration patterns) can also be very effective.

17.2.1.1 Species (or taxa) recognition

Although automated species recognition has not replaced experts (see why in Gaston and O'Neill, 2004), numerous studies developed routine IA identification, for various purposes.

Bacterial morphological diversity could be an indicator of dynamic ecological succession following a nutrient perturbation in bacterial communities (Liu et al., 2001). In microbial ecology, a major challenge is to develop reliable methods of computer assisted microscopy that can analyse digital images of complex microbial communities at the resolution of the single cell, and to compute useful quantitative characteristics of their organisation without the need for cultivation. Liu et al. (2001) described a computer-aided system (Figure 17.3), which extracted size and shape measurements of segmented, digital images of microorganisms and classified them into one of 11 predominant bacterial morphotypes (e.g. cocci, spirals, curved rods, ellipsoids). This shape classifier had an accuracy of 97% on a test set of 4,270 cells representing all these bacterial morphotypes, indicating that accurate classification of rich morphological diversity in microbial communities was possible using IA.

To complement shape recognition, bacterial cells can be classified in a given taxa (or group) after specific

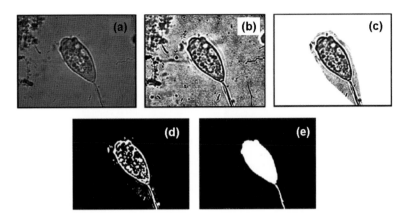

Figure 17.2 First steps of a classical image analysis: original image (a), pre-treated image (b), selection of the region of interest (c), binary image after segmentation (d) and final image (e). Reprinted from Analytica Chimica Acta, 595, Ginoris et al. Recognition of protozoa and metazoa using image analysis tools, discriminant analysis, neural networks and decision trees, 160–169, Copyright 2007, with permission from Elsevier.

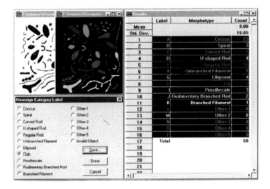

Figure 17.3 Interactive interface of the CMEIAS morphotype classifier. Shown (clockwise) are a portion of the analysed binary composite image overlaid with the corresponding classification result images in which the classified morphotype of each cell is noted by its unique pseudocolor assignment; the results window of classification data for each morphotype using the corresponding pseudocoloured text; and the interactive edit interface of 16 morphotype selections. With kind permission from Springer Science+Business Media: Microbial Ecology, CMEIAS: a computer-aided system for the image analysis of bacterial morphotypes in microbial communities, 41, 2001, 173–194, Liu et al.

colourations and/or excitations. From *in situ* PCR[2] followed by IA, Tani et al. (1998) detected *Escherichia coli* cells in polluted river water. They also showed that, under different techniques of excitation, some bacteria could exhibit blue or intense red fluorescence, allowing their automatic recognition. Within complex microbial

communities, Lee et al. (1999) simultaneously determined *in situ* the identities, activities, and specific substrate uptake profiles of individual bacterial cells, by combining fluorescent hybridisation and microautoradiography.

Phototrophic biofilms in aquatic environments represent an important carbon source for other trophic levels and affect mass transfer processes at the ecosystem scale. The description of biofilm structure in time and space is therefore very important when studying the functioning of aquatic ecosystems (Paterson et al., 2003). Recent IA advances have allowed the description of the dynamic spatial and temporal separation of diatoms, bacteria and organic and inorganic matter during the shift from a bacteria-dominated to a diatom-dominated phototrophic biofilm (Mueller et al., 2006). This approach facilitated the analyses of large amounts of multichannel confocal laser scanning microscopy data in an automated way. The different channels map individual biofilm components, describing various aspects of biofilm morphology (e.g. biovolume, substratum coverage, area to volume ratio and fractal dimensions).

Cyanobacteria occur in surface waters worldwide. Many of these produce peptides and/or alkaloids, which present potential risks for human and animal health. Effective risk assessment and management requires continuous and precise observation and quantification of cyanobacterial cell densities. Six cyanobacteria species (unstained cells) were classified from image processing techniques, with an error rate of approximately 3% (Walker and Kumagai, 2000). Walker et al. (2002) exploited the ability of microalgae to autofluoresce when exposed to epifluorescence illumination for detecting

---

[2] Polymerase Chain Reaction (PCR) is a technique used to replicate ADN fragments.

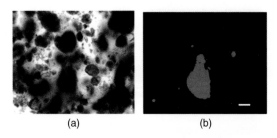

(a)   (b)

Figure 17.4 Image of Lake Biwa sediment, containing almost completely obscured microalgae specimens (a); corresponding fluorescence image (b). Scale bar = 100 Am. Reprinted from *Journal of Microbiological Methods*, 51, Walker et al., Fluorescence-assisted image analysis of freshwater microalgae, pp. 149–162. Copyright 2002, with permission from Elsevier.

(Figure 17.4) and analysing microalgae in sediment samples containing complex scenes. They quantitatively measured 120 characteristics of each object detected through fluorescence excitation, and used an optimised subset of these characteristics for later automated analysis and species classification. They succeeded in classifying two genera of microalgae (*Anabaena spp.* and *Microcystis spp.*) with accuracy higher than 97%.

Early *in situ* detection of algae species and the estimation of their potential to cause algal blooms is also possible from IA. For that purpose, an autonomous underwater vehicle equipped with a submersible microscope, video recording system and water quality monitoring sensors to detect the spatial structure of *Uroglena americana* (causing freshwater 'red tide') was developed (Ishikawa et al., 2005). Objects corresponding to the target species were detected and analysed by extracting 130 statistical and morphometrical features. The numbers of *U. Americana* objects per-unit-time were counted and combined with the recorded vehicle route trajectory data. Subsequently, colonies could be enumerated. In some cases (e.g. in palynology, see Weller et al., 2006), self-organised maps (a form of artificial neural networks), based on morphological and textural IA features, were used for image clustering of samples.

The discrimination of submerged macrophyte species from optical remote sensing is possible at a large scale by using appropriate spectral regions (see a review in Silva et al., 2008). As suggested by Rowlinson et al. (1999), an application of this method could be the detection of alien species in riparian zones (e.g. prior to their removing). However, water characteristics (such as turbidity and depth), the presence of epiphytes and physiological status of vegetation can be a source of variation in plant spectral signatures, and this variability can lead to poor results from simple automated classification procedures (Silva

et al., 2008). At a lower scale, automated plant recognition has almost exclusively only been undertaken for terrestrial vegetation species. Du et al. (2007) used 15 morphological features of leaves to classify 20 species. Prior to IA, chlorophyll *a* fluorescence induction curves had also been reliably used for automated identification of terrestrial plants (Keränen et al., 2003).

The characterisation of body ornamentation (e.g. pigmentation) can help to recognise and classify individuals into species. For example, shape differences among individual patches on the frontoclypeus can provide valuable information for rapid species identification of 10 *Hydropsyche* species (see Statzner and Mondy, 2009 for visual interpretation of digitalised images).

Among freshwater vertebrates, automated species recognition using IA has mainly been developed for fish. Recognition of fish species using imagery, for conservation (e.g. when they migrate using fishways) or commercial purposes (e.g. edible species), has been significantly enhanced over the last 20 years (see a review in Zion et al., 2007). Castignolles et al. (1994), achieved species classification in fishways by analysing morphometric features on multiple images of each fish as it swam across the passage. However, for real-time applications (i.e. when fish are lined up close to each other), multiple imaging tends to be impractical. At the same time as this development, Hatch et al. (1994) demonstrated that automated fish counting and speciation in fishways could also be performed from digitised sequences of video frames (the process determined the correct location and identification of 67 out of 70 fish for three salmonid species). Cadieux et al. (2000) used a set of infrared diodes and sensors that generate silhouettes as the fish swim between them. Once acquired, the data can be sent periodically to a computer using a direct link, a satellite link or a cellular phone. Cadieux et al. (2000) calculated some moment-invariants, Fourier descriptors of silhouette contours, and the geometric features described by Castignolles et al. (1994). A majority vote method (Xu et al., 1992) was used to classify images of five fish species with an overall accuracy of 78%. This system allows the operator to select the species of interest according to the fauna of the specified river. Tillett et al. (2000) segmented fish images by means of a modified point distribution model (PDM) which considered the strength of an edge and its proximity, to attract landmarks to edges. They estimated salmon length with an average accuracy of 95% when compared to manual measurement. However, their procedure required manual placement of the PDM in an initial position close to the centre of the fish, and some images could not be correctly fitted (e.g. neighbouring fish and fish whose orientation was very different from the initial PDM). Zion et al. (1999) extracted typical features from dead fish tails and used them for species identification. For three species (Common carp, *Cyprinus carpio*; St. Peter's fish, *Oreochromis* sp. and grey mullet, *Mugil cephalus*), the average identification accuracy was higher than 93%. More recently, Zion et al. (2007) improved their image-processing algorithms by extracting size- and orientation features from the fish silhouettes (Figure 17.5). The overall species recognition accuracy (from swimming fish) was about 98%.

Recent developments have improved the automated species recognition for species that have similar shape characters (e.g. seven salmonid species with similar morphology: Lee et al., 2004) or IA from low-quality images (i.e. images lacking distinctive or stable morphological features: Rova et al., 2007; detection of target species in turbid habitats from dual-frequency sonar techniques: Frias-Torres and Luo, 2009). White et al. (2006) developed a 'computer vision machine' for identifying and measuring different species of fish from 10 shape and 114 colour features. The fish were transported along a conveyor belt underneath a digital camera. The image processing algorithms determined the orientation of the fish, identified flat or round body shapes (with 100% accuracy), measured fish length (with a standard deviation of 1.2 mm) and differentiated between seven species (with up to 99.8% sorting reliability). This machine could theoretically process up to 30,000 fish per hour using a single conveyor belt based system.

Body shape descriptors are less sensitive to lighting variations and water quality and are generally preferred to colour features for species recognition. However, in some cases, colour restoration of underwater images can be achieved (Iqbal et al., 2007; Figure 17.6) and

Figure 17.6 Example of an underwater image before (left) and after (right) enhancement. Reproduced from Iqbal, K. et al. (2007) Underwater image enhancement using an integrated colour model. IAENG International Journal of Computer Science, 34, 2. © Copyright International Association of Engineers.

the subsequent restored images should give better results when displayed or processed (see an example of fish segmentation and feature extraction in Chambah et al., 2004).

Fish species recognition can also be performed by studying the shape of anatomical structures such as otoliths.[3] The general morphology of the saccular otoliths (i.e. the largest pair among three in teleosts) is usually species specific and has been used for species identification using IA (e.g. Tuset et al., 2003). Fourier analysis has traditionally been used to study otolith morphology, since it is an effective method for describing outline shapes, but it does not encourage intuitive understanding of the reason for subtle shape differences (Cadrin and Friedland, 1999). Studying numerous otolith shape descriptors (e.g. aspect ratio, compactness, eccentricity, ellipticity, bilateral symmetry), Tuset et al. (2006) demonstrated that species differentiation from otolith shape characterisation should be enhanced by standardising variables with respect to fish length and adding otolith weight in the analysis.

17.2.1.2 Stock differentiation

Variability in growth, development, and maturation creates a variety of body shapes within a species, and identifying discrete units of the stock is a basic requirement for fisheries science and management. Geographic variation in morphometry has been used to discriminate 'phenotypic stocks' of fish (defined as groups with similar growth, mortality, and reproductive rates, Cadrin, 2000) for over 130 years (Heinke, 1878 cited by Cadrin, 2000). For example, Corti et al. (1988) detected variations between six strains of common carp (*Cyprinus carpio*) from truss

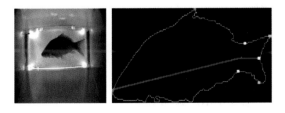

Figure 17.5 Left: a carp image acquired by a real-time underwater system, in a laboratory pool. The background light is reflected from the apparatus frame. Right: segmented contour and landmarks found by the system. Reprinted from Computers and Electronics in Agriculture, 56, Zion et al., Real-time underwater sorting of edible fish species, 34–45, Copyright 2007, with permission from Elsevier.

---

[3]Otoliths are aragonitic mineralisations positioned in the membraneous labyrinth of the inner ear of bony fishes and play an important role in the senses of hearing and balance (see Popper et al., 2005 for recent review).

network measurements. Meristic[4] features can also differ between stocks. For instance, Claytor and MacCrimmon (1988) visually counted vertebra and dorsal and anal fin rays from radiographs in order to differentiate regional stocks of Atlantic salmon (*Salmo salar*).

As for species recognition, subtle variations in size and contour of otoliths were used in the discrimination of individual populations/stocks (e.g. Campana and Casselman, 1993) and in the determination of phylogenetic lines (e.g. Gaemers, 1984). Furthermore, fish stock identification can rely on scale pattern analysis. Fish otoliths and scales grow by accretion, as more bone is periodically added along their periphery. Subsequently, their diameter increase reflects body growth and circuli[5] are closer together during periods of slow growth, such as winter in temperate regions. This growth pattern creates alternative dark and light bands (called annuli), which can be counted and analysed through IA. For diadromous species, the freshwater-ocean transition is usually associated with a change from thin, narrowly spaced circuli to thick, widely spaced ones (Bernard and Myers, 1996). Moreover, the freshwater and early marine portions of the scale were shown to differ respect to the origin of fish (i.e. hatchery *vs* wild specimens, see Davis and Light 1985). Using acetate impressions of scales and subsequent IA, Bernard and Myers (1996) showed that hatchery steelhead salmon (*Oncorhynchus mykiss*) had larger freshwater zones and more freshwater circuli than wild specimens. They conclude that this technique has good potential for estimating proportions of hatchery and wild steelhead in high-seas mixed-populations.

Farmed and wild fish may also differ in their general body shape. Using geometric morphometrics methods (Procrustes coordinates of landmarks and visualisation by thin-plate splines), captively reared adults were differentiated from wild ones by sharply reduced sexual dimorphism as well as numerous differences in body shape (Hard et al., 2000).

As for species recognition, colour features are less used than shape ones for stocks differentiation. However, Strachan and Kell (1995) used ten shape features and 114 colour features to discriminate between haddock (*Melanogrammus aeglefinus*) stocks from two different fishing regions, demonstrating that IA methods are efficient to segment fish images from colour information.

---

[4]Meristic features are all the variables that can be counted on a fish (e.g. number of fin rays, of vertebrae, of scales on the lateral line, etc.).
[5]Circuli are layers of roughly concentric circles of bone that appear on fish scales and otoliths.

17.2.1.3 Sexual dimorphism

Sexual dimorphism is common in animals where males and females have distinct roles in mating and courtship. Aquatic species are not an exception to this rule, and such a dimorphism can be found in a wide variety of freshwater taxa. Using Fourier analysis of outlines, Bertin et al. (2002) showed that three characters (pleotelson, paraeopods 4 and 5) differed significantly in shape between males and females of *Asellus aquaticus* (Crustacea). By comparing three species in the genus *Poecilia* (Pisces), Ptacek (1998) showed that differences in behaviour and morphometrics in males could play a role in female mate choice. Therefore, IA can rapidly sort individuals by gender, through morphological differences among sexes. For that purpose, Zion et al. (2008) used algorithms derived from shape (i.e. locating landmark positions on fish contours and extraction of shape-related features) and colour differences between female and male guppies (*Poecilia reticulata*) to classify them. Identification accuracy was approximately 90% using shape features, approximately 96% using colour features and was slightly improved when both colour and shape features were used. In a same way, studying skin colour in newts (*Notophthalmus viridescens viridescens*) through IA methods, Davis and Grayson (2007) showed that males were statistically greener than females, although this effect depended upon life-stage. If males and females often differ in external shape or colour features, sexual dimorphism can also be detected through analysis of histological images. Under a light microscope and using IA software, Hagen et al. (2006) showed that the total number of fast muscle fibres per trunk cross section was higher in females than males prior to sexual maturation in Atlantic halibut (*Hippoglossus hippoglossus*). These results illustrated a sexual dimorphism of muscle fibre recruitment patterns in some fish species.

17.2.2 Characterisation of life-history traits and ontogenetic stages

Characterisation of life-history traits is a very important task in population biology and evolution. For example, estimation of individual growth and reproductive investment is central in many studies (e.g. Jennings and Philipp, 1992). Moreover, despite the idiosyncratic nature of many invasions (Marchetti et al., 2004), it has been hypothesised that life-history traits such as early maturity, high fecundity and asexual reproduction are associated with successful invading species (Lodge, 1993). Therefore, numerous studies have characterised various life-history traits, sometimes using IA methods.

17.2.2.1 Egg size, fecundity

Manual measurements of oocyte number and/or sizes imply time-consuming work. Thorsen and Kjesbu (2001) used an IA system for estimating oocyte density, the 'auto-diametric fecundity method'. They determined the average diameter of oocytes in a sample, and this was converted into oocyte density using a calibration curve. Furthermore, they showed that accurate and precise measurement of oocyte size had practical implications for the assessment of maturity stage and predictions for the start of spawning. To estimate fecundity without sacrificing the fish, Will et al. (2002) assessed ovary volume from ultrasonic imaging, as in medical imagery. The total length of the ovary and maximum and mean cross-sectional ovary areas were measured. Oocyte number or sizes may also be measured *in situ*. MacInnis and Corkum (2000) examined nests of the invasive round goby (*Neogobius melanostomus*) through video recordings. The area of each egg mass or area covered by egg scars was subsequently measured using IA. Analysing the data in association with conventional methods (e.g. egg counting by hand and ovarian weighting) allowed the authors to conclude that the reproductive strategy of round gobies combined with its aggressive behavior may favour the species expansion throughout the Great Lakes.

17.2.2.2 Growth

Population studies depend upon correct age and growth estimates. There are a few ways to assess individual growth: by (1) analysing differential body lengths, (2) studying anatomical structures like shells, scales or otoliths and (3) quantifying chemical substances that accumulate within the body with age (e.g. lipofuscin). Monitoring the growth of live fish without manipulating specimens is a difficult task. Tillett et al. (2000) described an underwater stereo IA technique that offered the potential for estimating key dimensions of free-swimming fish. Comparing automatic measurements of fish dimensions with manual measurements demonstrated an average length error estimation of 5%.

In most cases, periodic growth increments have been measured to estimate the age. Tree rings are the archetypal ageing structure. In freshwater animals, growth increments of various anatomical structures (such as bivalve shells, tortoise scute, fish scales, vertebrae, fin rays, opercula and otoliths) are used to estimate age and reconstruct growth rate. In molluscans, shell increments contain information related to the evolution of the environment in which the organism grew during its biomineralisation.

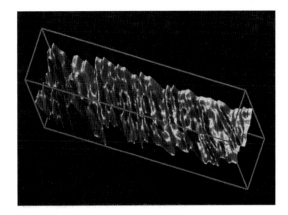

Figure 17.7 Visualisation of fine growth increments (seasonal and monthly increments) on a molluscan shell. Reprinted from Computers & Geosciences, 25(8), Toubin, M. et al., Multi-scale analysis of shell growth increments using wavelet transform, pp. 877–885, Copyright 1999, with permission from Elsevier.

To extract the information from variations in shell topography, Toubin et al. (1999) used a technique involving multi-scale analysis of the shell topography after a wavelet transform. An optical system, based on laser triangulation, mapped the shell surface (Figure 17.7). A multiscale representation allowed distinctions between growth increments of various orders (seasonal, monthly, daily), which were related to bivalve ontogenesis and environmental stress.

Furthermore, the analysis of molluscan shell growth may allow the reconstruction of environmental conditions *a posteriori*. Schöne et al. (2004) inferred summer air temperatures for each year over the period 1777–1993 from studying variations in annual shell growth of the freshwater pearl mussel *Margritifera margritifera* (using both live and collection specimens). Up to 55% in the variability of annual shell growth was explained by temperature changes. Such models can be used to test and verify other air temperature proxies and thus may help improve climate models. However, Dunca et al. (2005) underlined that shell growth does not co-vary with summer temperatures at polluted sites, suggesting that care is required when designing an appropriate sampling strategy when molluscan shells are used for climate reconstructions.

Ageing based on fish otoliths is often a subjective activity, based on experience (see Abecasis et al., 2007 for a comparison of aging fish from scales and otoliths). Evidently, although it is possible, it is more difficult to determine fish age at small time scales (e.g. age in days *vs* age in years). Over the last 35 years (following the seminal paper of Mason, 1974) improvements in (semi-) automatic measurements methods have made interpretation easier. Morales-Nin et al. (1998) used Fourier analysis and subsequent wavelet analysis for studying the periodic signals obtained from an otolith radius (Figure 17.8). This method allowed the analysis of signals having the potential to discriminate between the different time-scales (daily, weekly, monthly) interacting with the otolith growth rate.

Such methods were later complemented and somewhat enhanced by new algorithms. For instance, Cao and Fablet (2006) developed an automatic detection of the growth centre (i.e. otolith nucleus) to limit experimenter effects.

In some crustaceans, there is a relationship between age and lipofuscin concentrations in the brain (Sheehy, 1990a). Belchier et al. (1998) used confocal fluorescence microscopy and IA of histological sections to quantify lipofuscin in the crayfish *Pacifastacus leniusculus*. After the excitation of olfactory lobe sections (to reveal autofluorescence) and the optimisation of image catch (see Sheehy, 1990b for details), brightly autofluorescing lipofuscin granules were discriminated from the darker background of neurone somata using greyscale thresholding. The

outline of the cell-mass background in the image was traced manually and, finally, the total cross-sectional areas of both lipofuscin and the background cells were calculated by the software. Following this procedure, Belchier et al. (1998) showed that lipofuscin concentration was linearly associated with age ($r^2 = 0.92$) and produced much more accurate age estimates than conventional body size-based procedures.

Individual growth, rather than population growth rate (PGR), is central to the theory of population ecology. In some cases, IA offers potential tools for estimating PGR. In the Crustacean *Daphnia magna*, PGR was estimated as the ratio of population sizes at two different times, where size was measured by the sum of the individuals' surface areas (Hooper et al., 2006). The IA system proved reliable and reproducible in counting and estimating surface area of up to 440 individuals in 5 L of water.

17.2.2.3 Description of ontogenetic stages

For many species, successive ontogenetic stages differ in physiological, behavioural and morphological ways. Therefore, IA tools can be used to differentiate life stages,

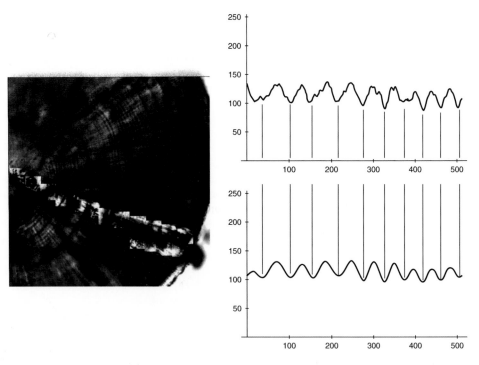

Figure 17.8 Left: otolith of Atlantic salmon (Salmo salar) showing the reading radius with enhanced increments. Right: signal vector corresponding to a grey level of enhanced increments, with noise (top), and the same signal once filtered by a Fast Fournier Transform (bottom). Reproduced with permission from Morales-Nin, B., Lombarte, A. and Japón, B. (1998) Approaches to otolith age determination: image signal treatment and age attribution. Scientia Marina 62, 247–256.

for subsequent biological or ecological considerations. For example, automatic or semi-automatic measurements of body length can give useful ecological information, such as the population structure in terms of length classes. Álvarez and Pardo (2005) estimated the body length of the Trichoptera *Agapetus quadratus* using IA. The stage-frequency size histograms suggested a trivoltine population, with an average cohort period of 4 months.

Slight morphological (or anatomical) differences assessed from IA can also help to distinguish between different life-stages of a given species. From Procrustes superimposition of 2D landmark data and subsequent analyses, morphological shifts in Eurasian perch (*Perca fluviatilis*) during its growth were demonstrated (Hjelm et al., 2001). By examining otoliths of the eastern rainbowfish (*Melanotaenia splendida splendida*) under cross-polarised light, Humphrey et al. (2003) showed that circuli discontinuities appeared at hatching and during yolk sac absorption. Moreover, among other features, accurate head measurements from microscopic views can be used to differentiate instars of insect larvae (Hanus et al., 2006).

17.2.3 Ecomorphological studies

Some morphological or anatomical characteristics of an organism may be related to its ecological role or preferences. Therefore, the power of IA to discriminate fine morphological features may be very useful in ecomorphological studies. Li and Yu (2009) scanned and subsequently analysed leaf outlines of *Ranunculus natans* and found that intra-specific foliar morphological variations were important and traduced functional responses to water-quantity and water-availability factors (e.g. altitude, pH). Sagnes et al. (2008) measured body lengths in numerous individuals of aquatic insect larvae, using semi-automated IA methods. They showed that, for a given species, habitat use could be different with respect to body size and concluded that future instream habitat models should consider the respective habitat requirements of different size classes of invertebrates.

Spatial segregation of species (or ontogenetic stages of a given species) living in the same system can be inferred from the ecomorphological study of specific functions, such as locomotion (Wood and Bain, 1995). After describing and analysing body lengths, widths and fin areas from photographs, Sagnes et al. (1997, 2000) demonstrated that the body shape of the European grayling (*Thymallus thymallus*) changed during its ontogenesis towards hydrodynamically efficient shape at highest velocities. Moreover, there was a relationship

between hydrodynamic abilities and the current velocities used by the successive morphological groups. The relationship between the hydrodynamic abilities of fishes and their utilisation of different flow velocities has been quantified and confirmed, using IA, for many riverine French species (Sagnes and Statzner, 2009).

Spatial segregation between species may also be due to feeding behaviour or kinematics. Maie et al. (2009) compared food capturing kinematics and performance of two Hawaiian stream fish species, *Awaous guamensis* and *Lentipes concolor*, using morphological data and high-speed video. They analysed in-lever and out-lever arms for both jaw opening and closing (from digital photographs) and digitally synchronised lateral and ventral views of feeding strikes (movies). Landmarks on the heads of the fishes were digitised for each frame (Figure 17.9). They concluded that an elevated suction pressure (due to jaw morphology) would enhance the ability of *L. concolor* to successfully capture food in the fast stream reaches it typically inhabits.

Body ornamentation (e.g. pigmentation) can also be used for ecomorphological purposes. After scanning individuals of freshwater isopods (*Asellus aquaticus*) and determining the optical density of the substrate, Hargeby et al. (2005) showed that *Asellus* pigmentation was correlated with substrate darkness across 29 localities. This differentiation suggested that habitat heterogeneity promotes genetic diversity. In newts (*N. viridescens viridescens*), Davis and Grayson (2007) showed that terrestrial post-breeding specimens were browner than aquatic ones (which were greener), postulating that the former developed greater skin granulation, for camouflage and resistance to desiccation. This example shows that IA of body ornamentations can give significant informations about individuals' way of life or strategies.

Architecture of animal building is poorly known for freshwater species (except for beaver), although some constructions may have important implication in ecosystems functioning. Using a variety of electron microscopic and computer image enhancement techniques, Schultze-Lam et al. (1992) showed the specific involvement of bacterial surfaces in natural mineral formation processes. From detailed ultrastructural studies of cell surfaces (Figure 17.10), they demonstrated that the S-layer[6] of a cyanobacteria (belonging to the *Synechococcus* group) acted as a template for fine-grain gypsum and calcite formation by providing discrete,

---

[6] An S-layer (surface layer) is a part of the cell envelope commonly found in bacteria. It consists of a monomolecular layer composed of identical proteins.

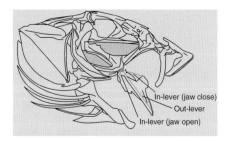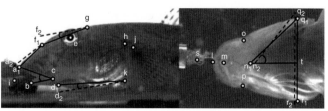

Figure 17.9 Left: Out-lever and in-lever arms for jaw opening and closing in the cranial skeleton of the Hawaiian gobiid fish Awaous guamensis. Right: Lateral and ventral views of Awaous guamensis, illustrating 11 lateral and eight ventral anatomical landmarks and angular excursions between vectors formed by landmark points. Dashed lines represent positions of corresponding lines (solid lines) when each element is further expanded toward full opening of the mouth. Reproduced from Maie, T. et al. (2009) with permission from John Wiley & Sons, Inc.

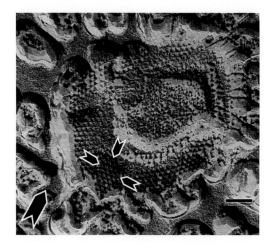

Figure 17.10 Freeze fractured and -etched cyanobacterial cell. A discontinuity in the S-layer lattice structure is visible (arrowheads). The large arrowhead indicates the shadowing direction. Scale bar = 200 nm. Reproduced from Schultze-Lam, S., Harauz, G., Beveridge, T.J. (1992) Journal of Bacteriology 174(24): 7971–7981, with permission from American Society for Microbiology.

regularly arranged nucleation sites for the critical initial events in the mineralisation process. Utilising automated IA, Statzner et al. (2009) described features of mineral grains in the pupal cases of lotic hydropsychids (e.g. number, area, shape and mass) and suggested that such characteristics could be related to forces deforming cases to fatal levels for the builder/occupant.

17.3 Abundance and biomass

Quantification of bacterial and viral cells is essential for understanding the role they play in diverse aquatic environments. Over time, many methods have been developed to enumerate microscopic organisms, of which epifluorescence microscopy has become the principle method for direct enumeration of bacteria without cultivation (Kepner and Pratt, 1994). Utilising this approach, bacteria can simply be differentiated from detritus by colour segmentation (Shopov et al., 2000 and see paragraph. 17.4.2). In some studies, the combination of epifluorescent microscopy and IA has been used to speed up sample processing and to support differentiation between heterotrophic and autotrophic filaments (see Massana et al. 1997 for a calibration of these techniques). Cynobacteria total filament length can be measured and Walsby and Avery (1996) developed an IA method for correcting errors arising from the orientation, crossing and overlapping of filaments. However, biovolume was shown to be a better measure of biomass than cell number or length. To estimate the biovolume of filtered bacterioplankton, Krambeck et al. (1981) developed a semi-automatic system that assisted cell size measurements on images: bacterial length and width were manually marked by a cursor on the image and coordinates were directly transferred to the computer. Fry and Davies (1985) stated that the optimum technique for measuring volumes of planktonic bacteria was to filter acridine orange stained bacteria through polycarbonate membrane filters, to photograph them with epifluorescence microscopy and then to estimate volumes from individual area and perimeter measurements (using IA). Following this technique, Bjørnsen (1986) revealed an empirical conversion factor from bacterioplankton biovolume to biomass.

During epifluorescence microscopy, halation can bias the automatic estimations of cells size and shape. To avoid this problem, Tani et al. (1996) suggested the use of a scanning electron microscopic for IA to measure bacterial biovolume. This method allowed the determination of biovolume-to-carbon and nitrogen conversion factors. They also showed that the average volume of bacterial cells was twice as high in polluted waters, while the bacterial biomass was 35 times higher than in unpolluted water. However, the combination of epifluorescent microscopy and IA is still utilised for studying the functioning of food webs and, subsequently, the recycling of nutrients (e.g. Das et al., 2007).

At a larger scale, macrophyte biomass can be estimated from reflectance measurements (see a review in Silva et al. 2008). Remote sensing has made it possible to assess the biomass of a given species from the relationships between reflectance values and ground estimations of the biomass of submerged vegetation (Valta-Hulkkonen et al., 2004). At a smaller scale, IA was shown to be a valuable tool for accurate estimations of surface areas and volumes of different plant parts (Gerber et al., 1994).

The estimation of invertebrate biomass (or secondary production) often required detailed and time consuming counting and measurement or organisms. IA can perform such tasks at a reasonable speed and level accuracy. Alver et al. (2007) designed a particle counter for making automatic measurements of rotifer densities. The instrument automatically extracted samples from tanks, and relied on a digital camera and image processing to measure density. Due to its autonomous nature, this apparatus could be used in monitoring and control processes. Bernardini et al. (2000) developed a method to quickly estimate individual biomass from video-recorded images of macroinvertebrates. The corresponding software automatically measured several body dimensions (area, perimeter, minor and major axes) on each individual and related these dimensions to dry weight using linear regressions. In this study, body area was found to be the best predictor of dry weight. In comparison, Davis and Grayson (2007) showed that total body surface area of newts (*N. viridescens viridescens*) provided a better correlation with newt mass than did body length using an automated IA; therefore total body surface area could potentially serve as an alternative to estimate individual mass or body condition.

Moreover, invertebrate biomass may be indirectly assessed from vegetation characteristics. McAbendroth et al. (2005) used fractal indices to describe the structural complexity of mixed stands of aquatic macrophytes, and these were employed to examine the effects of habitat complexity on the composition of invertebrate assemblages. After IA, and especially the quantification of the fractal dimension of both perimeter and area, fractal

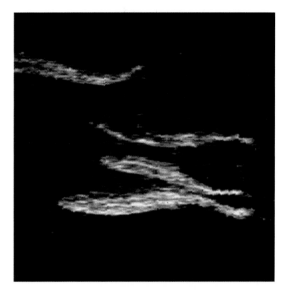

Figure 17.11 Image of swimming salmons, captured with an imaging SONAR. Imaged sourced from http://www .soundmetrics.com/industries-served/aquatic-life/fish-surveys, and reproduced courtesy of Sound Metrics Corporation.

indices were found to be significantly related to both invertebrate biomass scaled by body size and overall invertebrate biomass: more complex stands of macrophytes contained a greater number of small animals.

During automated species recognition in fishways specimens are usually counted (see 17.2.1.1 above). The most common automatic methods for fish enumeration involve acoustic techniques (Figure 17.11), which can be as accurate as visual counts (e.g. Holmes et al., 2006). If necessary, acoustic techniques can be coupled with imagery techniques (e.g. shape description of fish schools: Reid and Simmonds, 1993).

Enumeration of large animals (manatees, riverine dolphins, otters, flamingos ...) can also be carried out from air structures (e.g. planes, helicopters or fixed-wing platforms). Although these counts are commonly processed by eye, some low-resolution images (and/or high density populations) may need automatic procedures, for which classical image processing tools such as mathematical morphology are unsatisfactory. In this context, techniques combining geometric models and environmental constraints (colors, spatial repartition ...) were developed (Descamps et al., 2008; Figure 17.12). These techniques are less time consuming than classical ones, such as expert counting on some predefined small areas, and avoid the associated interpolation procedures.

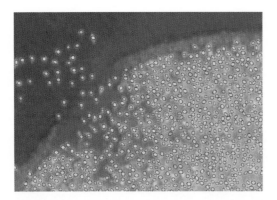

Figure 17.12 Automatic detection (before counting) of Greater Flamingos (*Phoenicopterus Roseus*) from a low-resolution aerial image (from Descamps et al., 2008). Each pink circle represents a specimen. Reproduced with permission from Descamps, S. et al. (2008) Automatic flamingo detection using a multiple birth and death process. In Proc. IEEE International Conference on Acoustics, Speech and Signal Processing (ICASSP), Las Vegas, March 2008. © INRIA/Tour du Valat.

As a complement to counting, imagery may also allow the estimation of fish biomass, which provides useful information for ecological or applied studies. Hockaday et al. (2000) demonstrated that individual fish biomass can be accurately estimated from truss dimensions. Ruff et al. (1995) described a non-invasive technology based on optical stereo visualisation and computer IA for continuous monitoring of size, position, shape, and spatial orientation of single fish among many others. Stereo IA requires two views of a fish, so that a point on the fish in one 2D image may be matched with the corresponding point in the second image. Given a calibration of the optical arrangement of the system, the coordinates of these two points may be used to directly estimate the 3D coordinates. Using this technique, fish dimensions could be measured to millimeter accuracy and fish could be tracked over limited time intervals to observe detailed movement. This underwater stereo IA technique was used to estimate key dimensions of free-swimming fish (Tillett et al., 2000), from which fish mass could be estimated (Lines et al., 2001).

17.4 Detection of stress and diseases

Visual inspection of the appearance and/or the activity of animals and plants can provide a variety of information regarding their health and development, or the presence of potential factors of stress in their environment.

17.4.1 Direct visualisation of stress (or its effects)

At trace concentrations, copper is an essential micronutrient for most living organisms, because it is important for many biologically important reactions. However, copper can become highly cytotoxic if accumulated in excess of cellular needs (e.g. causing oxidative damage). Nanoscale secondary-ion mass spectrometry has been able to detect and visualise the copper-ion distribution in microalgal cells exposed to copper (Slaveykova et al., 2009; Figure 17.13).

Variations of the physiological characteristics (e.g. under factors of stress) of macrophytes can be inferred from their spectral response during reflectance measurements, due to alterations in optically active substances (e.g. chlorophyll concentration, chemical composition, photosynthetic efficiency) under environmental pressures (see a review in Silva et al., 2008). At a smaller scale, plant damage due to herbivory (or other factors of stress) can be assessed from simple IA (e.g. quantification of leaf area removed or damaged plants: O'Neal et al., 2002) or more complex investigations (e.g. chlorophyll fluorescence or thermal imaging: Aldea et al. 2006).

The analysis of molluscan internal organs is widely utilized for biomonitoring purposes. For example, molluscan kidneys are able to excrete solids in the urine in the form of concretions and it is thought that increased formation of these concretions occur under stresses such as toxic exposure. Klobucar et al. (2001) examined the formation of such concretions in the kidney of the freshwater snail *Planorbarius corneus* experimentally exposed to pentachlorophenol (PCP). IA of the PCP-exposed individuals revealed significantly enhanced production of the kidney concretions (e.g. number of kidney concretions and area of the concretions) when compared to control individuals. After manual segmentation of microscopic images, Giambérini and Cajaraville (2005) showed the general activation of the lysosomal system, including an increase in both the number and the size of lysosomes, in the digestive gland cells of the zebra mussel (*Dreissena polymorpha*) experimentally exposed to cadmium. This type of IA can also be used for *in situ* experiments to examine the effects of toxic exposure on specific organs (e.g. Guerlet et al., 2006).

To enlarge the range of potential sentinel species, Guerlet et al. (2008), utilised histochemistry coupled to IA, to assess the spatial and temporal (monthly) morphological variations of four cellular compartments and their contents in the hepatopancreatic caeca of the freshwater

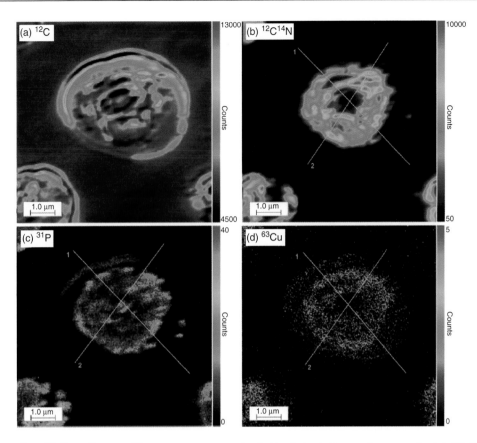

Figure 17.13 $^{12}C^-$, $^{12}C^{14}N^-$, $^{31}P^-$, and $^{63}Cu^-$ distributions in cells of *Chlorella kesslerii* exposed to $10\,\mu mol\,L^{-1}$ Cu(II). The straight lines represent the line scans made in two regions of interest. Different colours correspond to the different intensities of the signal, which increase from black to red. With kind permission from Springer Science+Business Media: *Analytical and Bioanalytical Chemistry, Dynamic NanoSIMS ion imaging of unicellular freshwater algae exposed to copper*, 393, 2009, 583–589, Slaveykova, V.I., Guignard, C., Eybe, T., Migeon, H.-N. and Hoffmann, L.

gammarid, *Dikerogammarus villosus*. They demonstrated that unsaturated neutral lipids were more abundant, the surface densities of the lysosomal and peroxisomal systems were, respectively, less and more important, and lipofuscin granules tended to accumulate in the amphipods from the most anthropised site. In a same way, morphometry and pigment composition of macrophage aggregates, which are thought to be dependent on the fish species, age, and health status, were studied in spleen and kidney of Blue Gourami (*Trichogaster trichopterus*) (Russo et al., 2007).

Hematological indices are also gaining general acceptance as valuable tools in monitoring various aspects of fish health when exposed to contaminants. Oliveira Ribeiro et al. (2006) examined the effects of methyl mercury (MeHg), inorganic lead (Pb2+), and tributyltin

(TBT) on abundance and morphometry of various blood cells of the top predator fish *Hoplias malabaricus* (after trophic exposure). Among other effects, they found differences in area, elongation, and roundness of erythrocytes for individuals exposed to Pb2+ and TBT. IA methods can also be used in histopathological studies, such as fish gills structure as indicators of water quality (Oropesa-Jimenez et al., 2005) or to detect environmental stresses from the quality of genetic structures (Hayashi et al., 1998).

Finally, fish parasites have generally been found to be useful indicative models for the indication of environmental disturbances (Koskivaara, 1992). By identifying the parasite species of chub (*Leuciscus cephalus*) using IA, Dušek et al. (1998) revealed a relationship between the diversity of parasites and the pollution level at the sites studied.

17.4.2 Activity of organisms as stress indicator

Bacteria play an important part in the functioning of aquatic ecosystems (e. g. crucial link in food webs, carbon cycling and degradation of heavy metals), and imagery methods for detecting physiologically active bacteria in natural microbial communities have been developed. Ogawa et al. (2003) presented a simple multicolor digital IA system, which differentiated between actively respiring bacteria and non-respiring bacteria based on distinctive color information using epifluorescence microscopic images. They also developed an algorithm to distinguish bacteria from detritus based on color segmentation, which produced bright fluorescence in different colors. Utilising a similar approach, a two-dye fluorescence bacterial viability kit rapidly distinguished between live and dead bacterial cells (Amalfitano et al., 2008).

Phosphorus concentrations, the main nutrient responsible for eutrophication, can be assessed indirectly by estimating the activity of various organisms. Using the 'easy image analyzer 2000 software', Kragh et al. (2008) analysed freshwater bacterial abundance, biovolume and growth efficiency. They were able to demonstrate experimentally that the availability of phosphorus has major implications in growing processes and, subsequently, for the quantitative transfer of carbon in microbial food webs. The regulated phosphatise activity of several phytoplankton species, based on external phosphorus concentration, were also demonstrated by fluorescence quantification (enzyme-labelled fluorescence technique) (Štrojsová et al., 2005). From fluorescence microscopy coupled with IA, Znachor and Nedoma (2008) quantified silicon (Si) deposition over time and distinguished between diatom cells that were actively depositing Si and those that were not. They showed that, at the water surface, silica deposition by diatoms was limited by phosphorus deficiency.

Burbank and Snell (1994) developed biomarkers of sublethal toxicity in the freshwater rotifer *Brachionus calyciflorus* based on the reduction of enzyme activity. Esterase and phospholipase A2 activity was quantified in single rotifers using IA and a fluorescence detection system. Quantification of enzyme activity demonstrated that toxic stress reduced rotifer activity in a dose-dependent manner. Johnson and Delaney (1998) showed that growth of the waterflea *Daphnia magna* may be a useful short-term (7 day) indicator of chronic toxicity (i.e. zinc and 3,4-dichloroaniline exposure) based on the body length measured on day 0, 2, 4, and 7 using an IA method. Population growth rate (PGR) of *D. magna* can also be used to estimate environment suitability (Hooper et al.,

2008). Moreover, for ecotoxicological purposes, IA may help to interpret behaviour patterns. Juchelka and Snell (1994) reported the effects of several types of toxicants on rotifer ingestion rate, by quantifying gut fluorescence (in this experiment, rotifers ingested fluorescently labeled latex spheres).

IA may also provide adequate tools to quantify motility of organisms, which is currently the most studied activity pattern for detecting various types of stress. The ability of the unicellular flagellate *Euglena gracilis* to oriente in the gravity field, its mobility and velocity were found to be sensitive to a number of toxic compounds. Therefore, a warning system for monitoring the quality of water has been developed using a real time IA system which quantified the movement parameters of this species (Tahedl and Häder, 1999). The software used the vectors of the tracks to calculate the number of motile cells, the percentage of cells moving upwards, their mean velocity and their precise orientation. Statistically significant changes in the parameters indicated variability in water quality, even over very short time periods. In similar studies, Unterstainer et al. (2003) showed that, after only 9 h of high copper exposure, a significant decrease of the average swimming velocity of *D. magna* was observed. In their experiment, they recorded the coordinates of daphnids from two frames and automatically reconstructed their trajectories. Lee et al. (2005) showed that individual medaka fish (*Oryzias latipes*), affected by diazinon (an insecticide) were less active, and their movement behaviour was more erratic than that of unaffected/control specimens. Finally, note that motility of gametes (a task that will not be developed here) may also help to assess, among other things, the potential hazards of environmental pollutants on reproduction (see a review in Kime et al., 2001, and further developments for fish in Van Look and Kime, 2003 and Marco-Jimenez et al., 2006).

17.4.3 Fluctuating asymmetry as stress indicator

Fluctuating asymmetry of bilateral traits (FA) refers to a population-level pattern of slight bilateral morphometric asymmetries observed in individuals, with no consistent directional pattern being observed in the population as a whole. Although a low FA is not the '*unambiguous measure of well being and good genes that has been claimed*' (Rasmuson, 2002), FA can increase under various stress conditions and, subsequently, is a potential indicator of past and present factors of stress. Servia et al. (2004) measured FA levels in various structures of the head capsule in fourth instar *Chironomus riparius* larvae (e.g. mentum width and antennal segment lengths), which were collected from various sampling sites subjected to different types and degrees of stress. The analysis grouped the sampling sites in a similar way to other measures of stress (e.g. head capsule deformities). However, FA as a stress indicator should be used carefully. Servia et al. (2004) suggested that several characters, not one, should be considered in such analyses. Davis and Grosse (2008) indicated that the magnitude of FA variation could differ between genders (e.g. higher among male compared to female turtles) and that asymmetry may vary naturally with increasing age.

17.5 Conclusion

With the emergence of new IA techniques, it is now possible to push the boundaries by examining objects either smaller or presenting more complex shapes. The development of these techniques has greatly improved accuracy and precision of various measurements and subsequent objectivity and coherence in decision making. The techniques for catching, correcting, sorting and segmenting images still increase in quality; new and powerful algorithms for extracting information from pictures and for analysing data (e.g. 3D and even 4D data, see Zhang et al., 2010) are still developed. Artificial intelligence, through artificial neural networks or fuzzy logic theory, is now being applied to image understanding (e.g. Hemanth et al., 2010). Although these recent developments often emerge from medical research, nothing prevents their use for ecological purposes. For instance, in aquatic environments, some current IA developments are oriented towards 1) a better detection of potential risks for human and animal health (e.g. through early detection of cyanobacteria or algae: Vardon et al., 2011); 2) a better detection of alien species, such as undesired macrophytes in riparian zones (Jones et al., 2011), that maybe conducted from satellite imagery (Dlamini, 2011); 3) a better automatic detection and classification of species (or stocks) from underwater videos (Spampinato et al., 2010), that can be of importance, for example, for the management of patrimonial species or the detection of invasive ones; 4) the use of non-invasive sex determination procedures (Ferreira et al., 2010), that can be of interest for the detection of environmental exposure to endocrine disruptive chemicals; 5) the documentation about altered morphological and anatomical features of organisms after an exposure to nanoparticles (Laban et al.,

2010), which is currently another major environmental problem. Obviously, the currently already amazing IA techniques will continue to grow in power and utility in applied and fundamental freshwater studies, towards actually unexpected applications.

References

Abecasis, D., Bentes, L., Coelho, R., Correia, C., Lino, P.G., Monteiro, P., Gonçalves, J.M.S., Ribeiro, J. and Erzini, K. 2007. Ageing seabreams: a comparative study between scales and otoliths. *Fisheries Research* 89, 37–48.

Aldea, M., Hamilton, J.G., Resti, J.P., Zangerl, A.R., Berenbaum, M.R., Frank, T.D. and DeLucia, E.H. 2006. Comparison of photosynthetic damage from arthropod herbivory and pathogen infection in understory hardwood saplings. *Oecologia* 149, 221–232.

Álvarez, M. and Pardo, I. 2005. Life history and production of *Agapetus quadratus* (Trichoptera: Glossosomatidae) in a temporary, spring-fed stream. *Freshwater Biology* 50, 930–943.

Alver, M.O., Tennøy, T., Alfredsen, J.A. and Øie, G. 2007. Automatic measurement of rotifer *Brachionus plicatilis* densities in first feeding tanks. *Aquacultural Engineering* 36, 115–121.

Amalfitano, S., Fazi, S., Zoppini, A., Barra Caracciolo, A., Grenni, P. and Puddu, A. 2008. Responses of benthic bacteria to experimental drying in sediments from Mediterranean temporary rivers. *Microbiology Ecology* 55, 270–279.

Belchier, M., Edsman, L., Sheehy, M.R.J. and Shelton, P.M.J. 1998. Estimating age and growth in long-lived temperate freshwater crayfish using lipofuscin. *Freshwater Biology* 39, 439–446.

Bernard, R.L. and Myers, K.W. 1996. The performance of quantitative scale pattern analysis in the identification of hatchery and wild steelhead (*Oncorhynchus mykiss*). *Canadian Journal of Fisheries and Aquatic Sciences* 53, 1727–1735.

Bernardini, V., Solimini, A.G. and Carchini, G. 2000. Application of an image analysis system to the determination of biomass (ash free dry weight) of pond macroinvertebrates. *Hydrobiologia* 439, 179–182.

Bertin, A., David, B., Cézilly, F. and Alibert, P. 2002. Quantification of sexual dimorphism in *Asellus aquaticus* (Crustacea: Isopoda) using outline approaches. *Biological Journal of the Linnean Society* 77, 523–533.

Bjørnsen, P.K. 1986. Automatic determination of bacterioplankton biomass by image analysis. *Applied and Environmental Microbiology* 51, 1199–1204.

Bookstein, F.L. 1986. Size and shape spaces for landmark data in two dimensions. *Statistical Science* 1, 181–222.

Burbank, S.E. and Snell, T.W. 1994. Rapid toxicity assessment using esterase biomarkers in *Brachionus calyciflorus* (Rotifera). *Environmental Toxicology and Water Quality* 9, 171–178.

Cadieux, S., Lalonde, F. and Michaud, F. 2000. Intelligent system for automated fish sorting and counting. In *International Conference on Intelligent Robots and Systems*, vol. 2, pp. 1279–1284. Takamatsu, Japan.

Cadrin, S.X. 2000. Advances in morphometric identification of fishery stocks. *Reviews in Fish Biology and Fisheries* 10, 91–112.

Cadrin, S.X. and Friedland, K.D. 1999. The utility of image processing techniques for morphometric analysis and stock identification. *Fisheries Research* 43, 129–139.

Campana, S.E. and Casselman, J.M. 1993. Stock discrimination using otolith shape analysis. *Canadian Journal of Fisheries and Aquatic Sciences* 50, 1062–1083.

Cao, F. and Fablet, R. 2006. Automatic morphological detection of otolith nucleus. *Pattern Recognition Letters* 27, 658–666.

Castignolles, N., Cattoen, M. and Larinier, M. 1994. Identification and counting of live fish by image analysis. In *Image and Video Processing II*, vol. 2182 eds. S. Rajala and R.L. Stevenson, pp. 200–209. San Jose, CA, USA: Proc. SPIE.

Chambah, M., Semani, D., Renouf, A., Courtellemont, P. and Rizzi, A. 2004. Underwater color constancy: enhancement of automatic live fish recognition. In *Electronic Imaging*, vol. 5293 (ed. SPIE). San Jose, CA, USA.

Claytor, R.R. and MacCrimmon, H.R. 1988. Morphometric and meristic variability among North American Atlantic salmon (*Salmo salar*). *Canadian Journal of Zoology* 66, 310–317.

Corti, M., Thorpe, R.S., Sola, L., Sbordoni, V. and Cataudella, S. 1988. Multivariate morphometrics in aquaculture: a case study of six stocks of the Common Carp (*Cyprinus carpio*) from Italy. *Canadian Journal of Fisheries and Aquatic Sciences* 45, 1548–1554.

Das, M., Royer, T.V. and Leff, L.G. 2007. Diversity of fungi, bacteria, and actinomycetes on leaves decomposing in a stream. *Applied and Environmental Microbiology* 73, 756–767.

Davis, A.K. and Grayson, K.L. 2007. Improving natural history research with image analysis: the relationship between skin color, sex, size and stage in adult red-spotted newts (*Notophthalmus viridescens viridescens*). *Herpetological Conservation and Biology* 2, 65–70.

Davis, A.K. and Grosse, A.M. 2008. Measuring fluctuating asymmetry in plastron scutes of yellow-bellied sliders: the importance of gender, size and body location. *American Midland Naturalist* 159, 340–348.

Davis, N.D. and Light, J.T. 1985. Steelhead age determination techniques. (Document submitted to annual meeting of the INPFC, Tokyo, Japan, November 1985.) Fisheries Research Institute, University of Washington, Seattle, Wash, 41 pp.

Descamps, S., Descombes, X., Béchet, A. and Zerubia, J. 2008. Automatic flamingo detection using a multiple birth and death process. *In Proc. IEEE International Conference on Acoustics, Speech and Signal Processing (ICASSP)*. Las Vegas: USA, March 2008.

Dlamini, W.M. 2011. Using probabilistic graphical models for invasive alien plant detection from Worldview-2 satellite imagery. *DigitalGlobe 8-Band Research Challenge*. Published online at: http://dgl.us.neolane.net/res/img/ 33b4ff59d197e9df3b8310515ae3af1e.pdf (accessed on 2011-XI-03).

Du, J.-X., Wang, X.-F. and Zhang, G.-J. 2007. Leaf shape based plant species recognition. *Applied Mathematics and Computation* 185, 883–893.

Dunca, E., Schöne, B.R. and Mutvei, H. 2005. Freshwater bivalves tell of past climates: But how clearly do shells from polluted rivers speak? *Palaeogeography, Palaeoclimatology, Palaeoecology* 228, 43–57.

Dušek, L., Gelnar, M. and Šebelová. 1998. Biodiversity of parasites in a freshwater environment with respect to pollution: metazoan parasites of chub (*Leuciscus cephalus* L.) as a model for statistical evaluation. *International Journal for Parasitology* 28, 1555–1571.

Ferreira, F., Santos, M.M., Reis-Henriques, M.A., Vieira, N.M. and Monteiro, N.M. 2010. Sexing blennies using genital papilla morphology or ano-genital distance. *Journal of Fish Biology* 77, 1432–1438.

Frias-Torres, S. and Luo, J. 2009. Using dual-frequency sonar to detect juvenile goliath grouper Epinephelus itajara in mangrove habitat. *Endangered Species Research* 7, 237–242.

Fry, J.C. and Davies, A.R. 1985. An assessment of methods for measuring volumes of planktonic bacteria, with particular reference to television image analysis. *Journal of Applied Bacteriology* 58, 105–112.

Gaemers, P.A.M. 1984. Taxonomic position of the Cichlidae (Pisces, Perciformes) as demonstrated by the morphology of their otoliths. *Netherlands Journal of Zoology* 34, 566–595.

Gaston, K.J. and O'Neill, M.A. 2004. Automated species identification: why not? *Philosophical Transactions of the Royal Society of London Series B Biological Sciences* 359, 655–667.

Gerber, D.T., Ehlinger, T.J. and Les, D.H. 1994. An image analysis technique to determine the surface area and volume for dissected leaves of aquatic macrophytes. *Aquatic Botany* 48, 175–182.

Giambérini, L. and Cajaraville, M.P. 2005. Lysosomal responses in the digestive gland of the freshwater mussel, *Dreissena polymorpha*, experimentally exposed to cadmium. *Environmental Research* 98, 210–214.

Ginoris, Y.P., Amaral, A.L., Nicolau, A., Coelho, M.A.Z. and Ferreira, E.C. 2007. Recognition of protozoa and metazoa using image analysis tools, discriminant analysis, neural networks and decision trees. *Analytica Chimica Acta* 595, 160–169.

Guerlet, E., Ledy, K. and Giambérini, L. 2006. Field application of a set of cellular biomarkers in the digestive gland of the freshwater snail *Radix peregra* (Gastropoda, Pulmonata). *Aquatic Toxicology* 77, 19–32.

Guerlet, E., Ledy, K. and Giambérini, L. 2008. Is the freshwater gammarid, *Dikerogammarus villosus*, a suitable sentinel species for the implementation of histochemical biomarkers? *Chemosphere* 72, 697–702.

Hagen, O., Solberg, C. and Johnston, I.A. 2006. Sexual dimorphism of fast muscle fibre recruitment in farmed

Atlantic halibut (*Hippoglossus hippoglossus* L.). *Aquaculture* 261, 1222–1229.

Hanus, R., Šobotník, J., Valterová, I. and Lukáš, J. 2006. The ontogeny of soldiers in *Prorhinotermes simplex* (Isoptera, Rhinotermitidae). *Insectes Sociaux* 53, 249–257.

Hard, J.J., Berejikian, B.A., Tezak, E.P., Schroder, S.L., Knudsen, C.M. and Parker, L.T. 2000. Evidence for morphometric differentiation of wild and captively reared adult coho salmon: a geometric analysis. *Environmental Biology of Fishes* 58, 61–73.

Hargeby, A., Stoltz, J. and Johansson, J. 2005. Locally differentiated cryptic pigmentation in the freshwater isopod *Asellus aquaticus*. *Journal of Evolutionary Biology* 18, 713–721.

Hatch, D.R., Pederson, D.R., Fryer, J.K., Schwartzberg, M. and Wand, A. 1994. The feasibility of documenting and estimating adult fish passage at large hydroelectric facilities in the Snake river using video technology, 131 pp. Portland: U.S. Department of Energy.

Hayashi, M., Ueda, T., Uyeno, K., Wada, K., Kinae, N., Saotome, K., Tanaka, N., Takai, A., Sasaki, Y.F., Asano, N., Sofuni, T. and Ojima, Y. 1998. Development of genotoxicity assay systems that use aquatic organisms. *Mutation Research* 399, 125–133.

Heinke, F. 1878. Die varietäten des herings I. *Jahresbuch, Kommission für die Untersuchungen der Deutschen Meere in Kiel* 4/6, 37–132.

Hemanth, J., Selvathi, D. and Anitha, J. 2010. Artificial Intelligence Techniques for Medical Image Analysis: Basics, Methods, Applications. Saarbrücken, Germany: VDM Verlag.

Hjelm, J., Svanbäck, R., Byström, P., Persson, L. and Wahlström, E. 2001. Diet-dependent body morphology and ontogenetic reaction norms in Eurasian perch. *Oikos* 95, 311–323.

Hockaday, S., Beddow, T.A., Stone, M., Hancock, P. and Ross, L.G. 2000. Using truss networks to estimate the biomass of *Oreochromis niloticus*, and to investigate shape characteristics. *Journal of Fish Biology* 57, 981–1000.

Holmes, J.A., Cronkite, G.M.W., Enzenhofer, H.J. and Mulligan, T.J. 2006. Accuracy and precision of fish-count data from a 'dual-frequency identification sonar' (DIDSON) imaging system. *ICES Journal of Marine Science* 63, 543–555.

Hooper, H.L., Connon, R., Callaghan, A., Fryer, G., Yarwood-Buchanan, S., Biggs, J., Maund, S.J., Hutchinson, T.H. and Sibly, R.M. 2008. The ecological niche of *Daphnia magna* characterized using population growth rate. *Ecology* 89, 1015–1022.

Hooper, H.L., Connon, R., Callaghan, A., Maund, S.J., Liess, M., Duquesne, S., Hutchinson, T.H., Moggs, J. and Sibly, R.M. 2006. The use of image analysis to estimate population growth rate in *Daphnia magna*. *Journal of Applied Ecology* 43, 828–834.

Humphrey, C., Klumpp, D.W. and Pearson, R.G. 2003. Early development and growth of the eastern rainbowfish, *Melanotaenia splendida splendida* (Peters) II. Otolith development, increment validation and larval growth. *Marine and Freshwater Research* 54, 105–11.

Iqbal, K., Salam, R.A., Osman, A. and Talib, A.Z. 2007. Underwater image enhancement using an integrated colour model. *IAENG International Journal of Computer Science* 34, 2.

Ishikawa, K., Kumagai, M. and Walker, R.F. 2005. Application of autonomous underwater vehicle and image analysis for detecting the three-dimensional distribution of freshwater red tide *Uroglena americana* (Chrysophyceae). *Journal of Plankton Research* 27, 1–6.

Jennings, M.J. and Philipp, D.P. 1992. Reproductive investment and somatic growth rates in longear sunfish. *Environmental Biology of Fishes* 35, 257–271.

Johnson, I. and Delaney, P. 1998. Development of a 7-Day *Daphnia magna* growth test using image analysis. *Bulletin of Environmental Contamination and Toxicology* 61, 355–362.

Jones, D., Pike, S., Thomas, M. and Murphy, D. 2011. Object-Based Image Analysis for Detection of Japanese Knotweed *s.l.* taxa (Polygonaceae) in Wales (UK). *Remote Sensing* 3, 319–342.

Juchelka, C.M. and Snell, T.W. 1994. Rapid toxicity assessment using rotifer ingestion rate. *Archives of Environmental Contamination and Toxicology* 26, 549–554.

Kepner, R.L. and Pratt, J.R. 1994. Use of fluorochromes for direct enumeration of total bacteria in environmental samples: past and present. *Microbiological Reviews* 58, 603–615.

Keränen, M., Aro, E.-M., Tyystjärvi, E. and Nevalainen, O. 2003. Automatic plant identification with chlorophyll fluorescence fingerprinting. *Precision Agriculture* 4, 53–67.

Kime, D.E., Van Look, K.J.W., McAllister, B.G., Huyskens, G., Rurangwa, E. and Ollevier, F. 2001. Computer-assisted sperm analysis (CASA) as a tool for monitoring sperm quality in fish. *Comparative Biochemistry and Physiology -C- Toxicology & Pharmacology* 130, 425–433.

Klobucar, G.I.V., Lajtner, J. and Erben, R. 2001. Increase in number and size of kidney concretions as a result of PCP exposure in the freshwater snail *Planorbarius corneus* (Gastropoda, Pulmonata). *Diseases of Aquatic Organisms* 44, 149–154.

Koskivaara, M. 1992. Environmental factors affecting monogeneans parasitic on freshwater fishes. *Parasitology Today* 8, 339–342.

Kragh, T., Søndergaard, M. and Tranvik, L. 2008. Effect of exposure to sunlight and phosphorus-limitation on bacterial degradation of coloured dissolved organic matter (CDOM) in freshwater. *Microbiology Ecology* 64, 230–239.

Krambeck, C., Krambeck, H.J. and Overbeck, J. 1981. Microcomputer-assisted biomass determination of plankton bacteria on scanning electron micrographs. *Applied and Environmental Microbiology* 42, 142–149.

Laban, G., Nies, L.F., Turco, R.F., Bickham, J.W. and Sepúlveda, M.S. 2010. The effects of silver nanoparticles on fathead minnow (*Pimephales promelas*) embryos. *Ecotoxicology* 19, 185–195.

Lee, D.-J., Schoenberger, R., Shiozawa, D., Xu, X. and Zhan, P. 2004. Contour matching for a fish recognition and migration monitoring system. In *Two and Three-Dimensional Vision*

Systems for Inspection, Control, and Metrology II, (ed. S.O. East). Philadelphia, PA, USA.

Lee, N., Nielsen, P.H., Andreasen, K.H., Juretschko, S., Nielsen, J.L., Schleifer, K.-H. and Wagner, M. 1999. Combination of fluorescent *in situ* hybridization and microautoradiography – a new tool for structure-function analyses in microbial ecology. *Applied and Environmental Microbiology* 65, 1289–1297.

Lee, S., Kim, J., Baek, J.-Y., Han, M.-W., Ji, C.W. and Chon, T.-S. 2005. Movement analysis of Medaka (*Oryzias latipes*) for an insecticide using decision tree. *Lecture Notes in Computer Science* 3735, 150–162.

Li, Z. and Yu, D. 2009. Factors affecting leaf morphology: a case study of *Ranunculus natans* C.A. Mey. (Ranunculaceae) in the arid zone of northwest China. *Ecological Research*, Online version, downloaded August 20, 2009.

Lines, J.A., Tillett, R.D., Ross, L.G., Chan, D., Hockaday, S. and McFarlane, N.J.B. 2001. An automatic image-based system for estimating the mass of free-swimming fish. *Computers and Electronics in Agriculture* 31, 151–168.

Liu, J., Dazzo, F.B., Glagoleva, O., Yu, B. and Jain, A.K. 2001. CMEIAS: a computer-aided system for the image analysis of bacterial morphotypes in microbial communities. *Microbial Ecology* 41, 173–194.

Lodge, D.M. 1993. Biological invasions: lessons for ecology. *Trends in Ecology and Evolution* 8, 133–137.

MacInnis, A.J. and Corkum, L.D. 2000. Fecundity and reproductive season of the round goby *Neogobius melanostomus* in the Upper Detroit River. *Transactions of the American Fisheries Society* 129, 136–144.

Maie, T., Wilson, M.P., Schoenfuss, H.L. and Blob, R.W. 2009. Feeding kinematics and performance of Hawaiian stream gobies, *Awaous guamensis* and *Lentipes concolor*: linkage of functional morphology and ecology. *Journal of Morphology* 270, 344–356.

Marchetti, M.P., Moyle, P.B. and Levine, R. 2004. Invasive species profiling? Exploring the characteristics of non-native fishes across invasion stages in California. *Freshwater Biology* 49, 646–661.

Marco-Jiménez, F., Pérez, L., Viudes de Castro, M.P., Garzón, D.L., Peñaranda, D.S., Vicente, J.S., Jover, M. and Asturiano, J.F. 2006. Morphometry characterisation of European eel spermatozoa with computer-assisted spermatozoa analysis and scanning electron microscopy. *Theriogenology* 65, 1302–1310.

Mason, J.E. 1974. A semi-automatic machine for counting and measuring circuli on fish scales. In *The ageing of fish*, (ed. T.B. Bagenal), pp. 87–102. Surrey, England: Unwin Brothers.

Massana, R., Gasol, J.M., Bjørnsen, P.K., Blackburn, N., Hagström, Å., Hietanen, S., Hygum, B.H., Kuparinen, J. and Pedrós-Alió, C. 1997. Measurement of bacterial size via image analysis of epifluorescence preparations: description of an inexpensive system and solutions to some of the most common problems. *Scientia Marina* 61, 397–407.

McAbendroth, L., Ramsay, P.M., Foggo, A., Rundle, S.D. and Bilton, D.T. 2005. Does macrophyte fractal complexity drive invertebrate diversity, biomass and body size distributions? *Oikos* 111, 279–290.

Morales-Nin, B., Lombarte, A. and Japón, B. 1998. Approaches to otolith age determination: image signal treatment and age attribution. *Scientia Marina* 62, 247–256.

Mueller, L.N., de Brouwer, J.F.C., Almeida, J.S., Stal, L.J. and Xavier, J.B. 2006. Analysis of a marine phototrophic biofilm by confocal laser scanning microscopy using the new image quantification software PHLIP. *BMC Ecology* 6, 1.

Ogawa, M., Tani, K., Yamaguchi, N. and Nasu, M. 2003. Development of multicolour digital image analysis system to enumerate actively respiring bacteria in natural river water. *Journal of Applied Microbiology* 95, 120–128.

Oliveira Ribeiro, C.A., Filipak Neto, F., Mela, M., Silva, P.H., Randi, M.A.F., Rabitto, I.S., Alves Costa, J.R.M. and Pelletier, E. 2006. Hematological findings in neotropical fish *Hoplias malabaricus* exposed to subchronic and dietary doses of methylmercury, inorganic lead, and tributyltin chloride. *Environmental Research* 101, 74–80.

O'Neal, M.E., Landis, D.A. and Isaacs, R. 2002. An inexpensive, accurate method for measuring leaf area and defoliation through digital image analysis. *Journal of Economic Entomology* 95, 1190–1194.

Oropesa-Jiménez, A.L., García-Cambero, J.P., Gómez-Gordo, L., Roncero-Cordero, V. and Soler-Rodríguez, F. 2005. Gill modifications in the freshwater fish *Cyprinus carpio* after subchronic exposure to simazine. *Bulletin of Environmental Contamination and Toxicology* 74, 785–792.

Paterson, D.M., Perkins, R., Consalvey, M. and Underwood, G.J.C. 2003. Ecosystem function, cell microcycling and the structure of transient biofilms. In *Fossil and recent biofilms – A natural history of life on Earth*, eds W.E. Krumbein D.M. Paterson and G.A. Zavarzin, pp. 47–63, Chapter 3. Dordrecht, The Netherlands: Kluwer Academic Publishers.

Popper, A.N., Ramcharitar, J. and Campana, S.E. 2005. Why otoliths? Insights from inner ear physiology and fisheries biology. *Marine and Freshwater Research* 56, 497–504.

Ptacek, M.B. 1998. Interspecific mate choice in sailfin and short-fin species of mollies. *Animal Behaviour* 56, 1145–1154.

Rasmuson, M. 2002. Fluctuating asymmetry – indicator of what? *Hereditas* 136, 177–183.

Reid, D.G. and Simmonds, E.J. 1993. Image analysis techniques for the study of fish school structure from acoustic survey data. *Canadian Journal of Fisheries and Aquatic Sciences* 50, 886–893.

Rova, A., Mori, G. and Dill, L.M. 2007. One fish, two fish, butterfish, trumpeter: recognizing fish in underwater video. In *Conference on Machine Vision Applications*, pp. 404–407. Tokyo, Japan.

Rowlinson, L.C., Summerton, M. and Ahmed, F. 1999. Comparison of remote sensing data sources and techniques for identifying and classifying alien invasive vegetation in riparian zones. *Water SA* 25, 497–500.

Ruff, B.P., Marchant, J.A. and Frost, A.R. 1995. Fish sizing and monitoring using a stereo image analysis system applied to fish farming. *Aquacultural Engineering* 14, 155–173.

Russo, R., Yanong, R.P.E. and Terrell, S.P. 2007. Preliminary morphometrics of spleen and kidney macrophage aggregates in clinically normal Blue Gourami *Trichogaster trichopterus* and freshwater Angelfish *Pterophyllum scalare*. *Journal of Aquatic Animal Health* 19, 60–67.

Sagnes, P., Champagne, J.-Y. and Morel, R. 2000. Shifts in drag and swimming potential during grayling ontogenesis: relations with habitat use. *Journal of Fish Biology* 57, 52–68.

Sagnes, P., Gaudin, P. and Statzner, B. 1997. Shifts in morphometrics and their relation to hydrodynamic potential and habitat use during grayling ontogenesis. *Journal of Fish Biology* 50, 846–858.

Sagnes, P., Mérigoux, S. and Péru, N. 2008. Hydraulic habitat use with respect to body size of aquatic insect larvae: Case of six species from a French Mediterranean type stream. *Limnologica* 38, 23–33.

Sagnes, P. and Statzner, B. 2009. Hydrodynamic abilities of riverine fish: a functional link between morphology and velocity use. *Aquatic Living Resources* 22, 79–91.

Schöne, B.R., Dunca, E., Mutvei, H. and Norlund, U. 2004. A 217-year record of summer air temperature reconstructed from freshwater pearl mussels (*M. margaritifera*, Sweden). *Quaternary Science Reviews* 23, 1803–1816.

Schultze-Lam, S., Harauz, G. and Beveridge, T.J. 1992. Participation of a cyanobacterial S Layer in fine-grain mineral formation. *Journal of Bacteriology* 174, 7971–7981.

Servia, M.J., Cobo, F. and González, M.A. 2004. Multiple-trait analysis of fluctuating asymmetry levels in anthropogenically and naturally stressed sites: a case study using *Chironomus riparius* Meigen, 1804 larvae. *Environmental Monitoring and Assessment* 90, 101–112.

Sheehy, M.R.J. 1990a. Potential of morphological lipofuscin age-pigment as an index of crustacean age. *Marine Biology* 107, 439–442.

Sheehy, M.R.J. 1990b. Individual variation in, and the effect of rearing temperature and body size on, the concentration of fluorescent morphological lipofuscin. *Comparative Biochemistry and Physiology -A- Molecular and Integrative Physiology* 96, 281–286.

Shopov, A., Williams, S.C. and Verity, P.G. 2000. Improvements in image analysis and fluorescence microscopy to discriminate and enumerate bacteria and viruses in aquatic samples. *Aquatic Microbial Ecology* 22, 103–110.

Silva, T.S.F., Costa, M.P.F., Melack, J.M. and Novo, E.M.L.M. 2008. Remote sensing of aquatic vegetation: theory and applications. *Environmental Monitoring and Assessment* 140, 131–145.

Slaveykova, V.I., Guignard, C., Eybe, T., Migeon, H.-N. and Hoffmann, L. 2009. Dynamic NanoSIMS ion imaging of unicellular freshwater algae exposed to copper. *Analytical and Bioanalytical Chemistry* 393, 583–589.

Spampinato, C., Giordano, D., Di Salvo, R., Chen-Burger, Y.-H.J., Fisher, R.B. and Nadarajan, G. 2010. Automatic fish classification for underwater species behavior understanding. *ARTEMIS '10, proceedings of the first ACM international workshop on Analysis and retrieval of tracked events and motion in imagery streams*. New York, USA: ACM, pp. 45–50.

Statzner, B., Dolédec, O. and Sagnes, P. 2009. Recent low-cost technologies to analyse physical properties of cases and tubes built by aquatic animals. *International Review of Hydrobiology* 94, 625–644.

Statzner, B. and Mondy, N. 2009. Variation of colour patterns in larval *Hydropsyche* (Trichoptera): implications for species identifications and the phylogeny of the genus. *Limnologica* 39, 177–183.

Strachan, N.J.C. and Kell, L. 1995. A potential method for the differentiation between haddock fish stocks by computer vision using canonical discriminant analysis. *ICES Journal of Marine Science* 52, 145–149.

Strauss, R.E. and Bookstein, F.L. 1982. The truss: body form reconstructions in morphometrics. *Systematic Zoology* 31, 113–135.

Štrojsová, A., Vrba, J., Nedoma, J. and Šimek, K. 2005. Extracellular phosphatase activity of freshwater phytoplankton exposed to different *in situ* phosphorus concentrations. *Marine and Freshwater Research* 56, 417–424.

Tahedl, H. and Häder, D.-P. 1999. Fast examination of water quality using the automatic biotest Ecotox based on the movement behavior of a freshwater flagellate. *Water Research* 33, 426–432.

Tani, K., Chen, J.M., Yamaguchi, N. and Nasu, M. 1996. Estimation of bacterial biovolume and biomass by scanning electron microscopic image analysis. *Microbes and Environments* 11, 11–17.

Tani, K., Kurokawa, K. and Nasu, M. 1998. Development of a direct *in situ* PCR method for detection of specific bacteria in natural environments. *Applied and Environmental Microbiology* 64, 1536–1540.

Thompson, D.W. 1917. On Growth and Form. London: Cambridge University Press.

Thorsen, A. and Kjesbu, O.S. 2001. A rapid method for estimation of oocyte size and potential fecundity in Atlantic cod using a computer-aided particle analysis system. *Journal of Sea Research* 46, 295–308.

Tillett, R., McFarlane, N. and Lines, J. 2000. Estimating dimensions of free-swimming fish using 3D point distribution models. *Computer Vision and Image Understanding* 79, 123–141.

Toubin, M., Dumont, C., Verrecchia, E.P., Laligant, O., Diou, A., Truchetet, F. and Abidi, M.A. 1999. Multi-scale analysis of shell growth increments using wavelet transform. *Computers and Geosciences* 25, 877–885.

Tuset, V.M., Lombarte, A., González, J.A., Pertusa, J.F. and Lorente, M.J. 2003. Comparative morphology of the sagittal otolith in *Serranus* spp. *Journal of Fish Biology* 63, 1491–1504.

Tuset, V.M., Rosin, P.L. and Lombarte, A. 2006. Sagittal otolith shape used in the identification of fishes of the genus *Serranus*. *Fisheries Research* 81, 316–325.

Untersteiner, H., Kahapka, J. and Kaiser, H. 2003. Behavioural response of the cladoceran *Daphnia magna* Straus to sublethal Copper stress – validation by image analysis. *Aquatic Toxicology* 65, 435–442.

Valta-Hulkkonen, K., Kanninen, A. and Pellikka, P. 2004. Remote sensing and GIS for detecting changes in the aquatic vegetation of a rehabilitated lake. *International Journal of Remote Sensing* 25, 5745–5758.

Van Look, K.J.W. and Kime, D.E. 2003. Automated sperm morphology analysis in fishes: the effect of mercury on goldfish sperm. *Journal of Fish Biology* 63, 1020–1033.

Vardon, D.R., Clark, M.M. and Ladner, D.A. 2011. The potential of laser scanning cytometry for early warning of algal blooms in desalination plant feedwater. *Desalination* 277, 193–200.

Walker, R.F., Ishikawa, K. and Kumagai, M. 2002. Fluorescence-assisted image analysis of freshwater microalgae. *Journal of Microbiological Methods* 51, 149–162.

Walker, R.F. and Kumagai, M. 2000. Image analysis as a tool for quantitative phycology: a computational approach to cyanobacterial taxa identification. *Limnology* 1, 107–115.

Walsby, A.E. and Avery, A. 1996. Measurement of filamentous cyanobacteria by image analysis. *Journal of Microbiological Methods* 26, 11–20.

Weller, A.F., Harris, A.J. and Ware, J.A. 2006. Artificial neural networks as potential classification tools for dinoflagellate cyst images: A case using the self-organizing map clustering algorithm. *Review of Palaeobotany and Palynology* 141, 287–302.

White, D.J., Svellingen, C. and Strachan, N.J.C. 2006. Automated measurement of species and length of fish by computer vision. *Fisheries Research* 80, 203–210.

Will, T.A., Reinert, T.R. and Jennings, C.A. 2002. Maturation and fecundity of a stock-enhanced population of striped bass in the Savannah River Estuary, U.S.A. *Journal of Fish Biology* 60, 532–544.

Wood, B.M. and Bain, M.B. 1995. Morphology and microhabitat use in stream fish. *Canadian Journal of Fisheries and Aquatic Sciences* 52, 1487–1498.

Xu, L., Krzyzak, A. and Suen, C.Y. 1992. Methods of combining multiple classifiers and their applications to handwriting recognition. *IEEE Transactions on Systems, Man, and Cybernetics* 22, 418–435.

Zelditch, M.L., Swiderski, D.L., Sheets, H.D. and Fink, W.L. 2004. Geometric Morphometrics for Biologists, a Primer. Amsterdam: Elsevier.

Zhang, D.S. and Lu, G. 2004. Review of shape representation and description techniques. *Pattern Recognition* 37, 1–19.

Zhang, H., Wahle, A., Johnson, R.K., Scholz, T.D. and Sonka, M. 2010. 4-D cardiac MR image analysis: left and right ventricular morphology and function. *IEEE Transactions on Medical Imaging* 29, 350–364.

Zion, B., Alchanatis, V., Ostrovsky, V., Barki, A. and Karplus, I. 2007. Real-time underwater sorting of edible fish species. *Computers and Electronics in Agriculture* 56, 34–45.

Zion, B., Alchanatis, V., Ostrovsky, V., Barki, A. and Karplus, I. 2008. Classification of guppies' (*Poecilia reticulata*) gender by computer vision. *Aquacultural Engineering* 38, 97–104.

Zion, B., Shklyar, A. and Karplus, I. 1999. Sorting fish by computer vision. *Computers and Electronics in Agriculture* 23, 175–187.

Znachor, P. and Nedoma, J. 2008. Application of the PDMPO technique in studying silica deposition in natural populations of *Fragilaria crotonensis* (bacillariophyceae) at different depths in a eutrophic reservoir. *Journal of Phycology* 44, 518–525.

18 Ground Imagery and Environmental Perception: Using Photo-questionnaires to Evaluate River Management Strategies

Yves-Francois Le Lay, Marylise Cottet, Hervé Piégay and Anne Rivière-Honegger
University of Lyon, CNRS, France

18.1 Introduction

In the previous chapters, the authors focused on the use of vertical images to characterise the biophysical structures of riverscapes, their spatial patterns and dynamics through time. For the purpose of diagnoses, ground or airborne oblique imagery is of increasing interest and opens new scientific and practical questions to solve. Oblique photos are valuable sources to assess landscape structures and to provide a view that is close to what the human eye sees. It is therefore possible to analyse the structure of the image itself and evaluate the perception of landscape. These aspects of environmental research are becoming key points for practitioners in charge of managing environmental features, because the effectiveness of environmental projects partly depends on the integration of the social demand. Riverscapes – i.e. landscapes that have a river as the focal point (Mosley, 1989) – are increasingly important components of everyday environments.

Literature has showed that observers cognitively differentiate landscapes with and without water (Herzog, 1985; Wherrett, 2000). Moreover, water appears to be a strong positive contributor to landscape attractiveness (Shafer et al., 1969; Kaplan, 1977a; Zube et al., 1982). The general public values waterscapes highly, not only for their visual aesthetics, but also for their beneficial psycho-physiological effects (Ulrich, 1981; Hartig et al., 1991; Parson, 1991). Water is one of the most desirable and preferred features in outdoor environments (Ulrich, 1981; Schroeder, 1982). In urban areas, water also enhances enjoyment; therefore landscape designers make use of water in plazas, parks and gardens (Whalley, 1988).

Many streams and rivers have been neglected after land-use changes in rural areas and degraded by 'over-engineering' in urban areas (Penning-Rowsell and Burgess, 1997). Hard engineering techniques were common to control risks and to promote water-related activities, such as navigation or agriculture. Such human interventions clearly manipulate river features and may reduce landscape attractiveness. Softer engineering techniques, such as restoration works, may also have some deleterious effects on 'a number of scenic components, particularly vegetation cover

Fluvial Remote Sensing for Science and Management, First Edition. Edited by Patrice E. Carbonneau and Hervé Piégay.
© 2012 John Wiley & Sons, Ltd. Published 2012 by John Wiley & Sons, Ltd.

of the surrounding landscape, vegetation cover along the banks (including bank protection), channel shape, and colour and turbidity' (Mosley, 1989, p. 11). Thus, the restoration of braided and meandering rivers aims to re-establish diverse in-stream and riparian habitats, involving river widening or relocation of flood levees. These river engineering operations induce changes in water quantity and quality, altering not only the forms of aquatic environments but also vegetation and wildlife. However, all the visible features, such as water depth, width, colour, odour, or movement, have implications for riverscape aesthetics, recreation purposes and other human activities.

Moreover, some of the rivers that are protected or proposed for conservation (notably braided rivers) may be less appreciated by the general public than familiar and accessible rivers – such as the rivers that run through urban and rural areas and are simply pleasing to the eye (Mosley, 1989). Reasons for the conservation of these rivers include factors that people generally do not take into account.

There is thus a need to identify the river-corridor features that the general public prefers and to include the public in the selection of river management plans. House and Sangster (1991) estimated that 'although the public may not possess formal knowledge of the best environmental options available for river works, they have a strong preference for certain environmental features' (p. 312). People have some very definite ideas of what they consider to be their ideal riverscape.

The use of photographs as a surrogate for landscape is well-established in environmental evaluation and preference surveys. This chapter focuses on defining valid methodological principles and describing some of the applications for environmental perception surveys. At first, a critical literature review indicates that several paradigms explore the assessment of public perception. In addition, it clarifies methodological issues and the implications for data analysis. Several examples of the evaluation of basic channel types corresponding to more or less humanised environments are presented. Different components of riverscape are then studied, and in particular water, gravel bars, and in-channel wood. These surveys were carried out in France and ten other countries, and designed to accomplish various objectives of landscape management. An effort was made to take account of public preferences when conceiving river operations, to improve the understanding of the motivations for maintaining watercourses and to show the influence of socio-cultural context on riverscape perception. Finally,

the advantages and limits of photo-questionnaires are demonstrated, and findings are discussed in terms of environmental education.

18.2 Conceptual framework

Several classifications present the numerous existing landscape studies in an orderly way. Turner (1975) distinguished three broad categories: (a) measurement techniques based on sophisticated statistical analysis that skilled professionals may achieve, (b) preference techniques derived from studies of perception and behavioural sciences, and (c) consensus approaches which are 'perhaps currently held in low esteem for their want of objective method' (p. 157). Gregory and Davis (1993) identified three types of research on the evaluation of riverscapes: (a) component or inventory approaches which provide an overall value judgment on river aesthetics (Morisawa, 1972), (b) approaches based upon uniqueness of specific areas (Leopold and Marchand, 1968; Leopold, 1969a and b; Williams, 1986), and (c) approaches which are more interested in public perception of riverscapes (Mosley, 1989; Gregory and Davis, 1993).

Studies on landscape perception and evaluation are diverse and research has evolved into different paradigms (Table 18.1). The most-cited classification and terminology were developed by Zube et al. (1982) and are used in Table 18.2, with a gradient in terms of respondent involvement (from the experiential to the expert paradigm).

The experiential paradigm is based on user-dependent methods that provide respondents with every opportunity to express freely their opinions, attitudes and feelings about their environment (Table 18.2). Coeterier (1996) used a verbal approach based on open, partially structured interviews. These are better suited to explore inhabitants' meanings of landscape than user-independent methods such as photo-questionnaires. He estimated that 'after six or seven interviews no new information about perceived qualities was added' (p. 30). Whereas photo-questionnaires involve the researcher taking or choosing the photographs, the photo-projective method asks individuals to take photographs of their surroundings and to describe each scene on site (Yamashita, 2002). However, the experiential paradigm has received relatively 'little attention in resource and environmental management because it has been considered to be idiosyncratic, individualistic and subjective' (Dakin, 2003, p. 190).

Table 18.1 Several classifications of landscape perception paradigms (Based on Dakin (2003)).

References	Approaches to Landscape Evaluation			
Dakin, 2003	Expert	Experimental		Experiential
Daniel and Vining, 1983	Ecological or formal aesthetic	Psychophysical	Psychological	Phenomenological
Porteous, 1982	Planner	Experimentalist		Humanist
Punter, 1982	Landscape perception	Landscape quality		Landscape interpretation
Zube et al., 1982	Expert/professional	Psychophysical	Cognitive	Experiential
Turner, 1975	Measurement techniques	Preference techniques		Consensus approaches

The expert paradigm involves the assessment of visual quality by trained and skilled observers working within a fine arts, landscape design, or ecosystem perspective (Table 18.2). It is assumed that experts are able to identify and objectively evaluate common features and relationships between landscape elements that contribute to the intrinsic aesthetic quality. These tasks are motivated by the pragmatic purposes of environmental planning, design, or management. They have produced rough categories of formal aspects (shapes, lines, colours, and the patterns combining these aspects) and formal qualities (balance, proportion, unity, and diversity). The expert approach has benefited from the tools and techniques of geomatics, such as remote sensing and geographic information systems. Penning-Rowsell and Hardy (1973) pointed out that this approach requires improvements to take into account 'the landscape user's preferences for certain landscape features' (p. 160).

The experimental paradigm is based on the responses of the selected groups of respondents, including non-expert judgments (Table 18.2). Observers make sense of surroundings according to their experience, expectations, or socio-cultural context. Nature does not 'possess normative proclivities in what it creates' (Ribe, 1982, p. 63). Appleton (1975b) underlined the lack of a convincing body of theoretical knowledge in landscape evaluation: 'we are perfectly entitled to reject the professional as an arbiter of excellence, if we have more confidence in ourselves than in him' (p. 122). As a consequence, Dearden (1981) considered some philosophical and pragmatic arguments supporting the inclusion of public participation in landscape evaluation techniques. The involvement of the public may improve the quality of the decision-making process.

Zube et al. (1982) distinguished two approaches. While the psychophysical paradigm postulates that individual evaluations and behaviour depend on the features of landscapes (considered to be a source of visual information to which observers respond), the cognitive paradigm postulates that human perceptions determine environmental preferences (Table 18.2). Both seek the relationships between value judgments or human meanings on the one hand, and the environment or landscape dimension on the other hand. Such measures have enabled the development of predictive models estimating landscape preferences, scenic beauty (Shafer and Mietz, 1970; Daniel and Boster, 1976), and riverscape attractiveness (Mosley, 1989; Gregory and Davis, 1993).

The measured landscape dimensions may be physical (e.g. topography or forest cover) or cognitive (naturalness or mystery). Following a neo-Darwinian conception, landscape preferences are motivated by the satisfaction of need. Appleton's theory (1975a) postulates that human preference for landscapes has its origins in the impulse to take the opportunities provided by habitats to see (prospect) without being seen (refuge). The environmental psychologists R. and S. Kaplan developed an informational approach to environmental preference. 'Like other animals (. . .), humans tend to evaluate the terrain and prefer habitats that are likely to offer safety and the resources they need' (p. 286). After Kaplan's works (e.g. 1989), Levin (1977), Lee (1978 and 1979) and Ellsworth (1979) attempted to predict the waterscape preferences by means of four concepts: (a) legibility, i.e. 'the perceptual establishment of relationships within and among elements of the visual display' (Lee, 1979, p. 574), (b) coherence, i.e. 'the degree of visual organization in the scene' (Kaplan, 1985, p. 167), (c) complexity, i.e. 'the number of independently perceived elements and their degree of dissimilarity' (Ulrich, 1981, p. 552), and (d) mystery, i.e. 'the extent to which the environment suggests that one could obtain new information

Table 18.2 Characteristics of landscape perception studies.

Broad Approaches (Zube et al., 1982)	Expert Paradigm	Experimental Paradigm — Psychophysical Trend	Experimental Paradigm — Cognitive Trend	Experiential Paradigm
Participants	Professionals	Sample groups of professionals and general public		Lay people
Subject involvement	Skilled and training observers	Respondents		Participants
Techniques of evaluation (Turner, 1975)	Measurement techniques	Preference techniques (relationship between physical features and value judgments)		Consensus approaches (in-depth interviews, semi-structured questionnaires)
Primary study object	Environment(s)	Human/environment interaction		Subjects
Methods	Reductionism	Systemics and complexity		Holism
Factors explicating riverscape value		Sociocultural characteristics (age, genre, income, profession, home, nationality)		
	Formal aesthetic, ecological and development principles	Biophysical features of riverscape (water color, clarity, depth, landforms)	Cognitive and affective dimensions (attractiveness, preference, aesthetics, naturalness, legibility, coherence, complexity, mystery)	Everyday experience (familiarity, sense of place, sociability)
Data collection	Structured questionnaires, photo-questionnaires, field measures			In-depth taped interviews, auto-directed photography, literature production
References	Griselin and Nageleisen, 2004; Cossin and Piégay, 2001; Williams, 1986; Pitt, 1976; Penning-Rowsell and Hardy, 1973; Leopold, 1968 and 1969b; Leopold and Marchand, 1968	Pflüger et al., 2010; Le Lay et al., 2008; Junker and Buchecker, 2008; Piégay et al., 2005; Meitner, 2004; Wilson et al., 1995; Gregory and Davis, 1993; House and Sangster, 1991; Mosley, 1989	Bulut et al., 2010; Ryan, 1998; Herzog, 1985; Ellsworth, 1982; Ulrich, 1981; Lee, 1979; Levin, 1977; Kaplan, 1977b	Dakin, 2003; Coeterier, 1996; Yamashita, 1992

if one were to travel deeper into it' (Herzog, 1985, p. 227). Legibility and coherence indicate the person's understanding of a scene, while complexity and mystery relate to the observer's involvement and interest in an environment.

The preference for river scenes is positively related to mystery (Levin, 1976; Ellsworth, 1982) and to coherence (Kaplan, 1977a). Herzog (1985) also found that the most preferred waterscapes received high rates in spaciousness, coherence, and mystery, but low rates in texture (that is, uneven terrain). Litton (1977) considered the whole river landscape and recognized the following elements of visual assessment: '(a) landforms, (b) vegetation patterns, (c) water presence and expression, (d) human use and impacts' (p. 46), (e) other natural influences 'including visual effects of climate, seasonal change, topographic orientation, and relative elevation' (p. 47). More recently, the results of the Ryan's study (1998) show that 'local residents see the river corridor in terms of four interconnected landscapes: the river, woods, farms, and built areas' (p. 236). The photographs of rivers were higher in preference than those of woods and grassland, of backyard, and of farm fields. In Western Europe, an inclination exists towards deep, slow-flowing mature rivers that are more than 4 meters wide (House and Sangster, 1991; House and Fordham, 1997). As curved lines create a sense of mystery, it appears that observers are particularly attracted to curving or meandering river scenes (Levin, 1977; Lee, 1979; Ellsworth, 1982; Kenwick et al., 2009).

The surveys detailed in this chapter are in line with the experimental paradigm and take advantage of both the psychophysical and cognitive approaches. They are based on a set of photographs and a series of questions concerning the views. The aim is to appreciate riverscape perceptions according to several criteria, to evaluate the consensual aspect of reactions, and to understand the underlying socio-cultural logics. Practically, such approaches are very efficient and can provide knowledge allowing to determine a management policy integrating social perception (the so called 'social engineering' approach), but they are also subject to criticism. Turner (1975) notably indicated for such an approach: 'Personally I am appalled at the prospect of planning landscape on the basis of public preference' (p. 160).

Whereas photographs are commonly recognised as an accurate indication of on-site conditions, a few studies have highlighted differences between static and dynamic representations in certain conditions (Brown and Daniel, 1991; Heft and Nasar, 2000; Huang, 2009). As human perception of the environment is multi-modal, many scientists acknowledge the role that both motion and sound play in the perception and evaluation of water landscapes. Moreover, intangible elements such as odours, temperature and humidity could also be important components of riverine environments, although the related impressions are only felt in the real world. Scientists attempted to use videotape images to represent waterscapes more realistically by conveying the dynamic qualities of water (Brown and Daniel, 1991; Hetherington et al., 1993). They found differences in the responses to water flow (water level, volume, and speed). The use of videos still remains limited however. A few studies have examined variations in perception depending on the types of stimulus (*in situ*, photo or video). The results are very variable, indicating that additional knowledge is needed to improve the understanding of when and why photos are an accurate medium or may produce biases (see detailed in Hetherington et al., 1993; Huang, 2000, 2004 and 2009). Anderson et al. (1983) indicated that realistic sound stimuli affected the aesthetic evaluation of outdoor settings. Motion, individually and jointly with sound, has been demonstrated to have important effects on human perception and evaluation of dynamic landscapes (Hetherington et al., 1993). Recently, Huang (2000, 2004 and 2009) investigated preferences for built waterscapes. Results showed that 'fast water movement, loud water sound, complex configurations of water movement, and complex profiles of containers and objects of waterscapes units were favored' (Huang, 2000, p. 11). Our own tests showed that photographs are a realistic medium compared to videos or in-situ conditions, notably along natural rivers and oxbows where motion is not a key point (the flow velocity is shown by turbulence) no more than odours or sounds.

18.3 The design of photo-questionnaires

18.3.1 The questionnaire and selection of photographs

Such an approach is considered as an experiment and as a consequence of this is performed in a clear scientific framework based on hypotheses, validation and inferential statistics. The aim is to validate or invalidate a hypothesis and the question that is asked is in strong interaction with the selected photo set and the submission of the view. As with every scientific experiment, there are simplifications of the reality. Nevertheless, unexpected perceptive aspects may contribute to the variability of answers and make the interpretation of results more complex. The questionnaire is then built to validate the hypothesis and cannot be used for multiple purposes. There are therefore clear constraints in building and submitting the questionnaire and in selecting the photographs to mitigate noise and maximise the validity of the experiment.

The choice of a method to submit the views to the people surveyed is a key issue in studies on landscape preferences. Only direct experience stimulates all of the senses of subjects in the real world. However, the requirements in terms of participants and landscape samples may create some insurmountable practical difficulties and arouse interest for environmental surrogates. Simple photographs (Ryan, 1998; Nasar and Lin, 2003; Bulut and Yilmaz, 2009), scanned and altered photographs (Wilson et al., 1995), colour slides (Herzog, 1985; Gregory and Davis, 1993; Le Lay et al., 2008), 360° panoramic views (Meitner, 2004), visual simulations (Kubota, 1997; Junker and Buchecker, 2008) and specially-drawn sketches (House and Fordham, 1997) have more or less frequently been used as visual media for waterscape assessment.

Photographs are two-dimensional images, the content of which does not integrally convey 'the temporal variety and visually dynamic qualities of many real-world nature scenes – e.g., such as moving water surfaces, wind-blown vegetation, and changes associated with seasons' (Ulrich, 1981, p. 551). Therefore, numerous studies have addressed the validity of photographs as a medium for presentation of the environment. Although surrogates may provoke perceptual distortion, individual responses to an actual physical setting are strongly and positively correlated to responses based on a comprehensive photograph of the same scene (Shafer and Richards, 1974; Daniel and Boster, 1976; Shuttleworth, 1980; Stewart et al., 1984; Nassauer, 1987; Trent et al., 1987; Zube et al., 1987; Brown et al., 1988; Stamps, 1990; Palmer and Hoffman, 2001). A meta-analysis of 11 papers yielded 'a combined correlation of 0.86 between preferences obtained *in situ* and preferences obtained through photographs' (Stamps, 1900, p. 907).

The great advantages of photo-based perceptual evaluation are the equivalence of presentation conditions to respondents (Shuttleworth, 1980), the low costs of (re)production and the easy access to and manipulations of the media (Huang, 2004), the possibility to present many identical landscapes to sample groups of participants (Gregory and Davis, 1993). 'The selection of particular stimulus photographs is critical for the evaluation of any hypothesis about the nature of people's landscape preferences' (Wilson et al., 1995, p. 53). Scientists, and not participants, choose what is presented: they take the photographs or at least select them. However, the reality is so complex that individual appreciations of the same view may derive from very different criteria. To reduce the bias of the observer and assess the importance of a specific feature, commonly-used measures of central tendency can be calculated using scores that are attributed to a scene. As nothing forces the analyst to consider each of the scenes individually, he may also distinguish different groups of photographs according to defined criteria.

The investigator has to take into account several constraints. First, photographs have to depict most of the variety in the outdoor scenes (Shafer and Richards, 1974). The slides are taken from the ground, without using telephoto lenses (Shafer and Brush, 1977). They should show an eye-level front view of the studied object, 'normally from an obvious viewpoint such as a riverside track, stopbank, or bridge parapet' (Mosley, 1989, p. 6). They are taken in clear weather and under similar lighting and sun angle conditions, using the same film and camera (Ulrich, 1981; Mosley, 1989). Lack of sunlight lowers scenic beauty (Brown and Daniel, 1991). No effort must be made to compose scenes. Aesthetically spectacular scenes must be avoided (Ulrich, 1981). No animals or people should be visible, and human influences should be minimal in any of the pictures (Ulrich, 1981; Herzog, 1985). The riverscape structures presented by the photographs must be similar in order to weaken the effect of factors such as depth of field, perspective, or relative importance of background.

It can be asked to what extent photographs are needed in studies on environmental perception and evaluation? No protocol has been fixed at this time and stimulus samples may vary considerably between 5 and 240 views, depending on the authors and the objects studied (Table 18.3). In order to diminish the specificity of each scene and to test precise hypotheses, it is necessary to present a reasonable number of photographs for each type of riverscape and to calculate the mean of the scores obtained. Two is a minimum, 4 to 5 should provide a good set for a given riverscape type, allowing for the assessment of the respective importance of the intra- and the inter-types of variability. This being said, Shafer and Brush (1977) suggested 'that 20 photographs probably would be adequate for any one management area' (p. 250). More than this makes the interview long and fastidious, and the order of submission of the photographs may influence the perception ranking. Twenty photos is a good compromise, thus indicating that not more than 4 to 5 types of scenes can be tested in a given survey.

18.3.2 The attitude scales

There is no general agreement on the metrics used for photo-questionnaires (Table 18.3). Many scales are commonly used in works of social psychology. Landscape evaluation is often based on bi-polar scales built width adjectives such as 'exciting', 'sad', or 'ugly'. Participants

Table 18.3 Sample sizes and attitudinal scales used in several waterscape perception studies.

Scenes (nb)	Respondents (nb)	Scales	Variables	References
25	120 university students	7-point scale	Preference and 5 descriptor variables (vividness, harmony, fascinaty, naturalness, being interesting)	Bulut et al., 2010
30	176 residents and 44 professional planners	5-point scale	Preference	Kenwick et al., 2009
8	1005 Swiss participants	7-point scale	Appeal, naturalness and satisfaction of needs	Junker and Buchecker, 2008
20	2200 students	Visual analog scale	Aesthetics, naturalness, dangerousness, and need for improvement	Piégay et al., 2005 ; Le Lay et al., 2008
47	118 undergraduates	10-point scale	Scenic beauty	Meitner, 2004
5	30 residents	7-point scale	Preference, calming, and excitement	Nasar and Lin, 2003
16	120 rural property owners	5-point scale	Preference	Ryan, 1998
16	105 members of an University community	7-point scale	Preference	Wilson et al., 1995
20	199 students	10-point scale	Attractiveness	Gregory and Davis, 1993
190	409 participants	10-point scale	Scenic beauty	Mosley, 1989
70	259 introductory psychology students	5-point scale	Spaciousness, texture, coherence, complexity, mystery, identifiability and preference	Herzog, 1985
60	98 college students	5-point scale	Legibility, coherence, complexity and mystery	Ellsworth, 1982
240	12 student judges	Semantic scale	Complexity, unity, beauty, and pleasantness. Dominance, wakefulness, attention/interest, and stability (affect)	Ulrich, 1981
60	54 subjects			
20	100 landscape architecture students	5-point scale	Legibility, spatial definition, complexity, and mystery	Lee, 1979
		7-point scale	Preference	
48	400 visitors	5-point scale	Preference	Hammitt, 1978

rate each photograph based on these differential semantic measurements. The so-called 5-point Likert scale seems the most conventional. Often, the mean score is calculated to evaluate, for instance, the preference for each scene, and to determine which landscapes tend to be preferred. The interpretation of the scale, however, remains problematic. Averages cannot be applied to categorical variables, thus complicating the appreciation of groups of photographs and the testing of hypotheses under relatively controlled experimental conditions. With such variables, it is more

difficult to predict reactions to external parameters using basic inferential tools.

A Visual Analog Scale (VAS) is a measure instrument that was built for the diagnostic of pain. Respondents indicate their level of agreement by marking a straight line between the two end-points of the scale (Gift, 1989). Although VAS is less conventional in landscape research, it outperforms the other discrete scales. Each photograph is individually evaluated on a continuous scale ranging from 0 (lowest degree of agreement) to 10 (highest degree of agreement), so that the averages are rightfully calculated and compared in various ways. Data are quantitative and continuous, and statistical models are more easily developed (Le Lay et al., 2008).

18.3.3 The selection of participant groups

Two questions are considered here: Who is surveyed and how many people? Each time, the answer depends on the scientific question asked. It may be focused on relative comparisons (e.g. a given group of actors compared to another one) or infer the opinion of a sample to an entire population, which is becoming complex because of the sample size and question of representativeness.

Several authors believe that there is a general consensus about landscape values and tastes in a society. Consequently, any scientist, as a member of society, can claim to represent its values. In the opinion of the philosopher Carlson (1977), any person who has good judgment and is conscious of public preferences can predict landscape evaluation. Therefore, managers can consider themselves as representatives of public opinion. Such practices may however be criticized (Wallace, 1974). Kroh and Gimblett (1992) consider that preferences derive from a framework of values, beliefs and experiences. Moreover, experiences accumulated during life partly determine the reactions to stimuli (Zube and Pitt, 1981). Few experimental studies focus on the observers' characteristics, but they are significant. Many variables have an effect upon descriptive and evaluative responses: (a) personality and socio-economic attributes (Zube and Pitt, 1981; Carp and Carp, 1982a et b); (b) profession and experience, in terms of resource and environmental management (Zube, 1973a; Buhyoff et al., 1978; Feimer, 1984; Gregory and Davis, 1993); and (c) familiarity with a particular environment or a type of environment (Pedersen, 1978; Zube and Pitt, 1981; Buhyoff et al., 1983).

The observer's characteristics have an influence on riverscape evaluation (Lee, 1979). The respondents can be categorised into several groups, based on the nature of their contact with rivers: contact sports (canoeists), non-contact activities (anglers and rowers), and remote-contact river users (walkers and picnickers) (House and Sangster, 1991; House and Fordham, 1997). However, the personal characteristics of the observer remain trivial compared to the environmental effects (Stamps, 1995 and 1999; Nasar and Lin, 2003). Moser (1984) found no relation between subjects' recreational activity and their perception of water quality. Swimmers and fishermen were not more sensitive than boaters and walkers.

Likewise, Gregory and Davis (1993) showed a considerable agreement between their three sample groups of respondents.

Few environmental studies provide statistical rationales for selecting the sizes of respondent samples, so that scientists use a wide variety of sample sizes (Table 18.3). Stamps (1992) applied a bootstrap procedure on a set of 200 student answers and indicated that about 25 to 30 respondents are needed to minimise the error estimate. Therefore, 'large respondent samples would probably not, in general, be a cost-effective protocol for person/environment research' (Stamps, 1992, p. 222). Nevertheless, the assessment of a relative opinion is not always the objective, notably when the managers may know the public opinion. When considering an inferential procedure therefore, the sample size is determined according to the traditional sampling strategy.

18.4 Applications with photo-questionnaires

18.4.1 From judgment assessment to judgment prediction

In the experiments that were performed, two issues were considered (Table 18.4):
• assessing a judgment in experimental conditions: riverscape with and without wood, riverscape before and after a restoration action, observed in summer or in winter, riverscapes ordered on a gradient from water to gravel dominance, and from a eutrophic to an oligotrophic state. The scoring is often combined with additional questions, allowing the respondents to account for their answers.
• linking the judgment with the structural organisation of the riverscape, in order to provide perception models that are potentially used in the decision-making and planning process.

For each of these issues, we worked on different sets of individuals depending on the question that was asked: first, a sample of the local population, in order to infer their opinions in terms of landscape judgments, and second a set of water/local actors, to evaluate perception in terms of both the differences and/or the consensus between the two groups. Most of the time, it is difficult to survey a sample that is representative of the population as this is so time consuming that the problem must be addressed in a comparative way, with inter-actor comparisons. The survey does not say

Table 18.4 Types of riverscape features and informal variables in the five case studies.

Case Studies	Types of Features	Judgment Items	Types of Respondents	Aim	Number of Photographs
1	Wood in streams and rivers	Aesthetic quality, naturalness, danger feeling, and management needs	2250 students from 10 countries	Test effect of wood presence on landscape value	20
2	Restoration works	Aesthetic quality, typicalness, danger feeling and strolling	118 children, local councilors and association members	Test effect of restoration actions on landscape value	24
3	Extent of gravel versus water area	Aesthetic quality, beneficial uses, and management needs	127 children, experts, river managers and basin inhabitants.	Test effect of water extent on landscape value	10
4	Riverscapes	Attractiveness	176 basin residents	Predict aesthetic perception from landscape characters	9
5	Water of floodplain lakes	Beauty	100 students	Predict aesthetic perception from landscape characters	34

what the local population thinks, but allows apprehending consensus and divergence of perception. Students were also surveyed for cross-cultural comparisons and methodological purposes (are there any differences in landscape perception, depending on whether the landscapes are submitted through photographs or videos). In terms of cross-cultural comparisons, student sets cover the same age range and are surveyed by disciplines, so that their type of training may not affect the perception evaluation.

In Figures 18.1 and 18.4, a few examples of photo sets are shown. The assessment of cross-cultural variations in the perception of riverscapes with and without wood was based on a set of 20 photographs, which represent watercourses running through various physical and humanised environments (see detail in Piégay et al. 2005, Chin et al., 2008; Le Lay et al., 2008). Half of the scenes are characterised by river and stream sections that are obstructed by wood, while the 10 others are free-flowing, without large pieces of wood. To evaluate the overall scenic attractiveness of each picture, respondents rated four different values perceived for each of the 20 colour photographs (namely aesthetics, naturalness, danger, and need for improvement) on Visual Analog Scales ranging from 0 to 10. The questionnaire also included two

qualitative variables for characterising the perception of danger and the motivation for improving riverscapes.

The perception of landscape changes, after proceeding with a restoration program on the Rhône River and its floodplain lakes, was studied within the reach of Pierre-Bénite, in the southern suburbs of the city of Lyon (France). Works were achieved in 1999. The survey aimed, initially, to compare 12 couples of scenes (before and after restoration works), presenting two distinct compartments of the river corridor (12 views of the channel and 12 views of the floodplain lakes). Second, it endeavoured to evaluate the possible effect of different seasons (12 photographs taken in summer, 12 others taken in winter). Three groups of respondents were surveyed: school children, association members, and local elected representatives.

Over the past century, the ecologically diverse, braided Magra River in Italy has narrowed, incised, and lost many gravel bars. These evolutions are due in part to the encroachment of the riparian vegetation, following the decrease in bedload supply, and channel degradation. Motivated by the European Water Framework Directive, river scientists and managers are beginning to plan projects to preserve and restore this dynamic mosaic of rare habitats and processes. To support this objective, a

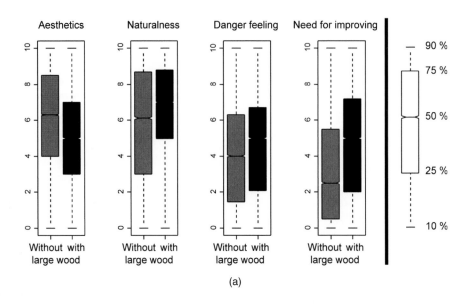

(a)

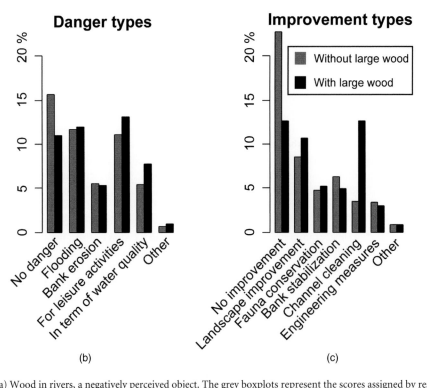

(b)   (c)

Figure 18.1 (a) Wood in rivers, a negatively perceived object. The grey boxplots represent the scores assigned by respondants to the riverscapes without wood and the black boxplots represent the scores assigned by respondants to the riverscapes with wood. (b) The percentage for each danger type rated by students from 20 riverscapes with and without wood. (c) The percentage for each improvement type rated by students from 20 riverscapes with and without wood.

study was conducted to assess the perception of braided patterns by different social groups. In June, 2006, 127 people were surveyed using a photo-questionnaire consisting of 10 photographs that depicted riverscapes with different proportions of water and gravel bar areas. Respondents were asked to score each photograph in terms of aesthetic value, beneficial uses, and river management needs.

In the Roubion catchment (near Montélimar, France), a survey focused on the local population and their preferred riverscape features. 176 individuals evaluated 9 photographs. Thus, 10% of the residents in the basin were surveyed, taking into account respondent's age and socio-professional group. The sample was representative of the public, so that findings are valuable in terms of public decision-making at a local scale. The participants were asked to consider each scene, to rank them according to their preference, and to justify and characterise their choice. The aesthetics scores were then linked to the structural characteristics of the riverscape in order to provide a predictive model. Similarly, scores for the perception of water in floodplain lakes was also related to the characteristics of the landscape. The Lower Ain and the Upper Rhône Rivers (Rhône-Alpes, France) are characterised by the existence of floodplain lakes caused by historical river migration and channel abandonment. The disappearance of these former channels over the last century led to numerous restoration projects. In this context, specific attention has been paid to the public perception of floodplain lakes, focusing on the appearance of the water. An internet survey using photo-questionnaires was conducted with more than 100 persons, who were asked to assess the aesthetics of the scene. The survey aims to build a model associating the visual physical characteristics of floodplain lakes and the public's aesthetics judgments.

18.4.2 Comparing reactions between scenes and between observers

Over the last three decades, there has been an increasing scientific interest in large pieces of wood (LW) and scientists have recognised the hydraulic, geomorphic and biological role of wood in temperate river systems. Although LW re-introduction has been promoted in different areas, such as North America, Australia, Switzerland or Germany, such measures are not commonly accepted by managers and users in other countries. In order to understand the reasons for some regional variations in LW public perception, we have analysed the social, cultural and historical context of the question. The wood perception study was performed in eleven geographical

areas (France, Poland, Sweden, India, Russia, Germany, Italy, Spain, China, Oregon (U.S.), Texas (U.S.)). These areas were selected for the potential diversity of socio-cultural environments that they represent. Owing to the low variability in age classes, the student community is a very interesting experimental population for international comparisons, and students' responses were presumed to represent the knowledge of non expert groups (Brown and Daniel, 1991). Similar disciplines were surveyed in each of the areas concerned. More information about methodological aspects is presented in Piégay et al. (2005).

The results show that the presence of in-channel LW modifies students' perceptions (Figure 18.1a). The respondents considered riverscapes with LW to be less aesthetically pleasing, more natural, more dangerous, and needing more improvement than those without LW. When considering the danger perceived (Figure 18.1b), the participants evaluate the scenes with wood as being more dangerous in terms of water quality and affecting leisure activities. Moreover, the students consider that the riverscapes with LW require much more improvement than those without (Figure 18.1c). With the presence of LW, there is an increase in the perceived need for cleaning the channel and improving the quality of the landscape. The comparative geographical analysis demonstrates some substantial cultural differences amongst the geographical areas (Figure 18.2a). Particularly, Asian students show a great motivation for improving watercourses, whereas respondents from other countries (German, Oregon, or Sweden) show a more conservationist attitude towards streams and rivers. Many factors can explain the geographical variability. For instance, the motivation for improvement (Figure 18.2b) seems to be linked to the familiarity with LW, the combination of forest cover and density of population, the history of land use (agricultural tradition in Western Europe or forestry in Oregon and Sweden), the technocratic management of watercourses (France and Russia) and the necessity of development (China and India).

In the survey that focused on the restored features of the Rhône, results show that the flowing water channels are preferred to their margins (Figure 18.3). The photographs of the Rhône River obtained higher aesthetic scores than those of the floodplain lakes. The participants considered the Rhône River to be more representative of the valley than the floodplain lakes. While the latter provoke a feeling of danger, the former seem to be considered suitable for recreational purposes. Moreover, riverscape evaluation is different before and after restoration (Figure 18.3). After river works, observers consider the photographs

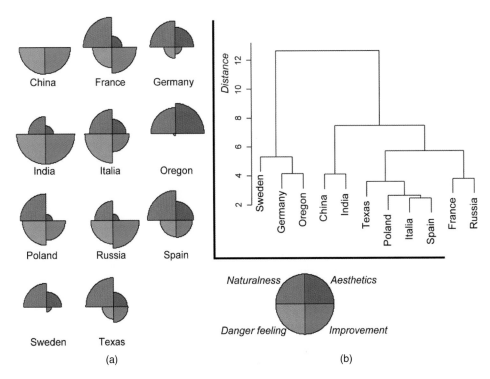

Figure 18.2 (a) Segment diagrams. Each diagram represents one geographical area. On each of them, four continuous variables (scores for aesthetics, naturalness, feeling of danger and need for improvement) are plotted by a radius whose length corresponds to the mean value attributed by the students of the given area. (b) Dendrogram. A hierarchical cluster analysis has been performed on the mean values of need for improvement (provided by students from 11 geographical areas who scored the 20 riverscapes).

to be more aesthetic, more typical of the valley, more reassuring, and more suitable for walks. The removal of trees and clearing of brushwood enhanced the feeling of safety on the margins of the restored floodplain lakes. Thus, the results demonstrate some positive effects of restoration works on riverscape perception. Further analyses highlighted clear differences between photographs taken in summer and in winter. Summery scenes are more appreciated, more typical of the valley, and more reassuring than wintry scenes. Thus the season effect is taken into account when selecting the sample of riverscapes. Overall responses appeared to be similar between groups of respondents. There was no statistical difference concerning the aesthetic quality and the motivation for strolling along a water body, however. School children tended to underestimate the typical characteristic of riverscapes and to be more sensitive to danger. By contrast, members of water-related associations felt danger less intensively, as if good knowledge of the river enhanced the feeling of control.

The third survey performed in the Magra region is based on a similar approach, focusing on the difference between water and sediment areas in riverscapes. There was a strong tendency towards a preference for landscapes that depicted high proportions of water, or conversely, low proportions of sediment (Figure 18.4). As an example, strong positive associations were demonstrated between the proportion of water depicted and the average responses for aesthetic value (r = 0.76; p < 0.015) and beneficial use (r = 0.87; p < 0.0015). Six significant negative correlations were also identified. Management needs are negatively correlated to the satisfaction of uses (r = −0.94; p < 0.0001), aesthetic value (r = −0.90; p < 0.0004), and proportion of water (r = −0.70; p < 0.025). Likewise, there were negative associations between the proportion of sediments and beneficial use (r = 0.80; p < 0.005) or aesthetic value (r = 0.70; p < 0.025). River landscapes characterised by a large proportion of sediment were recognised as less aesthetic, less usable, and led respondents to require an improvement. The size of

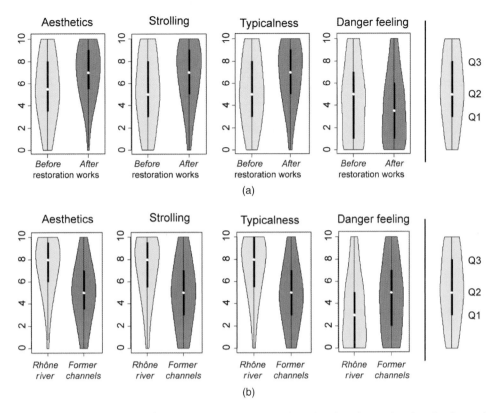

Figure 18.3 Perception of restoration works (a) and floodplain lakes, (b) The violin plots of scores (attributed to four evaluative variables) show the lower quartile (Q1), the median (Q2) and the upper quartile (Q3).

sediment particle also influenced landscape perception. For a given proportion of mineral cover, images with large-sized boulders were perceived as more attractive, more usable, and requiring less active management. However, these perceptions differed amongst groups of participants, reflecting different interests and objectives.

Public awareness of environmental stakes has increased very significantly in the most developed countries over the last decades. Nevertheless, values, feelings and beliefs may trigger the decision-making of LW or gravel removal throughout the world. This negative perception of LW or gravel may explain why conservation or restoration policies dealing with these features are not shared by local inhabitants or decision-makers. Does this mean that the management of riverine landscapes should acknowledge the commonly-held national perception of riverscape? There is a need to develop a greater appreciation of natural rivers that are scientifically and ecologically important. Environmental education should therefore be considered for river restoration purposes.

18.4.3 Linking judgments to environmental factors

Once affective responses are measured to evaluate attitudes towards various types of riverscape components, it is then possible to use these responses as independent variables to be predicted from landscape characteristics. Two approaches are detailed here.

The first approach was carried out within the Roubion catchment. Figure 18.5 shows the most and the least attractive riverscapes according to the population sample. Similarly to the Magra set, results underline some contrasts between the perception of experts, who evaluate braided rivers positively (a rare judgment on the French territory) and the perception of the general public. The identification of the least appreciated photograph was relatively consensual: scenes E and F received 69% of responses. Concerning the first scene, participants denounced the lack of water, the appearance of the channel, that seemed designed with a bulldozer, and

Figure 18.4 The influence of visible bed material on riverscape attractiveness. (a) Scores attributed by respondents in terms of aesthetics, satisfaction, and improvement. The size of sediment is indicated on the left side of the figure, and the areas occupied by water, vegetation and bed material are visible on the right side of the figure. (b) Set of the ten riverscapes. (c) A visual analog scale (VAS).

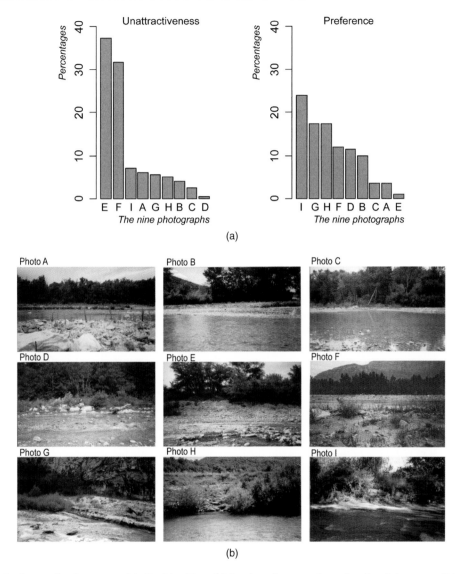

Figure 18.5 Preferences for riverscapes of the Roubion River. (a) Barplots of unattractive and preferred riverscapes. (b) Set of the nine landscapes of the Roubion River. Reproduced with permission from Le Lay et al. (2006) Les enquêtes de perception paysagère à l'aide de photographies : choix méthodologiques et exemples en milieu fluvial. In Septièmes Rencontres de Théo Quant à Besançon du 26 au 28 janvier. Fig. 5, p. 10.

the abundance of sediments, particularly gravel. The riverscape is perceived as deserted, barren, badly maintained, and disorderly. The local population preferred photograph I. Participants pointed out four characteristics: the abundance of flowing water, the green and well-maintained vegetation that provides shade for users, the presence of a sandy beach, and the feeling of calmness. Photograph G represents a typical gorge

waterscape. Sixty-five percent of respondents appreciated the dominant presence of rocks. Water was considered to be vivid and clear. The scene was perceived to be natural and wild. Participants evoked leisure activities, such as bathing or simply sitting near the water. Photograph H benefits from a balance between grass, shrub and trees, from the presence of calm and clear water, and from the glimpse of the surrounding mountains. Such a riverscape appears to be suitable for bathing, strolling, and children's play.

Following this first step, an analysis of the structure of the landscape was performed to improve the interpretation of these findings (Figure 18.6). The Roubion's riverscape patches were statistically analysed in order to distinguish different fluvial landscape types (Cossin and Piégay, 2001). A series of variables describe 34 waterscapes (Figure 18.6b). Each label identified the surface area occupied by a landscape features (e.g. B = block, LW = lotic water, and G = gravel...) in one portion of the photograph (1 for the foreground, 2 for the mid-ground...). A normalised Principal Component Analysis (nPCA) was performed to determine a smaller number of variables. The correlation circle of the first factorial plan provides a graphic synthesis of the results (Figure 18.6a). On the first component (F1), the open landscapes showing plains (Pl) tend to be opposed to the closed landscapes of gorge (Sl). The second component (F2) distinguishes on the one hand lentic sections (1LE) with meadows (2M) and semi-closed riverine forest (1TW and 2TW) and on the other hand lotic sections with a background of forest (3TU). Preferences for the nine waterscapes can be compared with the coordinates of axis F1 (Figure 18.6c). A clear relationship is then established between the aesthetic score and the first factorial coordinate. There is a strong preference for the closed waterscapes of gorges, contrary to the open waterscapes of plains. This conclusion supports the analysis of responses to the open questions of the questionnaire.

18.4.4 Modelling and predicting water landscape judgments

The last study is focused on modelling the public's aesthetic preferences of the waterbodies of floodplain lakes, within the context of the ecological restoration of the Ain River. It aims to predict the aesthetic assessments of different waterbodies from a set of qualitative visual variables. The choice of using visual variables is due to several operational stakes: it enables the model to be used by any practitioner, whether they are ecological experts or not. As in the previous study, the model consists in a factorial regression analysis: the dependent variable is the mean aesthetic grade given by the people surveyed; the independent variables are the physical visual characteristics of the waterbodies.

A photo-questionnaire survey was conducted in order to obtain data concerning the aesthetic preferences of the public: 100 students in geography were asked to assess the

aesthetics of 34 photographs of floodplain lakes. These photographs were sampled so that the diversity of the floodplain lakes of the Ain River was represented as much as possible. The visual variables characterising the physical aspects of the waterbodies were selected based on a previous study that focused on the perception of floodplain lakes (Cottet et al. 2010), as well on bibliography, enabling to select the most discriminating variables to distinguish positive and negative judgments. Finally, the modelling relies on six visual physical variables: (1) green dominance, (2) grey or brown dominance, (3) presence of warm and bright colours, (4) presence of a badly structured aquatic vegetation, (5) presence of sediments, and (6) muddy water. Each photograph was characterised according to these variables and a multiple correspondence analysis (MCA) was realised.

Strong correspondences are observed between variables (60% of the variance explained) (Figure 18.7). Axis F1 shows above all the information on the colour of the waterbodies; whereas axis F2 is structured by the surrounding objects (sediments, aquatic vegetation). The regression analysis was then based on the coordinates on the 2 axes of the 6 variables selected: y = 4.7 + 2 (F1 axis coord.) + 0.9 (F2 axis coord.). Several conclusions about perception mechanisms can be drawn from the resulting linear regression. The more the colour green dominates and the more the warm and bright colours are present, the more the waterbody is judged to be aesthetic. On the contrary, the muddier the water, the more the grey or brown colours dominate, and the more the badly structured aquatic vegetation are present, the less aesthetic the waterbody is judged. The influence of sediments on the perception is more uncertain. These results are rather encouraging: 2/3 of the preferences are explained by the model ($r^2 = 0.66$). Moreover the validation step, using the leave-one-out method, showed the robustness of the model (Figure 18.8): its power of generalisation can be considered further. Such a model may be an efficient tool in order to favour dialogue between stakeholders.

18.4.5 Photographs and landscape perception, a long history of knowledge production

The examples developed above illustrate a long collective effort of knowledge production in this domain. Different experiments have been published showing that humans react differently to the amount of water and its characteristics, but also to the type of waterscapes, the openness of the landscape or the riparian characteristics and the level of naturalness.

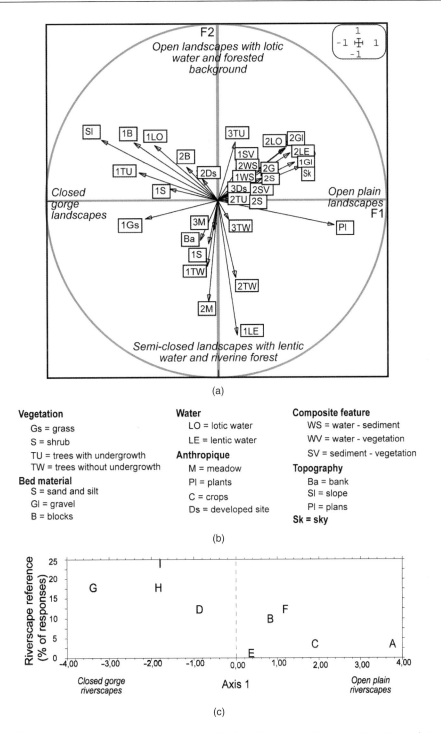

(a)

(b)

Vegetation
Gs = grass
S = shrub
TU = trees with undergrowth
TW = trees without undergrowth
Bed material
S = sand and silt
Gl = gravel
B = blocks

Water
LO = lotic water
LE = lentic water
Anthropique
M = meadow
Pl = plants
C = crops
Ds = developed site

Composite feature
WS = water - sediment
WV = water - vegetation
SV = sediment - vegetation
Topography
Ba = bank
Sl = slope
Pl = plans
Sk = sky

(c)

Figure 18.6 Unhabitant preferences and riverscape openness, the Roubion River, south France. (a) F1 × F2 correlation circle of the normalised principal component analysis concerning the structure of 34 riverscapes. (b) Labels of visual components. (c) Relation between riverscape preference (Figure 18.6) and openness (Figure 18.7a). Pictures A to I are shown on Figure 18.5b. Reproduced with permission from Le Lay et al. (2006) Les enquêtes de perception paysagère à l'aide de photographies : choix méthodologiques et exemples en milieu fluvial. In Septièmes Rencontres de Théo Quant à Besançon du 26 au 28 janvier. Fig. 6, p. 12.

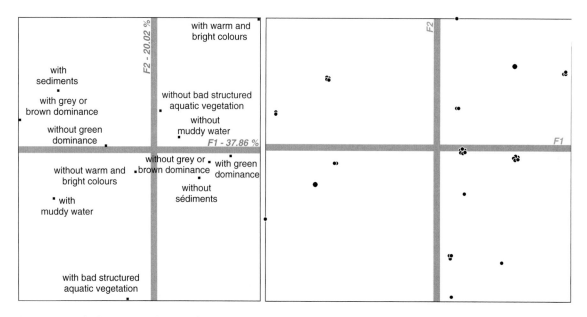

Figure 18.7 Multiple Correspondence Analysis (MCA) performed on six dichotomous variables describing waterscapes (green dominance, grey dominance, muddy water, warm or bright colour, presence of sediment, presence of badly structured aquatic vegetation) and 34 photographs. We showed here the position of the variables (a) and the photographs (b) on the first factorial map.

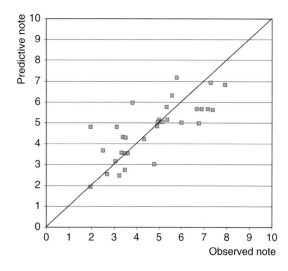

Figure 18.8 Scatter plot of the predictive versus the observed value of waterscape aesthetics. The predictive value is based on a multiple regression model performed on the F1 and F2 axes coordinates shown on Figure 18.7. It has been validated by a leave-one out procedure. Each grey square is a photo.

The perception of water therefore depends on the water level or water velocity (Burmil et al., 1999). Litton (1984) observed photographs taken at various flow levels along two California rivers and concluded that the aesthetic quality was diminished at both flood stage and lowest flow stages. Too much water masks the riffle-pool sequences, reduces the apparent differences of velocity, and drowns both islands and sandbars. Too little water gives the impression of a uniform, monotonous flow, and diminishes riverscape vividness because of the loss of white water. In shallow waters, the substrate and the aquatic vegetation are visible and make the signal received by the eye more complex (Smith et al., 1995b). Decreases in scenic beauty are linked to the percentage of the scene in exposed river bed (Brown and Daniel, 1991). Literature consistently indicates a nonlinear relationship of attractiveness to flow (Pflüger et al., 2010). On the Cache La Poudre River (in northern Colorado), Brown and Daniel (1991) estimated an optimum flow range (31–42 m³/s) for scenic beauty. They found an inverted U-shaped distribution: perception of scenic beauty is negatively affected by the lowest and highest flow events. These findings are consistent with economic studies that used contingent valuation techniques. Recreationists' willingness-to-pay for the maintenance of in-stream flow levels increases with flow up to a point, and then decreases for further increases in flow (Daubert and Young, 1981; Brown et al., 1991). The critical level differs by type of recreation activity. Anglers, for instance, prefer lower flow levels than floaters and streamside users (Loomis, 1987).

Perception is also affected by the water characteristics, as underlined by the former channel survey. Some archetypes explain why some types of landscape scenes are more highly regarded than others. Clean water is perceived as a vital fluid, a source of life, and a pre-eminently pure element, whereas LW evokes human death and is seen as an intolerable body polluting the maternal and divine element (Bachelard, 1942; Durand, 1969). Ordinary persons place a great importance upon the naturalness of river corridors, and particularly upon the quality of water (House and Fordham, 1997). When visually evaluating waterscapes, the lay public pays attention to a limited number of cues potentially affecting water quality (Table 18.5). Green and Tunstall (1992) suggested that a polluted river makes sense more easily than an unpolluted river. Individuals who judge the quality to be good use fewer features than those who judge the quality to be polluted. In the study of House and Sangster (1991), people perceived only three visual features to be indicators of good water quality: 'adults fishing', 'many fish' and where observers 'can see the river bottom'. By contrast, at the other end of the water quality spectrum, some criteria such as unusual color or smell, muddy water, presence of water plants and algae, green scum or foam on surface, protruding rubbish in river, or water movement and clarity, seem to indicate polluted water (Dinius, 1981; Moser, 1984; House and Sangster, 1991; Wilson et al., 1995; Smith et al., 1995b). These clues may detract from respondents' ratings of preference, and reduced appeal for particular recreational activities, such as boating, swimming, walking, and fishing. The public shows a rather optimistic attitude towards water quality. Indeed, 'pollution is generally judged to be less serious than it actually is, from the biological point of view' (Moser, 1984, p. 209). Clarity dominates if water is brown yet clear, and colour alters clarity perception if the water is turbid but blue. To a certain extent, the visual quality of yellow water seems more acceptable when perceived as

Table 18.5 Criteria of perceived water quality (After Dinius (1981); Moser (1984); House and Sangster (1991); Gregory and Davis (1993); Smith et al. (1995aa,b); House and Fordham (1997).

Criteria	Perceived Water Quality	
	Good Water Quality	*Polluted Water*
Color	Absence, blue and green	Grey, yellow and brown
Aquatic plants and algae	No water plants	Algae blooming
Floating debris	Absence	Litter, wooding debris,
Foam and scum	Absence	Presence
Odor	Absence	Unusual or bad smell
Movement	Flowing water	Standing water
Clarity	Clear	Muddy
Others	Presence of fish, visible bottom of the river	Rubbish

natural (Smith et al., 1995a). Turbid and brown waters, however, are considered 'unlikely to be highly regarded for bathing or aesthetic water use' (Smith et al., 1995b, p. 50). Some features evidently play a significant role in the discrimination of waterscapes, e.g. water movement and colour, spaciousness, or a mountainous setting. Rushing water received the higher preference rates, and swampy areas by far the lowest. Between these categories, the large bodies of water are preferred to rivers and lakes, because of their spaciousness, or 'a sand-and-surf mentality among [the] college-age raters' (Herzog, 1985, p. 237). Recently, Bulut and Yilmaz (2009) asked 128 university students to rate their visual preference for six waterscapes. In descending order of preference, the photographs were: (a) an urban pool with jet and still water, (b) a waterfall in forested setting, (c) standing water at the bottom of a deep valley with steep rocky walls, (d) a dam scene, (e) a wetland scene with forested background, and (f) a braided river scene.

Preference for riverscapes is more strongly affected by the environment of the river than by the characteristics of the river itself (Mosley, 1989; House and Sangster, 1991; Smith et al., 1995a; Kaplan et al., 1998). The waterscape is an 'enclosed landscape' (Litton, 1968). Landforms can be considered as a scene's backbone and thereby influence its perception. Three categories can be distinguished: mountainous, hilly, and flat plain (and plateau). For riverscapes, the flat surface of water maximises the contrast with bordering landscape elements framing the view (Lee, 1979). Shafer and Brush (1977) suggested that 'without the contrast of dark vertical masses of trees in the distance, the presence of water could actually diminish scenic quality' (p. 255). A too large part of water may make a scene appear monotonous and reduce its aesthetic value. Openness influences the comprehension of water views and the ability to project oneself into them. Riverscape openness is determined by the visible vertical elements: tall vegetation (such as alluvial forest) and buildings (bridges, dams, farmhouses) in the riverside. 'Landscapes that are too open offer little visual stimulation, and those that are too closed block one's view and cut off the possibility of more information' (Kelsch, 2000, p. 177). Spaciousness and mystery of waterscapes play an important role in riverscape preference. Mosley (1989) established that 'scenic attractiveness increases with the forest cover, the visible relief or grandeur, the area in alpine barrens visible, the area of water, and with the confinement of the river by overhanging vegetation or valley sides' (p. 10). There is a fascination for waters in mountainous settings that provide the opportunity to see unobstructed views into the distance (Shafer and Richards, 1974; Herzog, 1985).

In the same way as the landform enclosure and water elements, riparian vegetation can be very visually dominant. Vegetation can reduce the amount of light and alter the vision of landscape elements. This mysteriousness creates a sense of interest and attracts the observers insofar as they wish to obtain additional information (Lee, 1979). Some of the most highly preferred scenes include river views through open trees (Levin, 1977). Concerning marshes, the preference rates are higher for those with foreground vegetation and clear water reflections (Ellsworth, 1982). However, the presence of foreground vegetation may detract from the visual quality of scenes where the intermediate zone is largely forested (Shafer and Brush, 1977). The observer negatively perceives tangled vegetation, for it 'does not allow for visual penetration or physical penetration into the shoreline environment' (Lee, 1979, p. 578).

Whereas river managers have removed trees from banks because of the possible negative effects on flood control, a percentage of the scene in trees is associated with increases in scenic beauty (Brown and Daniel, 1991). The desire for trees may turn out to be so overwhelming that

vegetation is more influential than the details of the river form in assessing water landscape preference (House and Sangster, 1991). 'The viewer interprets a tree standing on a waterfront as an imaginal or alternative self' (Kubota, 1997, p. 184). In New Zealand, Mosley established that the percentage of native forest in a scene was the best singular variable to predict riverscape attractiveness. On Japanese urban riverfronts, the preference is for cherry trees (particularly in full bloom), followed by willow (with the branches descending near the water surface) and other deciduous species (Kubota, 1997). A linear arrangement of cherry trees with a 10 m interval is highly appreciated. Likewise, in Great Britain, the public's ideal river setting includes an open deciduous forest (House and Sangster, 1991). Indeed, the fleetingness of flower blooming and leaf falling mirrors the life of human beings. The authors also identified a strong preference for vegetational diversity, 'a mixture of grass and plants or grass and trees' (House and Fordham, 1997, p. 37) either overhanging or lining both sides of the river, but 'without an overabundance of vegetation in the water' (House and Sangster, 1991, p. 315).

Therefore, increasing the amount of vegetation at the river's edge – and not only the cosmetic placement of occasional trees at a suitable distance from the river bank – has the potential to enhance riverscape aesthetics and attractiveness, insofar as access to the bank and glimpses of water remain possible between open trees. The findings show substantial support for management strategies that include riparian forest and meandering channels.

Research has consistently shown preference for landscapes that are perceived as 'natural' and 'less built' (Kaplan et al., 1998; Van den Berg et al., 2003). Hodgson and Thayer (1980) demonstrated that the scenes labelled as natural (e.g. lakes) were appreciated more highly than the same scenes that were labelled as artificial (e.g. reservoirs). People prefer a 'shallow pool with still water' to 'the same shaped pool drained of water and covered with a glass surface' (Nasar and Li, 2004, p. 234). The natural condition appeared as more attractive than the artificial material. Calvin et al. (1972) studied preferences for natural landscapes and found that the scenes of waterfalls and rapids are regarded as higher in natural scenic beauty than a photograph depicting an algae bloom. Likewise, photographs of swampy regions were rated relatively low in preference by 49 participants in a wilderness outing program (Kaplan, 1984). In the Ellsworth study (1982), all the river scenes (n = 40) were rated higher in preference than the marsh slides (n = 20). Preferences for marshes and swampy areas are negatively correlated to coherence and spaciousness, and positively correlated to mystery and texture (Ellsworth, 1982; Herzog, 1985). There is substantial preference for natural channels and sinuous banks, with randomly-located trees (House and Sangster, 1991; House and Fordham, 1997). Indeed, straight lines and rigid rows of uniform elements do not seem natural. Along suburban waterways, Kenwick et al. (2009) revealed a slightly higher preference for earthen and vegetated banks than for banks made of stone or concrete. Likewise, according to Gregory and Davis (1993), the assessment of water colour (perceived to be an indicator of water quality) and the percentage of channelised bank shown in the scene were the two best single predictors of the riverscape aesthetics. The preference was therefore 'for riverscapes which have clean water, are relatively deep and are being actively eroded' (p. 181). By contrast, the scenes of urban reaches and channelised concrete banks attracted the lowest attractiveness scores. These findings raise the question as to whether the evaluation of river landscape demonstrates that the public's greatest preference is for the channels which have been the least affected by men. In fact, there is definite inclination and desire to maintain a certain degree of management, at least in urban areas (Asakawa, 2004). The idealised river would have naturally-worn paths following its course and short mown grass rather than uncut long grass (House and Sangster, 1991). Mosley (1989) found that a number of familiar, orderly and heavily modified river scenes in parklands or pleasant residential areas received ratings that were higher in scenic beauty than others in wilderness settings.

18.5 Conclusions and perspectives

Besides aerial imagery, ground photographs may also be useful to extract the structural elements of landscapes. Assessment of perception is truly a challenging issue for river managers when they are implementing actions, notably when such actions are not well perceived by the population. It is therefore interesting to discover why, and adjust the program accordingly or develop communication tools to explain the different options to the population. It is also interesting to obtain information about perception before the implementation of actions, so that the development of predictive models of riverscape perception can be used effectually for river management purposes.

The general public has 'a more or less clear image of which elements belongs to a certain type of landscape and which elements do not' (Coeterier, 1996, p. 31). Stranger elements may corrupt a landscape. As shown by previous authors, many factors interact to explain why a landscape is appreciated or not. Point elements, such as distinctive vegetation, specimen trees, snags, rocks, and mineral bars may attract the observer's eye (Lee, 1979). This means that superficial features can have less impact than small, localised, and specific elements on the pictures, opening another point of discussion dealing with the parameters to be extracted from a photograph and the link existing between their impact on perception and their extent on the photograph. Different techniques exist to evaluate what the eye is looking at on a picture to improve the general understanding of such a question.

By providing quantitative information on the different aspects of social perception, photo-questionnaires are a powerful approach combining social and ecological data in order to provide management scenarios, identifying for example the most valuable ecological and social reaches to be preserved. Research perspectives are still wide in this domain with the increasing interest of virtual reality, allowing to evaluate the perception of different kinds of 3D riverscapes but also considering odour and sound, as well as the view, as a part of the perception evaluation. Movement is also a parameter of interest in riverscape; so that both video and photography are valuable techniques for surveys.

For river management purposes and notably for river restoration, the question of what landscape we want is a critical question, for which perception surveys can be a valuable tool to accompany the technical work that is made in such operations. There is a strong discrepancy between the perceptions of nature of various actors. What natural landscape do people really want? Inhabitants may perceive naturalness in a very different manner from experts (Coeterier, 1996). 'The public appears to want a degree of naturalness and visual diversity that the river engineer – or at least some engineers – want to discard in favour of a more ordered, more controlled scene' (House and Fordham, 1997, p. 41). For the same river, different principles apply to different kinds of riverscape. Three views on nature can be distinguished (Swart et al., 2001; Van der Windt et al., 2007): (a) wilderness is essentially conceived as a self-regulating entity where human activities are absent or rare and do not disturb physical and biological processes; (b) arcardian nature refers to cultural ecosystems where human influence contributes to the harmony of landscapes and enhances biological values; (c) functional nature is strongly anthropocentric and adapted to human uses. An integrative and sustainable management of streams and rivers must reconcile different objectives – such ecologic restoration, flood protection, or human recreation – the importance of which varies according to river sections. The watercourse multifunctionality does not imply that all uses must be supported everywhere. There is a need to divide management practises into sectors (Asakawa et al., 2004; Dufour and Piégay, 2009). River management must involve the lay public to ascertain the key variables that are considered by the silent majority when evaluating riverine environments and establish its preferences for river corridor features. The inclusion of public participation in river management and planning could reduce the public's possible resistance to change. However, if the enhancement of public acceptance and support is the target, it must be noted that the public is not a homogeneous group, but encompasses people of diverse interests (House and Fordham, 1997). Public involvement also aims to identify misperceptions and to expand knowledge regarding ecosystem functioning. There is 'a need to develop a greater appreciation of unique and scientifically important landscapes' (Gregory and Davis, 1993, p. 172). Environmental information campaigns should explain the ecological benefits of these unique landscapes, such as, for example, in-channel woody debris or gravel-bed rivers.

Acknowledgements

The research presented in this chapter is supported by the CNRS-funded interdisciplinary project "Ingénierie écologique", by the ANR-funded project "Gestrans", and by the "Accord cadre ZABR/Agence de l'eau RMC". Many thanks to Mélanie Cossin who contributed to the survey in the Roubion catchment and the preliminary discussions. Comments by Patrice Carbonneau and two external reviewers helped to improve this chapter.

References

Anderson, L.M., Mulligan, B.E., Goodman, L.S., and Regen, H.Z. 1983. 'Effects of sounds on preferences for outdoor setting'. *Environment and Behavior*, 15, p. 539–566.

Appleton, J. 1975a. *The experience of landscape*. New York, Wiley, 293 p.

Appleton, J. 1975b. 'Landscape evaluation: The theoretical vacuum'. *Transactions of the Institute of British Geographers*, 66, p. 120–123.

Asakawa, S., Yoshida, K., and Yabe, K. 2004. 'Perceptions of urban stream corridors within the greenway system of Sapporo, Japan'. *Landscape and Urban Planning*, 68, p. 167–182.

Bachelard, G. 1983. *L'eau et les rêves. Essai sur l'imagination de la matière*. Paris: José Corti, 265 p.

Brown, T.C. and Daniel, T.C. 1991. 'Landscape aesthetics of riparian environments: relationship of flow quantity to scenic quality along a wild and scenic river'. *Water Resources Research*, 27, p. 1787–1795.

Brown, T.C. and Daniel, T.C. 1986. 'Predicting scenic beauty of timber stands'. *Forest Science*, 32, p. 471–487.

Brown, T.C., Daniel, T.C., Richards, M.T., and King, D.A. 1988. 'Recreation participation and the validity of photo-based preference judgments'. *Journal of Leisure Research*, 20, p. 40–60.

Brown, T.C., Taylor, T.C., and Shelby, B. 1992. 'Assessing the direct effects of streamflow on recreation: a literature review'. *Water Resources Bulletin*, 27, p. 979–989.

Buhyoff, G.J., Wellman, J.D., Koch, N.E., Gauthier, L., and Hultman, S. 1983. 'Landscape preference metrics: An international comparison'. *Journal of Environmental Management*, 16, p. 181–190.

Buhyoff, G.J., Wellman, J.D., Harvey, H., and Fraser, R.A. 1978. 'Landscape architects' interpretations of people's landscape preferences'. *Journal of Environmental Management*, 6, p. 255–262.

Bulut Z., Karahan F., and Sezen I. 2010. 'Determining visual beauties of natural waterscape: A case study for Tortum Valley (Erzurum/Turkey)'. *Scientific Research and Essay*, 5, p. 170–182.

Bulut, Z. and Yilmaz, H. 2009. 'Determination of waterscape beauties through visual quality assessment method'. *Environmental Monitoring and Assessment*, 154, p. 459–468.

Burmil, S., Daniel, T.C., and Hetherington, J.D. 1999. 'Human values and perceptions of water in arid landscapes'. *Landscape and Urban Planning*, 44, p. 99–109.

Calvin, J.S., Dearinger, J.A., and Curtain, M.E. 1972. 'An attempt at assessing preferences for natural landscapes'. *Environment and behavior*, 4, 447–470.

Carlson, A.A. 1977. 'On the possibility of quantifying scenic beauty'. *Landscape Planning*, 4, p. 131–172.

Carp, F.M. and Carp, A. 1982a. 'A role for technical environment assessment in perceptions of environmental quality and well-being'. *Journal of Environmental Psychology*, 2, p. 171–192.

Carp, F.M. and Carp, A. 1982b. 'Perceived environmental quality of neighbourhoods: development of assessment scales and their relation to age and gender'. *Journal of Environmental Psychology*, 3, p. 295–312.

Chin, A., Daniels, M.D., Urban, M.A., Piégay, H., Gregory, K.J., Bigler, W., Butt, A., Grable, J., Gregory, S.V., Lafrenz, M., Laurencio, L.R., and Wohl, E. 2008. 'Perceptions of wood in rivers and challenges for stream restoration in the United States'. *Environmental Management*, 41, p. 893–903.

Coeterier, J.F. 1996. 'Dominant attributes in the perception and evaluation of the Dutch landscape'. *Landscape and Urban Planning*, 34, p. 27–44.

Cossin, M. and Piégay, H. 2001. 'Les photographies prises au sol, une source d'information pour la gestion des paysages riverains des cours d'eau'. *Cahiers de Géographie du Québec*, 45, p. 37–62.

Cottet, M.L., Honegger, A., and Piégay, H. 2010. Mieux comprendre la perception des paysages de bras morts en vue d'une restauration écologique: quels sont les liens entre les qualités esthétique et écologique perçues par les acteurs ? Norois. 3: 85–103

Dakin, S. 2003. 'There's more to landscape than meets the eye: Toward inclusive landscape assessment in resource and environmental management'. *The Canadian Geographer*, 47, p. 185–200.

Daniel, T.C. and Boster, R.S. 1976. *Measuring landscape esthetics: The scenic beauty estimation method.* Res. Pap. RM-167. Fort Collins, Colo.: U.S. Dept. of Agriculture, Forest Service, Rocky Mountain Forest and Range Experiment Station. 66 p.

Daubert, J.T. and Young, R.A. 1981. 'Recreational demands for maintaining instream flows: a contingent valuation approach'. *American Journal of Agricultural Economics*, 63, p. 666–676.

Dearden, P. 1981. 'Public participation and scenic quality analysis'. *Landscape Planning*, 8, p. 3–19.

Dinius, S.H. 1981. 'Public perception in water quality evaluation'. *Water Resources Bulletin*, 17, p. 116–121.

Dufour, S. and Piégay, H. 2009. 'From the myth of a lost paradise to targeted river restoration: forget natural references and focus on human benefits'. *River Research and Applications*, 25, p. 568–581.

Durand, G. 1969. *Les structures anthropologiques de l'imaginaire (Introduction à l'archétypologie générale).* Paris: Bordas, 536 p.

Ellsworth, J.C. 1982. *Visual assessment of rivers and marshes: an examination of the relationship of visual units, perceptual variables and preference.* Master's Thesis, Utah State University.

Feimer, N.R. 1984. 'Environmental perception: the effects of media, evaluative context and observer sample'. *Journal of Environmental Management*, 4, p. 61–80.

Gift, A. 1989. 'Visual analog scales: measurement of subjective phenomenon'. *Nursing Research*, 38, p. 286–288.

Green, C.H. and Tunstall, S.M. 1992. 'The amenity and environmental value of river corridors in Britain'. *In* Boon, P.J., Calow, P., and Petts, G.E. (eds), *River conservation and management.* Chichester, Wiley, p. 425–441.

Gregory, S.V., Boyer, K.L., and Gurnell, A M. 2003. *The Ecology and Management of Wood in World Rivers.* Bethesda, American Fisheries Society, 444 p.

Gregory, K.J. and Davis, R.J. 1993. 'The perception of riverscape aesthetics: an example from two Hampshire rivers'. *Journal of Environmental Management*, 39, p. 171–185.

Griselin, M. and Nageleisen, S. 2004. '"Quantifier" le paysage au long d'un itinéraire à partir d'un échantillonnage photographique au sol'. *Cybergeo*, http://www.cybergeo.eu/index3684.html.

Hammitt, W.E. 1978. *Visual and user preference for a bog environment.* Ph.D. dissertation, Univ. Michigan, Ann Arbor., 159 p.

Hartig, T., Mang, M., and Evans, G.W. 1991. 'Restorative effects of natural environment experiences'. *Environment and Behavior*, 23, p. 3–26.

Heft, H. and Nasar, J.L. 2000. 'Evaluating environmental scenes using dynamic versus static displays'. *Environment and Behavior*, 32, p. 301–322.

Herzog, T.R. 1985. 'A cognitive analysis of preference for waterscapes'. *Journal of Environmental Psychology*, 5, p. 225–241.

Hetherington, J., Daniel, T.C., and Brown, T.C. 1993. 'Is motion more important than it sounds?: the medium of presentation in environment perception research'. *Journal of Environmental Psychology*, 13, p. 283–291.

Hodgson R.W. and Thayer R.L. 1980. 'Implied human influence reduces landscape beauty'. *Landscape Planning*, 7, p. 171–179.

House, M. and Fordham, M. 1997. 'Public perceptions of river corridors and attitudes towards river works'. *Landscape Research*, 22, p. 25–44.

House, M. and Sangster, E.K. 1991. 'Public perception of river-corridor management'. *Journal of the Institution of Water and Environmental Management*, 5, p. 312–317.

Huang, S.-C.L. (2009). 'The validity of visual surrogates for representing waterscapes'. *Landscape Research*, 34, p. 323–335.

Huang, S.-C.L. (2004). 'An exploratory approach for using videos to represent dynamic environments'. *Landscape Research*, 29, p. 205–218.

Huang, S.-C.L. and Tassinary, L. 2000. 'A study of people's perception of waterscapes in built environment'. *Journal of Public Affair Review*, 1, p. 1–19.

Junker, B. and Buchecker, M. 2008. 'Aesthetic preferences versus ecological objectives in river restorations'. *Landscape and Urban Planning*, 85, p. 141–154.

Kaplan, R. 1985. 'The analysis of perception via preference: a strategy for studying how the environment is experienced'. *Landscape Planning*, 12, p. 161–176.

Kaplan, R. 1984. 'Wilderness perception and psychological benefits: an analysis of a continuing program'. *Leisure Sciences*, 6, p. 271–290.

Kaplan, R. 1977a. 'Down by the riverside: Informational factors in waterscape preference. In River recreation management and research symposium'. USDA Forest Service General Technical Report NC-28, p. 285–289.

Kaplan, R. 1977b. 'Preference and everyday nature: method and application'. *In* Stokols D. (Ed), *Perspectives on environment and behavior.* New York, Plenum, p. 235–250.

Kaplan, R. and Kaplan, S. 1989. *The experience of nature.* Cambridge, Cambridge University Press, 340 p.

Kaplan, R., Kaplan, S., and Ryan, R.L. 1998. *With people in mind. Design and management for everyday nature.* Washington, Island Press, 239 p.

Kelsch, P. 2000. 'Constructions of American forest: four landscapes, four readings'. In Conan, M., (ed), *Environmentalism in landscape architecture*. Washington, Dumbarton Oaks, p. 163–186.

Kenwick, R.A., Shammin, Md R., and Sullivan, W.C. 2009. 'Preferences for riparian buffers'. *Landscape and Urban Planning*, 91, p. 88–96.

Kroh D.P. and Gimblett R.H. 1992. 'Comparing live experience with pictures in articulating landscape preference'. *Landscape Research*, 17, p. 58–69.

Kubota, Y. 1997. 'Preference for trees on urban riverfronts'. *In* Wapner, S., Demick, J., Yamamoto, T., and Takahashi, T. (Eds), *Handbook of Japan-United States environment-behavior research. Toward a transactional approach*. New York: Plenum Press, p. 183–198.

Lee, M.S. 1979. 'Landscape preference assessment of Louisiana river landscapes: a methodological study'. *In*: Elsner, Gary H., and Richard C. Smardon, technical coordinators. 1979. *Proceedings of our national landscape: a conference on applied techniques for analysis and management of the visual resource* [Incline Village, Nev., April 23-25 1979]. Gen. Tech. Rep. PSW-GTR-35. Berkeley, CA. Pacific Southwest Forest and Range Exp. Stn., Forest Service, U.S. Department of Agriculture: p. 572–580.

Lee, M.S. 1978. *Visual quality assessment of Louisiana river landscapes: A methodological study*. Master's Thesis, Louisiana State University.

Le Lay, Y.-F., Piégay, H., Gregory, K., Chin, A., Dolédec, S., Elosegi, A., Mutz, M., Wy ga, B., and Zawiejska, J. 2008. 'Variations in cross-cultural perception of riverscapes in relation to in-channel wood'. *Transactions of the Institute of British Geographers*, 33, p. 268–287.

Leopold, L.B. 1969a. *Quantitative comparison of some aesthetic factors among rivers*. U.S. Geological Survey Circular 620, 14 p.

Leopold, L.B. 1969b. 'Landscape Esthetics. How to quantify the scenics of a river valley?' *Natural History*, October, p. 37–45.

Leopold, L.B. and Marchand, M. O. 1968. 'On the quantitative inventory of the riverscape'. *Water Resources Research*, 4, p. 709–717.

Levin, J.E. 1977. *Riverscape preference: on site photographic reactions*. Master's Thesis, University of Michigan, 114 p.

Litton, R.B. 1984. 'Visual fluctuations in river landscape quality'. *In* Popadic, J. S., Butterfield, D. I., Anderson, D. H., and Popadic, M. R. *National river recreation symposium proceedings*, Baton Rouge, Louisiana State University, p. 369–384.

Litton, R.B. 1977. 'River landscape quality and its assessment'. USDA Forest Service General Technical Report NC-28, p. 46–54.

Litton, R.B. Jr 1968. *Forest landscape description and inventories – a basis for landplanning and design*. Res. Paper PSW-RP-049. Albany, CA: Pacific Southwest Research Station, Forest Service, U.S. Department of Agriculture; 88 p.

Loomis, J. 1987. 'The economic value of instream flow: methodology and benefit estimated for optimum flows'. *Journal of Environmental Management*, 24, p. 169–179.

Meitner, M.J. 2004. 'Scenic beauty of river views in the Grand Canyon: relating perceptual judgments to locations'. *Landscape and Urban Planning*, 68, p. 3–13.

Morisawa, M.E. 1972. 'A methodology for watershed evaluation'. *In* Csallany, S. C., McLaughlin, T.G., and Striffler, W. D. (eds), *Watersheds in transition: national symposium for the international hydrological decade*. Fort Collins, Colorado State University Press, p. 153–158.

Moser, G. 1984. 'Water quality perception, a dynamic evaluation'. *Journal of Environmental Psychology*, 4, p. 201–210.

Mosley, M.P. 1989. 'Perceptions of New Zealand river scenery'. *New Zealand Geographer*, 45, p. 2–13.

Mutz, M., Piégay, H., Gregory, K.J., Borchardt, D., Reich, M., and Schmieder, K. 2006. 'Perception and evaluation of dead wood in streams and rivers by German students'. *Limnologica – Ecology and Management of Inland Waters*, 36, p. 110–118.

Nasar, J.L. and Li, M. 2004. 'Landscape mirror: the attractiveness of reflecting water'. *Landscape and Urban Planning*, 66, p. 233–238.

Nasar, J.L. and Lin, Y.-H. 2003. 'Evaluative responses to five kinds of water features'. *Landscape Research*, 28, p. 441–450.

Nausser, J.I. 1982. 'Framing the landscape in photographic simulation'. *Journal of Environmental Management*, 17, p. 1–16.

Palmer, J.F. and Hoffman, R.E. 2001. 'Rating reliability and representation validity in scenic landscape assessments'. *Landscape and Urban Planning*, 54, p. 149–161.

Parson, R. 1991. 'The potential influences of environmental perception on human health'. *Journal of Environmental Psychology*, 11, p. 1–23.

Pedersen, D.M. 1978. 'Relationship between environmental familiarity and environmental preference'. *Perceptual and Motor Skills*, 47, p. 739–743.

Penning-Rowsell, E. and Burgess, J. 1997. 'River landscapes: changing the concrete overcoat?'. *Landscape Research*, 22, p. 5–11.

Penning-Rowsell, E.C. and Hardy, D.I. 1973. 'Landscape evaluation and planning policy: a comparative survey in the Wye valley area of outstanding natural beauty'. *Regional Studies*, 7, p. 153–160.

Pflüger, Y., Rackham, A. and Larned, S. 2010. 'The aesthetic value of river flow: an assessment of flow preferences for large and small rivers'. *Landscape and Urban Planning*, 95, p. 68–78.

Piégay, H., Gregory, K.J., Bondarev, V., Chin, A., Dahlström, N., Elosegi, A., Gregory, S., Joshi, V., Mutz, M., Rinaldi, M., Wyzga, B., and Zawiejska, J. 2005. 'Public perception as a barrier to introducing wood in rivers for restoration purposes'. *Environmental Management*, 36, p. 665–674.

Pitt, D.G. 1976. 'Physical dimensions of scenic quality in streams'. *In* Zube, E. H. (ed.) *Studies in landscape perception*. Amherst, Institute for Man and Environment, p. 143–161.

Ribe, R.G. 1982. 'On the possibility of quantifying scenic beauty – a response'. *Landscape Planning*, 9, p. 61–75.

Ryan, R.L. 1998. 'Local perceptions and values for a Midwestern river corridor'. *Landscape and Urban Planning*, 42, p. 225–237.

Schroeder, H.W. 1982. 'Preferred features of urban parks and forests'. *Journal of Arboriculture*, 8, p. 317–322.

Shafer, E. Jr and Brush, R.O. 1977. 'How to measure preferences for photographs of natural landscapes'. *Landscape Planning*, 4, p. 237–256.

Shafer, E. Jr and Mietz, J. 1970. *It seems possible to quantify scenic beauty in photographs*. Res. Pap. Ne-162. Upper Darby, PA: U.S. Department of Agriculture, Forest Service, Northeastern Forest Experiment Station, 12 p

Shafer, E. Jr, Hamilton, J.F., and Schmidt, E.A. 1969. 'Natural landscape preferences'. *Journal of Leisure Research*, 1, p. 1–19.

Shafer, E. Jr and Richards, T.A. 1974. *A comparison of viewer reactions to outdoor scenes and photographs of those scenes*. Res. Pap. Ne-302. Upper Darby, PA: U.S. Department of Agriculture, Forest Service, Northeastern Forest Experiment Station, 26 p.

Shuttleworth, S. 1980. 'The use of photographs as an environment presentation medium in landscape studies'. *Journal of Environmental Management*, 11, p. 61–76.

Smith, D.G., Croker, G.F., and McFarlane, K. 1995a. 'Human perception of water appearance. 1. Clarity and colour for bathing and aesthetics'. *New Zealand Journal of Marine and Freshwater Research*, 29, p. 29–43.

Smith, D.G., Croker, G.F., and McFarlane, K. 1995b. 'Human perception of water appearance. 2. Colour judgment, and the influence of perceptual set on perceived water suitability for use'. *New Zealand Journal of Marine and Freshwater Research*, 29, p. 45–50.

Stamps, A.E. III 1999. 'Demographic effects in environmental preferences: a meta-analysis'. *Journal of Planning Literature*, 14, p. 155–175.

Stamps, A.E. III 1995. 'Stimulus and respondent factors in environmental preference'. *Perceptual and Motor Skills*, 80, p. 668–670.

Stamps, A.E. III 1992. 'Bootstrap investigation of respondent sample size for environmental preference'. *Perceptual and Motor Skills*, 75, p. 220–222.

Stamps, A.E. III 1990. 'Use of photographs to simulate environments: a meta-analysis'. *Perceptual and Motor Skills*, 71, p. 907–913.

Stewart, T.R., Middleton, P., Downton, M., and Ely, D. 1984. 'Judgments of photographs vs. field observations in studies of perception and judgment of the visual environment'. *Journal of Environmental Psychology*, 4, p. 283–302.

Swart, J.A.A., Van der Windt, H.J., and Keulartz, J. 2001. 'Valuation of nature in conservation and restoration'. *Restoration Ecology*, 9, p. 230–238.

Trent, R.B., Neumann, E., and Kvashny, A. 1987. 'Presentation mode and question format artifacts in visual assessment research'. *Landscape and Urban Planning*, 14, p. 225–235.

Turner, J.R. 1975. 'Applications of landscape evaluation: A planner's view'. *Transactions of the Institute of British Geographers*, 66, p. 156–161.

Ulrich, R.S. 1981. 'Natural versus urban scenes: some psychophysiological effects'. *Environment and Behavior*, 13, p. 523–556.

Van den Berg, A.E., Koole, S.L., and Van der Wulp, N.Y. 2003. 'Environmental preference and restoration: (how) are they related?' *Journal of Environmental Psychology*, 23, p. 135–146.

Van der Windt, H.J., Swart, J.A.A., and Keulartz, J. 2007. 'Nature and landscape planning: exploring the dynamics of valuation, the case of the Netherlands'. *Landscape and Urban Planning*, 79, p. 218–228.

Wallace, B.C. 1974. 'Landscape evaluation and the Essex coast'. *Regional Studies*, 8, p. 299–305.

Whalley, J.M. 1988. 'Water in the landscape'. *Landscape and Urban Planning*, 16, p. 145–162.

Wherrett, J.R. 2000. 'Creating landscape preference models using internet survey techniques'. *Landscape Research*, 25, p. 79–96.

Williams, A.T. 1986. 'Landscape aesthetics of the River Wye'. *Landscape Research*, 11, 25–30.

Wilson, M.I., Robertson, L.D., Daly, M., and Walton, S.A. 1995. 'Effects of visual cues on assessment of water quality'. *Journal of Environmental Psychology*, 15, p. 53–63.

Wyzga, B., Zawiejska, J., and Le Lay, Y.-F 2009. 'Influence of academic education on the perception of wood in watercourses'. *Journal of Environmental Management*, 90, 587e603

Yamashita, S. 2002. 'Perception and evaluation of water in landscape: use of Photo-Projective Method to compare child and adult residents' perceptions of a Japanese river environment'. *Landscape and Urban Planning*, 62, p. 3–17.

Zube, E.H. 1973. 'Rating everyday rural landscapes of the Northeastern United States'. *Landscape Architecture*, 63, p. 370–375.

Zube, E.H. and Pitt, D.G. 1981. 'Cross-cultural perceptions of scenic and heritage landscapes'. *Landscape Planning*, 8, p. 69–87.

Zube, E.H., Sell, J.L., and Taylor, J.G. 1982. 'Landscape perception: research, application and theory'. *Landscape Planning*, 9, p. 1–33.

Zube, E.H., Simcox, D.E., and Law, C.S. 1987. 'Perceptual landscape simulations: history and prospect'. *Landscape Journal*, 6, p. 62–80.

19 Future Prospects and Challenges for River Scientists and Managers

Patrice E. Carbonneau[1] and Hervé Piégay[2]

[1]Department of Geography, Durham University, Science site, Durham, UK
[2]University of Lyon, CNRS, France

We hope to have demonstrated in this edited volume that fluvial remote sensing is now capable of delivering unprecedented data to river scientists and managers. Starting from basic principles, the volume has illustrated how fluvial remote sensing has evolved into a self-contained discipline. We have discussed hyperspectral imagery, thermal imagery and visible imagery (from the ground, the air or space) all of which offer new ways of imaging rivers capable of resolving fine spatial and spectral details. We have also discussed new LiDAR approaches, both terrestrial and airborne, which offer topographic data of unprecedented quality. Finally, we hope to have shown the value of this data with some emerging applications which covered the biotic, abiotic and even social aspects of river sciences. Where is the new emerging discipline of fluvial remote sensing heading? If the pace of technical progress is maintained, it would seem that there are few fundamental limitations impeding the improvement of spatial and spectral resolutions. Furthermore, the temporal resolution of fluvial remote sensing datasets, which is currently impeded mostly by cost and logistic issues, is expected to improve markedly in the next decade owing to reductions in costs of classic airborne platforms and to progress in the areas of terrestrial fluvial remote sensing and UAV technology. As a result, our ability to produce vast hyperspatial (see Chapter 8 for a definition) and hyperspectral (see Chapter 4 for a

definition) datasets can only improve. When compared to the datasets that led to major and significant river sciences contributions such as Hydraulic Geometry (Leopold and T., 1953) or the network dynamics hypothesis (Benda et al., 2004), fluvial remote sensing datasets are bigger by orders of magnitude. But does this over-abundant nirvana of data necessarily lead to improved science, new knowledge and better management? Our answer here is a cautious *maybe*.

This move from a data-sparse to a data-rich situation is a fundamental change which has not been fully appreciated by the river sciences community. A key distinction must be made between data and knowledge. A vast data set might contain vast amounts of information but extracting this information and transforming it into knowledge via the scientific method is a more difficult task when compared to smaller, sparse datasets. Unfortunately, somewhat less attention has been paid to analysis methods and conceptual frameworks underpinning these hyper-large datasets. As authors who have worked extensively with large fluvial remote sensing datasets (e.g. Carbonneau et al., 2004; Alber et Piégay, 2011; Carbonneau et al., 2011; Wawrzyniak et al., 2011), we, the editors of this volume, are well aware of the difficulties posed by their analysis. Further progress and reflection focussed on the meaningful analysis of these datasets is now a crucial challenge for the future. For example, Carbonneau et al.

Fluvial Remote Sensing for Science and Management, First Edition. Edited by Patrice E. Carbonneau and Hervé Piégay.
© 2012 John Wiley & Sons, Ltd. Published 2012 by John Wiley & Sons, Ltd.

(2012) present a comprehensive dataset comprised of particle sizes, channel widths, channel depths and slopes all sampled continuously at metric resolutions for an entire river. Faced with the complexity of the data the authors have proposed new analysis methods aimed at synoptic views and multiscale analysis of river variable response along both the length and width of the river (see also Figure 9.15 in Chapter 9 of this volume). One of the observations of Carbonneau et al. (2012) is a frequent lack of agreement between the data and established theories, the authors therefore argue that such hyperspatial data challenges current paradigms in river sciences. Continuous data collection over the fluvial continuum provides exceptional information to explore the network structure in a multi-scale context which opens discussions on the 'scalar dissonance', the idea that the different scale levels are not nested in each other but disjointed; the boundaries of units at each scale level not being overlain (Leviandier et al., 2012). The hierarchical theory as described by Frissell et al. (1986) can be now explored, validated or, if warranted by the evidence of real data, reformulated.

Such a widening application of fluvial remote sensing to a greater number of catchments and rivers highlights another challenge currently faced by the discipline: widening the user base and moving this technology beyond the quasi-exclusive use of its developers. If we examine the fluvial remote sensing literature discussed in Chapter 1, we find that the vast majority of contributions were made by the scientists responsible at least partially for the methodological development. Rarely do we see a contribution where an explicit thematic question is solved using established remote sensing methods. Most of the time, contributions explored the method itself or the capacity of a given technology or a specific type of data to answer the question. Moreover, most of the river management initiative in Europe, following the implementation of the Water Framework Directive, are based on field measures, sometimes combined with GIS information but never with Remote Sensing information (e.g. The River Habitat Survey in the UK or the QualPhy or SEQphy methods in France). This is representative of a slightly worrying gap between the developers and proponents of this new technology and the potential users. Indeed, the potential of fluvial remote sensing in river management is still underexplored because practitioners still consider these techniques as complex approaches which are not traditionally taught and thus reserved to academic circles. However, we note that even within academic circles, lotic ecologists rarely use fluvial remote sensing methods

as data sampling method. It is therefore becoming clear that the dissemination of fluvial remote sensing methods within a broader community of users is an urgent matter. This is certainly not an impossible tasks since the raw cost of imagery is at an all-time low and many fluvial remote sensing tools are quite compatible with GIS packages already widely used. For example, the freely available 'River Bathymetry Toolkit' offers users of the EAARL system for bathymetric river topography measurements i.e. (McKean et al., 2009) a set of tools which are seamlessly integrated into the popular ArcGIS software. The river bathymetry toolbox also comes with a range of web based tutorial services such as YouTube videos and html help files. Such open source approaches are needed in order to facilitate the teaching and dissemination of fluvial remote sensing to academics outside the remote sensing community and river managers. Nevertheless, we can argue this evolution is ongoing with the emergence of private companies providing more and more frequent airborne imagery campaigns but also LIDAR surveys to help designing river management plans.

The wider availability of fluvial remote sensing to a broad range of users would be a very powerful tool. It could allow for the development of new procedures and methods for automating and analysis data for describing the state of biotic and abiotic features and for providing query procedures vitally needed in order to target management actions in a period where maximum impact must be reached at minimal costs. These tools could also be used to predict the evolution of the managed system thus allowing for risk analysis (e.g. assess sensitivity of the system to change, assess potential effects of given actions or pressures on channel network characters), such as shown in Chapters 11 and 12, and for an a priori evaluation of long-term costs. Furthermore, these methods could provide crucial support to monitoring efforts needed to survey fish populations, restoration work evolution, pollution, natural hazards, ice cover and even invasive species. Additionally, as demonstrated by Chapter 18, river imagery has a role to play in the social aspect of river restoration and could thus contribute to a better integration of the socio-economic aspects and the traditional biophysical aspects of river sciences. In such context, the development of the 3D approaches which allow people to interact and evaluate river aesthetics in a virtual environment should provide also new indicators and new opportunities to view the river corridors providing a bridge between biophysical characterisation and social perceptions.

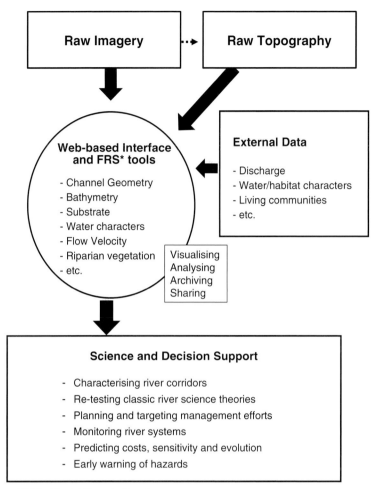

Figure 19.1 Flow chart describing future fluvial remote sensing framework for supporting management decisions and scientific challenges.

In closing, we propose the following flow chart describing what we perceive as an idealised future state for fluvial remote sensing (Figure 19.1). At the core of the flowchart is the idea that fluvial remote sensing will be adopted by a very wide user group of non-remote-sensing-specialists in a trajectory similar to that of traditional remote sensing. The history of western science, going as far back as Da Vinci and other Renaissance thinkers, has clearly shown that our ability to understand and conceptualise physical phenomena is linked to our ability to measure them. Fluvial remote sensing therefore offers us an enormous opportunity to push river sciences forward by fundamentally changing our perception and by allowing us to thoroughly investigate river catchments in a manner which was previously impossible. We therefore hope that this volume will serve as a useful reference and encourage new users to adopt these methods in order to further their respective fields of investigation.

References

Alber, A. and Piégay, H. 2011. Spatial aggregation procedure for characterizing physical structures of fluvial networks: applications to the Rhône basin. *Geomorphology*. 125(3): 343–360.

Benda, L., Poff, N.L., Miller, D., Dunne, T., Reeves, G., Pess, G., and Pollock, M. 2004. The network dynamics hypothesis: How channel networks structure riverine habitats. *Bioscience*, **54**(5), 413–427. 10.1641/0006-3568(2004)054[0413:tndhhc]2.0.co;2.

Carbonneau, P.E., Fonstad, M.A., Marcus, W.A., and Dugdale, S.J. 2012. Making riverscapes real. *Geomorphology*, **137**(1), 74–86. DOI:10.1016/j.geomorph.2010.09.030.

Carbonneau, P.E., Lane, S.N., and Bergeron, N.E. 2004. Catchment-scale mapping of surface grain size in gravel bed rivers using airborne digital imagery. *Water Resources Research*, **40**(7). DOI:10.1029/2003WR002759.

Leopold, L.B. and T., M. 1953. "The hydraulic geometry of stream channels and some physiographic implications." U.S.G.S., ed., United States Government Printing Office, Washington, DC., 64 pp.

Leviandier, T., Alber, A., Le Ber, F., and Piégay, H. 2012. Comparison of statistical algorithms for detecting homogeneous river reaches along a longitudinal continuum. *Geomorphology*, **138**(1), 130–144. DOI:10.1016/j.geomorph.2011.08.031.

McKean, J., Nagel, D., Tonina, D., Bailey, P., Wright, C.W., Bohn, C., and Nayegandhi, A. 2009. Remote Sensing of Channels and Riparian Zones with a Narrow-Beam Aquatic-Terrestrial LIDAR. *Remote Sensing*, **1**(4), 1065–1096.

Wawrzyniak, V., Piégay, H., Poirel, A., 2011. Longitudinal and Temporal Thermal Patterns of the French Rhône River using Landsat ETM+ Thermal Infrared (TIR) Images. Aquatic Sciences. DOI: 10.1007/s00027-011-0235-2. On line.

Index